# Advances in Experimental Medicine and Biology

Volume 1058

**Editorial Board:**
IRUN R. COHEN, *The Weizmann Institute of Science, Rehovot, Israel*
ABEL LAJTHA, *N.S.Kline Institute for Psychiatric Research, Orangeburg, NY, USA*
JOHN D. LAMBRIS, *University of Pennsylvania, Philadelphia, PA, USA*
RODOLFO PAOLETTI, *University of Milan, Milan, Italy*
NIMA REZAEI, *Tehran University of Medical Sciences, Children's Medical Center Hospital, Tehran, Iran*

More information about this series at http://www.springer.com/series/5584

J. Miguel Oliveira • Sandra Pina • Rui L. Reis
Julio San Roman
Editors

# Osteochondral Tissue Engineering

Nanotechnology, Scaffolding-Related Developments and Translation

*Editors*
J. Miguel Oliveira
3B's Research Group – Biomaterials,
Biodegradables and Biomimetics
University of Minho
Headquarters of the European Institute
of Excellence on Tissue Engineering
and Regenerative Medicine
Barco, Guimarães, Portugal

Sandra Pina
3B's Research Group – Biomaterials,
Biodegradables and Biomimetics
University of Minho
Headquarters of the European Institute
of Excellence on Tissue Engineering
and Regenerative Medicine
Barco, Guimarães, Portugal

Rui L. Reis
3B's Research Group – Biomaterials,
Biodegradables and Biomimetics
University of Minho
Headquarters of the European Institute
of Excellence on Tissue Engineering
and Regenerative Medicine
Barco, Guimarães, Portugal

Julio San Roman
Polymeric Nanomaterials and Biomaterials
Department
Institute of Polymer Science and Technology
Spanish Council for Scientific Research (CSIC)
Madrid, Spain

ISSN 0065-2598     ISSN 2214-8019   (electronic)
Advances in Experimental Medicine and Biology
ISBN 978-3-319-76710-9     ISBN 978-3-319-76711-6   (eBook)
https://doi.org/10.1007/978-3-319-76711-6

Library of Congress Control Number: 2018938100

© Springer International Publishing AG, part of Springer Nature 2018
This work is subject to copyright. All rights are reserved by the Publisher, whether the whole or part of the material is concerned, specifically the rights of translation, reprinting, reuse of illustrations, recitation, broadcasting, reproduction on microfilms or in any other physical way, and transmission or information storage and retrieval, electronic adaptation, computer software, or by similar or dissimilar methodology now known or hereafter developed.
The use of general descriptive names, registered names, trademarks, service marks, etc. in this publication does not imply, even in the absence of a specific statement, that such names are exempt from the relevant protective laws and regulations and therefore free for general use.
The publisher, the authors and the editors are safe to assume that the advice and information in this book are believed to be true and accurate at the date of publication. Neither the publisher nor the authors or the editors give a warranty, express or implied, with respect to the material contained herein or for any errors or omissions that may have been made. The publisher remains neutral with regard to jurisdictional claims in published maps and institutional affiliations.

Printed on acid-free paper

This Springer imprint is published by the registered company Springer International Publishing AG, part of Springer Nature.
The registered company address is: Gewerbestrasse 11, 6330 Cham, Switzerland

# Preface

In the last few years, osteochondral tissue engineering has shown an increasing development in advanced tools and technologies for damaged underlying subchondral bone and cartilage tissue repair and regeneration. Considering the limitation of articular cartilage to heal and self-repair, new therapeutic options are essential to develop approaches based on suitable strategies made of appropriate engineered biomaterials. This book reviews the most recent developments in the field of osteochondral tissue engineering and presents challenges and strategies being developed that not only face bone and cartilage regeneration but also establish osteochondral interface formation, in order to translate it into a clinical setting. Topics include biomaterials advances in osteochondral tissue engineering, namely natural, synthetic, and bioceramics-based materials, nanotechnology approaches, as well as advanced processing methodology underlying tissue-engineered scaffolding development, such as 3D bioprinting, electrospinning, and supercritical fluid technology. Hydrogel systems for osteochondral applications are also detailed thoroughly. It also maximizes the reader insights into translational research and turning research into products, clinical trials and management of osteochondral lesions, and commercially available products. This is an ideal book for biomedical engineering students and a wide range of established researchers and professionals working in the orthopedic field.

| | |
|---|---|
| Barco, GMR, Portugal | J. Miguel Oliveira |
| Barco, GMR, Portugal | Sandra Pina |
| Barco, GMR, Portugal | Rui L. Reis |
| Madrid, Spain | Julio San Roman |

# Contents

**Part I  Biomaterials Advances in Osteochondral Tissue**

1 **Natural Origin Materials for Osteochondral Tissue Engineering** ... 3
Walter Bonani, Weerasak Singhatanadgige, Aramwit Pornanong, and Antonella Motta

2 **Synthetic Materials for Osteochondral Tissue Engineering** ........ 31
Antoniac Iulian, Laptoiu Dan, Tecu Camelia, Milea Claudia, and Gradinaru Sebastian

3 **Bioceramics for Osteochondral Tissue Engineering and Regeneration** ......................................... 53
Sandra Pina, Rita Rebelo, Vitor Manuel Correlo, J. Miguel Oliveira, and Rui L. Reis

**Part II  Nanotechnology Approaches for Osteochondral Tissue Engineering**

4 **Nanomaterials/Nanocomposites for Osteochondral Tissue**......... 79
Ohan S. Manoukian, Connor Dieck, Taylor Milne, Caroline N. Dealy, Swetha Rudraiah, and Sangamesh G. Kumbar

5 **Nanofibers and Microfibers for Osteochondral Tissue Engineering**. 97
Zaida Ortega, María Elena Alemán, and Ricardo Donate

6 **Micro/Nano Scaffolds for Osteochondral Tissue Engineering** ...... 125
Albino Martins, Rui L. Reis, and Nuno M. Neves

**Part III  Osteochondral Tissue Scaffolding**

7 **Mimetic Hierarchical Approaches for Osteochondral Tissue Engineering** ......................................... 143
Ivana Gadjanski

8  Porous Scaffolds for Regeneration of Cartilage, Bone
   and Osteochondral Tissue .................................. 171
   Guoping Chen and Naoki Kawazoe

9  Layered Scaffolds for Osteochondral Tissue Engineering ......... 193
   Diana Ribeiro Pereira, Rui L. Reis, and J. Miguel Oliveira

**Part IV  Advanced Processing Methodology**

10 Preparation of Polymeric and Composite Scaffolds
   by 3D Bioprinting ......................................... 221
   Ana Mora-Boza and María Luisa Lopez-Donaire

11 The Use of Electrospinning Technique on Osteochondral Tissue
   Engineering ............................................... 247
   Marta R. Casanova, Rui L. Reis, Albino Martins,
   and Nuno M. Neves

12 Supercritical Fluid Technology as a Tool to Prepare Gradient
   Multifunctional Architectures Towards Regeneration
   of Osteochondral Injuries................................... 265
   Ana Rita C. Duarte, Vitor E. Santo, Manuela E. Gomes,
   and Rui L. Reis

**Part V  Hydrogels Systems for Osteochondral Tissue Applications**

13 Gellan Gum-Based Hydrogels for Osteochondral Repair........... 281
   Lígia Costa, Joana Silva-Correia, J. Miguel Oliveira, and Rui L. Reis

14 Silk Fibroin-Based Hydrogels and Scaffolds
   for Osteochondral Repair and Regeneration .................... 305
   Viviana P. Ribeiro, Sandra Pina, J. Miguel Oliveira, and Rui L. Reis

15 In Situ Cross-Linkable Polymer Systems and Composites
   for Osteochondral Regeneration ............................. 327
   María Puertas-Bartolomé, Lorena Benito-Garzón,
   and Marta Olmeda-Lozano

**Part VI  Translation of Osteochondral Tissue Products**

16 Stem Cells in Osteochondral Tissue Engineering ............... 359
   Eleonora Pintus, Matteo Baldassarri, Luca Perazzo, Simone Natali,
   Diego Ghinelli, and Roberto Buda

17 Osteochondral Tissue Engineering: Translational Research
   and Turning Research into Products.......................... 373
   Victoria Spencer, Erica Illescas, Lorenzo Maltes, Hyun Kim,
   Vinayak Sathe, and Syam Nukavarapu

| | | |
|---|---|---|
| 18 | **Clinical Trials and Management of Osteochondral Lesions** ........ | 391 |
| | Carlos A. Vilela, Alain da Silva Morais, Sandra Pina, J. Miguel Oliveira, Vitor M. Correlo, Rui L. Reis, and João Espregueira-Mendes | |
| 19 | **Commercial Products for Osteochondral Tissue Repair and Regeneration** ........................................... | 415 |
| | Diana Bicho, Sandra Pina, Rui L. Reis, and J. Miguel Oliveira | |
| **Index**............................................................. | | 429 |

# Part I
# Biomaterials Advances in Osteochondral Tissue

# Chapter 1
# Natural Origin Materials for Osteochondral Tissue Engineering

**Walter Bonani, Weerasak Singhatanadgige, Aramwit Pornanong, and Antonella Motta**

**Abstract** Materials selection is a critical aspect for the production of scaffolds for osteochondral tissue engineering. Synthetic materials are the result of man-made operations and have been investigated for a variety of tissue engineering applications. Instead, the products of physiological processes and the metabolic activity of living organisms are identified as natural materials. Over the recent decades, a number of natural materials, namely, biopolymers and bioceramics, have been proposed as the main constituent of osteochondral scaffolds, but also as cell carriers and signaling molecules. Overall, natural materials have been investigated both in the bone and in the cartilage compartment, sometimes alone, but often in combination with other biopolymers or synthetic materials. Biopolymers and bioceramics possess unique advantages over their synthetic counterparts due similarity with natural extracellular matrix, the presence of cell recognition sites and tunable chemistry. However, the characteristics of natural origin materials can vary considerably depending on the specific source and extraction process. A deeper understanding of the relationship between material variability and biological activity and the definition of standardized manufacturing procedures will be crucial for the future of natural materials in tissue engineering.

**Keywords** Proteins · Polysaccharides · Polyhydroxyalkanoates · Bioceramics · Osteochondral tissue engineering

---

W. Bonani (✉) · A. Motta
BIOtech Research Center and Department of Industrial Engineering,
European Institute of Excellence on Tissue Engineering and Regenerative Medicine,
University of Trento, Trento, Italy
e-mail: walter.bonani@unitn.it

W. Singhatanadgige
Department of Orthopedic Surgery, Faculty of Medicine, Chulalongkorn University
and King Chulalongkorn Memorial Hospital, Thai Red Cross Society, Bangkok, Thailand

P. Aramwit
Bioactive Resources for Innovative Clinical Applications Research Unit
and Department of Pharmacy Practice, Faculty of Pharmaceutical Sciences,
Chulalongkorn University, Bangkok, Thailand

## 1.1 Introduction

Even if synthetic materials can offer multiple choices in terms of adaptability to specific physical requirements, they do not generally possess biomolecular recognition features that are essential for the induction of the regenerative pathway. Instead, natural materials possess unique properties like the ability to interact with the biological environment, particularly with the regeneration process. In this context, natural materials display important bioactive properties, having been designed and fabricated from nature to fulfill specific functions. Biopolymers and bioceramics can be extracted from natural sources and manipulated in terms of composition and structure to obtain multifunctional systems, specifically designed to direct host cell activity and tissue responses. This concept has been extensively studied and applied in the last years to the design and fabrication of scaffolds for tissue engineering (TE) strategies.

### 1.1.1 Naturally Derived Materials for Osteochondral Tissue Engineering Applications

Natural materials for TE scaffolds can be divided in two major classes: biopolymers (proteins, polysaccharides, and polymers derived from bacterial fermentation) and bioceramics. Recently, proteins such as collagen, silk, fibrin, and keratin, as well as polysaccharides like chitosan, alginate, and hyaluronic acid have been investigated in osteochondral tissue engineering (OCTE) [1–5]. In general, the sources of natural materials are plants, animals, or microorganisms [6, 7]. In Table 1.1, we briefly summarized the most important natural sources used to extract proteins and polysaccharides currently used in OCTE, along with materials properties relevant to OC regeneration. Polymeric materials like polyesters produced by bacterial fermentation have been also proposed for engineered degradable scaffolds [8]. In the last decade, innovative material sources have also been considered [6, 9, 10]. For example, collagen, typically isolated from pig and cow, is now isolated from squid, jellyfish, and fish bone and scales [6]. In addition, marine organisms can provide bioactive ceramics such as natural aragonite, calcium phosphates, and biogenic silica derived from corals, sponges, and diatomaceous deposits [11, 12]. Biopolymers and bioceramics can be processed and assembled into 3D constructs with different morphology, architecture and presentation. For example, porous scaffolds (sponges), hydrogels, membranes, and fibrous structures have been developed [9]. Similarly, cross-linkable hydrogels prepared from water-based processes in relatively mild conditions have been applied to cell encapsulation, bioprinting, and microfabrication [13–15]. In OCTE, natural materials are seldom used alone. Multicomponent and multilayered scaffolds are usually designed combining different biopolymers to meet the requirements of both subchondral and cartilage regeneration. In addition, natural materials can be combined with synthetic polymers to modulate mechanical

**Table 1.1** Summary of naturally derived biopolymers: their main sources, relevant characteristics, and applications proposed in OCTE

| Material | Main sources | Relevant properties in OCTE | Material presentation | OCTE applications |
|---|---|---|---|---|
| **Proteins** | | | | |
| Collagen [160] | Rat tail, Bovine tendon, Fish bone, scales and skin, jellyfish | Main component of OC ECM, biodegradable, bioresorbable, biorecognition, optimal cell adhesive properties, support pluripotent cell differentiation | Sponge, Hydrogel | Multilayered scaffold for OC regeneration [161], Bilayered scaffold with HA and HAp [4], Scaffold for chondrogenic differentiation of AdMSCs [162] – In situ gelling cell carrier [120] |
| Gelatin [44] | Porcine/calf skin, bovine tendon, Marine organisms | Biodegradable, bioresorbable, biorecognition, optimal adhesive properties, low antigenicity compared to collagen | Sponge, Hydrogel, Signaling molecule | GFs-loaded chitosan-gelatin gel for OCTE [163], Cross-linked gel for cell/GFs delivery [48], Cell adhesive coating [47] |
| Silk fibroin [52] | Silkworm cocoons, Recombinant bacteria | Versatile processing methods, slow degradation in vivo, good adhesive properties, support pluripotent cells differentiation | Sponge, Hydrogel | Bilayered scaffold for OCTE in vivo [59], Trilayered scaffold with HAp for OCTE [60], Cell-laden hydrogel for OC interphase [2] |
| Silk Sericin [61] | Silkworm cocoons | Cell adhesive properties, ability to induce nucleation of HAp crystals, mitogenic effect | Hydrogel, Signaling molecule | Additive to induce HAp deposition and BMSCs osteogenic differentiation [65] |
| Fibrin [26] | Bovine/human plasma | Partially controllable degradability, cell compatibility and flexibility, hemostatic effect, bioactivity and biomimetic properties | Hydrogel, Coating, Signaling molecule | Cell-seeded hydrogel for cartilage TE [36], Carrier for chondrogenic cells [69], Biomimetic cell-laden coating [72] |

(continued)

**Table 1.1** (continued)

| Material | Main sources | Relevant properties in OCTE | Material presentation | OCTE applications |
|---|---|---|---|---|
| Keratin [76] | Animal/human hair, horns, claws, hooves | Large availability, minimal inflammatory reaction, excellent cell adhesive properties, ability to support vascularization, slow degradation | Sponge Film | Osteoinductive porous scaffold [82] |
| Polysaccharides | | | | |
| Chitosan [83] | Partial acetylation of chitin from crustaceous exoskeletons | Similarity to GAGs in articular cartilage, antibacterial properties, support wound healing | Hydrogel Sponge | Bilayer scaffold with HAp [91] Porous scaffold for OC [164] Cell-laden injectable hydrogel [165] |
| Alginate [97] | Brown seaweed algae | Nontoxic, mild cross-linking conditions, hydrophilicity (hydrogel absorb considerable amount of water and body fluids) | Hydrogel Sponge | Injectable hydrogel for cartilage defects [166] Porous scaffold with TGF-β/HAp for OC [110] Chondrogenic differentiation of AdMSCs [106] |
| Hyaluronic acid (HA) [111] | Rooster combs Microbial fermentation | Main component of hyaline cartilage ECM, water soluble and versatile cross-linking methods, easy chemical modification, degradability | Hydrogel | MSCs and GFs delivery for cartilage repair [167] Cell laden hydrogel for OC bioprinting [168] Injectable gel for MSCs delivery [75] |
| Gellan gum (GG) [126] | Bacterial fermentation (*Sphingomonas* group) | Similarity to GAGs in articular cartilage, GG forms thermoreversible hydrogels with tunable mechanical properties, injectable hydrogels, | Hydrogel | Cell-laden hydrogel for chondral region [132] Bilayered scaffold for OC regeneration [134] Injectable cell-laden hydrogel [169] |

behavior or degradation kinetics. Ceramics have found application both as porous scaffolds for bone regeneration and as fillers/additives to improve mechanical performances and bioactivity of hydrogel constructs [16]. Recently, a variety of natural origin bioceramics have raised great interest particularly in bone TE [17], and are expected to receive larger and larger attention in the coming years due their inherent morphological and chemical advantages over their synthetic counterparts [10]. Synthetic materials for OCTE are reviewed elsewhere in this book. Here we present biopolymers (proteins and polysaccharides) and bioceramics that have been significantly investigated for OCTE approaches. Topics such as natural sources and extraction methods, physicochemical structure and bioactivity of the different materials are the focus of this chapter. For each material we concisely highlight properties relevant to OCTE, chemical modifications, area of application (bone TE, OC regeneration, or cartilage TE), scaffold presentation, and possible clinical impact. We report protein-based materials such as collagen/gelatin, silk proteins, fibrin, and keratin, along with polysaccharides like chitosan, alginate, hyaluronic acid (HA), and gellan gum (GG). Bacterial-derived polyhydroxyalkanoates (PHAs) are also briefly discussed. In the last section, we introduce materials like coral-derived aragonite, natural calcium phosphates, and hydroxyapatite (HAp). In all cases representative studies on scaffolds for OCTE will be underlined.

## 1.2 Biopolymers

### 1.2.1 Proteins

#### 1.2.1.1 Collagen and Gelatin

Collagen is the most abundant protein in mammals, but it is present throughout the entire animal kingdom, including birds and fishes. Collagen fundamental building block is a right-handed triple-helix consisting of three polypeptide chains, called $\alpha$-chains, held together by hydrogen bonding [18]. The characteristic repeating unit of $\alpha$-chains is glycine-X-Y, where X and Y are often proline and hydroxyproline, respectively. Collagen fibrils are maintained in position within a collagen fiber by interfibrillar proteoglycan bridges [19]. To date, at least 29 variants of collagen have been identified which differ for amino acid composition of the $\alpha$-chains and for the nature of nonhelical proteins, resulting in significant differences in structures and functions [20]. The majority of these collagens, especially the fibril forming collagens (type I, II, III, and V) is commonly found in vertebrates and has excellent biocompatibility and biodegradability. The most abundant type of collagen is collagen type I, whose triple helix is a heteropolymer consisting of two $\alpha$1-chains and one $\alpha$2-chain [21]. Collagen type I is a structural protein that can be found in tendon, skin and bone, and largely determines the mechanical behavior of these connective tissues. Collagen type II, V, VI, IX, X, and XI are normally found in hyaline cartilage tissue [22], while only collagen type I and V are consistently present in the underlying subchondral bone tissue [23]. The bone-to-cartilage transition region is usually characterized by collagen type II and X [24, 25]. Both type I and type II collagen scaffolds can facilitate cartilaginous tissue formation; however, collagen type I collagen is known to induce chondrocytes dedifferentiation [26]. Common sources of collagen for TE include bovine tendons and skin, rat-tail and porcine skin, but collagen scaffolds can also be prepared from allogeneic or xenogeneic decellularized tissues.

Recently, marine-derived collagen has attracted much attention as an alternative to mammalian collagen [27]. Within marine resources, collagen is commonly isolated from fish skins, jellyfish, sea sponges, echinoderms, and cephalopods [28]. Generally, marine origin collagen extracted from fish skins, scales and bones is considered to be consistent with collagen type I, while collagens derived from jellyfish and sea sponges are consistent with type II and type IV collagen, respectively [29]. Sources, extraction methods, and pretreatments affect the final characteristics of collagen, such as composition, rheological properties, solubility, and thermal stability which consequently affect its biological activity [30]. Scaffolds prepared from fibrillized jellyfish collagen *(Rhopilema esculentum)* showed good biocompatibility and the ability to preserve the chondrogenic phenotype of porcine chondrocytes [31, 32]. Furthermore, jellyfish collagen–alginate hybrid systems demonstrated to support human MSCs chondrogenic differentiation [33].

Mammalian collagen has been used as the main component of a variety of OC scaffolds, both for chondral and bone regeneration [4]. Several studies investigated the use of collagen from different sources in combination with other biomaterials for OCTE applications [34–36]; some examples are summarized in Table 1.2. Collagen type I/III membranes stabilized by fibrin glue (Chondro-Gide®), have shown promising results for the treatment of knee OC lesions with autologous matrix-induced chondrogenesis (AMIC) treatment [37, 38]. MaioRegen® is a commercially available, trilayered scaffold with graded composition consisting of equine collagen type I (in the chondral region) and magnesium-doped HAp (Mg-HAp, predominant in the subchondral region) [39]. Success of this product in case of severe and large chondral/OC lesions has been demonstrated in some literature [40, 41]. Nevertheless, incomplete cartilage repair and poor subchondral bone integration was found in some cases at the 1- and 2.5-year follow-ups [42]. Recently, a similar bilayered scaffold composed of cross-linked collagen and Mg-HAp demonstrated initial formation of new bone and chondral tissue in vivo [43].

Gelatin is the result of non-reversible thermal denaturation or hydrolysis of collagen. The triple helix structure of the collagen is lost at relatively mild temperatures (40 °C) due to the disruption of hydrogen bonds between α-chains. Hydrolytic denaturation can occur either in alkaline or acidic conditions and it is induced by cleavage of covalent bonds along the peptide backbone. Overall, the properties of gelatin materials are strongly dependent from collagen type and source as well as from the denaturation process. In solution, gelatin acts as a thermally responsive protein that can form reversible gels at temperature around 35 °C. As a result, gelatin hydrogel is not stable around body temperature limiting the possibility to implement pure gelatin scaffolds. However, gelatin can be chemically cross-linked to increase stability, mechanical properties and delay degradation in in vivo conditions [44]. Moreover, gelatin can be combined with other materials such as starch or synthetic polymers to improve the mechanical properties of the resulting scaffolds [45, 46]. In OCTE field, gelatin is frequently used to increase bioactivity and biocompatibility of other natural or synthetic materials. For example, gelatin chemically

**Table 1.2** Selected examples of collagen-based scaffolds for OCTE applications obtained by combination of collagen with other natural derived or synthetic biomaterials

| Combined biomaterial | Material presentation | Testing | Results |
|---|---|---|---|
| Hydroxyapatite/ alginate (HAp/ alginate) [170] | Hydrogel | In vitro (chondrocytes from articular cartilage) | Enhancement of tensile and compressive moduli, increase of cell viability, upregulation of cell proliferation and hyaline cartilage marker production |
| β-tricalcium phosphate (β-TCP) [171] | Ceramic sponge | In vitro (BMSCs from new-born NZ white rabbit) | (β-TCP) scaffold was fabricated by ceramic stereolithography with collagen incorporation. Efficient cell adhesion, migration, and distribution in 3D |
| Polyvinyl alcohol (PVA) [172] | Nanofibers and sponge | In vitro (primary porcine chondrocytes) | Activation of chondrocytes' synthesis of glycosaminoglycan and type II collagen in all structure considered. Cells organized in lacunae 3 weeks after seeding |
| Glycosaminoglycans (GAGs) [34] | Hydrogel | In vitro (porcine AdMSCs and BMSCs) | Cell adhesion, proliferation, and migration in floating and static conditions. No cells adhesion when cultured in dynamic roller condition. |
| (GAGs) [173] | Biphasic scaffold | In vivo (acute OC defect in sheep model) | Improvement in osteochondral repair with collagen/GAG scaffold loaded with fibroblast growth factor-18 |
| (GAGs)/calcium phosphate [174] | Sponge | In vitro (primary human MSCs) | Incorporation of a dual chondrogenic and osteogenic GFs system (BMP-9/ MC-GAG). Up-regulation of chondrogenic markers and sulfated GAGs. No effect on ALP, mineralization, collagen I production |
| Hydroxyapatite/ hyaluronic acid (HAp/HA) [4] | Multilayered scaffold | In vivo OC critical-sized defect in rabbit knee | Increased levels of repair in presence of the multilayer scaffold. Diffuse host cellular infiltration, with a zonal organization, formation of a cartilaginous layer with evidence of an intermediate area |
| Magnesium-doped HAp crystals [43] | Bilayered scaffold | In vitro (hMSCs) in vivo (nude mice) -loaded with hMSCs | Improvement of cells attachment and proliferation, chondrogenic and osteogenic differentiation, and synthesis of ECM molecules. New tissue growth (bone and chondral), neoangiogenesis activation starting at 4 weeks |

(continued)

Table 1.2 (continued)

| Combined biomaterial | Material presentation | Testing | Results |
|---|---|---|---|
| Hyaluronate/TCP [35] | Sponge | In vitro: Rabbit MSCs In vivo (ectopic implantation rabbit knee) | Ability to act as a stem cell carrier; chondrogenesis in vitro and in vivo. Good cartilage regeneration and integration. Chondroinductive properties due to the presence of hyaluronate |
| HAp [175] | Sponge | In vivo (OC defects in the patellar groove of Japanese white rabbits) | Bone repair properties, high affinity for BMP-2. Effective for rigid subchondral bone repair |
| HAp [161] | Multilayered scaffold | Clinical study (30 patients with knee chondral or OC lesions) | Follow-up at 2 years. Slower recovery for all patients considered (patients with adverse events, old, previous surgery, with patellar lesions). A faster recovery was observed in active patients. Safety and potential clinical benefit of the biomimetic OC scaffold, able to promote bone and cartilage tissue restoration |

attached to calcium alginate porous scaffolds largely improved cell adhesion and proliferation, and supported differentiation of MSCs into chondrogenic and osteogenic lineages [47]. A furfurylamine-conjugated gelatin hydrogel cross-linked by visible light has been used as scaffold for BMSCs and collagen-binding GFs in a rabbit OC defect model; the system demonstrated regeneration of novel articular cartilage-like tissue and integrated subchondral bone [48]. Gelatin/chitosan hydrogels loaded with TGF-β1 exhibited low cytotoxicity the good repair potential for OC defect in vivo [49].

### 1.2.1.2 Silk Proteins

Silk is natural protein-based fiber secreted by arthropods like silkworms and spiders. Fibers are produced by a spinning process where an aqueous protein solution is converted in an insoluble filament [50]. The amino acid composition, structure, and mechanical properties of silks can be extremely different depending on animal species and specific silk function. Silk from silkworm cocoons have been used for many years as suture filaments, more recently acquiring novel attention for other applications in medicine and particularly in tissue engineering [51, 52]. Silkworm silk is composed of fibroin, the structural core component, and sericin, the hydrophilic protein coating. The primary structure of silk fibroin proteins is mainly composed of glycine and sericin amino acids, with heavy (~350 kDa) and light (~25 kDa) chain peptides connected by disulfide bond [53]. Silk can be processed into versatile formats, with tunable properties such as degradation kinetic and bioactivity, for a diversity of medical applications responding to precise physical and biological requirements. Many of these techniques are water based and take inspiration from the natural silk spinning process.

Highly porous 3D silk fibroin scaffolds can be fabricated using several techniques such as porogen-leaching [53], freeze-drying [54], or electrospinning technology [55]. Silk fibroin scaffolds combined with human MSCs, dexamethasone, and TGF-β3 were tested in cell culture and chondrogenesis of MSCs was assessed for cartilage-specific ECM gene markers. After a 3-week culture, histological analysis showed a spatial cell arrangement and collagen type-II distribution of human mesenchymal stem cells comparable with articular cartilage tissue [56]. Chemical modifications with cell binding domains and growth factors where proposed in order to improve the interactions between chondrocytes/MSCs and fibroin. Arg-Gly-Asp-Ser (RGDS)-modified fibroin demonstrated the ability to enhance mRNA expression levels of integrin α5β1 (an integrin that binds to matrix macromolecules and proteinases and thereby stimulates angiogenesis) and aggrecan (cartilage-specific proteoglycan core protein (CSPCP) or chondroitin sulfate proteoglycan 12 h after seeding [57]. It was also suggested that RGDS induced moderate chondrocyte adhesion to fibroin while maintaining the chondrogenic phenotype and facilitated chondrogenesis. Fibroin was also modified using the diazonium-coupling chemistry to control protein structure and overall hydrophilicity to direct encapsulated MSCs toward enhanced osteogenic differentiation [58]. Chemically modified fibroin hydrogels showed the ability to effectively interact with chondrocytes and MSCs due to the immobilized cell binding domains and growth factors. The constructs maintained the chondrogenic/osteogenic phenotype without extra growth factors [2]. Recently, fibroin scaffolds from mulberry (*Bombyx mori*) and non-mulberry (*Antheraea mylitta*) silkworms loaded with TGF-β3 and BMP-2 were compared in vitro and in vivo [59]. The study showed species-specific ECM deposition and envisioned interspecies fibroin blends or multilayered scaffolds with a combination of mulberry and non-mulberry silks for the regeneration of OC defects. Ding et al. developed a biomimetic trilayered scaffold using silk fibroin and HAp by combining paraffin leaching and thermally induced phase separation techniques. The construct supported simultaneous differentiation of AdMSCs toward chondrocytic/osteoblastic phenotype in the chondral and subchondral region in vitro [60]. Chondral and subchondral repair in a rabbit femur model was achieved with fibroin/chitosan/nano-HAp layered scaffolds [33]. A bilayered scaffold consisting of integrated fibroin layer and a silk–calcium phosphate layer supported cartilage regeneration and subchondral bone ingrowth and angiogenesis in a rabbit knee model [32].

Sericin is obtained as a by-product in the silk industry from the degumming process of cocoon silk fibers. Soluble sericin extracted from native silk fibers was considered mainly in wound dressing but also in bone TE applications [61–63]. For OCTE, sericin could find application as bioactive signaling molecule for the regeneration of subchondral bone. For example, silk sericin extracted from non-mulberry silkworm cocoon (*Antheraea pernyi*) can mediate the formation of HAp crystals in simulated body fluid [64]. The so formed HAp crystals have been shown to stimulate cell adhesion and proliferation but also promote osteogenic differentiation of human BMSCs [65].

### 1.2.1.3 Fibrin

Fibrin is a major component of the blood clot and it is formed by polymerization of fibrinogen mediated by thrombin enzymatic activity [66]. Fibrin clot acts as a temporary template guiding the process of wound healing and the formation of new tissue, and is later degraded by plasmin. Fibrin and fibrinogen have a pivotal role in a number of biological processes, such as blood clotting, inflammatory response, cell–cell and cell–matrix interaction, wound healing, endothelial cells recruitment and angiogenesis, and regulation of fibroblasts activity [67]. Fibrinogen is a soluble molecule consisting of two sets of three polypeptide chains, namely, A$\alpha$-, B$\beta$- and $\gamma$-chains, jointed by interchain disulfide bridges. The thrombin-mediated cleavage of the N-terminal fibrino-peptide segments in the fibrinogen A$\alpha$- and B$\beta$-chains initiates the spontaneous assembly of polypeptides into a network fibrin fibrils [26]. Fibrin network is further stabilized by physical and chemical cross-links between $\gamma$-chains (or between one A$\alpha$-chain and one $\gamma$-chain) initiated by Factor XIIIa.

Thanks to its hemostatic effects, fibrin has a long history as hemostatic agent, sealant, and tissue adhesive [66]. In addition, fibrin has been proposed as a carrier for the delivery of antibiotics, chemotherapeutic drugs, GFs, and engineered plasmids for GF expression in gene therapy [68]. In general, the degradation profile and mechanical properties of fibrin gels depend on the fibrinogen concentration, calcium concentration and local pH [69]. Due to the characteristic properties of biocompatibility, partially controllable degradability, cell compatibility, and flexibility fibrin hydrogels have been proposed in a number of TE applications. The major disadvantages of fibrin hydrogels include poor mechanical properties and rapid enzyme-catalyzed degradation in vivo, which can be however controlled by chemical cross-linking [70]. In addition, mechanical properties and degradation kinetics can be improved by blending fibrin gels with other synthetic or natural polymers. For example, Filova et al. developed a composite hydrogel consisting of collagen type I, hyaluronan, and fibrin with better mechanical properties and stability, when compared to pure fibrin gel; such composite hydrogel, seeded with autologous chondrocytes, showed the formation of new hyaline cartilage in rabbits [36]. Fibrin hydrogels seeded with chondrocytes demonstrated to promote the regeneration of cartilaginous matrix and new cartilage tissue both in vitro and in vivo [26]. Moreover, fibrin gel was infiltrated into porous scaffolds to increase the bioactivity and biomimetic properties of synthetic materials [71, 72]. Wang et al. demonstrated that PLGA porous scaffold filled with fibroin gel could guide the regeneration of a well-integrated neocartilage in full-thickness rabbit defect [73]. A biphasic scaffold consisting of platelet-rich fibrin gel combined with HAp was considered to support xenogenic transplantation of differentiated hMSCs in rabbit model [74]. Recently, an injectable hydrogel consisting of fibrin and chemically modified HA was designed to delivery of BMSCs for articular cartilage repair in osteoarthritis therapy. The composite hydrogel showed to promote BMSCs proliferation and possess chondrogenesis potential in vitro [75].

#### 1.2.1.4 Keratin

The term "keratin" defines a family of structural, filament-forming proteins found in epidermal and corneous tissues like hair, nails, horns, claws, hooves, turtle scute, whale baleen, beaks, and feathers [76, 77]. In these materials, keratin proteins are assembled in a complex hierarchical structure. Being insoluble and heavily crosslinked via sulfur bridges, some keratin-based structures are between of the toughest biological materials [78]. Keratins extracted from hair and wool represents a whole new family of biomaterials for applications like TE scaffolds, drug delivery, and wound healing [79, 80]. Keratin biopolymers carry cell-binding domains such as arginine–glycine–aspartic acid (RGD) and leucine–aspartic acid–valine (LDV) and exhibit excellent cell adhesion properties [81]. Lately, keratin-based materials have gained some attention in bone TE applications [80], and are considered promising candidates for the subchondral region of bilayered scaffolds. Keratin porous scaffolds were demonstrated to elicit minimal inflammatory reaction and to support tissue healing and neovascularization [7]. Similarly, human hair keratin/jellyfish collagen/eggshell-derived HAp scaffolds supported self-differentiation of human AdMSCs into osteogenic lineage without additional induction agents [82].

### 1.2.2 Polysaccharides

#### 1.2.2.1 Chitin and Chitosan

Chitin is a linear polysaccharide made of N-acetylglucosamine residues linked through β-(1,4)-glycosidic bonds. Chitin is the second most abundant natural polymer in nature after cellulose, and is present in many biological structures like crustaceous exoskeletons, fungal cell walls, and insect cuticles [83]. The presence of strong intermolecular bonds prevents chitin solubility, thus limiting de facto the possibility of manipulation and development of chitin-based products. Chitosan is produced by partial deacetylation of chitin and consists of D-glucosamine (deacetylated unit) and N-acetyl-D-glucosamine (acetylated unit) randomly distributed within the polymer and linked by β-(1–4)-glycosidic bonds [84]. Chitosan materials are considered to be nontoxic, non-allergenic and to elicit minimal foreign body reaction. Moreover, chitosan demonstrated also good antithrombogenic and hemostatic potential, bacteriostatic, antifungal activity, analgesic effect, and positive interaction with wound healing progression [85, 86]. Chitosan is only soluble in acidic solutions, while it remains insoluble in neutral and alkaline conditions. However, water-soluble alternatives can be obtained by reducing chitosan molecular weight and by chemical modification. Methods like grafting with glycolic acid, enzymatic degradation and hydrolysis with hydrogen peroxides have been developed to obtain water-soluble chitosan [7, 83, 87]. Chitosan is degraded in vivo by lysosome-mediated hydrolysis producing nontoxic saccharide by-products. Degradation rate can be modulated acting on the degree of deacetylation; as a general rule, the rate of degradation decreases as the degree of deacetylation increases [88]. Chitosan can be chemically functionalized thanks to the presence of reactive side groups. Chitin and chitosan have been successfully used to produce scaffolds for TE applications in different presentations like hydrogels, nanofibers, beads, micro/nanoparticles, membranes, and sponges [83]. Thanks to the chemical similarity to some components of natural cartilage like GAGs, HA and chondroitin sulfate, chitosan has received much attention as an alternative material for articular cartilage repair. Chitosan hydrogels prepared by enzymatic cross-linking were shown to support the proliferation of chondrocytes and MSCs, maintain the chondrogenic phenotype and morphology, and improve the deposition of cartilaginous ECM in vitro [89]. Long-term subcutaneous implantation demonstrated a robust chondrogenic potential of MSCs-laden chitosan hydrogels with accumulation of high levels of aggrecan and deposition of collagen type II. At front of the deposition of a large amount of cartilaginous ECM, neither vascularization nor endochondral ossification were observed in vivo [90]. As a consequence, chitosan finds applications in the chondral region of biphasic scaffolds for OCTE. For example, a macroporous HAp/chitosan bilayered scaffold was developed combining a sintered HAp construct (bone layer) with a chitosan sponge (cartilage-like layer) obtained by freeze-drying. Such bilayered scaffold supported goat BMSC attachment, proliferation, and selective differentiation in to osteoblasts and chondrocytes in the two layers [91]. Chitosan is often blended with other polysaccharides like CS, HA, and alginate to create blended hydrogels for OCTE, to encapsulate and deliver pluripotent cells and preserve chondrogenic phenotype [92–95]. Scaffolds based on fumarate–vinyl acetate copolymer and chitosan supported bone marrow progenitor cell osteogenic development, primary chondrocyte growth and extracellular matrix deposition [96].

### 1.2.2.2 Alginate

Alginic acid, or alginate, is a natural anionic polysaccharide found in the cell walls of brown seaweeds of the class *Phaeophyceae*. From a chemical point of view, alginates are a family of unbranched block copolymers of $(1\rightarrow4')$-linked β-D-mannuronic acid (M) and α-L-guluronic acid (G) units. The copolymer can consist of pure M-blocks, pure G-block, alternating GM-blocks or random GM blocks with variable length and arrangement [97]. The relative amount of each group as well as the block sequence can vary with the alginate source, harvesting time and extraction procedures [98]. Alginate molecules can be physically cross-linked in presence of divalent cations, typically $Ca^{2+}$, $Mg^{2+}$, or $Ba^{2+}$, that cooperatively interact with G-blocks to form ionic interchain bridges; thus leading to physical gelation of the solution (ionotropic gelation) [99]. Due to the large availability in nature, mild gelation conditions, limited toxicity, low immunogenicity and cost, alginate has been widely proposed for drug release and for cells encapsulation applications [100, 101]. Moreover, delayed alginate gelation was exploited to develop in situ crosslinkable injectable hydrogels for cell delivery [102]. Due to the high water content and the lack of cell binding sites, alginate hydrogels have limited protein adsorption and cell adhesion ability. However, cell compatibility can be improved by blending alginate with pro-adhesive proteins (gelatin, collagen, and laminins) and/or by chemical modifications with active species like RGD-containing peptides [103]. Physically cross-linked alginate hydrogels are relatively stable in water; however, with time the leaching of divalent cations can undermine long-term stability and mechanical strength, particularly in physiological conditions [2]. Hydrogel stability can be greatly improved using methacrylated alginate to form covalent cross-linking bonds.

In the field of OCTE, alginate-based hydrogels were proposed both in the cartilage compartment and in the subchondral bone region. In particular, alginate hydrogels were used to deliver bone progenitor cells, including MSCs, for bone regeneration [104]. Pure alginate and RGD-modified alginate gels were shown to support a complete regeneration of critical-sized bone defects in various animal models. In addition, alginate was combined with inorganic materials like HAp and TCP to develop scaffold with interconnected porosity with a potential in bone tissue engineering applications [97]. On the other side, alginate was used to encapsulate and delivery chondrocytes to support chondrogenic phenotype and to promote the synthesis of cartilage-specific macromolecules such as proteoglycans and collagen [26]. It has been shown that the fate of adult stem cells could be controlled in alginate hydrogels with a proper mix of soluble factors and physical interaction with the 3D environment [105]. Chondrogenic differentiation of BMSCs and AdMSCs in alginate hydrogels was repeatedly confirmed [106, 107]. Alginate hydrogels was also combined with synthetic degradable polyesters to prepare multilayered scaffolds loaded with GFs for the treatment of full-thickness OC defects in rabbits with encouraging results [108, 109]. Overall, alginate scaffolds with different presentations and compositions could be used both in the bone phase and the chondral phase of multilayered OC scaffolds. Recently, Colaccino et al. fabricated a mechanically competent and cell compatible scaffold for OCTE based on freeze-dried alginate gels. The scaffold presented a bilayered structure, with the combination of alginate-sulfate and TGF-β1 in the chondral layer and a alginate/HAp composite in the bone layer [110].

### 1.2.2.3 Hyaluronic Acid

Hyaluronic acid (HA) or hyaluronan is a fundamental component of the human connective tissue. HA can be found in the ECM of the skin, hyaline cartilage, vitreous humor, and nucleus polposus and is present at elevated concentrations in the synovial fluid. HA is critical for the correct lubrication of arthritic joints and mechanical behavior of soft tissues; furthermore, it influences several biological functions such as cell motility, organization, and cell–ECM interaction [111]. HA is a glycosaminoglycan (GAG) composed of repeating disaccharide units, namely, N-acetyl-α-D-glucosamine and β-D-glucuronic acid linked by alternate β(1,3)- and β(1,4)-glycoside bonds [7]. Due to a high density of negative charges along the polymer chain, HA is very hydrophilic and adopts extended random coil conformation in solution. HA has been used for wound healing, tissue engineering, ophthalmic surgery, arthritis treatments [111–113]. HA hydrogel mechanical properties and degradation kinetics can be controlled by chemical cross-linking [114, 115].

In hyaline cartilage, HA plays a central role regulating protein adsorption and providing adhesion sites for chondrocytes in ECM [116]. HA-based scaffolds are usually designed to promote and maintain the chondrocytic phenotype and promote chondrogenesis in vitro and in vivo [117]. The implantation of autologous chondrocytes using a HA matrix has been used to treat OC defects resulting in the formation of hyaline-like cartilage [118, 119]. Chemically modified HA hydrogels, alone or in combination with other biopolymers, were used to develop injectable in situ forming hydrogel in cartilage regeneration or OCTE applications [75, 93, 120]. For example, the ability of chitosan/HA hydrogels to retain large amount of water, support encapsulated cells and maintain chondrogenic cell phenotype make them ideal candidates for cartilage repair [94]. Some examples of HA-based scaffolds for OCTE are reported in Table 1.3.

**Table 1.3** Selected examples of hyaluronic acid-based scaffolds for OCTE applications obtained by combination with other biomaterials or by chemical modifications

| Combined material/ chemical modification | Material presentation | Testing | Results |
|---|---|---|---|
| Chondroitin sulfate (CS) [176] | Biphasic scaffold | In vitro (rat BMSCs) In vivo (Rat OC defect model | Elevated expression of osteogenic markers (bone sialoprotein, runt related transcription factor-2, and BMP-2) in the bone side and chondrogenic markers (collagen type II, aggrecan) in the cartilage. Good alignment of collagen type II fibrils and aggrecan |
| Sodium alginate [177] | Sponge | In vitro (chondrocytes embedded into scaffold) | Proteoglycan and collagen synthesis. Cells showed an evident spherical shape and a non-oriented and disperse actin microfilament network. Production of collagen II |
| Cross-linked hyaluronan, benzylated hyaluronan [178] | Sponge | In vivo (rabbit articular cartilage defect) | Cross-linked hyaluronan was more effective than benzylated hyaluronan in the treatment of articular cartilage defect at 12 weeks after implantation |
| Type I and type II collagen [4] | Multilayered scaffold | In vivo (OC defects in rabbit model) | Ability to guide host reparative response. Tissue regeneration with zonal organization, repair of the subchondral bone, formation of an overlying cartilaginous layer, presence of an intermediate tidemark |

(continued)

1 Natural Origin Materials for Osteochondral Tissue Engineering

**Table 1.3** (continued)

| Combined material/ chemical modification | Material presentation | Testing | Results |
|---|---|---|---|
| Gelatin/CS [179] | Bilayered scaffold | In vivo (OC defects in rabbit) preloaded with chondrocytes/ BMSCs) | After 6 and 12 weeks: Hyaline-like cartilage formation, with collagen II synthesis. Scaffold replaced mainly by new bone, with little remained in the underlying cartilage. 36 weeks: Scaffold completely resorbed |
| Collagen type I/ fibrinogen [180] | Composite hydrogel | In vivo (OC defects in rabbit loaded with synovium MSCs) | 24 weeks after transplantation: Hyaline cartilage-like tissue production into the defect, with high content of GAGs and collagen II |
| Fibronectin coating [181] | Sponge | In vivo (OC defects in rabbit) | 24 weeks of implantation: The defects filled with new bone with a upper layer of cartilage, evident integration with the adjacent cartilage |
| Chitosan [182] | Hydrogel fibers by wet spinning | In vivo (OC defects in rabbit) | 12 weeks after implantation: Production of mechanically competent hyaline-like cartilage into the defect with integration with the adjacent native cartilage. Physiological reconstitution of subchondral bone |

### 1.2.2.4 Gellan Gum

Gellan gum (GG) is an anionic extracellular polysaccharide secreted by fermentation of carbohydrates by different bacteria from the *Sphingomonas* genus [121]. Properties and purity of the gellan gum materials depend on bacterial population, dietary pattern, and extraction procedure (i.e., temperature, pH, and purification cycles) [122]. GG is a water-soluble heterosaccharide based on D-glucose, D-glucuronic acid, and L-rhamnose repeating subunits [123]. GG is a FDA-approved compound used as stabilizer and gelling agent in food and cosmetic industry, but also commercialized for clinical applications. Typically, GG can be prepared in either low or high acylated forms; the deacylated form is more commonly used in TE applications. GG molecules can be homogeneously dispersed in hot aqueous conditions, but undergo gelation upon cooling below the sol–gel transition temperature (between 40 and 50 °C). The gelation is initiated by GG chains self-assembling of into paired helical structures; in a second stage, helices further aggregate forming cluster junctions, eventually leading to the gelation of the system [124]. The junction zones of the paired GG helical chains can be stabilized in presence of divalent cations (typically $Ca^{2+}$ and $Mg^{2+}$), thus resulting in irreversible gelation [125].

Over the past 20 years, GG-based hydrogels were taken into consideration for OCTE applications thanks to their biocompatibility, biodegradability, and injectability [126]. In fact, hydrogels can be easily produced by thermal gelation and in presence of cations without toxic chemical reagents. However, the temperature of spontaneous gelation temperature is generally too high, and the resulting gel is considerably more fragile and weaker when compared to natural cartilage. In addition, GG hydrogels tends to reduce mechanical properties in vivo due to divalent cations exchange with body fluids. These factors represent critical limitations for TE clinical applications. A number of strategies have been devised to decrease gelation temperature, to improve injectability, and to increase mechanical performances in vivo—i.e., by GG oxidation, by optimizing cations content, and by blending low and high acyl GG in the right proportion [15, 127, 128]. In alternative, GG can be covalently modified to improve cell adhesion and mechanical properties of the hydrogel. Most notably, GG methacrylation has been implemented to obtain modified GG species with photo-induced cross-linking abilities and improved reactivity [129–131]. Cell-laden GG hydrogels were studied for *in vitro* chondrogenesis and long-term cartilage regeneration [15, 132]. Injectable GG hydrogels loaded with autologous AdMSCs predifferentiated toward chondrogenic lineage were used to successfully treat full-thickness articular cartilage defects in a rabbit model [133]. Pereira et al. presented a monolithic bilayered scaffold with distinct cartilage-like and bone-like zones and cohesive interface [134].

## *1.2.3 Polyhydroxyalkanoates*

Polyhydroxyalkanoates (PHAs) are a family of linear polyesters synthetized by a variety of bacterial species through the fermentation of lipids, sugars, alkanes, alkenes, and alkanoic acids under controlled environmental and feeding conditions [135]. PHAs can be retrieved from bacterial cell cytoplasm in form of water-insoluble spherical granules. The physicochemical properties of PHAs compounds are strictly dependent on the monomer chemical structure, which can be tuned acting directly on the composition of the growing broth. Since 1927, when Maurice Lemoigne first discovered that poly(3-hydroxybutyrate) could be recovered from *Bacillus megaterium* cells, more than 100 different hydroxy acid monomers have been studied [136]. The PHAs are generally classified in 2 different groups depending on the number of carbon atoms in the monomeric unit; short-chain-length PHAs and medium-chain-length PHAs are based on monomers with 3–5 and 6–16 carbon atoms, respectively. Medium-chain-length PHAs have an elastomeric behavior, while short-chain-length PHAs are usually stiffer and more brittle due to a higher degree of crystallinity. These bacterial synthetized biopolymers are degraded primarily by hydrolysis via surface erosion both *in vitro* and *in vivo* [137]. Mechanical performances and resistance to degradation can be modulated acting on monomer length and side groups, blending different PHAs and by additional chemical modifications. As their synthetic counterparts PGA, PLA, and PLGA, PHAs lack of cell-recognition sites. Therefore, cell compatibility of PHAs-based scaffolds have been improved by blending or surface modification with natural biopolymers such as collagen, gelatin, chitosan and RGD peptides [8, 138, 139]. In addition, surface hydrophobicity can be reduced by plasma treatment [140]. Recently, PHAs have attracted great interest in biomedical field thanks to unlimited availability, chemical variability, tunable mechanical and degradation behavior, and optimal processability. PHAs can be formed as films, hydrogels, porous scaffolds, and nanofibrous meshes with applications in many TE areas [141]. A clear effect of composition and surface properties of different PHAs blend on chondrocytes behavior and osteochondral ossification was demonstrated as early as in 2005 [142]. PHA/Bioglass® scaffolds were developed promoting the formation of a thick layer of mechanically competent cartilage-like tissue in rabbit model [143]. Chondrogenic differentiation of BMSCs was obtained on PHA scaffolds coated with PHA granules binding protein PhaP and fused with RGD peptide, suggesting that such scaffolds could support cartilage regeneration [144].

## 1.3 Bioceramics

### 1.3.1 Natural Aragonite

Aragonite is one of the polymorphs of calcium carbonate mineral ($CaCO_3$) along with calcite and vaterite. Inorganic aragonite is formed in high-pressure metamorphic rocks and by precipitation of seawater in oceans and submerged caves. Nevertheless, aragonite is also the main component of most mollusk shells and coralline exoskeletons. Corals are a large family of marine invertebrates of the Anthozoa class, including more than 7000 species; however, species considered for medical applications are mainly of the genera *Porites* and *Acropora* [145]. Coral aragonite has long been considered in orthopedic and reconstructive surgery due to the analogies with cancellous bone in terms of 3D structure, pore size and interconnectivity, mechanical performances and calcium carbonate crystalline structure [146]. Thanks to its osteoconductive properties [147], marine-derived aragonite has been proposed as subchondral graft to support the regeneration of OC defects. Kon et al. studied a bilayered aragonite-based scaffold with HA impregnation in the chondral phase for the treatment of critical OC defect in a goat model. The construct induced both subchondral bone regeneration and consistent repair of the articular hyaline cartilage [12]. Recently, such aragonite-based scaffold showed a potential for the treatment OC lesions in humans, with encouraging clinical improvement 12 months after implantation [148].

### 1.3.2 Natural Calcium Phosphates and Hydroxyapatite

Calcium phosphate-based ceramics (CaPs) are a large class of bioactive materials that has been used as bone grafts and bone substitutes for as long as a century [149]. CaPs have been widely investigated in TE due to their excellent biocompatibility, bioactivity, and osteoconductivity [150]. The CaP crystalline forms most commonly used in TE are hydroxyapatite (HAp) and α-/β-tricalcium phosphates (TCP). Nanometric HAp is characterized by remarkable chemical and structural similarity to the natural apatite in bones. HAp substrates favor adhesion, proliferation, and osteogenic differentiation of MSCs in vitro. Furthermore, HAp bioactivity and osteoconductivity were shown to promote rapid bone formation in vivo [151]. Due to low mechanical strength and limited fracture toughness, the use of pure HAp scaffolds is hindered in load bearing applications, and HAp is often used in combination with TCP. Moreover, HAp has been used as functional coating or filler for polymeric porous scaffolds and hydrogels [152]. Natural HAp can be obtained from natural sources, such as mammalian bones, eggshells, fish scales, seashells, and marine food wastes via hydrothermal conversion of calcium carbonate structures [153, 154]. HAp extracted from natural sources has a calcium-deficient nonstoichiometric composition and incorporates other ions like $Na^+$, $K^+$, $Mg^{2+}$, $Sr^{2+}$, $Zn^{2+}$, $Al^{3+}$, $Cl^-$, $SO_4^{2-}$, and $CO_3^{2-}$. As a result, the composition of natural HAp is more similar to bone than stoichiometric HAp produced by synthetic methods [155]. It has been shown that nonstoichiometric Ca/P ratio and the presence of different ions strongly enhance the dissolution rate and bioactivity of natural HAp [156]. Several studies agreed that natural HAp from natural sources has a great potential for bone TE applications [157–159], and it should be considered in OCTE for the regeneration of subchondral bone.

## 1.4 Conclusion

Natural materials represent a unique opportunity for the design and fabrication of bioactive scaffolds for OCTE. Biopolymers and bioceramics can be successfully combined with synthetic materials as well as with growth factors and cells. Natural materials can be isolated from several sources, but depending on the organism considered and the extraction process used, chemistry and bioactivity can sensibly change. On the contrary, recombinant biopolymers produced by microbial fermentation can have consistent and tailorable composition and are emerging as a class of next generation biomaterials. However, the development of recombinant DNA technology is still limited by high production cost, infrastructural problems, and legal restrictions. Processing wastes from food industry, agriculture, sericulture, fishing, and textile industry can be also exploited for the production of natural materials with obvious environmental, ethical, and economical benefits, particularly in developing countries. To this regard, biomaterials and tissue engineering industry should capitalize on renewable sources and address issues related to environmental sustainability of the raw materials.

Moreover, the definition of common protocols for source selection, material extraction and purification as well as for scaffold fabrication and validation is now critical for the advancement of the field. This standardization process should aim to obtain materials with consistent and reproducible properties, to minimize interlab variability and batch-to-batch variations. Therefore, the improvement of quality assurance policies and standardization of manufacturing procedures are essential to pass regulatory approval process and for translation of naturally derived tissue engineered products to the clinic .

## References

1. Mano JF, Silva GA, Azevedo HS et al (2007) Natural origin biodegradable systems in tissue engineering and regenerative medicine: present status and some moving trends. J R Soc Interface 4:999–1030
2. Yang J, Zhang YS, Yue K, Khademhosseini A (2017) Cell-laden hydrogels for osteochondral and cartilage tissue engineering. Acta Biomater 57:1–25
3. Yan LP, Oliveira JM, Oliveira AL, Reis RL (2013) Silk fibroin/nano-CaP bilayered scaffolds for osteochondral tissue engineering. Key Eng Mater 587:245–248
4. Levingstone TJ, Thompson E, Matsiko A, Schepens A, Gleeson JP, O'Brien FJ (2016) Multi-layered collagen-based scaffolds for osteochondral defect repair in rabbits. Acta Biomater 32:149–160
5. Nooeaid P, Salih V, Beier JP, Boccaccini AR (2012) Osteochondral tissue engineering: scaffolds, stem cells and applications. J Cell Mol Med 16:2247–2270
6. Silva TH, Alves A, Ferreira BM, Oliveira JM, Reys LL, Ferreira RJF, Sousa RA, Silva SS, Mano JF, Reis RL (2012) Materials of marine origin: a review on polymers and ceramics of biomedical interest. Int Mater Rev 57:276–306
7. Ige OO, Umoru LE, Aribo S (2012) Natural products: a minefield of biomaterials. ISRN Mater Sci 2012:1–20
8. Ali I, Jamil N (2016) Polyhydroxyalkanoates: current applications in the medical field. Front Biol (Beijing) 11:19–27
9. Malafaya PB, Silva GA, Reis RL (2007) Natural-origin polymers as carriers and scaffolds for biomolecules and cell delivery in tissue engineering applications. Adv Drug Deliv Rev 59:207–233
10. Baino F, Novajra G, Vitale-Brovarone C (2015) Bioceramics and scaffolds: a winning combination for tissue engineering. Front Bioeng Biotechnol. https://doi.org/10.3389/fbioe.2015.00202
11. Wang X, Schröder HC, Grebenjuk V, Diehl-Seifert B, Mailänder V, Steffen R, Schloßmacher U, Müller WEG (2014) The marine sponge-derived inorganic polymers, biosilica and polyphosphate, as morphogenetically active matrices/scaffolds for the differentiation of human multipotent stromal cells: potential application in 3D printing and distraction osteogenesis. Mar Drugs 12:1131–1147
12. Kon E, Filardo G, Shani J, Altschuler N, Levy A, Zaslav K, Eisman JE, Robinson D (2015) Osteochondral regeneration with a novel aragonite-hyaluronate biphasic scaffold: up to 12-month follow-up study in a goat model. J Orthop Surg Res 10:81
13. Markstedt K, Mantas A, Tournier I, Martínez Ávila H, Hägg D, Gatenholm P (2015) 3D bioprinting human chondrocytes with nanocellulose–alginate bioink for cartilage tissue engineering applications. Biomacromolecules 16:1489–1496
14. Bartnikowski M, Akkineni A, Gelinsky M, Woodruff M, Klein T (2016) A hydrogel model incorporating 3D-plotted hydroxyapatite for osteochondral tissue engineering. Materials (Basel) 9:285
15. Gong Y, Wang C, Lai RC et al (2009) An improved injectable polysaccharide hydrogel: modified gellan gum for long-term cartilage regeneration in vitro. J Mater Chem 19:1968
16. Salinas AJ, Vallet-Regí M (2013) Bioactive ceramics: from bone grafts to tissue engineering. RSC Adv 3:11116–11131
17. Sprio S, Sandri M, Ruffini A, Adamiano A, Iafisco M, Dapporto M, Panseri S, Montesi M, Tampieri A (2017) Tissue engineering and biomimetics with bioceramics. In: Adv Ceram Biomater. Elsevier, pp 407–432
18. Ramachandran GN, Kartha G (1954) Structure of collagen. Nature 174:269–270
19. Puxkandl R, Zizak I, Paris O, Keckes J, Tesch W, Bernstorff S, Purslow P, Fratzl P (2002) Viscoelastic properties of collagen: synchrotron radiation investigations and structural model. Philos Trans R Soc B Biol Sci 357:191–197

20. Pal GK, Suresh PV, Kalal KM, Laxman RS, Saxena RK, Okabe A, FitzGerald RJ, Nasri M, Zhang YZ, Murayama K (2016) Microbial collagenases: challenges and prospects in production and potential applications in food and nutrition. RSC Adv 6:33763–33780
21. Krishnamoorthi J, Ramasamy P, Shanmugam V, Shanmugam A (2017) Isolation and partial characterization of collagen from outer skin of Sepia pharaonis (Ehrenberg, 1831) from Puducherry coast. Biochem Biophys Reports 10:39–45
22. Mayne R (1989) Cartilage collagens. What is their function, and are they involved in articular disease? Arthritis Rheum 32:241–246
23. Niyibizi C, Eyre DR (1994) Structural characteristics of cross-linking sites in type V collagen of bone. Chain specificities and heterotypic links to type I collagen. Eur J Biochem 224:943–950
24. Walker GD, Fischer M, Gannon J, Thompson RC, Oegema TR (1995) Expression of type-X collagen in osteoarthritis. J Orthop Res 13:4–12
25. Wong MWN, Qin L, Lee KM, Leung KS (2009) Articular cartilage increases transition zone regeneration in bone-tendon junction healing. Clin Orthop Relat Res 467:1092–1100
26. Zhao W, Jin X, Cong Y, Liu Y, Fu J (2013) Degradable natural polymer hydrogels for articular cartilage tissue engineering. J Chem Technol Biotechnol 88:327–339
27. Tongnuanchan P, Benjakul S, Prodpran T (2012) Properties and antioxidant activity of fish skin gelatin film incorporated with citrus essential oils. Food Chem 134:1571–1579
28. Yousefi M, Ariffin F, Huda N (2017) An alternative source of type I collagen based on by-product with higher thermal stability. Food Hydrocoll 63:372–382
29. Miles CA, Avery NC, Rodin VV, Bailey AJ (2005) The increase in denaturation temperature following cross-linking of collagen is caused by dehydration of the fibres. J Mol Biol 346:551–556
30. Pati F, Adhikari B, Dhara S (2010) Isolation and characterization of fish scale collagen of higher thermal stability. Bioresour Technol 101:3737–3742
31. Sewing J, Klinger M, Notbohm H (2017) Jellyfish collagen matrices conserve the chondrogenic phenotype in two- and three-dimensional collagen matrices. J Tissue Eng Regen Med 11:916–925
32. Hoyer B, Bernhardt A, Lode A, Heinemann S, Sewing J, Klinger M, Notbohm H, Gelinsky M (2014) Jellyfish collagen scaffolds for cartilage tissue engineering. Acta Biomater 10:883–892
33. Pustlauk W, Paul B, Gelinsky M, Bernhardt A (2016) Jellyfish collagen and alginate: combined marine materials for superior chondrogenesis of hMSC. Mater Sci Eng C 64:190–198
34. Womack SA, Milner DJ, Weisgerber DW, Harley BAC, Wheeler MB (2017) Behavior of porcine mesenchymal stem cells on a collagen-glycosaminoglycan hydrogel scaffold for bone and cartilage tissue engineering. Reprod Fertil Dev 29:205
35. Meng F, Zhang ZZ, Huang G, Chen W, Zhang ZZ, He A, Liao W (2016) Chondrogenesis of mesenchymal stem cells in a novel hyaluronate-collagen-tricalcium phosphate scaffolds for knee repair. Eur Cells Mater 31:79–94
36. Filová E, Jelínek F, Handl M, Lytvynets A, Rampichová M, Varga F, Činátl J, Soukup T, Trč T, Amler E (2008) Novel composite hyaluronan/type I collagen/fibrin scaffold enhances repair of osteochondral defect in rabbit knee. J Biomed Mater Res Part B Appl Biomater 87B:415–424
37. Benthien JP, Behrens P (2010) Autologous matrix-induced Chondrogenesis (AMIC): combining microfracturing and a collagen I/III matrix for articular cartilage resurfacing. Cartilage 1:65–68
38. Piontek T, Bąkowski P, Ciemniewska-Gorzela K, Naczk J (2015) Arthroscopic treatment of chondral and osteochondral defects in the ankle using the autologous matrix induced chondrogenesis technique. Arthrosc Tech 4:e463–e469
39. Kon E, Delcogliano M, Filardo G et al (2009) Orderly osteochondral regeneration in a sheep model using a novel nano-composite multilayered biomaterial. J Orthop Res 28:116–124

40. Delcogliano M, de Caro F, Scaravella E, Ziveri G, De Biase CF, Marotta D, Marenghi P, Delcogliano A (2013) Use of innovative biomimetic scaffold in the treatment for large osteochondral lesions of the knee. Knee Surgery, Sport Traumatol Arthrosc 22:1260–1269
41. Kon E, Perdisa F, Filardo G, Marcacci M (2014) MaioRegen: Our experience. In: Tech. Cartil. Repair Surg. Springer Berlin Heidelberg, Berlin, Heidelberg, pp 81–95
42. Christensen BB, Foldager CB, Jensen J, Jensen NC, Lind M (2016) Poor osteochondral repair by a biomimetic collagen scaffold: 1- to 3-year clinical and radiological follow-up. Knee Surgery, Sport Traumatol Arthrosc 24:2380–2387
43. Sartori M, Pagani S, Ferrari A, Costa V, Carina V, Figallo E, Maltarello MC, Martini L, Fini M, Giavaresi G (2017) A new bi-layered scaffold for osteochondral tissue regeneration: in vitro and in vivo preclinical investigations. Mater Sci Eng C 70:101–111
44. Echave MC, Burgo LS, Pedraz JL, Orive G (2017) Gelatin as biomaterial for tissue engineering. Curr Pharm Des 23:3567–3584
45. Van Nieuwenhove I, Salamon A, Adam S, Dubruel P, Van Vlierberghe S, Peters K (2017) Gelatin- and starch-based hydrogels. Part B: in vitro mesenchymal stem cell behavior on the hydrogels. Carbohydr Polym 161:295–305
46. Zhang S, Chen L, Jiang Y et al (2013) Bi-layer collagen/microporous electrospun nanofiber scaffold improves the osteochondral regeneration. Acta Biomater 9:7236–7247
47. Petrenko YA, Ivanov RV, Petrenko AY, Lozinsky VI (2011) Coupling of gelatin to inner surfaces of pore walls in spongy alginate-based scaffolds facilitates the adhesion, growth and differentiation of human bone marrow mesenchymal stromal cells. J Mater Sci Mater Med 22:1529–1540
48. Mazaki T, Shiozaki Y, Yamane K et al (2015) A novel, visible light-induced, rapidly cross-linkable gelatin scaffold for osteochondral tissue engineering. Sci Rep 4:4457
49. Han F, Yang X, Zhao J, Zhao Y, Yuan X (2015) Photocrosslinked layered gelatin-chitosan hydrogel with graded compositions for osteochondral defect repair. J Mater Sci Mater Med 26:160
50. Sutherland TD, Young JH, Weisman S, Hayashi CY, Merritt DJ (2010) Insect silk: one name, many materials. Annu Rev Entomol 55:171–188
51. Altman GH, Diaz F, Jakuba C, Calabro T, Horan RL, Chen JS, Lu H, Richmond J, Kaplan DL (2003) Silk-based biomaterials. Biomaterials 24:401–416
52. Kundu B, Rajkhowa R, Kundu SC, Wang X (2013) Silk fibroin biomaterials for tissue regenerations. Adv Drug Deliv Rev 65:457–470
53. Kim U-J, Park J, Joo Kim H, Wada M, Kaplan DL (2005) Three-dimensional aqueous-derived biomaterial scaffolds from silk fibroin. Biomaterials 26:2775–2785
54. Sangkert S, Kamonmattayakul S, Chai WL, Meesane J (2017) Modified porous scaffolds of silk fibroin with mimicked microenvironment based on decellularized pulp/fibronectin for designed performance biomaterials in maxillofacial bone defect. J Biomed Mater Res Part A 105:1624–1636
55. Singh BN, Pramanik K (2017) Development of novel silk fibroin/polyvinyl alcohol/sol–gel bioactive glass composite matrix by modified layer by layer electrospinning method for bone tissue construct generation. Biofabrication 9:15028
56. Wang Y, Kim U-J, Blasioli DJ, Kim H-J, Kaplan DL (2005) In vitro cartilage tissue engineering with 3D porous aqueous-derived silk scaffolds and mesenchymal stem cells. Biomaterials 26:7082–7094
57. Kambe Y, Yamamoto K, Kojima K, Tamada Y, Tomita N (2010) Effects of RGDS sequence genetically interfused in the silk fibroin light chain protein on chondrocyte adhesion and cartilage synthesis. Biomaterials 31:7503–7511
58. Murphy AR, John PS, Kaplan DL (2008) Modification of silk fibroin using diazonium coupling chemistry and the effects on hMSC proliferation and differentiation. Biomaterials 29:2829–2838
59. Saha S, Kundu B, Kirkham J, Wood D, Kundu SC, Yang XB (2013) Osteochondral tissue engineering in vivo: a comparative study using layered silk fibroin scaffolds from mulberry and non-mulberry silkworms. PLoS One 8:e80004

60. Ding X, Zhu M, Xu B et al (2014) Integrated trilayered silk fibroin scaffold for osteochondral differentiation of adipose-derived stem cells. ACS Appl Mater Interfaces 6:16696–16705
61. Lamboni L, Gauthier M, Yang G, Wang Q (2015) Silk sericin: A versatile material for tissue engineering and drug delivery. Biotechnol Adv 33:1855–1867
62. Siritientong T, Aramwit P (2012) A novel silk sericin/poly (vinyl alcohol) composite film crosslinked with genipin: fabrication and characterization for tissue engineering applications. Adv Mater Res 506:359–362
63. Nayak S, Kundu SC (2014) Sericin-carboxymethyl cellulose porous matrices as cellular wound dressing material. J Biomed Mater Res Part A 102:1928–1940
64. Jiayao Z, Guanshan Z, Jinchi Z, Yuyin C, Yongqiang Z (2017) Antheraea pernyi silk sericin mediating biomimetic nucleation and growth of hydroxylapatite crystals promoting bone matrix formation. Microsc Res Tech 80:305–311
65. Yang M, Shuai Y, Zhang C, Chen Y, Zhu L, Mao C, OuYang H (2014) Biomimetic nucleation of hydroxyapatite crystals mediated by Antheraea pernyi silk sericin promotes osteogenic differentiation of human bone marrow-derived mesenchymal stem cells. Biomacromolecules 15:1185–1193
66. Mosesson MW (2005) Fibrinogen and fibrin structure and functions. J Thromb Haemost 3:1894–1904
67. Laurens N, Koolwijk P, De Maat MPM (2006) Fibrin structure and wound healing. J Thromb Haemost 4:932–939
68. Spotnitz WD, Burks S (2010) State-of-the-art review: hemostats, sealants, and adhesives II: update as well as how and when to use the components of the surgical toolbox. Clin Appl Thromb 16:497–514
69. Eyrich D, Brandl F, Appel B, Wiese H, Maier G, Wenzel M, Staudenmaier R, Goepferich A, Blunk T (2007) Long-term stable fibrin gels for cartilage engineering. Biomaterials 28:55–65
70. Schek RM, Michalek AJ, Iatridis JC (2011) Genipin-crosslinked fibrin hydrogels as a potential adhesive to augment intervertebral disc annulus repair. Eur Cell Mater 21:373–383
71. Li B, Yang J, Ma L, Li F, Tu Z, Gao C (2013) Fabrication of poly(lactide-co-glycolide) scaffold filled with fibrin gel, mesenchymal stem cells, and poly(ethylene oxide)-b-poly(L-lysine)/TGF-β1 plasmid DNA complexes for cartilage restoration in vivo. J Biomed Mater Res Part A 101:3097–3108
72. Pei M, He F, Boyce BM, Kish VL (2009) Repair of full-thickness femoral condyle cartilage defects using allogeneic synovial cell-engineered tissue constructs. Osteoarthr Cartil 17:714–722
73. Wang W, Li B, Yang J, Xin L, Li Y, Yin H, Qi Y, Jiang Y, Ouyang H, Gao C (2010) The restoration of full-thickness cartilage defects with BMSCs and TGF-beta 1 loaded PLGA/fibrin gel constructs. Biomaterials 31:8964–8973
74. Jang K-M, Lee J-H, Park CM, Song H-R, Wang JH (2014) Xenotransplantation of human mesenchymal stem cells for repair of osteochondral defects in rabbits using osteochondral biphasic composite constructs. Knee Surgery, Sport Traumatol Arthrosc 22:1434–1444
75. Snyder TN, Madhavan K, Intrator M, Dregalla RC, Park D (2014) A fibrin/hyaluronic acid hydrogel for the delivery of mesenchymal stem cells and potential for articular cartilage repair. J Biol Eng 8:10
76. Wang B, Yang W, McKittrick J, Meyers MA (2016) Keratin: structure, mechanical properties, occurrence in biological organisms, and efforts at bioinspiration. Prog Mater Sci 76:229–318
77. Bragulla HH, Homberger DG (2009) Structure and functions of keratin proteins in simple, stratified, keratinized and cornified epithelia. J Anat 214:516–559
78. Wegst UGK, Ashby MF (2004) The mechanical efficiency of natural materials. Philos Mag 84:2167–2186
79. Rouse JG, Van Dyke ME (2010) A review of keratin-based biomaterials for biomedical applications. Materials (Basel) 3:999–1014

80. Dias GJ, Peplow PV, McLaughlin A, Teixeira F, Kelly RJ (2010) Biocompatibility and osseointegration of reconstituted keratin in an ovine model. J Biomed Mater Res Part A 92A:513–520
81. Tachibana A, Furuta Y, Takeshima H, Tanabe T, Yamauchi K (2002) Fabrication of wool keratin sponge scaffolds for long-term cell cultivation. J Biotechnol 93:165–170
82. Arslan YE, Sezgin Arslan T, Derkus B, Emregul E, Emregul KC (2017) Fabrication of human hair keratin/jellyfish collagen/eggshell-derived hydroxyapatite osteoinductive biocomposite scaffolds for bone tissue engineering: from waste to regenerative medicine products. Colloids Surfaces B Biointerfaces 154:160–170
83. Elieh-Ali-Komi D, Hamblin MR (2016) Chitin and chitosan: production and application of versatile biomedical nanomaterials. Int J Adv Res 4:411–427
84. Muzzarelli C, Muzzarelli RAA (2002) Natural and artificial chitosan-inorganic composites. J Inorg Biochem 92:89–94
85. Cho YW, Cho YN, Chung SH, Yoo G, Ko SW (1999) Water-soluble chitin as a wound healing accelerator. Biomaterials 20:2139–2145
86. Mi F-L, Wu Y-B, Shyu S-S, Schoung J-Y, Huang Y-B, Tsai Y-H, Hao J-Y (2002) Control of wound infections using a bilayer chitosan wound dressing with sustainable antibiotic delivery. J Biomed Mater Res 59:438–449
87. Xia Z, Wu S, Chen J (2013) Preparation of water soluble chitosan by hydrolysis using hydrogen peroxide. Int J Biol Macromol 59:242–245
88. Tomihata K, Ikada Y (1997) In vitro and in vivo degradation of films of chitin and its deacetylated derivatives. Biomaterials 18:567–575
89. Jin R, Moreira Teixeira LS, Dijkstra PJ, Karperien M, van Blitterswijk CA, Zhong ZY, Feijen J (2009) Injectable chitosan-based hydrogels for cartilage tissue engineering. Biomaterials 30:2544–2551
90. Sheehy EJ, Mesallati T, Vinardell T, Kelly DJ (2015) Engineering cartilage or endochondral bone: a comparison of different naturally derived hydrogels. Acta Biomater 13:245–253
91. Oliveira JM, Rodrigues MT, Silva SS, Malafaya PB, Gomes ME, Viegas CA, Dias IR, Azevedo JT, Mano JF, Reis RL (2006) Novel hydroxyapatite/chitosan bilayered scaffold for osteochondral tissue-engineering applications: scaffold design and its performance when seeded with goat bone marrow stromal cells. Biomaterials 27:6123–6137
92. Sechriest VF, Miao YJ, Niyibizi C, Westerhausen-Larson A, Matthew HW, Evans CH, Fu FH, Suh JK (2000) GAG-augmented polysaccharide hydrogel: a novel biocompatible and biodegradable material to support chondrogenesis. J Biomed Mater Res 49:534–541
93. Tan H, Chu CR, Payne KA, Marra KG (2009) Injectable in situ forming biodegradable chitosan–hyaluronic acid based hydrogels for cartilage tissue engineering. Biomaterials 30:2499–2506
94. Mohan N, Mohanan P, Sabareeswaran A, Nair P (2017) Chitosan-hyaluronic acid hydrogel for cartilage repair. Int J Biol Macromol 104:1936–1945
95. Reed S, Lau G, Delattre B, Lopez DD, Tomsia AP, Wu BM (2016) Macro- and microdesigned chitosan-alginate scaffold architecture by three-dimensional printing and directional freezing. Biofabrication 8:15003
96. Lastra ML, Molinuevo MS, Cortizo AM, Cortizo MS (2017) Fumarate copolymer-chitosan cross-linked scaffold directed to osteochondrogenic tissue engineering. Macromol Biosci. https://doi.org/10.1002/mabi.201600219
97. Lee KY, Mooney DJ (2012) Alginate: properties and biomedical applications. Prog Polym Sci 37:106–126
98. Tønnesen HH, Karlsen J (2002) Alginate in drug delivery systems. Drug Dev Ind Pharm 28:621–630
99. Rowley JA, Madlambayan G, Mooney DJ (1999) Alginate hydrogels as synthetic extracellular matrix materials. Biomaterials 20:45–53
100. Goh CH, Heng PWS, Chan LW (2012) Alginates as a useful natural polymer for microencapsulation and therapeutic applications. Carbohydr Polym 88:1–12

101. Gasperini L, Maniglio D, Migliaresi C (2013) Microencapsulation of cells in alginate through an electrohydrodynamic process. J Bioact Compat Polym 28:413–425
102. Bidarra SJ, Barrias CC, Granja PL (2014) Injectable alginate hydrogels for cell delivery in tissue engineering. Acta Biomater 10:1646–1662
103. Alsberg E, Anderson KW, Albeiruti A, Franceschi RT, Mooney DJ (2001) Cell-interactive alginate hydrogels for bone tissue engineering. J Dent Res 80:2025–2029
104. Comisar WA, Kazmers N, Mooney DJ, Linderman J (2007) Engineering RGD nanopatterned hydrogels to control preosteoblast behavior: a combined computational and experimental approach. Biomaterials 28:4409–4417
105. Guilak F, Cohen DM, Estes BT, Gimble JM, Liedtke W, Chen CS (2009) Control of stem cell fate by physical interactions with the extracellular matrix. Cell Stem Cell 5:17–26
106. Herlofsen SR, Küchler AM, Melvik JE, Brinchmann JE (2011) Chondrogenic differentiation of human bone marrow-derived mesenchymal stem cells in self-gelling alginate discs reveals novel chondrogenic signature gene clusters. Tissue Eng Part A 17:1003–1013
107. Kim D-H, Kim D-D, Yoon I-S (2013) Proliferation and chondrogenic differentiation of human adipose-derived mesenchymal stem cells in sodium alginate beads with or without hyaluronic acid. J Pharm Investig 43:145–151
108. Wayne JS, McDowell CL, Shields KJ, Tuan RS (2005) In vivo response of polylactic acid–alginate scaffolds and bone marrow-derived cells for cartilage tissue engineering. Tissue Eng 11:953–963
109. Reyes R, Delgado A, Sánchez E, Fernández A, Hernández A, Evora C (2012) Repair of an osteochondral defect by sustained delivery of BMP-2 or TGF-β1 from a bilayered alginate-PLGA scaffold. J Tissue Eng Regen Med 8:521–533
110. Coluccino L, Stagnaro P, Vassalli M, Scaglione S (2016) Bioactive TGF-β1/HA alginate-based scaffolds for osteochondral tissue repair: design, realization and multilevel characterization. J Appl Biomater Funct Mater 14:0–0
111. Hemshekhar M, Thushara RM, Chandranayaka S, Sherman LS, Kemparaju K, Girish KS (2016) Emerging roles of hyaluronic acid bioscaffolds in tissue engineering and regenerative medicine. Int J Biol Macromol 86:917–928
112. Zhang H, Zhang K, Zhang X et al (2015) Comparison of two hyaluronic acid formulations for safety and efficacy (CHASE) study in knee osteoarthritis: a multicenter, randomized, double-blind, 26-week non-inferiority trial comparing Durolane to Artz. Arthritis Res Ther 17:51
113. Sparavigna A, Fino P, Tenconi B, Giordan N, Amorosi V, Scuderi N (2014) A new dermal filler made of cross-linked and auto-cross-linked hyaluronic acid in the correction of facial aging defects. J Cosmet Dermatol 13:307–314
114. Burdick JA, Chung C, Jia X, Randolph MA, Langer R (2005) Controlled degradation and mechanical behavior of photopolymerized hyaluronic acid networks. Biomacromolecules 6:386–391
115. Nettles DL, Vail TP, Morgan MT, Grinstaff MW, Setton LA (2004) Photocrosslinkable hyaluronan as a scaffold for articular cartilage repair. Ann Biomed Eng 32:391–397
116. Knudson CB, Nofal GA, Pamintuan L, Aguiar DJ (1999) The chondrocyte pericellular matrix: a model for hyaluronan-mediated cell-matrix interactions. Biochem Soc Trans 27:142–147
117. Bian L, Zhai DY, Tous E, Rai R, Mauck RL, Burdick JA (2011) Enhanced MSC chondrogenesis following delivery of TGF-β3 from alginate microspheres within hyaluronic acid hydrogels in vitro and in vivo. Biomaterials 32:6425–6434
118. Marcacci M, Berruto M, Brocchetta D et al (2005) Articular cartilage engineering with Hyalograft C: 3-year clinical results. Clin Orthop Relat Res 436:96–105
119. Kon E, Filardo G, Berruto M, Benazzo F, Zanon G, Della Villa S, Marcacci M (2011) Articular cartilage treatment in high-level male soccer players. Am J Sports Med 39:2549–2557
120. Kontturi L-S, Järvinen E, Muhonen V, Collin EC, Pandit AS, Kiviranta I, Yliperttula M, Urtti A (2014) An injectable, in situ forming type II collagen/hyaluronic acid hydrogel vehicle for chondrocyte delivery in cartilage tissue engineering. Drug Deliv Transl Res 4:149–158

121. Fialho AM, Martins LO, Donval M-L, Leitao JH, Ridout MJ, Jay AJ, Morris VJ, Sa-Correia I (1999) Structures and properties of gellan polymers produced by Sphingomonas paucimobilis ATCC 31461 from lactose compared with those produced from glucose and from cheese whey. Appl Environ Microbiol 65:2485–2491
122. Prajapati VD, Jani GK, Zala BS, Khutliwala TA (2013) An insight into the emerging exopolysaccharide gellan gum as a novel polymer. Carbohydr Polym 93:670–678
123. Chandrasekaran R, Radha A (1995) Molecular architectures and functional properties of gellan gum and related polysaccharides. Trends Food Sci Technol 6:143–148
124. Yuguchi Y, Urakawa H, Kajiwara K (1997) Structural characteristics of crosslinking domain in gellan gum gel. Macromol Symp 120:77–89
125. Morris ER, Nishinari K, Rinaudo M (2012) Gelation of gellan – a review. Food Hydrocoll 28:373–411
126. Stevens LR, Gilmore KJ, Wallace GG et al (2016) Tissue engineering with gellan gum. Biomater Sci 4:1276–1290
127. Ferris CJ, Gilmore KJ, Wallace GG, Panhuis M (2013) Modified gellan gum hydrogels for tissue engineering applications. Soft Matter 9:3705
128. Lee H, Fisher S, Kallos MS, Hunter CJ (2011) Optimizing gelling parameters of gellan gum for fibrocartilage tissue engineering. J Biomed Mater Res Part B Appl Biomater 98B:238–245
129. Shin H, Olsen BD, Khademhosseini A (2012) The mechanical properties and cytotoxicity of cell-laden double-network hydrogels based on photocrosslinkable gelatin and gellan gum biomacromolecules. Biomaterials 33:3143–3152
130. Pacelli S, Paolicelli P, Pepi F, Garzoli S, Polini A, Tita B, Vitalone A, Casadei MA (2014) Gellan gum and polyethylene glycol dimethacrylate double network hydrogels with improved mechanical properties. J Polym Res 21:409
131. Pacelli S, Paolicelli P, Dreesen I, Kobayashi S, Vitalone A, Casadei MA (2015) Injectable and photocross-linkable gels based on gellan gum methacrylate: a new tool for biomedical application. Int J Biol Macromol 72:1335–1342
132. Oliveira JT, Santos TC, Martins L, Silva MA, Marques AP, Castro AG, Neves NM, Reis RL (2009) Performance of new gellan gum hydrogels combined with human articular chondrocytes for cartilage regeneration when subcutaneously implanted in nude mice. J Tissue Eng Regen Med 3:493–500
133. Oliveira JT, Gardel LS, Rada T, Martins L, Gomes ME, Reis RL (2010) Injectable gellan gum hydrogels with autologous cells for the treatment of rabbit articular cartilage defects. J Orthop Res 28:1193–1199
134. Pereira DR, Canadas RF, Silva-Correia J, Marques AP, Reis RL, Oliveira JM (2013) Gellan gum-based hydrogel bilayered scaffolds for osteochondral tissue engineering. Key Eng Mater 587:255–260
135. Philip S, Keshavarz T, Roy I (2007) Polyhydroxyalkanoates: biodegradable polymers with a range of applications. J Chem Technol Biotechnol 82:233–247
136. Marchesini S, Erard N, Glumoff T, Hiltunen JK, Poirier Y (2003) Modification of the monomer composition of polyhydroxyalkanoate synthesized in Saccharomyces Cerevisiae expressing variants of the beta-oxidation-associated multifunctional enzyme. Appl Environ Microbiol 69:6495–6499
137. Shishatskaya EI, Volova TG, Gordeev SA, Puzyr AP (2005) Degradation of P(3HB) and P(3HB-co-3HV) in biological media. J Biomater Sci Polym Ed 16:643–657
138. Baek J-Y, Xing Z-C, Kwak G, Yoon K-B, Park S-Y, Park LS, Kang I-K (2012) Fabrication and characterization of collagen-immobilized porous PHBV/HA nanocomposite scaffolds for bone tissue engineering. J Nanomater 2012:1–11
139. Peschel G, Dahse H-M, Konrad A, Wieland GD, Mueller P-J, Martin DP, Roth M (2008) Growth of keratinocytes on porous films of poly(3-hydroxybutyrate) and poly(4-hydroxybutyrate) blended with hyaluronic acid and chitosan. J Biomed Mater Res Part A 85A:1072–1081

140. Wang Y, Lu L, Zheng Y, Chen X (2006) Improvement in hydrophilicity of PHBV films by plasma treatment. J Biomed Mater Res Part A 76A:589–595
141. Chen G, Wang Y (2013) Medical applications of biopolyesters polyhydroxyalkanoates. Chinese J Polym Sci 31:719–736
142. Zheng Z, Bei F-F, Tian H-L, Chen G-Q (2005) Effects of crystallization of polyhydroxyalkanoate blend on surface physicochemical properties and interactions with rabbit articular cartilage chondrocytes. Biomaterials 26:3537–3548
143. Wu J, Xue K, Li H, Sun J, Liu K (2013) Improvement of PHBV scaffolds with bioglass for cartilage tissue engineering. PLoS One 8:e71563
144. You M, Peng G, Li J, Ma P, Wang Z, Shu W, Peng S, Chen G-Q (2011) Chondrogenic differentiation of human bone marrow mesenchymal stem cells on polyhydroxyalkanoate (PHA) scaffolds coated with PHA granule binding protein PhaP fused with RGD peptide. Biomaterials 32:2305–2313
145. Vago R (2008) Beyond the skeleton: cnidarian biomaterials as bioactive extracellular microenvironments for tissue engineering. Organogenesis 4:18–22
146. Demers C, Hamdy CR, Corsi K, Chellat F, Tabrizian M, Yahia L (2002) Natural coral exoskeleton as a bone graft substitute: a review. Biomed Mater Eng 12:15–35
147. Viateau V, Manassero M, Sensébé L, Langonné A, Marchat D, Logeart-Avramoglou D, Petite H, Bensidhoum M (2016) Comparative study of the osteogenic ability of four different ceramic constructs in an ectopic large animal model. J Tissue Eng Regen Med 10:177–187
148. Kon E, Robinson D, Verdonk P, Drobnic M, Patrascu JM, Dulic O, Gavrilovic G, Filardo G (2016) A novel aragonite-based scaffold for osteochondral regeneration: early experience on human implants and technical developments. Injury 47:27–32
149. Damien CJ, Parsons JR (1991) Bone graft and bone graft substitutes: a review of current technology and applications. J Appl Biomater 2:187–208
150. Samavedi S, Whittington AR, Goldstein AS (2013) Calcium phosphate ceramics in bone tissue engineering: a review of properties and their influence on cell behavior. Acta Biomater 9:8037–8045
151. Habraken W, Habibovic P, Epple M, Bohner M (2016) Calcium phosphates in biomedical applications: materials for the future? Mater Today 19:69–87
152. Venkatesan J, Kim S-K (2014) Nano-hydroxyapatite composite biomaterials for bone tissue engineering: a review. J Biomed Nanotechnol 10:3124–3140
153. Vecchio KS, Zhang X, Massie JB, Wang M, Kim CW (2007) Conversion of bulk seashells to biocompatible hydroxyapatite for bone implants. Acta Biomater 3:910–918
154. Ivankovic H, Tkalcec E, Orlic S, Gallego Ferrer G, Schauperl Z (2010) Hydroxyapatite formation from cuttlefish bones: kinetics. J Mater Sci Mater Med 21:2711–2722
155. Zhou H, Lee J (2011) Nanoscale hydroxyapatite particles for bone tissue engineering. Acta Biomater 7:2769–2781
156. Ivankovic H, Orlic S, Kranzelic D, Tkalcec E (2010) Highly porous hydroxyapatite ceramics for engineering applications. Adv Sci Technol 63:408–413
157. Pon-On W, Suntornsaratoon P, Charoenphandhu N, Thongbunchoo J, Krishnamra N, Tang IM (2016) Hydroxyapatite from fish scale for potential use as bone scaffold or regenerative material. Mater Sci Eng C 62:183–189
158. Mondal S, Pal U, Dey A (2016) Natural origin hydroxyapatite scaffold as potential bone tissue engineering substitute. Ceram Int 42:18338–18346
159. Zhang X, Vecchio KS (2013) Conversion of natural marine skeletons as scaffolds for bone tissue engineering. Front Mater Sci 7:103–117
160. Dong C, Lv Y (2016) Application of collagen scaffold in tissue engineering: recent advances and new perspectives. Polymers (Basel) 8:42–62
161. Kon E, Delcogliano M, Filardo G, Busacca M, Di Martino A, Marcacci M (2011) Novel nano-composite multilayered biomaterial for osteochondral regeneration. Am J Sports Med 39:1180–1190

162. Calabrese G, Forte S, Gulino R et al (2017) Combination of collagen-based scaffold and bioactive factors induces adipose-derived mesenchymal stem cells chondrogenic differentiation in vitro. Front Physiol 8:1–27
163. Chen J, Chen H, Li P, Diao H, Zhu S, Dong L, Wang R, Guo T, Zhao J, Zhang J (2011) Simultaneous regeneration of articular cartilage and subchondral bone in vivo using MSCs induced by a spatially controlled gene delivery system in bilayered integrated scaffolds. Biomaterials 32:4793–4805
164. Abarrategi A, Lópiz-Morales Y, Ramos V, Civantos A, López-Durán L, Marco F, López-Lacomba JL (2010) Chitosan scaffolds for osteochondral tissue regeneration. J Biomed Mater Res Part A 95A:1132–1141
165. Naderi-Meshkin H, Andreas K, Matin MM, Sittinger M, Bidkhori HR, Ahmadiankia N, Bahrami AR, Ringe J (2014) Chitosan-based injectable hydrogel as a promising in situ forming scaffold for cartilage tissue engineering. Cell Biol Int 38:72–84
166. Liao J, Wang B, Huang Y, Qu Y, Peng J, Qian Z (2017) Injectable alginate hydrogel crosslinked by calcium gluconate-loaded porous microspheres for cartilage tissue engineering. ACS Omega 2:443–454
167. Fisher MB, Belkin NS, Milby AH et al (2016) Effects of mesenchymal stem cell and growth factor delivery on cartilage repair in a mini-pig model. Cartilage 7:174–184
168. Park JY, Choi J-C, Shim J-H, Lee J-S, Park H, Kim SW, Doh J, Cho D-W (2014) A comparative study on collagen type I and hyaluronic acid dependent cell behavior for osteochondral tissue bioprinting. Biofabrication 6:35004
169. Oliveira JT, Santos TC, Martins L, Picciochi R, Marques AP, Castro AG, Neves NM, Mano JF, Reis RL (2010) Gellan gum injectable hydrogels for cartilage tissue engineering applications: in vitro studies and preliminary in vivo evaluation. Tissue Eng Part A 16:343–353
170. Zheng L, Jiang X, Chen X, Fan H, Zhang X (2014) Evaluation of novel in situ synthesized nano-hydroxyapatite/collagen/alginate hydrogels for osteochondral tissue engineering. Biomed Mater 9:65004
171. Bian W, Li D, Lian Q, Li X, Zhang W, Wang K, Jin Z (2012) Fabrication of a bio-inspired beta-Tricalcium phosphate/collagen scaffold based on ceramic stereolithography and gel casting for osteochondral tissue engineering. Rapid Prototyp J 18:68–80
172. Lin H-Y, Tsai W-C, Chang S-H (2017) Collagen-PVA aligned nanofiber on collagen sponge as bi-layered scaffold for surface cartilage repair. J Biomater Sci Polym Ed 28:664–678
173. Getgood A, Henson F, Skelton C, Brooks R, Guehring H, Fortier LA, Rushton N (2014) Osteochondral tissue engineering using a biphasic collagen/GAG scaffold containing rhFGF18 or BMP-7 in an ovine model. J Exp Orthop 1:13
174. Ren X, Weisgerber DW, Bischoff D, Lewis MS, Reid RR, He T-C, Yamaguchi DT, Miller TA, Harley BAC, Lee JC (2016) Nanoparticulate mineralized collagen scaffolds and BMP-9 induce a long-term bone cartilage construct in human mesenchymal stem cells. Adv Healthc Mater 5:1821–1830
175. Taniyama T, Masaoka T, Yamada T et al (2015) Repair of osteochondral defects in a rabbit model using a porous hydroxyapatite collagen composite impregnated with bone morphogenetic protein-2. Artif Organs 39:529–535
176. Lee P, Tran K, Zhou G, Bedi A, Shelke NB, Yu X, Kumbar SG (2015) Guided differentiation of bone marrow stromal cells on co-cultured cartilage and bone scaffolds. Soft Matter 11:7648–7655
177. Miralles G, Baudoin R, Dumas D, Baptiste D, Hubert P, Stoltz JF, Dellacherie E, Mainard D, Netter P, Payan E (2001) Sodium alginate sponges with or without sodium hyaluronate: in vitro engineering of cartilage. J Biomed Mater Res 57:268–278
178. Solchaga LA, Yoo JU, Lundberg M, Dennis JE, Huibregtse BA, Goldberg VM, Caplan AI (2000) Hyaluronan-based polymers in the treatment of osteochondral defects. J Orthop Res 18:773–780
179. Deng T, Lv J, Pang J, Liu B, Ke J (2012) Construction of tissue-engineered osteochondral composites and repair of large joint defects in rabbit. J Tissue Eng Regen Med 8:546–556

180. Lee J-C, Lee SY, Min HJ, Han SA, Jang J, Lee S, Seong SC, Lee MC (2012) Synovium-derived mesenchymal stem cells encapsulated in a novel injectable gel can repair osteochondral defects in a rabbit model. Tissue Eng Part A 18:2173–2186
181. Solchaga LA, Gao J, Dennis JE, Awadallah A, Lundberg M, Caplan AI, Goldberg VM (2002) Treatment of osteochondral defects with autologous bone marrow in a hyaluronan-based delivery vehicle. Tissue Eng 8:333–347
182. Kasahara Y, Iwasaki N, Yamane S, Igarashi T, Majima T, Nonaka S, Harada K, Nishimura S-I, Minami A (2008) Development of mature cartilage constructs using novel three-dimensional porous scaffolds for enhanced repair of osteochondral defects. J Biomed Mater Res Part A 86A:127–136

# Chapter 2
# Synthetic Materials for Osteochondral Tissue Engineering

**Antoniac Iulian, Laptoiu Dan, Tecu Camelia, Milea Claudia, and Gradinaru Sebastian**

**Abstract** The objective of an articular cartilage repair treatment is to repair the affected surface of an articular joint's hyaline cartilage. Currently, both biological and tissue engineering research is concerned with discovering the clues needed to stimulate cells to regenerate tissues and organs totally or partially. The latest findings on nanotechnology advances along with the processability of synthetic biomaterials have succeeded in creating a new range of materials to develop into the desired biological responses to the cellular level. 3D printing has a great ability to establish functional tissues or organs to cure or replace abnormal and necrotic tissue, providing a promising solution for serious tissue/organ failure. The 4D print process has the potential to continually revolutionize the current tissue and organ manufacturing platforms. A new active research area is the development of intelligent materials with high biocompatibility to suit 4D printing technology. As various researchers and tissue engineers have demonstrated, the role of growth factors in tissue engineering for repairing osteochondral and cartilage defects is a very important one. Following animal testing, cell-assisted and growth-factor scaffolds produced much better results, while growth-free scaffolds showed a much lower rate of healing.

**Keywords** Biomaterial · Cartilage · Scaffold · 4D printing · Meniscal lesion

## 2.1 Introduction

Tissue engineering is the partial or total replacement of biological tissues by means of combinations of cells, engineering methods, biomaterials, and biochemical and physical–chemical factors. This engineering is based on the use of a scaffold that

---

A. Iulian (✉) · T. Camelia · M. Claudia
University Politehnica of Bucharest, Bucharest, Romania

L. Dan · G. Sebastian
University of Medicine and Pharmacy C.Davila Bucharest, Bucharest, Romania

© Springer International Publishing AG, part of Springer Nature 2018
J. M. Oliveira et al. (eds.), *Osteochondral Tissue Engineering*,
Advances in Experimental Medicine and Biology 1058,
https://doi.org/10.1007/978-3-319-76711-6_2

will later contribute to the formation of new viable tissue for medical purposes. Although it has been considered a branch of biomaterials, tissue engineering is of major importance in the treatment of certain diseases that cannot be treated on the basis of biomaterials, for example osteochondral and cartilaginous tissue engineering. Due to this important vital but also the very large domain, tissue engineering can be considered a domain itself. In osteochondral tissue engineering, the restoration of the osteochondral defect is based on scaffolding.

Scaffolds are biomaterials that have been specially designed to achieve certain cellular interactions in order to form new biological tissues; these cells are seeded into material structures that are capable of supporting 3D tissue formation. Because of tissue engineering studies, these scaffolds mimic the extracellular matrix of the target tissue, thereby mimicking the in vivo environment and allowing cells to influence their own micromedia.

The objective of an articular cartilage repair treatment is to repair the affected surface of an articular joint's hyaline cartilage. Over the years, both researchers and surgeons have been trying to solve this problem and innovate the old surgical cartilage repair interventions. Even if these solutions, for now, cannot restore the entire articular cartilage, the newest techniques and the biomaterials used have started to look very promising in repairing cartilage from traumatic injuries or others chondropathies.

## 2.2 Articular Cartilage Tissue Engineering

Tissue engineering is the study based on the principles of both engineering and biology, designed to innovate biological substitutes with the ability to restore the functions of the replaced tissues [1]. This discipline is based on an association of biomaterials, cells and biological or environmental factors, also known as the "tissue engineering triad" (Fig. 2.1).

In recent years, great progress has been made with osteochondral engineering and cartilage. These advances are in terms of the biomaterials used and the understanding of the role of growth factors and stem cells in tissues [2].

## 2.3 Biomaterials

In tissue engineering, the main objective is to replicate the tissue-target ECM characteristics; this is done through scaffolds made of biomaterials and seeded with cells. Through this replication of ECM functions, biomaterials make for cells used, an environmentally capable structure to provide viability, proliferation, and secretory cell activities. Thus, a large number of ECM-like scaffolding types exist and have been used for osteochondral and cartilaginous tissue engineering. These biomaterials can be classified as synthetic (Fig. 2.2) or natural matrices, in which we can distinguish those based on proteins or polysaccharides [3].

2 Synthetic Materials for Osteochondral Tissue Engineering

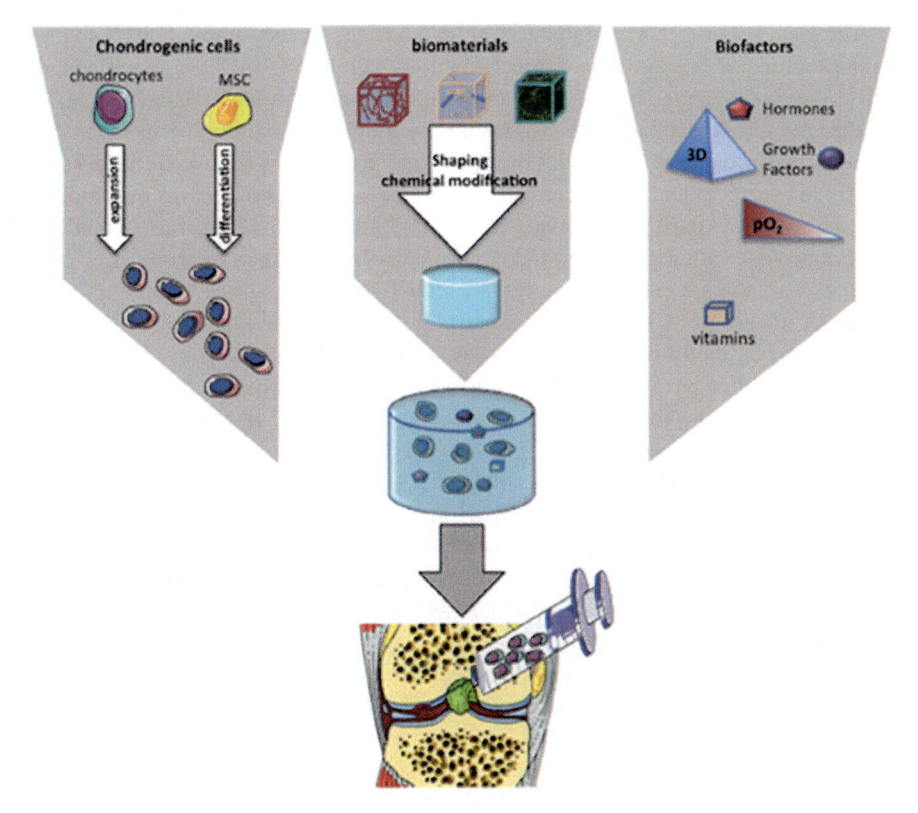

**Fig. 2.1** The tissue engineering triad used for articular cartilage repair. The combination of chondrogenic cells (expanded chondrocytes or differentiated MSCs) with biomaterials and biofactors is crucial for the development of cartilage tissue-engineering strategies [2]

**Fig. 2.2** Principal synthetic matrices used in osteochondral and cartilage engineering

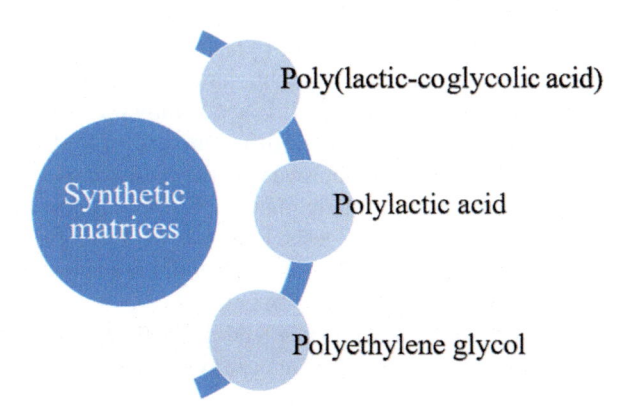

The ideal features that a biomaterial should present are biocompatibility—to prevent inflammatory and immunological reactions, adhesion—to allow attachment of cells to the lesion and to provide a favorable environment for the maintenance of 3D of the chondrocyte phenotype. Additionally, they must exhibit permeability to achieve the diffusion of molecules, nutrients and growth factors, and must be sufficiently biodegradable to be integrated into the physiological processes of remodeling the tissue. Apart from these, an ideal feature of biomaterials would be injectability—to allow implantation by minimally invasive intervention [2].

Because there are currently no percutaneously injected matrices, although there is a great deal of preclinical and clinical studies related to biomaterials used in the design of matrices for osteochondral tissue and cartilage engineering. For this reason, some studies have been developed in this direction, i.e., the development of self-reinforced and injectable biomaterials that can be used in percutaneous transplantation of chondrogenic cells [2].

Some of the biomaterials whose properties and structures make them ideal for injectable implantation, namely minimally invasive surgery, are hirdogels [4]. These are made up of chains of synthetic or natural macromolecules that have the ability to form hydrogels after physical, ionic or covalent crosslinking [5]. Hydrogels have a hydration level close to that of articular cartilage, allowing them to imitate the 3D environment of chondrocytes [6] (Table 2.1).

The use of biphasic biomaterials has become fashionable. These bilayered scaffolds consist of two different materials: a lower ceramic-like layer, which can be fitted as a plug into the subchondral bone, and an upper one, intended for the cartilaginous compartment of osteochondral defects. Different combinations of ceramic, synthetic polymers and natural materials such as collagen or hydroxyapatite in bilayered scaffolds have been investigated in clinical trials [12].

## 2.4 Other Examples of Synthetic Biomaterials Used as Matrices

*Polylactic acid (PLGA) Poly-L-lactic acid-co-glycolic acid (PLGA)* is a synthetic polymer with structural versatility and mechanical properties that can be manipulated, this being a feature favorable for regeneration of cartilaginous and osteochondral tissue. Uematsu et al. [13] in their study on the PLGA scaffold found that the biomaterial had infiltration and differentiation of MSC in vivo [13, 14]. This biomaterial, PLGA, has been approved by the FDA for use in medical applications, this being a rarity for a synthetic material. In order to limit the degree of degeneration and to support chondrogenesis, Fan et al. have demonstrated in their study that by combining PLGA with another polymer or even two a hybrid will be produced with the ability to meet these requirements [15].

*Polyvinyl alcohol (PVA)* is a biodegradable and biocompatible polymer that has been deployed to repair cartilage and osteochondral defects due to its notable water

Table 2.1 The main biomaterials used in the manufacture of matrices used in osteochondral and cartilaginous tissue engineering

| Matrices type | Component | Commercial product name | Area of use | Composition | References |
|---|---|---|---|---|---|
| Protein | Collagen | MACI1, Maix1, Atelocollagen1, MaioRegen1 | Consists of deantigenated type I equine collagen and magnesium enriched-hydroxyapatite | Promising initial results that are to be supported by further studies | Filardo et al. [7] |
|  | Fibrin | Tissucol kit1 |  |  |  |
|  | Silk |  |  |  |  |
| Polysaccharides | Hyaluronic acid | HYAFF-111 | Consists of autologous chondrocytes associated with a hyaluronic acid matrix | Clinical improvements in articular cartilage in humans | Stelzeneder et al. [8] |
|  |  |  |  |  | Nehrer et al. [9] |
|  | Chitosan | BST-CarGel1 | Made of chitosan and b-glycerophosphate | Developed for use as part of the MACI procedure | Steinwachs et al. [10] |
|  | Cellulose |  |  |  |  |
|  | Alginate |  |  |  |  |
| Synthetic | Poly(lactic-coglycolic acid) | Bio-Seed1-C | Is a porous biomaterial based on polyglycolic acid (PGA), polylactic acid (PLA) and polydioxa-none, on which chondrocytes suspended in fibrin glue are deposited | Enables the formation of hyaline cartilage and leads to clinical improvement in joint function | Ossendorf et al. [11] |
|  |  |  |  |  | Seo S-J et al. [12] |
|  | Polylactic acid |  |  |  |  |
|  | Polyethylene glycol |  |  |  |  |

content and hydrophilic behavior, in addition to its elastic and compressive properties [16]. Tadavarthy et al. [17] have conducted a study on the biocompatibility of this biomaterial so that the water content of PVA gels (on average 80–90% of their weight) was implanted both intramuscularly and subcutaneously in a rabbit in order to restore cartilage tissue [17]. This biomaterial has been studied for its use in the cartilage and osteochondral tissue engineering since 1970, but there are numerous other studies demonstrating that PVA can be used successfully in repairing these tissues [18–21]. The PVA hydrogel can be prepared at different polymer concentrations and a number of cycles tested to have a tensile strength in the cartilage range of 1–17 MPa [22], as well as a modulus of elastic range between 0.0012 and 0.85 MPa [23]. In addition, due to the fact that PVA has a low rate of degradation, its mechanical properties can be preserved, while preserving yet another chondrogenic phenotype, as a scaffold for sufficient time until the neocartilage tissue restores [14, 24].

## 2.5 Current Strategies and Challenges for Biomaterials Used in Osteochondral Tissue Engineering

The defect management and repair of osteochondral tissue have been an extremely important issue in orthopedic surgery. Osteochondral defects present damage to both the articular cartilage as well as the underlying subchondral bone. In other words, for a complete restoration of an osteochondral defect, we must take into account three needs: the bone's needs, the cartilage, and the bone–cartilage interface. In order to meet these needs, tissue engineering has evolved as an alternative that can be used to regenerate the bone, the cartilage and the bone–cartilage interface. Several scaffolding strategies, such as single-phase, layered and newly classified structures, have been developed and evaluated to repair osteochondral defects [25].

The patients quality of life may suffer due to joint pain and impaired joint function, all of which are caused by osteochondral defects resulting from trauma, joint disease, or aging [26, 27]. Treatment of osteochondral tissue lesions remains a challenge for orthopedic surgeons due to the fact that this cartilage is avascular, and has a low cell density and metabolic activity [28, 29].

As a potential source of cells, differentiated tissue cells and progenitor cells are closely observed not only in cell culture models, but also in osteochondral scaffolds in vitro and in vivo. Newly developed strategies, including the single factor, multifactor or factor release in a controlled manner helps to regenerate bone and cartilage, and to set up the formation of the osteochondral interface.

Over the last two decades, the tissue engineering methods for articular cartilage restoration, like autologous chondrocyte implantation (ACI) and matrix assisted chondrocyte implantation (MACI) have been in the clinical stage. Based on this accumulated knowledge, the area of application of tissue engineering has increased to osteochondral damage repair, which is composed of engineering articular cartilage, subchondral bone, and a smooth cartilage–bone interface [25].

In fact, the complex hierarchical structure of articular cartilage and subchondral bone is another major challenge in the reconstruction of osteochondral tissue. Nowadays, several types of techniques are used to treat osteochondral defects, including autograft transplantation [30, 31], autologous chondrocyte transplantation [32, 33] and marrow stimulation, such as subchondral drilling [34] and microfracture [29, 35].

Osteochondral defects cannot be reversed by themselves for the following reasons: the complicated hierarchical structure, and the lack of blood supply to the cartilage. With this in mind, the major challenge in this area remains the structural design of a biomimetic scaffold that meets all the specific requirements for osteochondral restoration. To meet these requirements, a selective laser sintering technique (SLS), i.e., a multilayered osteochondral scaffolding consisting of poly (ε-caprolactone) (PCL) and hydroxyapatite (HA) /PCL microspheres, has been implemented [29].

Initially, single phase scaffolds were used as the standard method, but later studies on bilayered scaffolds demonstrated that they ultimately support bone and cartilage growth on individual layers to form a tissue very similar to osteochondral tissue [25]. The scaffolds obtained by SLS showed excellent biocompatibility to support cell adhesion and proliferation in vitro. The implantation of multilayer acellular scaffolds into osteochondral defects of a rabbit model could demonstrate the desired effects. Yingying Du et al. [29] have shown that multilayered scaffolds were able to induce articular cartilage formation by accelerating early subchondral bone regeneration, and thereby newly formed tissues could integrate with native tissues. The aforementioned researchers demonstrated that a biomimetic osteochondral scaffold with continuous multistrand architecture and gradient composition from the articular cartilage layer to the subchondral bone layer was fabricated using a microsphere based SLS technique. Their results demonstrated that the resultant multilayer scaffold featured highly interconnected porosity and desirable mechanical properties, as well as excellent biocompatibility. In vivo, animal evaluation further verified that the multilayer scaffold could successfully induce osteochondral repair, and that the newly formed tissue manifested multiple tissues types including articular cartilage and subchondral bone [29].

Osteochondral tissue could thus be obtained in vitro and in vivo on the basis of biodegradable scaffolds and differentiated cells for each type of tissue (osteoblasts for bone tissue, chondroblasts or chondrocytes for cartilage) or undifferentiated cells (adult stem cells, embryonic stem cells or induced pluripotent stem cells).

Current research including the gradual development of scaffolds that have a structure similar to the natural structure of an osteochondral tissue and the controlled release methods of the factor promises to favor the formation of the osteochondral interface.

A significant issue is the challenge of developing scaffolds or factors that can later regenerate osteochondral tissue. Thus, it is desired that the osteochondral tissue thus formed is very similar to the natural one, both structure- and function-wise, to form a smooth interface between the articular cartilage and the bone surface, and also to prevent the phenotypic drift in the regenerated cartilage. In addition, advancing these

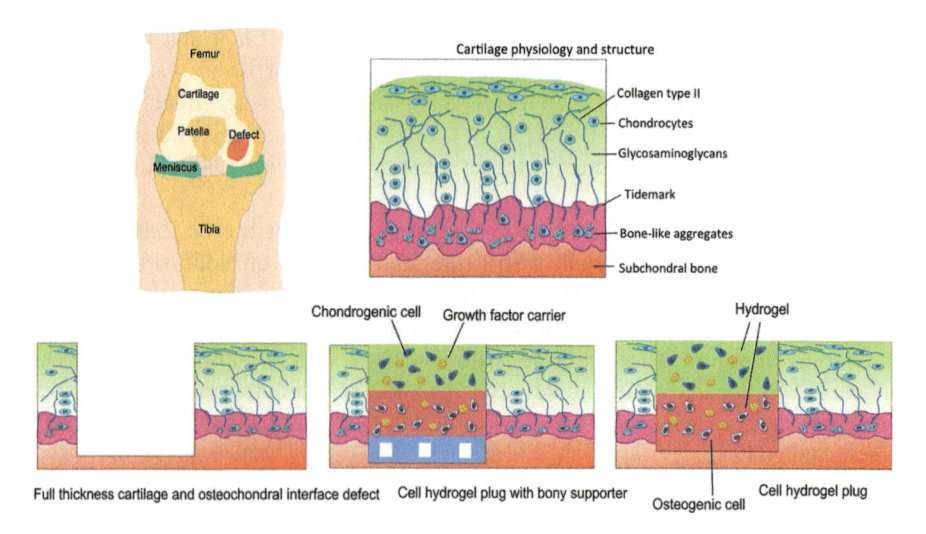

**Fig. 2.3** Tissue-engineering strategy for the treatment of osteochondral interface and full-thickness cartilage defects with cell-laden hydrogel constructs [36]

complex technologies may contain not only the biodegradable scaffolds, but also cells and growth factors as another important factor in osteochondral tissue engineering [25] (Fig. 2.3).

## 2.6 Hydrogels, Cells, and Growth Factors Used for Osteochondral Tissue Engineering (OTE) and Cartilage Tissue Engineering (CTE)

Since the biggest challenge was to find a biomaterial that can solve both the complex hierarchical structure of the articular cartilage and of the subchondral bone, hydrogels are emerging to be a promising class of biomaterials. They can be used for the regeneration of soft tissue and hard tissue as well. By proper selection of the material and chemistry, many critical properties of hydrogels (i.e., elasticity, bioactivity, degradation, mechanical stiffness, and water content) can be rationally and conveniently designed for these kinds of needs. Particularly, advances in the development of cell-loaded hydrogels have opened up new possibilities for cell therapy. Jingzhou Yang et al. [36] described the specific problems in this domain and have also described the latest findings on hydrogel-cell hybrids that can be used to restore osteochondral tissue. They focused on the efficiency of osteogenesis and chondrogenesis due to the type of hydrogel chosen, the cell and the growth factor [36] (Table 2.2).

Over time, the hydrogels have become more known and developed a series of hydrogels obtained from naturally derived polymers, chemically modified natural polymers, synthetic polymers, and their combinations for regenerating cartilage and

**Table 2.2** Hydrogels, cells and growth factors used for osteochondral tissue engineering (OTE) and cartilage tissue engineering (CTE) [36]

| Hydrogel type | Application (OTE/CTE) | Cell type | Growth factor | Crosslinking/gelation method |
|---|---|---|---|---|
| Alginate | OTE CTE | Chondrocytes MSCs IPSCs | TGF-β3 | Physical (ionic interaction) |
| Agarose | CTE | Chondrocytes MSCs | TGF-β3 | Physical (temperature change) |
| Collagen | OTE CTE | MSCs Chondrocytes | TGF-β1–3 BMP-2-7 FGF-1 | Chemical |
| Chitosan | CTE | Chondrocytes MSC | TGF-β1 | Chemical |
| Gellan gum | CTE | Chondrocytes | – | Physical (ionic interaction and temperature change) |
| Gelatin | OTE | MSCs | BMP-4 | Chemical |
| Fibroin | CTE | Chondrocytes | – | Physical (ionic interaction) |
| HA | CTE | Chondrocytes MSCs | TGF-β | Chemical (UV photopolymerization) |
| PVA | OTE | Osteoblasts Chondrocytes | – | Chemical (UV photopolymerization) |
| PEG | CTE OTE | MSCs Chondrocytes ESCs | TGF-β3 | Chemical (UV photopolymerization) |
| OPF | OTE | MSCs | TGF-β1 | Chemical |
| PDMMAAm | OTE | Chondrocytes | – | Chemical |
| Peg-PCL | CTE | Chondrocytes | – | Physical (temperature change) |
| GelMA | CTE | Chondrocytes | TGF-β1 | Chemical (UV photopolymerization) |
| – | OTE | PBMCs | – | – |

*HA* Hydroxyapatite, *PVA* Poly(vinyl alcohol), *PEG* Poly(ethylene glycol), *PVA* Polyvinyl alcohol, *OPF* Oligo(poly(ethylene glycol)fumarate), *PDMAAm* Poly(N,N-dimethylacrylamide), *PEG–PCL* Poly(ethylene glycol)–poly(e-caprolactone), *GelMA* Gelatin methacryloyl, *BMP* Bone morphogenetic protein, *PBMCs* Peripheral blood mononuclear cells [36]

osteochondral tissues. Studies have shown that some types of hydrogels can support proliferation, growth and cellular spread in the area where they are introduced, in this case being osteochondral cells, and they are also able to maintain the structure and phenotype of the tissue.

It has been found that naturally derived hydrogels have a high biocompatibility and at the same time support cellular viability, while synthetic hydrogels have mechanical and biodegradable properties that can be easily varied, making them vital for proper cartilage regeneration and clinical application. Rationally designed composite hydrogels could thus combine the advantages of natural, modified, and synthesized polymers [36].

In conclusion, hydrogels have certain high biodegradability and biocompatibility properties; they can also be easily molded for OTE and CTE applications. Although they are well suited to osteochondral applications, the disadvantages encountered may be related to insufficient mechanical stiffness, osteoconductivity, osteoinductivity, injectability, or printability. Thus, the inorganic particles contained in hybrid hydrogels composites place them among the biomaterials used to repair osteochondral tissues. Because of excellent osteoconductivity and osteoconductivity properties, the most favorable inorganic particles are phosphates and silicate minerals and bioactive glass [37, 38].

There is a variety of types of stem cells that may be able to replace osteochondral and cartilaginous tissue, but most of them are stem cells with a capacity for chondrogenic and osteogenic differentiation. These are best suited for such applications as they have an abundant source of supply. As Jingzhou Yang et al. [36] sustain in their review, different types of stem cells encapsulated in hydrogels were shown to differentiate into chondrocytes or osteoblasts induced by growth factors (e.g., the family of TGFs or BMPs) and to promote chondrogenesis and osteogenesis both in vitro and in vivo. Despite these advantages, engineered cartilage/osteochondral constructs combined with stem cells may readily become hypertrophic and undergo endochondral ossification, which hinders the formation of effective functional chondrogenesis and osteochondrogenesis [36].

For controlled release of the differentiation factor, strategies such as the development of the microcarrier growth factor, the covalent linkage of the hydrogel network, and the gene delivery have been designed. New bioactive factors have also been developed, with small molecules like kartogenin [39] or t-butyl methacrylate [40], and are used for inducing chondrogenic or osteogenic differentiation of stem cells.

Even after various research activities, a perfect restoration of osteochondral and cartilaginous tissue from the viewpoint of composition, structure and function of the area remains very hard to achieve. It is very difficult to make a hydrogel having a biodegradation rate directly proportional to the rate of cartilage and bone growth, making it a real challenge to find the ideal rate of degradation of hydrogels depending on the target tissue. Due to the properties of hydrogels suitable for the recovery of osteochondral tissue such as mechanical strength, osteoconductivity, ease of injection and printing, hybrid hydrophilic composites with inorganic particles and stem cells are very suitable as biomaterials for repairing osteochondral defects and cartilage thickness [36].

## 2.7 Composites Used as Synthetic Materials for Osteochondral Tissue Restoration

Composites are also used in tissue engineering strategies to repair osteochondral defects, when the subchondral bone as well as the cartilage is damaged, as a result of degenerative diseases such as osteoarthritis [41]. In this case, a simultaneous regeneration of both cartilage and subchondral bone is desired, using biphasic (or layered) composite scaffolds to guide the simultaneous regeneration of both tissues [42].

## 2.8 Stem Cell- and Scaffold-Based Tissue Engineering

To achieve a scaffold that delivers MSCs to defective areas in the osteochondral tissue, biomaterials must meet certain conditions, as shown in Fig. 2.4. The right choice of scaffolding design is very much about knowing the endogenous MSC activity in skeletogenesis, which includes cellular processes and cell sensitivity to biochemical and mechanical stimuli. To determine which stimuli to present to cells, bioengineering strategies can benefit significantly from endogenous examples of skeletogenesis. As an example of developmental skeletogenesis, the developing

**scaffold considerations**

choice of biomaterial & biocompatibility

geometry & architecture

porosity

mechanical properties

degradation rate

biochemical stimuli

**biological requirements**

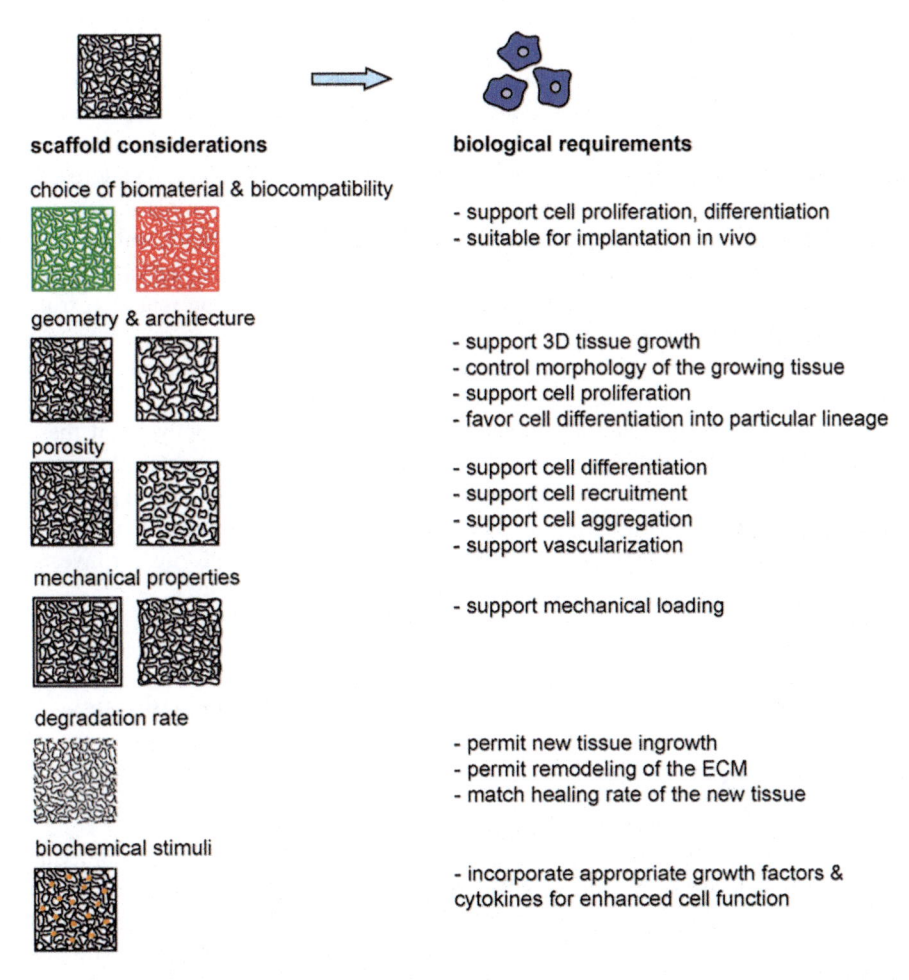

- support cell proliferation, differentiation
- suitable for implantation in vivo

- support 3D tissue growth
- control morphology of the growing tissue
- support cell proliferation
- favor cell differentiation into particular lineage

- support cell differentiation
- support cell recruitment
- support cell aggregation
- support vascularization

- support mechanical loading

- permit new tissue ingrowth
- permit remodeling of the ECM
- match healing rate of the new tissue

- incorporate appropriate growth factors & cytokines for enhanced cell function

**Fig. 2.4** Biologically informed design specifications for biomaterials in tissue engineering. Scaffold properties such as material biocompatibility, geometry, porosity, mechanical strength, degradation rate, and incorporation of signaling molecules can be optimized to address various physiological requirements of an engineered tissue [43]

limb bud serves as an excellent model system of how osteochondral structures form from undifferentiated precursor cells [43]. Sundelacruz et al. [43] present their observations on a particular biomaterial that has proved promising in supporting osteochondral growth in vitro and in vivo: silk fibroin from the silkworm *Bombyx mori* [44, 45].

Currently, both biological and tissue engineering research is concerned with discovering the clues needed to stimulate cells to regenerate tissues and organs totally or partially. For this to be possible, both approaches should take into account the progenitor behavior of stem cells, scaffolding and extracellular matrix (ECM), but also that of signaling molecules. Based on the biological research on cells that assist the formation of new tissues, tissue engineers have identified adult mesenchymal stem cells as promising cell sources that exhibit the plasticity required to perform similar cell functions in vitro and in vivo. This accumulated knowledge can be used by tissue engineers to help incorporate biomolecules chosen in systems, whether they are later used as scaffolding biomaterials or as a released factor. On the other hand, because tissue engineering is an inverse approach to regenerative medicine, scaffold-based tissue formation is an important challenge that cannot be addressed in developmental biology. The properties of the scaffold biomaterials dictate the microenvironment of the regenerated tissue. This physical microenvironment can adjust a wide range of parameters, for example cell proliferation, cell differentiation, healing mode, in vivo scaffolding stability, and release or presentation from delivered growth factors.

Through the mutual efforts of these two areas, progress can be made towards finding the appropriate balance between biochemical and physical cues for tissue formation and for the final adjustment of the spatial–temporal delivery of these broad scale tissue modeling indexes. Such knowledge is very important for engineering complex tissues in vitro. Functionally produced tissues have great potential in channeling efforts in the field of regenerative medicine, both by addressing the current clinical need for tissue replacement and by providing platforms for investigating treatment strategies to stimulate tissue regeneration [43].

## 2.9 Chemical, Topological, Mechanical, and Structural Cues of Biomaterials Used to Evoke Biological Responses on Cells

Even though many biomaterials have been developed to repair osteochondral tissue (where a key design requirement is the ability to repair both bone and cartilage in the osteochondral unit), so far few studies have examined the role of structure and composition biomaterials on the properties of the repair tissue. The mechanistic role of the structure and composition of biomaterials has been studied in many systems, and the extracellular matrix regeneration is believed to be dependent on a complex interaction of mechanical and chemotactic signals [46, 47]. Since it was assumed that cell recognition in this biochemical micro medium is sufficient to induce cell adhesion, migration and differentiation, much of the research has been based on the use of decellularized and naturally derived scaffolds, or of those made from synthetically produced or purified protein products constituting a cartilage ECM (or combinations thereof). The first approved procedure in the USA in the treatment of osteochondral defects involving biological materials was the use of collagen flaps as support matrix for autologous chondrocyte transplantation or for attachment and differentiation during microfracture procedures [47].

Niemeyer et al. [49] and Moseley et al. [48]—in their studies on the long-term outcome of these clinical treatments—have shown that there is a limited improvement in tissue functionality, resulting in the need for more complex materials that can orchestrate the development and cartilage tissue homeostasis either endogenously derived cells (MSC or chondrocytes) or allogeneic cells [48, 49]. Technological biofunctionality in synthetic materials has been seen as a promising alternative to current clinical treatments [47].

The latest findings on nanotechnology advances along with the processability of synthetic biomaterials have succeeded in creating a new range of materials to develop into the desired biological responses to the cellular level. In this way, their adhesion, migration and differentiation by integrin binding and activation of critical determinations of signaling pathways such as those regulated by Rac and Rho, protein kinase C and MAP kinase [50, 51] can be controlled. The design of biomaterials to improve biofunctionality can be achieved by using a wide range of tests involving the introduction of chemical, topographic or mechanical indices through top-down or bottom-up approaches (Fig. 2.5) [52].

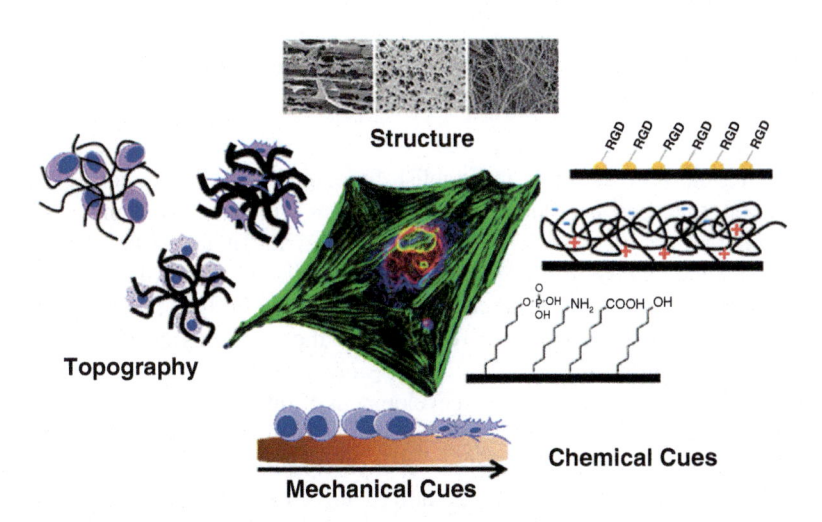

**Fig. 2.5** Schematic representation of common chemical, topological, mechanical, and structural cues used to evoke biological responses on cells [52]

## 2.10 Osteochondral Defect Repair Using a Biphasic Scaffold Based on a Collagen–Glycosaminoglycan Biopolymer

Getgood et al. [47] compared in their study the performance of a biphasic scaffold based on a CaP-mineralized collagen–GAG biopolymer [53–55] with a synthetic PLGA-PGA-CaS composite. In this study we can see that the scaffolds made of both biomaterials have the capacity to support the repair of the osteochondral tissue, but there is also a tendency towards the phenotype of the tissue improved by adding the natural collagen material [47].

The use of the PLGA material and the incidence of cystic change is consistent with some existing clinical literature [56, 57]. Because these porous biphasic biomaterials allow the defect to be linked to sub-congenital bone marrow (containing nutrients and mesenchymal stem cells), they can be seeded with cells that have the ability to differentiate into relevant phenotypes and help synthesize new osteochondral and cartilaginous tissues [58]. In particular, porous collagen-GAG biomaterials have shown they minimize the formation of fibrous scar tissue [59–61] and are considered promising tissue repair mediators.

The study authored by M. J. Getgood et al. [47] has shown that a scaffolding microporous material can be used to provide a three-dimensional environment that encourages the development of a hyaline-like cartilage structure between 12 and 26 week points compared to an empty defect.

In conclusion, the studies demonstrated that the biphasic biomaterial from which the scaffolding was designed can provide a viable alternative for osteochondral autograft or allograft materials used to repair osteochondral defects associated with bone disorders under the condyle [47].

## 2.11 Application of 4D Printing in Osteochondral Tissue Regeneration

3D printing has a great ability to establish functional tissues or organs to cure or replace abnormal and necrotic tissue, providing a promising solution for serious tissue/organ failure [62, 63]. 3D printing is evolving into an unparalleled biomanufacturing technology [64, 65]. 4D printing, based on the advanced 3D print features, is a dynamic, time-dependent process in manufacturing design. The 4D print process has the potential to continually revolutionize the current tissue and organ manufacturing platforms [66].

A new active research area is the development of intelligent materials with high biocompatibility to suit 4D printing technology. For example, 3D STLs, scaffolds capable of supporting the growth of mesenchymal stem cells from the multipotent bone marrow (hMSC) with acrylic epoxidized soybean have been printed [67]. The laser frequency and print speed have shown that they can have particular effects on the superficial structures of the epoxidized acrylate polymerized soybean oil shown in Fig. 2.6.a. These scaffolds, from a temporary set at 18 ° C, can completely recover

2 Synthetic Materials for Osteochondral Tissue Engineering 45

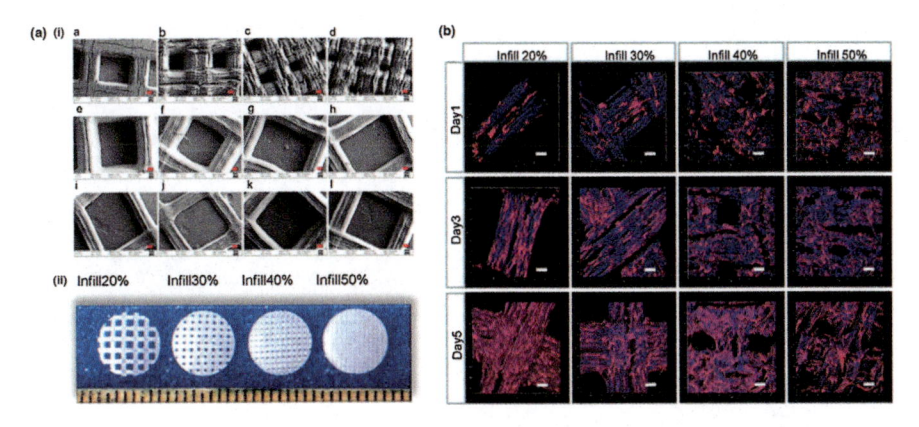

**Fig. 2.6** Shape memory biomedical scaffolds. (**a**) (i) SEM images of printed scaffolds, red scale bar 100 mm. (ii) The photos are of the printed scaffolds. (**b**) Confocal images of hMSCs spreading on printed scaffolds. Scale bars are 100 mm [66]

their original shape at the temperature of the human body (37 ° C). Thus, this novel material showed a significantly higher adhesion and proliferation of hMSC than polyethylene glycol diacetate (PEGDA) and did not have statistical differences between PLA and PCL. This increase in hMSC on printed patterns can be seen in Fig. 2.6.b. Given the multipotent nature of hMSCs, these scaffolds have great potential applications for the engineering of osteochondral and neuronal tissues [66].

## 2.12 Medical Results

One of the most common diseases related to osteochondral tissue is the meniscal lesion. The purpose of treating this area is to preserve as much tissue as possible, but when the lesions are very high, the treatment chosen is meniscectomy, but this procedure can lead to osteoarthritis. Following this procedure, the treated area is subjected to a much higher pressure, so coarse tissue will subsequently require multiple surgeries.

Research has led to the emergence of various therapeutic strategies for this condition, one of the most used being the replacement of the meniscus by a meniscal transplant, which later led to the development of artificial implantable scaffolds made of natural or synthetic polymers. The only scaffolds approved so far are Actifit® and Menaflex® (which can only be used in patients who have had partial meniscectomy) [68] (Fig. 2.7).

Following the research, treatment of this meniscal zone has evolved, but the total meniscectomy procedure has been abandoned due to the link between this procedure and the occurrence of early osteoarthritis. However, there is currently partial meniscectomy or special sutures, but the emphasis is on tissue engineering, by developing an artificial meniscus used to restore the function and structure of the meniscal fibrocartilage [69] (Fig. 2.8).

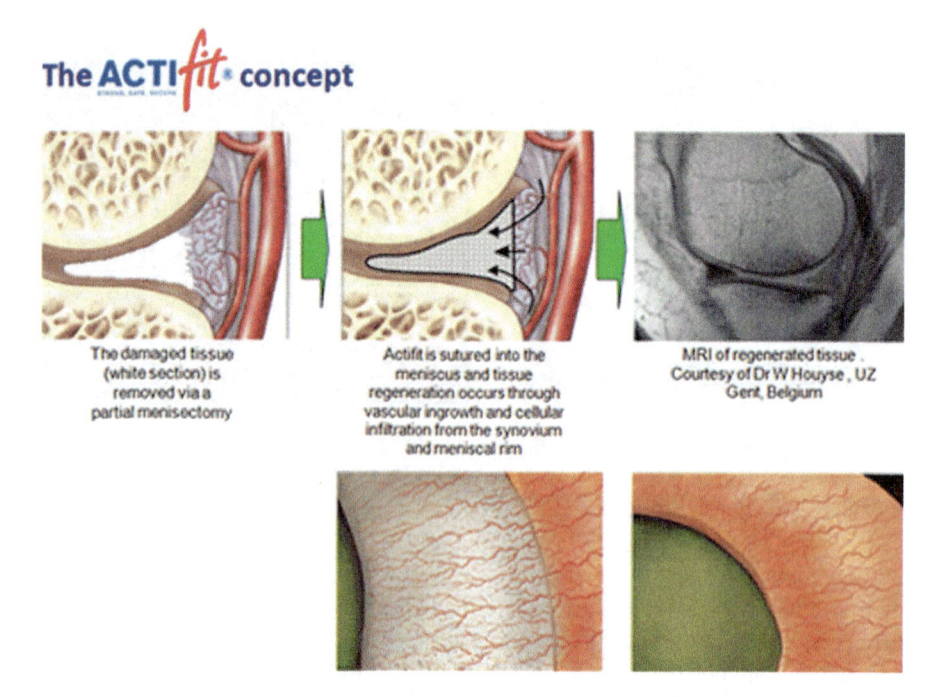

**Fig. 2.7** How Actifit® works [68]

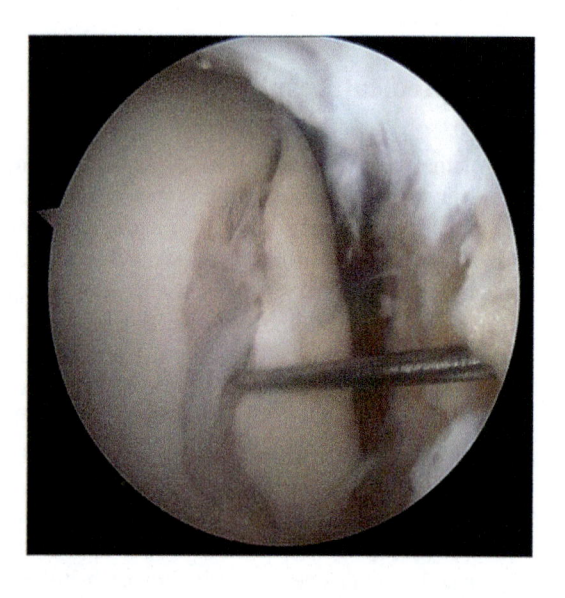

**Fig. 2.8** Chondral lesion on femoral condyle [70]

The basic cells that make up the meniscal tissue are fibrohydrocytes that create two main areas (i.e., superficial and deep) at the microscopic level. They closely resemble phenotypically chondrocytes and also have the ability to establish and develop a fibrous tertiary matrix [71]. Following modern testing, it has been demonstrated that at the time of compression of the tibiofemoral joint, a vertical force and a radial component appear on the meniscus surface. This external force moving the menus during the movement is limited by the rigid attachment of the meniscus to the horns that produces a circumferential force and a stretching force of the chain. All these forces are further distributed in tissue to the collagen fibers in the deep meniscus layer [72].

## 2.13 Meniscal Substitutes

Meniscal allografts come from multiorgan donors. There are different ways of processing and storing allografts: fresh, fresh-frozen, cryopreserved, and lyophilized. For scaffolding, synthetic polymeric biomaterials can be easily designed in any shape, with a particular geometric structure, porosity and biomechanical properties. They may also include a specific rate of biodegradation of the scaffold in nontoxic products [68, 73]. The scaffolds used for substitution of artificial meniscus may be synthetic or natural, as shown in Table 2.3.

## 2.14 Surgical Technique of Implantation

The length of the defect is measured using a malleable ruler (Figs. 2.9, 2.10, 2.11, and 2.12; Table 2.4). So that the implant would fit perfectly into place, 10 percent is added to the length prior read [68].

Table 2.3 Features that a biomaterial used for scaffolding must be/have

| No. | Characteristic | Description | References |
|-----|----------------|-------------|------------|
| 1. | "Cell-instructive" | The ability to induce (or to support) the host cell migration and differentiation | [69, 74] |
| 2. | "Biomimetic" | Similar to the structure and the mechanical characteristics of the native meniscus | |
| 3. | Resistance | Resilient and resistant to withstand the forces that appear in the tibiofemural joint | |
| 4. | Biocompatibility | Tolerated by the host and lacking the immune reactions to the degradation products that result over time | |
| 5. | Biodegradability | Allowing replacement by normal meniscus tissue | |
| 6. | Porosity | Being porous, allowing nutrients and catabolic substances to flow through | |
| 7. | Flexibility | Easily handled by the orthopedic surgeon | |

**Fig. 2.9** Arthroscopic technique 1st step measuring [68]

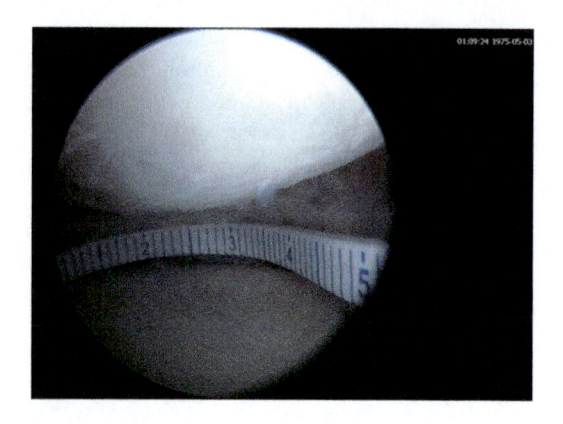

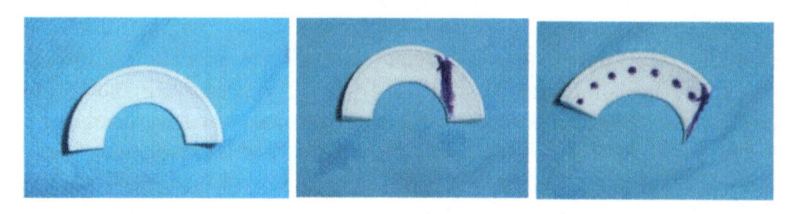

**Fig. 2.10** Preparing the scaffold [68]

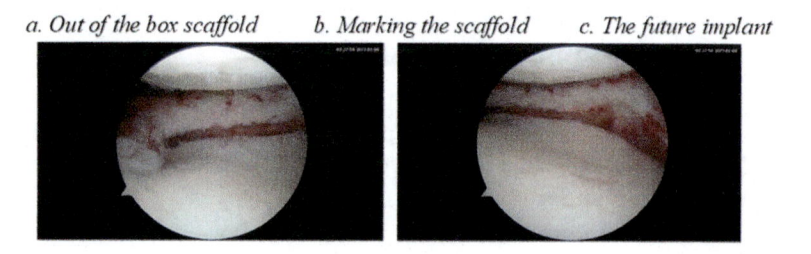

*a. Out of the box scaffold*   *b. Marking the scaffold*   *c. The future implant*

**Fig. 2.11** Partial meniscectomy [68]

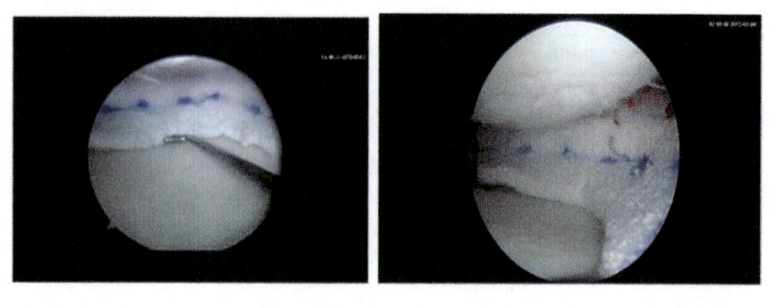

**Fig. 2.12** Final results after implantation [68]

**Table 2.4** The Biomechanical, Biochemical and Histological Changes of Ligament Scars Compared with Normal Ligaments Approximately 1 Year after Injury [70]

| | |
|---|---|
| Biomechanical changes | Weaker |
| | Inferior material quality |
| | Larger |
| | Greater creep |
| Biochemical changes | Increased type V collagen |
| | Decreased hydroxypyridinium cross-links |
| | Increased glycosaminoglycans |
| Histologic changes | "Flaws" in matrix |
| | Abnormal collagen fibril diameter distribution |

## 2.15 Summary

Articular osteochondral defect repair involves two types of distinct tissues: articular cartilage and subchondral bone. A biomaterial is any substance that has been designed to interact with biological systems for medical purposes—either therapeutic (treating, augmenting, repairing, or replacing a tissue function of the body) or diagnostic. When a non-body material is placed in the human body, the biological tissue reacts in different ways, depending on the type of material. The mechanism of tissue interaction depends on the tissue's response to the surface of the implant. From the results obtained during the studies carried out over time, it has been noticed that the biomaterials under study show all similarities and differences in their way of restoration of the osteochondral tissue [2, 6, 7, 27, 29, 30, 74]. As various researchers and tissue engineers have demonstrated, the role of growth factors in tissue engineering for repairing osteochondral and cartilage defects is a very important one. Following animal testing, cell-assisted and growth-factor scaffolds produced much better results, while growth-free scaffolds showed a much lower rate of healing.

## References

1. Langer R, Vacanti JP (1993) Tissue engineering. Science 260:920–926
2. Vinatier C, Guicheux J (2016) Cartilage tissue engineering: from biomaterials and stem cells to osteoarthritis treatments. Ann Phys Rehabil Med 59:139–144
3. Vinatier C, Bouffi C, Merceron C et al (2009) Cartilage tissue engineering: towards a biomaterial-assisted mesenchymal stem cell therapy. Curr Stem Cell Res Ther 4:318–329
4. Vinatier C et al (2006) Cartilage and bone tissue engineering using hydrogels. Biomed Mater Eng 16:S107–S113
5. Drury JL, Mooney DJ (2003) Hydrogels for tissue engineering: scaffold design variables and applications. Biomaterials 24:4337
6. Vinatier C, Gauthier O, Fatimi A et al (2009) An injectable cellulose-based hydrogel for the transfer of autologous nasal chondrocytes in articular cartilage defects. Biotechnol Bioeng 102:1259–1267

7. Filardo G, Kon E, Perdisa F et al (2013) Osteochondral scaffold reconstruction for complex knee lesions: a comparative evaluation. Knee 20(6):570
8. Brix MO et al (2013) Cartilage repair of the knee with Hyalograft C1: magnetic resonance imaging assessment of the glycosaminoglycan content at midterm. Int Orthop 37:39–43
9. Nehrer S et al (2006) Three-year clinical outcome after chondrocyte transplantation using a hyaluronan matrix for cartilage repair. Eur J Radiol 57:3–8
10. Steinwachs MR, Waibl B, Mumme M (2014) Arthroscopic treatment of cartilage lesions with microfracture and BST-CarGel. Arthrosc Tech 3:399–402
11. Ossendorf C et al (2007) Treatment of posttraumatic and focal osteoarthritic cartilage defects of the knee with autologous polymer-based three-dimensional chondrocyte grafts: 2-year clinical results. Arthritis Res Ther 9:R41
12. Seo S-J, Mahapatra C, Singh RK, Knowles JC, Kim H-W (2014) Strategies for osteo-chondral repair: focus on scaffolds. J Tissue Eng 5:204173141454185
13. Uematsu K, Hattori K, Ishimoto Y, Yamauchi J, Habata T, Takakura Y et al (2005) Cartilage regeneration using mesenchymal stem cells and a three-dimensional polylactic- glycolic acid (PLGA) scaffold. Biomaterials 26(20):4273–4279
14. Jazayeri HE et al (2017) A current overview of materials and strategies for potential use in maxillofacial tissue regeneration. Mater Sci Eng C 70:913–929
15. Fan H, Hu Y, Zhang C, Li X, Lv R, Qin L et al (2006) Cartilage regeneration using mesenchymal stem cells and a PLGA–gelatin/chondroitin/hyaluronate hybrid scaffold. Biomaterials 27(26):4573–4580
16. Baker MI, Walsh SP, Schwartz Z, Boyan BD (2012) A review of polyvinyl alcohol and its uses in cartilage and orthopedic applications. J Biomed Mater Res B Appl Biomater 100(5):1451–1457
17. Tadavarthy SM, Moller JH, Amplatz K (1975) Polyvinyl alcohol (Ivalon)—a new embolic material. Am J Roentgenol 125(3):609–616
18. Bray JC, Merrill EW (1973) Poly (vinyl alcohol) hydrogels for synthetic articular cartilage material. J Biomed Mater Res 7(5):431–443
19. Kobayashi M, Chang Y-S, Oka M (2005) A two year in vivo study of polyvinyl alcoholhydrogel (PVA-H) artificial meniscus. Biomaterials 26(16):3243–3248
20. Kobayashi M, Toguchida J, Oka M (2003) Preliminary study of polyvinyl alcoholhydrogel (PVA-H) artificial meniscus. Biomaterials 24(4):639–647
21. Bodugoz-Senturk H, Macias CE, Kung JH, Muratoglu OK (2009) Poly (vinyl alcohol)–acrylamide hydrogels as load-bearing cartilage substitute. Biomaterials 30(4):589–596
22. Stammen JA, Williams S, Ku DN, Guldberg RE (2001) Mechanical properties of a novel PVA hydrogel in shear and unconfined compression. Biomaterials 22(8):799–806
23. Holloway JL et al (2011) Analysis of the in vitro swelling behavior of poly (vinyl alcohol) hydrogels in osmotic pressure solution for soft tissue replacement. Acta Biomater 7(6):2477–2482
24. Shokrgozar MA, Bonakdar S, Dehghan MM, Emami SH, Montazeri L, Azari S et al (2013) Biological evaluation of polyvinyl alcohol hydrogel crosslinked by polyurethane chain for cartilage tissue engineering in rabbit model. J Mater Sci Mater Med 24(10):2449–2460
25. Nukavarapu SP, Dorcemus DL (2013) Osteochondral tissue engineering: current strategies and challenges. Biotechnol Adv 31:706–721
26. Mandelbaum BR, Browne JE, Fu F, Micheli L, Mosely JB, Erggelet C et al (1998) Articular cartilage lesions of the knee. Am J Sports Med 26:853
27. Hunziker EB (2002) Articular cartilage repair: basic science and clinical progress. A review of the current status and prospects. Osteoarthr Cartil 10:432
28. Huey DJ, Hu JC, Athanasiou KA (2012) Unlike bone, cartilage regeneration remains elusive. Science 338:917
29. Du Y et al (2017) Selective laser sintering scaffold with hierarchical architecture and gradient composition for osteochondral repair in rabbits. Biomaterials 137:37
30. Smith GD, Knutsen G, Richardson JB. A clinical review of cartilage repair techniques. J Bone Jt Surg Br. 2005; 87.

31. Marcacci M et al (2007) Arthroscopic autologous osteochondral grafting for cartilage defects of the knee: prospective study results at a minimum 7-year follow-up. Am J Sports Med 35:2014e2021
32. Salzmann GM et al (2011) The dependence of autologous chondrocyte transplantation on varying cellular passage, yield and culture duration. Biomaterials 32:5810
33. Filardo G et al (2013) Matrix-assisted autologous chondrocyte transplantation for cartilage regeneration in osteoarthritic knees: results and failures at midterm follow-up. Am J Sports Med 41:95
34. Eldracher M, Orth P, Cucchiarini M, Pape D, Madry H (2014) Small subchondral drill holes improve marrow stimulation of articular cartilage defects. Am J Sports Med 42:2741e2750
35. Gobbi A, Karnatzikos G, Kumar A (2014) Long-term results after microfracture treatment for full-thickness knee chondral lesions in athletes. Knee Surg Sports Traumatol Arthrosc 22:1986e1996
36. Yang J, Zhang YS, Yue K, Khademhosseini A (2017) Cell-laden hydrogels for osteochondral and cartilage tissue engineering. Acta Biomater 57:1–25
37. Gantar A et al (2014) Nanoparticulate bioactive-glass-reinforced gellan-gum hydrogels for bone-tissue engineering. Mater Sci Eng C Mater Biol Appl 43:27–36
38. Xavier JR et al (2015) Bioactive nanoengineered hydrogels for bone tissue engineering: a growth-factor-free approach. ACS Nano 9(3):3109–3118
39. Johnson K et al (2012) A stem cell-based approach to cartilage repair. Science 336(6082):717–721
40. Benoit DS, Schwartz MP, Durney AR, Anseth KS (2008) Small functional groups for controlled differentiation of hydrogel-encapsulated human mesenchymal stem cells. Nat Mater 7(10):816–823
41. Mano JF, Reis RL (2007) Osteochondral effects: present situation and tissue engineering approaches. J Tissue Eng Regenerative Med 1:281–287
42. Chen Q, Roether JA, Boccaccini AR. Tissue engineering scaffolds from bioactive glass and composite materials. 2008; 4, chapter 6.
43. Sundelacruz S, Kaplan DL (2009) Stem cell- and scaffold-based tissue engineering approaches to osteochondral regenerative medicine. Semin Cell Dev Biol 20:646–655
44. Karageorgiou V, Meinel L, Hofmann S et al (2004) Bone morphogenetic protein-2 decorated silk fibroin films induce osteogenic differentiation of human bone marrow stromal cells. J Biomed Mater Res A 71:528–537
45. Sofia S, McCarthy MB, Gronowicz G, Kaplan DL (2001) Functionalized silk-based biomaterials for bone formation. J Biomed Mater Res 54:139–148
46. Ghosh K, Ingber DE (2007) Micromechanical control of cell and tissue development: implications for tissue engineering. Adv Drug Deliv Rev 59(13):1306–1318
47. Getgood MJ, Simon JK, Brooks R, Aberman H, Simon T, Lynn AK, Rushton N (2012) Evaluation of early-stage osteochondral defect repair using a biphasic scaffold based on a collagen–glycosaminoglycan biopolymer in a caprine model. Knee 19:422–430
48. Moseley JB et al (2010) Long-term durability of autologous chondrocyte implantation a multicenter, observational study in US patients. Am J Sports Med 38:238–246
49. Niemeyer P et al (2014) Long-term outcomes after first-generation autologous chondrocyte implantation for cartilage defects of the knee. Am J Sports Med 42:150–157
50. Humphries JD, Byron A, Humphries MJ (2006) Integrin ligands at a glance. J Cell Sci 119:3901–3903
51. Prowse ABJ, Chong F, Gray PP, Munro TP (2011) Stem cell integrins: implications for ex-vivo culture and cellular therapies. Stem Cell Res 6:1–12
52. Camarero-Espinosaa S, Whitea JC (2017) Tailoring biomaterial scaffolds for osteochondral repair. Int J Pharm 523:476–489
53. Harley BA, Lynn AK et al. Design of a multiphase osteochondral scaffold III: fabrication of layered scaffolds with continuous interfaces. J Biomed Mater Res 2010; 92A(3):1078–93
54. Yannas IV, Tzeranis DS, Harley BA (2010) Biologically active collagen-based scaffolds: advances in processing and characterization. Philos Trans Royal Society A Mathemat Phys Eng Sci 368(1917):2123–2139

55. Harley BA, Lynn AK, et al. Design of a multiphase osteochondral scaffold. II. Fabrication of a mineralized collagen glycosaminoglycan scaffold. J Biomed Mater Res 2010; 92A(3):1066–77
56. Khan WS, Johnson DS, MR, et al. Delayed incorporation of a TruFit plug: perseverance is recommended. Arthroscopy 2009; 25(7):810–4
57. Kerker JT, Leo AJ, Sgaglione NA (2008) Cartilage repair: synthetics and scaffolds: basic science, surgical techniques, and clinical outcomes. Sports Med Arthrosc 16(4):208–216
58. Khan WS, Johnson DS, Hardingham TE (2010) The potential of stem cells in the treatment of knee cartilage defects. Knee 17(6):369–374
59. Yannas IV, Lee E, Orgil DP, Skrabut EM, Murphy GF (1989) Synthesis and characterization of a model extracellular matrix that induces partial regeneration of adult mammalian skin. PNAS 86(3):933–937
60. Chamberlain LJ, Yannas IV, Hsu HP, Strichartz G, Spector M (1998) Collagen–GAG substrate enhances the quality of nerve regeneration through collagen tubes up to level of autograft. Exp Neurol 154(2):315–329
61. Yannas IV (2001) Tissue and organ regeneration in adults. Springer, New York
62. Cui H, et al. Adv Healthc Mater 2016;6
63. Murphy SV et al (2014) Nat Biotechnol 32(8):773
64. Holmes B et al (2016) Nanotechnology 27(6):064001
65. Guo T et al (2016) Tissue Eng. B: Rev
66. Miao S, Castrol N, Nowicki M, et al., 4D printing of polymeric materials for tissue and organ regeneration, Materials Today 2017; 00(00).
67. Miao S, et al. Sci Rep 2016;6.
68. Codorean IB, Tanase S, et al. Current strategies and advanced materials for the treatment of injured meniscus.
69. Celeste S, Hirschmann MT, Antinnolfi P, Martin I, Peretti GM (2013) Meniscus repair and regeneration: review on current methods and research potential. Eur Cells Mater 26:150–170
70. Codorean IB, Tănase S, Diaconu F, et al. The treatment of articular cartilage lesions using two polymer scaffolds.
71. Makris EA, Hadidi P, Athannasiou KA (2011) The knee meniscus: structure-function, pathophysiology, current repair techniques and prospects for regeneration. Biomaterials 32(30):7411–7431
72. Kawamura S, Lotito K, Rodeo SA (2003) Biomechanics and healing response of the meniscus. Oper Tech Sports Med 11:68–76
73. Schoenfeld AJ, Landis WJ, Kay DB (2007) Tissue- engineered meniscal constructs. Am J Orthop 36:614–620
74. Haut Donahue TL, Hull ML, Rashid MM, Jacobs CR (2003) How the stiffness of meniscal attachments and meniscal material properties affect tibio-femoral contact pressure computed using a validated finite element model of the human knee joint. J Biomech 36:19–34

# Chapter 3
# Bioceramics for Osteochondral Tissue Engineering and Regeneration

Sandra Pina, Rita Rebelo, Vitor Manuel Correlo, J. Miguel Oliveira, and Rui L. Reis

**Abstract** Considerable advances in tissue engineering and regeneration have been accomplished over the last decade. Bioceramics have been developed to repair, reconstruct, and substitute diseased parts of the body and to promote tissue healing as an alternative to metallic implants. Applications embrace hip, knee, and ligament repair and replacement, maxillofacial reconstruction and augmentation, spinal fusion, bone filler, and repair of periodontal diseases. Bioceramics are well-known for their superior wear resistance, high stiffness, resistance to oxidation, and low coefficient of friction. These specially designed biomaterials are grouped in natural bioceramics (e.g., coral-derived apatites), and synthetic bioceramics, namely bioinert ceramics (e.g., alumina and zirconia), bioactive glasses and glass ceramics, and bioresorbable calcium phosphates-based materials. Physicochemical, mechanical, and biological properties, as well as bioceramics applications in diverse fields of tissue engineering are presented herein. Ongoing clinical trials using bioceramics in osteochondral tissue are also considered. Based on the stringent requirements for clinical applications, prospects for the development of advanced functional bioceramics for tissue engineering are highlighted for the future.

---

[$]The authors contributed equally to this work.

S. Pina (✉) · R. Rebelo
3B's Research Group – Biomaterials, Biodegradables and Biomimetics, University of Minho, Headquarters of the European Institute of Excellence on Tissue Engineering and Regenerative Medicine, Barco, Guimarães, Portugal

ICVS/3B's - PT Government Associate Laboratory, Barco, Guimarães, Portugal
e-mail: sandra.pina@dep.uminho; rita.rebelo@dep.uminho.pt

V. M. Correlo · J. M. Oliveira · R. L. Reis
3B's Research Group – Biomaterials, Biodegradables and Biomimetics, University of Minho, Headquarters of the European Institute of Excellence on Tissue Engineering and Regenerative Medicine, Barco, Guimarães, Portugal

ICVS/3B's - PT Government Associate Laboratory, Barco, Guimarães, Portugal

The Discoveries Centre for Regenerative and Precision Medicine, Headquarters at University of Minho, Barco, Guimarães, Portugal

**Keywords** Bioceramics · Natural and synthetic bioceramics · Calcium phosphates · Clinical trials · Osteochondral regeneration

## 3.1 Introduction

Over the last century, new biomaterials have considerably changed the lives of millions of patients. Biomaterials have made an important contribution to modern health care and will expand further, especially for osteoporosis, osteoarthritis, and fragility fractures, increasing with elderly population. Each biomaterial has specific physicochemical, mechanical, and biological characteristics which can originate variations in host/material response when applied for healthcare.

Bioceramics can be classified as inorganic and non-metallic ceramics used for the repair and regeneration of diseased and damaged parts of the musculoskeletal system and periodontal anomalies [1]. Bioceramics are known to promote biomineralization with excellent osteoconductivity, chemical corrosion resistance, and a hard brittle surface. However, limitations include brittleness, poor fracture toughness, very low elasticity, and extremely high stiffness [2]. Bioceramics are categorized depending on their ability to bond with living tissues after implantation, as: (a) bioinert ceramics (e.g., alumina and zirconia) has no interaction with its surrounding tissue after implantation. They have a reasonable fracture toughness, and resistance to corrosion and wear, (b) bioactive ceramics (e.g., bioglasses and glass-ceramics) bond directly with living tissues, with the pattern of bonding osteogenesis, and (c) bioresorbable ceramics (e.g., calcium phosphates (CaPs), calcium phosphate cements (CPCs), calcium carbonates, and calcium silicates) are gradually absorbed in vivo and is replaced by bone with time.

Considering their unique properties, bioceramics are commonly used in tissue engineering (TE) and biomedical applications, particularly for developing 3D–based scaffolds able to mimic the native tissues [3, 4]. Bioceramics are stronger under compression and weak under tension, important facts to have into account in particular biomedical application. Natural and synthetic bioceramics have been proposed to be used in the processing TE scaffolding considering specific composition, microstructure, and long-term reproducibility. Natural bioceramics include coral-derived materials, sponges, nacres, and animal (fish and chicken) bones, and offer an abundant source of calcium compounds (e.g., calcium carbonate and calcium phosphate) [5]. Synthetic bioceramics embrace alumina and zirconia, bioactive porous glasses and glass-ceramics, and CaPs-based materials in the form of sintered ceramics, coatings and cement pastes [6, 7].

Fabrication methodology available for bioceramics production are wet precipitation, hydrolysis, sol–gel synthesis, hydrothermal synthesis, mechanochemical synthesis, microwave processing, and spray drying methods. Among them, wet precipitation method has the benefit on the homogeneity of the final product, and the easiness of controlling certain parameters, such as temperature, pH, and the presence of additives, during the synthesis [8].

Many studies are devoted to bioceramics incorporating ionic elements (e.g., strontium, zinc, magnesium, manganese, and silicon) that would be released during bone graft resorption, and hence can influence bone health and enhance biocompatibility, while strengthening the mechanical properties of the implants [9–13]. Besides, minerals and traces of metal elements may provide physicochemical modifications in the produced materials, which can accelerate bone formation and resorption in vivo [14, 15].

This chapter presents a concise overview of natural and synthetic bioceramic materials for bone, cartilage, and OC tissue applications. A variety of materials are considered, from bioinert to bioactive and bioresorbable ceramics. It is presented their physicochemical, mechanical and biological properties. Clinical trials involving bioceramics, challenges and future prospects of research in this field, are also underlined.

## 3.2 Bioceramic Materials and Properties

### 3.2.1 Natural Bioceramics

Naturally derived bioceramics can offer an abundant source of inorganic materials (e.g., calcium carbonate and CaPs) with high applicability for tissue replacement and regeneration [5]. Emphasis are put on the ones from marine origin, such as natural corals, nacres (or mollusc shells), sponges, and fish bones (Fig. 3.1I).

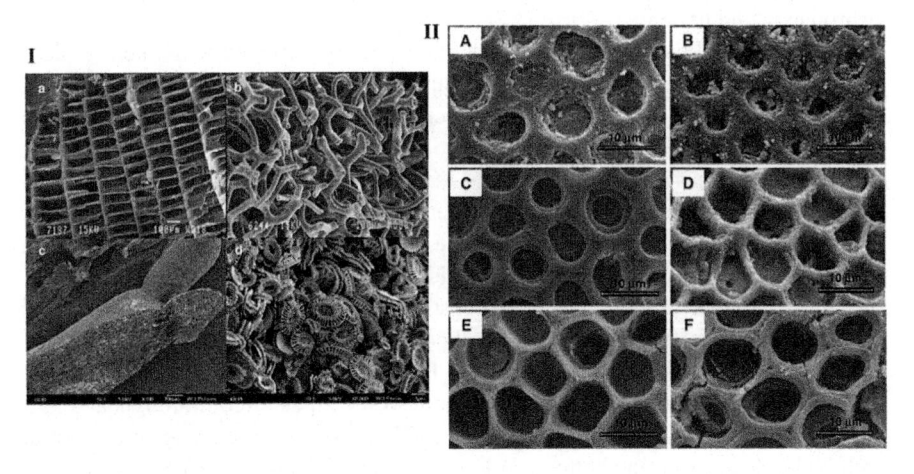

**Fig. 3.1** I—Scanning electron micrographs of (a) cuttlefish (*Sepia officinalis*), (b) sponge (*Spongia agaricina*), (c) red algae (*Corallina officinalis*), and (d) coccolithophores (*Emiliania huxleyi*) demonstrating a range of macroporous and microporous structures. Reprinted with permission [19]. Copyright 2011, Elsevier. II—Scanning electron micrographs of *Coralline officinallis*: (A) as received, (B) after 6 h at 400 °C, and after different chemical treatments with (C) $Na_4P_2O_7 \cdot 10H_2O$, (D) $H_3PO_4$, (E) $H_3PO_4 + (NH_4)_2HPO_4$, and (F) HF 48%. Reprinted with permission [17]. Copyright 2006, Elsevier

Calcium carbonate (aragonite or calcite forms) can be found in many of these marine organisms, and then converted to CaPs for the biomedical field, owing their unique structure, architecture, and mechanical properties.

Corals have been the most widely investigated as scaffolds, since they combine a multiscale porosity, and interconnected pores (100–500 μm diameter) and channels, crucial for healthy bone replacement [16]. Also, coral skeletons hold in situ resorption and high versatility, thus be capable of genes and bioactive factors delivering. Our group reported the use of *Coralline officinallis* to be useful as bone fillers targeting its repair and regeneration [17, 18]. Calcium carbonate skeletons of *C. officinallis* were converted into CaPs with hydroxyapatite (HAp) nanocrystallites, by combining a thermal and chemical treatment (Fig. 3.1II) [17]. Results showed that the coralline particulates preserved their morphology, after heat treatment and by soaking in different solutions. Furthermore, it was demonstrated that it was possible to tailor the microstructure of coralline, as well as the bioactivity and degradation profile.

## 3.2.2 Synthetic Bioceramics

### 3.2.2.1 Bioinert Materials

Alumina ($Al_2O_3$) and Zirconia ($ZrO_2$) are chemically inert and have high mechanical resistance, high hardness, and are resistant to cracking and corrosion. They are bioinert ceramics, successfully used in orthopedics, specifically for total hip/knee arthroplasty and in dentistry (Fig. 3.2) [20, 21].

Alumina-based bioceramics were the first to be available in the market, for dental implants and acetabular cup replacements in total hip prostheses [22]. Alumina positively combines good flexural and mechanical strength, excellent resistance to dynamic and fatigue, and high resistance to abrasion. As a result, alumina has been effectively used as synthetic bone grafts and as reinforcement agents for ceramics, or even as porous prosthetic devices, by means of a biomimetic coating on alumina, to afford a stable bond with the host tissue. Other clinical applications of alumina prostheses include bone screws, alveolar ridge (jaw bone) and maxillofacial reconstruction, ossicular (middle ear) bone substitutes, corneal replacements, segmental bone replacements, and blade, screw, and post-type dental implants [23]. However, alumina ceramics have a low toughness to fracture. This disadvantage can be overcome if zirconia is added to alumina ceramics (known as zirconia-toughened alumina, ZTA, or alumina-toughened zirconia, ATZ), resulting in a composite material with higher toughness and better tribological properties [24, 25]. ZTA contains alumina (70–95%) as the matrix phase and zirconia polycrystals (TZP, 5–30%) as the secondary phase, combining the positive properties of monolithic alumina with zirconia. Moreover, the wear properties and low susceptibility to stress-assisted degradation of alumina ceramics are also preserved in ZTA ceramics, reducing the risk of impingement and dislocation, and improving stability [25].

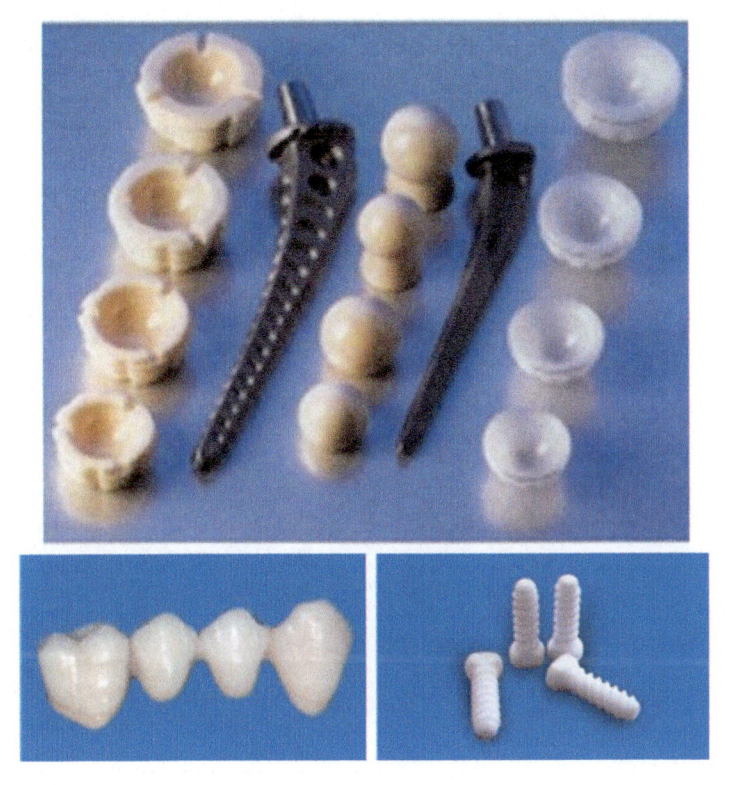

3 Bioceramics for Osteochondral Tissue Engineering and Regeneration

**Fig. 3.2** Bioceramics of alumina and zirconia used for orthopedics and dentistry applications [30, 31]

Zirconia has a polymorphic crystalline structure depending on the temperature, i.e., monoclinic at temperatures <1170 °C, tetragonal at 1170 °C which is stable up to 2370 °C, and cubic structure at 2370 °C. However, this phase transformation also occurs at the surface of the ceramic when present in body fluid, producing an aging which compromises the lifetime of zirconia implants. As a result zirconia-based bioceramics, especially tetragonal TZP, have been widely used in bone TE, due to their excellent toughness to fracture, high strength, high elastic modulus, wear resistance, and low temperature degradation [26]. For example, partially stabilized zirconia (with yttria, CaO, and MgO) materials are recognized for their flexural strength (higher than 1.0 GPa) and fracture toughness (above 8 MPam$^{1/2}$) [27, 28]. Besides its mechanical properties, zirconia promotes cell proliferation and differentiation in osteogenic pathways and osseointegration. Also, as it is radiopaque, it helps in the monitoring of radiographs [29]. Zirconia has often been used in dentistry since it can be colored to match the shade of existing teeth.

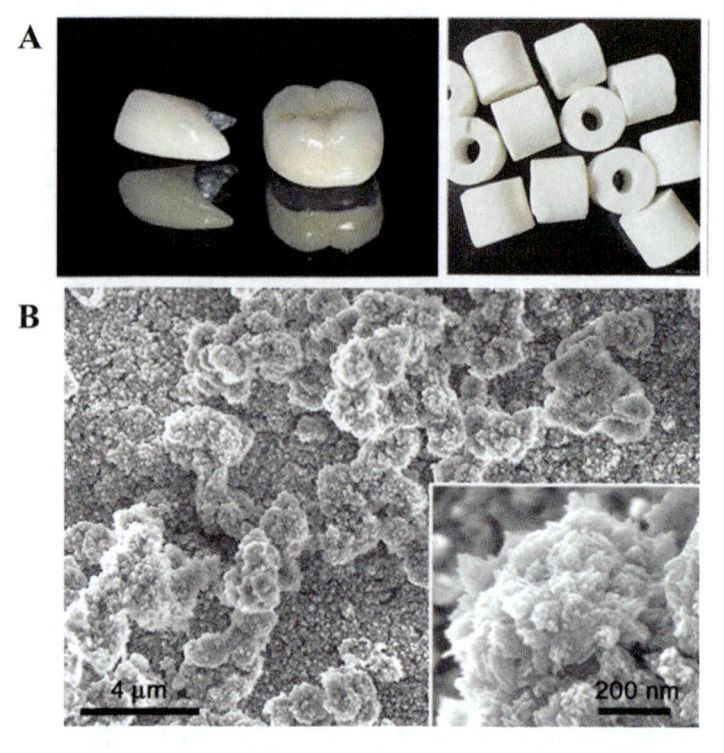

**Fig. 3.3** (**a**) Bioactive glass and glass-ceramics applied in the biomedical field. (**b**) Micrograph image of 45S5 Bioglass® after immersion in simulated body fluid [49, 50]

### 3.2.2.2 Bioactive Glasses and Glass-Ceramics

Bioactive glasses and glass-ceramics have been developed, in dense and porous form, for TE applications in orthopedics and dentistry (Fig. 3.3) [32–34]. Heat treated glasses result in crystalline glasses with higher strength and toughness, elastic modulus, and wear resistance.

Bioactive glasses have demonstrated their appropriateness to form a bond with the living bone tissue more rapidly than other bioceramics. They are converted into an amorphous CaP or apatite material after implantation. Moreover, it has also been reported that the ions Si, Ca, P, and Na, released during dissolution of certain bioactive glasses, seem to activate the expression of osteogenic genes and to stimulate neovascularization and angiogenesis, enzymatic activity, and differentiation of mesenchymal stem cells (MSCs) [35–37].

In the 1970s, Larry Hench [23] undertook the pioneering work in the field of bioactive glasses for biomedical applications with the development of 45S5 Bioglass®. It is a silica-based bioactive glass in the $Na_2O–CaO–SiO_2–P_2O_5$ system with a composition near the ternary eutectic in the $Na_2O–CaO–SiO_2$ diagram. Upon implantation, this unique biomaterial releases soluble $Si^{2+}$, $Ca^{2+}$, and P ions into solution, forming a hydroxycarbonate surface layer through a biochemical transformation. The dissolution of the ions of the bioactive glasses stimulates the genes responsible for osteoblast differentiation and proliferation [34, 38, 39]. Besides, it also combines the advantage of having an antimicrobial (ions increase the pH resulting in an osmotic effect), and an angiogenic activity, as well as stimulates the release of angiogenetic growth factors [33]. For example, when 45S5 Bioglass® was used in medium conditioned for fibroblasts there was an increase in tubule branching and the development of a complex network [40].

Besides the conventional silicate glasses, other types of bioactive glasses developed for biomedical applications include borate-based and phosphate-based glasses. Borate-based glasses, in the $B_2O_3$–$Na_2O$–$CaO$–$P_2O_5$ system, have fast degradation rates and are able to be completely converted into apatite when immersed in an aqueous phosphate solution, following a similar process of Bioglass®, but without the formation of a silica-rich layer [41, 42]. Borate glasses have also been used as drug release systems in the treatment of bone infection [43]. A disadvantage of this type of glasses is the toxicity of boron, which is released in the solution as borate ions; this disadvantage can be overturned in in vitro dynamic culture conditions [44]. Phosphate bioactive glasses, in the $Na_2O$–$CaO$–$P_2O_5$ system, have faster dissolution rates in aqueous fluids than silica glasses, which is a useful property for the healing of chronic wounds and as carriers in drug delivery, such as antibacterial ions and complex organic molecules for chemotherapy applications [45, 46]. The incorporation of metal oxides such as $TiO_2$, $Al_2O_3$, and $B_2O_3$ into the composition of phosphate glasses can stabilize the glass network, resulting in a slower degradation of the glass [24].

The common synthesis methods for bioactive glasses include the conventional melt-quenching, sol–gel process, flame spray synthesis, and microwave irradiation [47, 48].

### 3.2.2.3 Calcium Phosphates

Calcium phosphates (CaPs) are naturally found in the body and are bone-like materials proposed for a broad range of orthopedic and dental applications owing the similarity with the mineral component of major normal calcified tissues [51–53]. These types of bioceramics possess an outstanding biocompatibility, osteoconductivity, and bioresorbability, thus integrating into living tissues by the same processes active in bone remodeling. This phenomenon occurs when part of CaPs is dissolved into the microenvironment, and once the liberated ions are released, protein adsorption and precipitation of the biological apatite crystals takes place creating a layer on the surface of the biomaterial. Besides, CaPs are easy to obtain with a low cost, and can be relatively easily certified as medical grade. The most known CaPs comprise compounds with different chemical compositions, solubility, and properties (Table 3.1). CaPs can be ordered by in situ degradation rate as: MCPM > TTCP ≈ α-TCP > DCPD > OCP > β-TCP > HAp [54]. Differences in dissolution behavior of CaPs are related to changes in the

**Table 3.1** Main calcium phosphates compounds used in the biomedical field [3, 52]

| Calcium phosphate | Formula | Ca/P | pH stability range in aqueous solutions at 25 °C | Properties |
|---|---|---|---|---|
| Monocalcium phosphate monohydrate (MCPM) | $Ca(H_2PO_4)_2 \cdot H_2O$ | 0.5 | 0.0–2.0 | Water-soluble compound. Not biocompatible |
| Monocalcium phosphate anhydrous (MCPA) | $Ca(H_2PO_4)_2$ | 0.5 | Stable at T > 100 °C | Not biocompatible |
| Dicalcium phosphate dihydrate (DCPD or brushite) | $Ca(HPO_4) \cdot 2H_2O$ | 1.0 | 2.0–6.0 | Biocompatible, biodegradable and osteoconductive |
| Dicalcium phosphate anhydrous (DCPA) | $Ca(HPO_4)$ | 1.0 | Stable at T > 100 °C | Less soluble than DCPD |
| Amorphous calcium phosphate (ACP) | $Ca_xH_y(PO_4)_z \cdot nH_2O$ (n = 3–4.5) | 1.2–2.2 | ~ 5–12 (always metastable) | Lacks long range order |
| Octacalcium phosphate (OCP) | $Ca_8(HPO_4)_2(PO_4)_4 \cdot 5H_2O$ | 1.33 | 5.5–7.0 | Metastable precursor of CaPs that transforms into HAp |
| β-Tricalcium phosphate (β-TCP) | $\beta\text{-}Ca_3(PO_4)_2$ | 1.5 | Cannot be precipitated from aqueous solutions | Biodegradable. Fast resorption rate |
| α-Tricalcium phosphate (α-TCP) | $\alpha\text{-}Ca_3(PO_4)_2$ | 1.5 | Cannot be precipitated from aqueous solutions | Biodegradable. Fast resorption rate |
| Calcium-deficient hydroxyapatite (CDHA) | $Ca_9(HPO_4)(PO_4)_5(OH)$ | 1.5–1.67 | 6.5–9.5 | Poorly crystalline |
| Hydroxyapatite (HAp) | $Ca_{10}(PO_4)_6(OH)_2$ | 1.67 | 9.5–12 | Osteoconductive and osteoinductive |
| Tetracalcium phosphate (TTCP) | $Ca_4(PO_4)_2O$ | 2.0 | Cannot be precipitated from aqueous solutions | Biocompatible but poorly biodegradable |

chemical composition, as well as, the microporosity, and pore size, and have been connected with the process of osteoinduction in vivo [55]. Studies have shown the importance of the sintering process as a way to control the pore size and porosity, thus rendering an osteoinductive ceramic [56].

Despite that, CaPs are limited to load-bearing applications due to their poor mechanical properties, namely, strength and fatigue resistance, and for this reason they are mostly used as coatings and as fillers [51, 57]. However, CaPs bioceramics

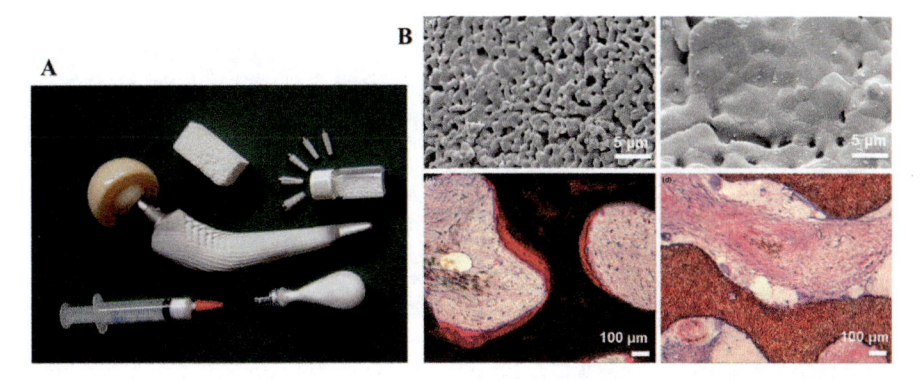

**Fig. 3.4** (**a**) Calcium phosphates-based materials for the biomedical field, such as porous blocks, powders and granules, injectable compositions, and coatings on metallic implants to help fixation into bone. Reprinted with permission from [52]; (**b**) Biphasic calcium phosphate ceramic with small pore size and microporosity (a) induces bone formation after 12 week implantation in a goat (c), in contrast to its non-osteoinductive counterpart with larger grains and less micropores (b) that is only infiltrated by fibrous tissue (d). Reprinted with permission from [71]

varied from thin coatings on metallic implants to help fixation into bone, to dense or porous blocks to be used as bone grafts, or even as injectable compositions (Fig. 3.4). Custom-designed forms as wedges for tibial opening osteotomy, cones for spine and knee, and inserts for vertebral cage fusion, are also available. CaPs are used in alveolar ridge augmentation, tooth replacement, maxillofacial reconstruction, orbital implants, increment of the hearing ossicles, spine fusion, and repair of bone defects [58].

Among CaPs most commonly investigated for biomedical purposes are $\alpha$- and $\beta$-TCP, CDHA, HAp, and biphasic CaPs which is the mixture of HAp and TCP [51, 59]. HAp is crystalline and is the most stable and least soluble CaPs in an aqueous solution below pH 4.2 [51]. HAp can be produced through wet methods, such as precipitation method, hydrothermal synthesis and solid-state reaction above 1200 °C of, for example, MCPM, DCPA, DCPD, OCP [60–63]. $\beta$-TCP is a high temperature phase of CaPs, obtained by thermal decomposition at temperatures above 800 °C. TCP can occur under three recognized polymorphs, such as $\beta$-TCP stable below 1120 °C, $\alpha$-TCP stable between 1120 °C and 1470 °C, and $\alpha$'-TCP above 1470 °C. Generally, $\beta$-TCP densification is difficult because the low temperature of $\beta \rightarrow \alpha$ phase transformation does not permit the sintering to high temperature. However, doping $\beta$-TCP with magnesia or calcium pyrophosphate can stabilize this $\beta \rightarrow \alpha$ transition at high temperatures. $\beta$-TCP is biodegradable and has been extensively investigated as bone substitute, either as granules or blocks, or even in CaPs-based bone cements [52]. $\alpha$-TCP is usually prepared from $\beta$-TCP phase, and quenching it prevents the reverse transformation $\alpha \rightarrow \beta$ [64]. $\alpha$-TCP is biocompatible, and more biodegradable and reactive than $\beta$-TCP [65]. It has been reported that the biological resorption capability of $\beta$-TCP and HAp is different though their similarity in terms of chemical composition. HAp has a slow resorption rate and

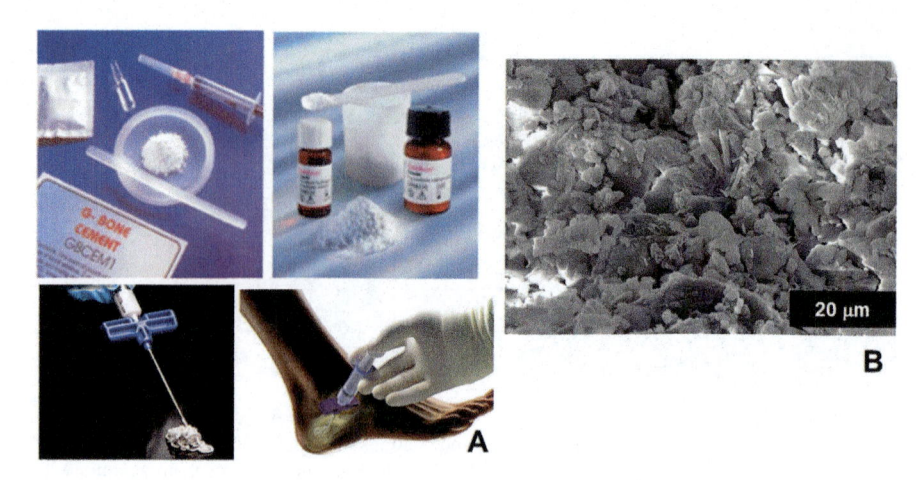

**Fig. 3.5** (**a**) Calcium phosphate cements resultant from the mixture of CaPs powders and aqueous solutions to be further injected into the bone defects; and (**b**) Brushite cement microstructure after hardening, showing entangled growing crystals, which provides the mechanical stiffness to the cement [80, 81]. Reprinted with permission from [81]

may remain integrated into the regenerated bone tissue after implantation, whereas β-TCP is completely reabsorbed [66, 67]. Hence, biomaterials for clinical applications have been performed combining HAp and β-TCP, for the bioresorbability and strength improvement of the implants [59, 63, 68]. CDHA is obtained by precipitation in an aqueous solution above a pH 7 [51]. Their crystals are in general poorly crystalline and of submicron dimensions. The solubility of CDHA increases with a decrease of Ca–P molar ratio, crystallinity, and size. CDHA can decompose into β-TCP, into a mixture of HAp and β-TCP or into pure HAp, when heating above 700 °C [59, 69]. As a first approximation, CDHA may be considered as HAp with some ions missing [70].

Calcium phosphates-based cements (CPCs) are mixture of one or several CaPs and an aqueous solution, which then precipitate into a less soluble CaP and sets by the entanglement of the growing crystals, providing mechanical stiffness to the cement. Once placed into the bone defect, the paste hardens in situ, at body temperature, and then displays limited solubility (Fig. 3.5) [72]. CPCs relevant features are excellent biocompatibility and resorbability, bioactivity, non-cytotoxicity, development of osteoconductive pathways, and sufficient compressive strength for a number of applications [51, 66, 73, 74]. CPCs are mechanically much stronger in compression than in tension or shear, because entangled crystals are not well bonded. Compressive strength values are typically 5–10 times superior to that of tensile. The main advantages of the CPCs include fast setting, excellent moldability, and manipulation. Hence, these bioceramics are commonly used to fill bone defects and trauma surgeries as moldable paste-like bone substitute materials. Besides, like any other bioceramics, CPCs provide the opportunity for bone grafting using alloplastic materials, which are unlimited in quantity and provide no risk of infectious diseases.

CPCs can be classified according to their end product into apatite (AP) cements and dicalcium phosphate dehydrate (DCPD or brushite) cements, upon the pH value of a cement paste after setting. AP is formed above pH 4.2, whereas brushite is preferentially formed when pH value of the paste is <4.2, although it may grow even up to pH 6.5, due to kinetics reasons [75, 76]. Brushite cements have raised interest due to their higher solubility and resorbability in vivo much faster than AP cements. Although AP cements show higher mechanical strength, they have slow in vivo resorption rates that interfere with the bone regeneration process [77, 78]. Moreover, brushite-based cements possess faster setting reactions [9, 79].

## 3.3 Applications of Bioceramics in Osteochondral Tissue Engineering

Current clinical use of bioceramics for bone, cartilage, and OC repair include, bone grafting, microfracture, arthroscopic mosaicplasty, periosteal and perichondrial transplantation, autologous chondrocyte implantation, drug delivery, and gene transfection [82–84]. Despite the fact that autologous grafts are the most ideal treatment, the rate of morbidity and the difficulty in trimming and grafting for the desired shape are important drawbacks of this technique [85–87].

Growth factors integration, like bone morphogenic proteins (BMPs), into scaffold, through structural entrapment or surface complexes have been widely reported for bone growth and healing, for instance in long bone defects, for their osteoinduction ability. Growth factors play a major role in cellular guiding and control. The incorporation of stem cells, like bone marrow stromal stem cells (BMSCs) or mesenchymal stem cells (MSCs), in bioceramics scaffolds for OC defects repairs, have been studied, since these cells had demonstrated promising results in bone recovery [88, 89]. For example, Lv and Yu [90] studied the viability of a composite lamellar scaffold made of nano-β-TCP/collagen type I and type II with BMSCs, for the articular OC defects repair in canine knee joints. In articular OC defects, subchondral bone plays an important role, once it is responsible for the formation of bones outline shape and provides the biomechanical needed environment cartilage differentiation and development. Thus, the biphasic composite scaffold used consisted of a mineralized collagen type I/β-TCP scaffold for bone regeneration and a non-mineralized collagen type II/β-TCP scaffold for cartilage regeneration. That study showed a gradual degradation and absorption of the scaffolds, while new cartilage tissue was formed. After 24 weeks of implantation, the defect space was fulfilled with new cartilage tissue integrated in the surrounding cartilage.

Several bioceramics and polymer composites have been developed and reported as an attractive solution for the repair of OC injuries. Xue et al. [91] evaluated the use of a poly-(lactide-co-glycolide) (PLGA)/nano HAp scaffold for potential use in cartilage tissue engineering applications. For that, MSCs were seeded in scaffolds and their efficacy was evaluated on a rat model. After 12 weeks after implantation,

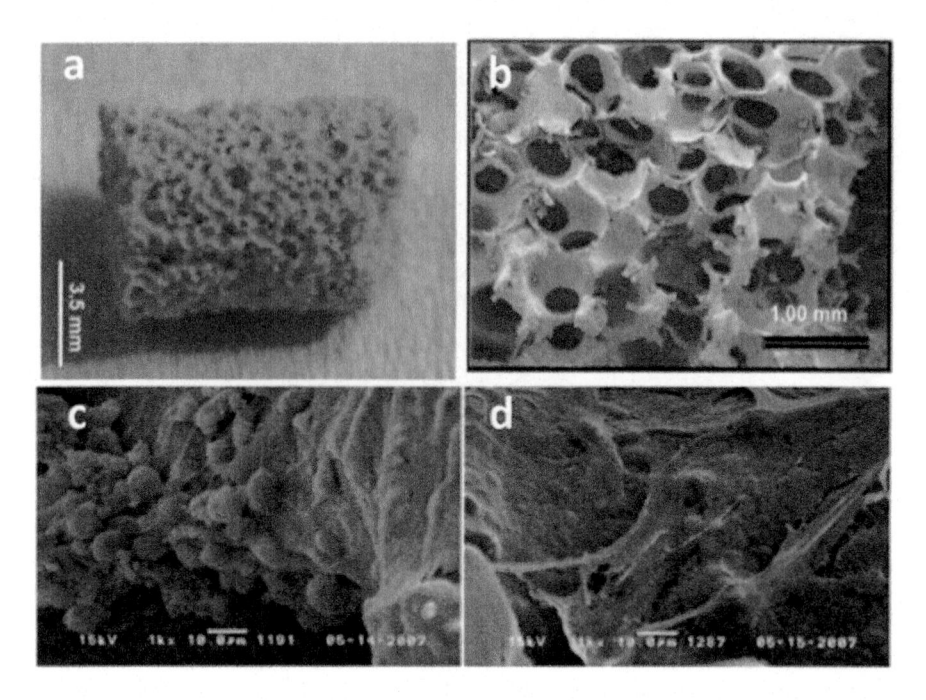

**Fig. 3.6** HAp porous scaffold: (**a**) macroscopic image, (**b**) microstructure, (**c**) macroporous HAp scaffolds seeded with RBMSCs at a cell density of $1 \times 10^4$ cells per scaffold after culturing for 24 h, and (**d**) 7 days. Reprinted with permission from [92]. Copyright 2008, Wiley

it was possible to observe that OC defects in rat knees were filled with smooth and hyaline-like cartilage with glycosaminoglycan and collagen type II deposition. Results were compared to those obtained only for PLGA scaffolds and, PLGA/nano HAp hybrid scaffold facilitated more significantly the cartilage repair, and provided a higher viability and proliferation of MSCs. In another study by Oliveira et al. [92] was reported the development of a porous HAp scaffolds with high interconnectivity, using an organic sacrifice template, for bone regeneration/repair. The scaffolds were tested in vitro, using rat bone marrow stromal cells (rBMSCs) confirmed that the cells adhered, proliferated well and remained viable (Fig. 3.6).

Zylinska et al. [93] also took advantaged from polymers and HAp and evaluated the applicability of a poly-L/D-lactide (PLDLA)/nano HAp composite scaffold enriched with sodium alginate in OC lesions of rabbit femoral trochlea. The use of sodium alginate is limited due to its low mechanical strength and fast degeneration. Thus, combining sodium alginate with PLDLA/nano HAp scaffold overcome these limitations, and the incorporation of the nano-sized HAp provides bioactivity, osteoinductivity and osteoconductivity, which facilitated new bone tissue regeneration. The bioactivity of the composites is low on the initial phase, increasing over time, due to its biodegradation.

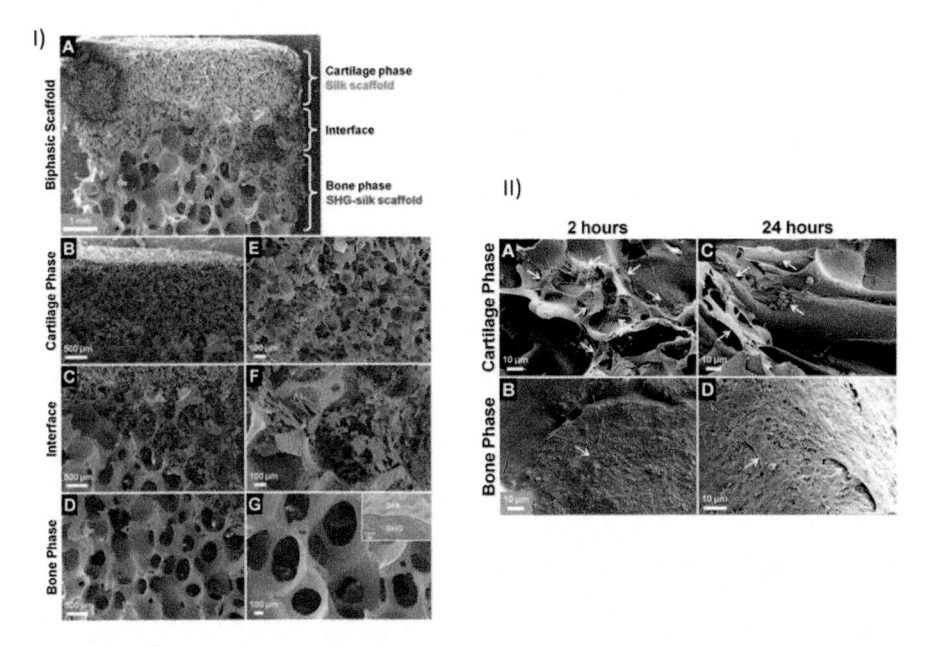

**Fig. 3.7** Biphasic scaffold for OC repair/regeneration: I) microscopic morphology showing complete scaffold (**A**), cartilage phase (**B**), interface region (**C**), bone phase (**D**), and structural features at higher magnifications (**E–G**); II) attachment and morphology of hMSCs cultured on the scaffold after (**A**, **B**) 2 hours and (**C**, **D**) 24 hours. Arrows indicate attached cells on the scaffold surface. Adapted with permission [94]. Copyright 2015, RSC

Taking advantage of naturally derived polymers, a study reported the development of a biphasic scaffold using silk fibroin and strontium-hardystonite-gahnite ceramic with stratified structure composed of distinct cartilage and bone phases which were well-integrated at a continuous interface, to satisfy the complex and diverse regenerative requirements of OC tissue [94]. Microstructure analysis showed that the cartilage phase had pores highly interconnected with sizes of 100–120 μm, while the bone phase had large pore sizes of 400–500 μm, along with interconnectivity (Fig. 3.7 I). In vitro behaviour of human mesenchymal stem cells (hMSCs) cultured in the scaffolds indicated that the cells infiltrated throughout its entire structure by allowing cell migration within and between phases. The SEM images showed that the scaffold was biocompatible and provided favourable substrates for cell attachment in its cartilage and bone phases, as well as a continuous interface which allowed cell migration and interaction between phases (Fig. 3.7 II).

In Table 3.2 are summarized diverse bioceramics materials used for bone, cartilage, and OC applications.

**Table 3.2** Summary of bioceramics in OC applications

| Application | Bioceramic materials | Function | References |
|---|---|---|---|
| Cartilage | PLGA/nano HAp | Investigate OC repair potential | [91] |
| Cartilage | PLDLA/nano HAp enriched with sodium alginate | Articular cartilage treatment | [93] |
| Cartilage | β-TCP scaffolds seeded with autologous chondrocytes | Repair of OC defects | [95] |
| Cartilage | β-TCP/collagen type I | Improve articular cartilage restoration and reconstruction | [96] |
| Joint arthroplasty | Alumina, zirconia, bioglasses, and HAp coatings for acetabular cup | Osteoconduction and osteointegration of prosthetic devices | [97] |
| Knee joint | Scaffold of nano-β-TCP/collagen type I and type II | Repair of articular OC defects of the canine knee joint | [90] |
| Bone defects and diseases | Bioglass, CaPs and CPCs | Repair and regeneration of defects and damaged bone | [48, 98, 99] |
| Deep OC defects | Bilayered implant of β-TCP and fibrous collagen type I and type III | Bone reconstruction and cartilage regeneration | [100] |
| OC defect regeneration | Bilayered scaffolds of silk fibroin and silk-nano CaP | OC defect regeneration | [94] |
| OC tissue engineering | Bilayered chitosan /HAp-based scaffolds | OC tissue engineering | [101] |

## 3.4 Clinical Trials of Bioceramics for Osteochondral Regeneration

Human clinical trials are research studies worldwide which evaluate the safety and effectiveness of a medical strategy, treatment, or device for humans. These studies follow strict scientific standards and are only performed after the approval of the health ethics committee. These standards are designed to protect patients and help to produce reliable study results. Those results are only obtained after a long and careful process which begins in the laboratory, following by animal tests and, as final stage come to clinical trials [102]. In Table 3.3 are reported the completed and ongoing clinical trials of using different types of bioceramics for OC applications.

Before their commercialization, implantable devices went through a rigorous, long and detailed process, involving several stages of R&D under restrict guidance of FDA. During this process, the safety of the medical device is ensured, and validated by scientific evidences, and after the approval, they are classified according with the associated risk. For instance, medium risk Class II devices include fracture fixation devices, while devices for organs replacement are in high risk Class III [102].

3 Bioceramics for Osteochondral Tissue Engineering and Regeneration 67

**Table 3.3** List of completed and ongoing clinical trials of bioceramics for OC applications

| NCT number | Date and status | Study | Patients age | Follow-up | Procedure |
|---|---|---|---|---|---|
| NCT0128034 | 2011–2016 completed | Chondral and Osteochondral lesions: Marrow stimulation techniques vs Maioregen (bioceramic scaffold) | 18–60 yrs | 24 mths | Marrow stimulation—Drilling or microfractures |
| NCT00841152 | 2009–2018 | Comparison of two ceramic bone graft substitutes, bioactive glass (BAG) and beta-tricalcium phosphate (TCP), in filling of contained bone defect | Adult, senior | 12 mths | Bone defects |
| NCT00900822 | 2005–2008 completed | Comparison of Straumann bone ceramic and BioOss as bone grafting materials for bone augmentation in the posterior upper jaw | Adult, senior | 3 yrs | Bone augmentation |
| NCT02128256 | 2014–2017 phase 4 | CERAMENT™|G—bone healing and reinfection prophylaxis | Adult, senior | 12 mths | Device absorption and bone in-growth of CERAMENT™|G |
| NCT00200603 | 2005 | Autograft versus calcium phosphate macroporous bioceramics as bone substitute for Tibial valgus osteotomy | Adult, senior | 3, 6, 12 and 24 mths | Tibial valgus osteotomy |
| NCT01813188 | 2011–2014 completed | Noninferiority and lower morbidity of the use of bone marrow mononuclear cells seeded onto a porous matrix of calcium phosphate, for the consolidation of tibial bone defects (pseudoarthrosis), compared with autologous bone graft | 18–75 yrs | 6 mths | Autologous bone marrow cells seeded onto a porous tricalcium phosphate ceramic and demineralized bone matrix |

(continued)

**Table 3.3** (continued)

| NCT number | Date and status | Study | Patients age | Follow-up | Procedure |
|---|---|---|---|---|---|
| NCT01824706 | 2012–2016 completed | A prospective, multicenter observational study evaluating the long term safety in terms of Explantation rate and number of infections of the custom-made bioceramic implant CustomBone™ | Child, adult, and senior | 2 yrs | Craniectomy |
| NCT01742260 | 2013–2017 phase 1 | A pilot study to demonstrate safety and feasibility of cranial reconstruction using Mesenchymal stromal cells and Resorbable biomaterials | 18–80 yrs | n.d. | Repair of cranial defects |
| NCT02910232 | 2014–2016 completed | In vivo clinical trial of porous starch–hydroxyapatite composite biomaterials for bone regeneration | 20–60 yrs | 6 mths | Bone void filler of foot fracture |
| NCT03302520 | 2017–2020 | Comparison of bioactive glass ceramics spacer and PEEK cages in posterior lumbar Interbody fusion | 30–80 yrs | n.d. | Posterior lumbar interbody fusion (PLIF) surgery |
| NCT02503891 | 2015–2019 | AL-2 MP-1 (polyimide) Acetabular liner | 21–90 yrs | 2 yrs | Polymer on ceramic articulation system |
| NCT01751841 | 2012–2020 | Providing compelling evidence for the efficacy of Si-CaP in terms of resulting in satisfactory fusion | Child, adult, senior | 12 mths | Spinal fusion |
| NCT03302429 | 2018–2020 | Evaluation of platelet rich fibrin / biphasic calcium phosphate effect versus autogenous bone graft on reconstruction of alveolar cleft | Child (>8 yrs), adult, senior | n.d. | Reconstruction of alveolar cleft defect in maxillary arch |
| NCT01409447 | 2009–2011 | Repair of articular Osteochondral defect | 18–60 yrs | 12 mths | Biphasic osteochondral composite implantation |

(continued)

Table 3.3 (continued)

| NCT number | Date and status | Study | Patients age | Follow-up | Procedure |
|---|---|---|---|---|---|
| NCT01183637 | 2010–2014 terminated | Evaluation of an Acellular Osteochondral graft for cartilage lesions | ≥ 21 yrs | 24 mths | Microfracture |
| NCT01209390 | 2010–2016 terminated | A prospective, post-marketing registry on the use of ChondroMimetic® for the repair of Osteochondral defects | 18–65 yrs. | 36 mths | Chondromimetic |
| NCT01282034 | 2011–2015 completed | Study for the treatment of knee Chondral and Osteochondral lesions | 18–60 yrs | 24 mths | Marrow stimulation—drilling or microfractures |
| NCT01471236 | 2011–2018 | Evaluation of the Agili-C biphasic implant in the knee joint | 18–55 yrs | 24 mths | Agili-C biphasic implantation and mini-arthrotomy or arthroscopy |

Information obtained from: http://clinicaltrials.gov
*n.d.* not defined, *yrs.* years, *mths* months

There are already some engineered bioceramic based materials and scaffolds regulatory approved as: (a) bone grafts substitutes, namely CERAMENT™G, Bonalive (Vivoxid Ltd), NovoMax® (BioAlpha Inc.,), ChronOs (DePuySynthes), Straumann® BoneCeramic™, and Geistlich Bio-Oss®; (b) cartilage repair, namely Cartilage Repair Device (Kensey Nash Corporation), and (c) OC defects such as ChondroMimetic ®, MaioRegen®, and Agili-CTM.

## 3.5 Concluding Remarks and Future Outlook

Bioceramics have demonstrated very important successes for applications in orthopedic and dental surgery. They are, however, potentially suitable for a wide range of essential TE purposes, namely, to restore the natural state and function of damaged OC tissue. Advanced strategies present some of the current challenges in this field, and may constitute a major step forward in the future. Bioceramics offer desirable characteristics such as biocompatibility, chemical inertness in biological medium, and hardness, but they have low resistance to traction. Ongoing research involves the chemistry, composition, and microstructure and nanostructure of the materials to improve their mechanical integrity upon implantation, and appropriate porosity for the cellular adhesion, proliferation, and differentiation. Although there have been significant advances in engineer new tissues, developments aimed at designing materials perfectly matching their biomedical purposes are necessary. Strategies should be devoted on the clear understanding of the bioceramics–tissue interactions, and hierarchical structure for long-term service, and the related mechanical strength, especially the fatigue limit under periodic external stress.

**Acknowledgments** The authors acknowledge the project FROnTHERA (NORTE-01-0145-FEDER-000023), supported by Norte Portugal Regional Operational Programme (NORTE 2020), under the PORTUGAL 2020 Partnership Agreement, through the European Regional Development Fund (ERDF). Also, H2020-MSCA-RISE program, as this work is part of developments carried out in BAMOS project, funded from the European Union's Horizon 2020 research and innovation program under grant agreement N° 734156. The financial support from the Portuguese Foundation for Science and Technology for the funds provided under the program Investigador FCT 2012, 2014, and 2015 (IF/00423/2012, IF/01214/2014, and IF/01285/2015) is also greatly acknowledged.

# References

1. Salinas AJ, Vallet-Regi M (2013) Bioactive ceramics: from bone grafts to tissue engineering. RSC Adv 3(28):11116–11131. https://doi.org/10.1039/C3RA00166K
2. Hasan MS, Ahmed I, Parsons AJ, Rudd CD, Walker GS, Scotchford CA (2013) Investigating the use of coupling agents to improve the interfacial properties between a resorbable phosphate glass and polylactic acid matrix. J Biomater Appl 28(3):354–366. https://doi.org/10.1177/0885328212453634
3. Pina S, Oliveira JM, Reis RL (2015) Natural-based Nanocomposites for bone tissue engineering and regenerative medicine: a review. Adv Mater 27(7):1143–1169. https://doi.org/10.1002/adma.201403354
4. Yan LP, Silva-Correia J, Correia C, Caridade SG, Fernandes EM, Sousa RA, Mano JF, Oliveira JM, Oliveira AL, Reis RL (2013) Bioactive macro/micro porous silk fibroin/nano-sized calcium phosphate scaffolds with potential for bone-tissue-engineering applications. Nanomedicine (Lond) 8(3):359–378. https://doi.org/10.2217/nnm.12.118
5. Silva TH, Alves A, Ferreira BM, Oliveira JM, Reys LL, Ferreira RJF, Sousa RA, Silva SS, Mano JF, Reis RL (2012) Materials of marine origin: a review on polymers and ceramics of biomedical interest. Int Mater Rev 57(5):276–306. https://doi.org/10.1179/1743280412Y.0000000002
6. Oliveira J, Costa S, Leonor I, Malafaya P, Mano J, Reis R (2009) Novel hydroxyapatite/carboxymethylchitosan composite scaffolds prepared through an innovative "autocatalytic" electroless coprecipitation route. J Biomed Mater Res A 88:470–480
7. Oliveira JM, Kotobuki N, Tadokoro M, Hirose M, Mano JF, Reis RL, Ohgushi H Ex vivo culturing of stromal cells with dexamethasone-loaded carboxymethylchitosan/poly(amidoamine) dendrimer nanoparticles promotes ectopic bone formation. Bone 46(5):1424–1435. doi:https://doi.org/10.1016/j.bone.2010.02.007
8. Fomin A, Barinov S, Ievlev V, Smirnov V, Mikhailov B, Belonogov E, Drozdova N (2008) Nanocrystalline hydroxyapatite ceramics produced by low-temperature sintering after high-pressure treatment. Doklady Chem 418:22–25
9. Pina S, Ferreira J (2010) Brushite-forming Mg-, Zn- and Sr-substituted bone cements for clinical applications. Materials 3:519–535
10. Tomoaia G, Mocanu A, Vida-Simiti I, Jumate N, Bobos LD, Soritau O, Tomoaia-Cotisel M (2014) Silicon effect on the composition and structure of nanocalcium phosphates: in vitro biocompatibility to human osteoblasts. Mater Sci Eng C Mater Biol Appl 37:37–47. https://doi.org/10.1016/j.msec.2013.12.027
11. Vallet-Regi M, Arcos D (2005) Silicon substituted hydroxyapatites. A method to upgrade calcium phosphate based implants. J Mater Chem 15(15):1509–1516

12. Kose N, Otuzbir A, Peksen C, Kiremitci A, Dogan A (2013) A silver ion-doped calcium phosphate-based ceramic nanopowder-coated prosthesis increased infection resistance. Clin Orthop Relat Res 471(8):2532–2539. https://doi.org/10.1007/s11999-013-2894-x
13. LeGeros RZ, Kijkowska R, Bautista C, Retino M, LeGeros JP (1996) Magnesium incorporation in apatites: effect of CO3 and F. J Dent Res 75:60–60
14. Mestres G, Le Van C, Ginebra M-P (2012) Silicon-stabilized α-tricalcium phosphate and its use in a calcium phosphate cement: characterization and cell response. Acta Biomater 8(3):1169–1179. https://doi.org/10.1016/j.actbio.2011.11.021
15. Pina S, Vieira SI, Rego P, Torres PMC, Goetz-Neunhoeffer F, Neubauer J, da Cruz e Silva OAB, da Cruz e Silva EF, Ferreira JMF (2010) Biological responses of brushite-forming Zn- and ZnSr-substituted β-TCP bone cements. Eur Cells Mater (in press) 20:162–177
16. Green DW, Ben-Nissan B, Yoon KS, Milthorpe B, Jung H-S (2017) Natural and synthetic coral biomineralization for human bone revitalization. Trends Biotechnol 35(1):43-54. doi:10.1016/j.tibtech.2016.10.003
17. Oliveira JM, Grech JMR, Leonor IB, Mano JF, Reis RL (2007) Calcium-phosphate derived from mineralized algae for bone tissue engineering applications. Mater Lett 61:3495–3499
18. Correlo VM, Oliveira JM, Mano JF, Neves NM, Reis RL (2011) Chapter 32 - Natural origin materials for bone tissue engineering—properties, processing, and performance A2 - Atala, Anthony. In: Lanza R, Thomson JA, Nerem R (eds) Principles of regenerative medicine (second edition). Academic, San Diego, pp 557–586. doi:https://doi.org/10.1016/B978-0-12-381422-7.10032-X
19. Clarke SA, Walsh P, Maggs CA, Buchanan F (2011) Designs from the deep: marine organisms for bone tissue engineering. Biotechnol Adv 29(6):610–617. doi:https://doi.org/10.1016/j.biotechadv.2011.04.003
20. Maccauro G, Iommetti PR, Raffaelli L, Manicone PF (2011) Alumina and zirconia ceramic for orthopaedic and dental devices. In: Biomaterials applications for nanomedicine. InTech
21. Ghaemi MH, Reichert S, Krupa A, Sawczak M, Zykova A, Lobach K, Sayenko S, Svitlychnyi Y (2017) Zirconia ceramics with additions of Alumina for advanced tribological and biomedical applications. Ceramics Int 43(13):9746-9752. doi:https://doi.org/10.1016/j.ceramint.2017.04.150
22. Kolos E, Ruys A (2015) Biomimetic coating on porous alumina for tissue engineering: characterisation by cell culture and confocal microscopy. Materials 8(6):3584
23. Greenspan DC (2016) Glass and medicine: the Larry Hench story. Int J Appl Glas Sci 7(2):134–138. https://doi.org/10.1111/ijag.12204
24. Biamino S, Fino P, Pavese M, Badini C (2006) Alumina–zirconia–yttria nanocomposites prepared by solution combustion synthesis. Ceram Int 32(5):509–513. https://doi.org/10.1016/j.ceramint.2005.04.004
25. Kurtz SM, Kocagöz S, Arnholt C, Huet R, Ueno M, Walter WL (2014) Advances in zirconia toughened alumina biomaterials for total joint replacement. J Mech Behav Biomed Mater 31:107–116. https://doi.org/10.1016/j.jmbbm.2013.03.022
26. Pieralli S, Kohal RJ, Jung RE, Vach K, Spies BC (2016) Clinical outcomes of zirconia dental implants: a systematic review. J Dent Res 96(1):38–46. https://doi.org/10.1177/0022034516664043
27. Nakamura K, Adolfsson E, Milleding P, Kanno T, Örtengren U (2012) Influence of grain size and veneer firing process on the flexural strength of zirconia ceramics. Eur J Oral Sci 120(3):249–254. https://doi.org/10.1111/j.1600-0722.2012.00958.x
28. Benzaid R, Chevalier J, Saâdaoui M, Fantozzi G, Nawa M, Diaz LA, Torrecillas R (2008) Fracture toughness, strength and slow crack growth in a ceria stabilized zirconia–alumina nanocomposite for medical applications. Biomaterials 29(27):3636–3641. https://doi.org/10.1016/j.biomaterials.2008.05.021
29. Afzal A (2014) Implantable zirconia bioceramics for bone repair and replacement: a chronological review. Mater Express 4(1):1–12. https://doi.org/10.1166/mex.2014.1148

30. http://www.cinn.es/mathys-orthopaedics-files-an-opposition-to-ceramtecs-patent-on-medical-use-of-biolox-delta/
31. http://www.nevz-ceramics.com/en/produktyi-i-materialyi/biokeramika.html
32. Rawlings RD (1993) Bioactive glasses and glass-ceramics. Clin Mater 14(2):155–179. https://doi.org/10.1016/0267-6605(93)90038-9
33. Rahaman MN, Day DE, Sonny Bal B, Fu Q, Jung SB, Bonewald LF, Tomsia AP (2011) Bioactive glass in tissue engineering. Acta Biomater 7(6):2355–2373. https://doi.org/10.1016/j.actbio.2011.03.016
34. Jones JR (2013) Review of bioactive glass: from Hench to hybrids. Acta Biomater 9(1):4457–4486. https://doi.org/10.1016/j.actbio.2012.08.023
35. Lobel KD, Hench LL (1996) In-vitro protein interactions with a bioactive gel-glass. J Sol-Gel Sci Technol 7(1–2):69–76. https://doi.org/10.1007/BF00401885
36. Gorustovich AA, Roether JA, Boccaccini AR (2010) Effect of bioactive glasses on angiogenesis: a review of in vitro and in vivo evidences. Tissue Eng Part B Rev 16(2):199–207. https://doi.org/10.1089/ten.teb.2009.0416
37. Xynos ID, Edgar AJ, Buttery LDK, Hench LL, Polak JM (2000) Ionic products of bioactive glass dissolution increase proliferation of human osteoblasts and induce insulin-like growth factor II mRNA expression and protein synthesis. Biochem Biophys Res Commun 276(2):461–465. https://doi.org/10.1006/bbrc.2000.3503
38. Hench LL (1998) Bioceramics. J Amer Ceram Soc 81:1705–1728
39. Huang W, Day D, Kittiratanapiboon K, Rahaman M (2006) Kinetics and mechanisms of the conversion of silicate (45S5), borate, and borosilicate glasses to hydroxyapatite in dilute phosphate solutions. J Mater Sci Mater Med 17(7):583–596. https://doi.org/10.1007/s10856-006-9220-z
40. Leu A, Leach JK (2008) Proangiogenic potential of a collagen/bioactive glass substrate. Pharm Res 25(5):1222–1229. https://doi.org/10.1007/s11095-007-9508-9
41. Fu Q, Rahaman MN, Fu H, Liu X (2010) Silicate, borosilicate, and borate bioactive glass scaffolds with controllable degradation rate for bone tissue engineering applications. I. Preparation and in vitro degradation. J Biomed Mater Res A 95(1):164–171. https://doi.org/10.1002/jbm.a.32824
42. Knowles JC (2003) Phosphate based glasses for biomedical applications. J Mater Chem 13(10):2395–2401. https://doi.org/10.1039/B307119G
43. Xie Z, Cui X, Zhao C, Huang W, Wang J, Zhang C (2013) Gentamicin-loaded borate bioactive glass eradicates osteomyelitis due to Escherichia Coli in a rabbit model. Antimicrob Agents Chemother 57(7):3293–3298. https://doi.org/10.1128/AAC.00284-13
44. Brown RF, Rahaman MN, Dwilewicz AB, Huang W, Day DE, Li Y, Bal BS (2009) Effect of borate glass composition on its conversion to hydroxyapatite and on the proliferation of MC3T3-E1 cells. J Biomed Mater Res A 88A(2):392–400. https://doi.org/10.1002/jbm.a.31679
45. Marikani A, Maheswaran A, Premanathan M, Amalraj L (2008) Synthesis and characterization of calcium phosphate based bioactive quaternary P2O5–CaO–Na2O–K2O glasses. J Non-Cryst Solids 354(33):3929–3934. https://doi.org/10.1016/j.jnoncrysol.2008.05.005
46. Pickup DM, Newport RJ, Knowles JC (2010) Sol–gel phosphate-based glass for drug delivery applications. J Biomater Appl 26(5):613–622. https://doi.org/10.1177/0885328210380761
47. Kashif I, Soliman AA, Sakr EM, Ratep A (2012) Effect of different conventional melt quenching technique on purity of lithium niobate (LiNbO3) nano crystal phase formed in lithium borate glass. Results Phys 2(0):207–211. https://doi.org/10.1016/j.rinp.2012.10.003
48. Balamurugan A, Rebelo A, Kannan S, Ferreira JMF, Michel J, Balossier G, Rajeswari S (2007) Characterization and in vivo evaluation of sol–gel derived hydroxyapatite coatings on Ti6Al4V substrates. J Biomed Mater Res B Appl Biomater 81B(2):441–447. https://doi.org/10.1002/jbm.b.30682
49. Brunner TJ, Stark WJ, Grass RN (2006) Glass and bioactive glass Nanopowders by flame synthesis. AIChE Annual Meeting, Hilton San Francisco

50. http://ceramics.org/ceramic-tech-today/biomaterials/glass-scaffolds-help-heal-bone
51. Bohner M (2000) Calcium orthophosphates in medicine: from ceramics to calcium phosphate cements. Injury Int J Care Injured 31:37–47
52. Dorozhkin S (2009) Calcium orthophosphates in nature, biology and medicine. Materials 2:399–498
53. Dorozhkin SV (2007) Calcium orthophosphates. J Mater Sci 42(4):1061–1095. https://doi.org/10.1007/s10853-006-1467-8
54. Le Geros RZ, Le Geros JP (2003) Calcium phosphate bioceramics: past, present and future. Key Eng Mater 3:240–242
55. Yuan H, Fernandes H, Habibovic P, de Boer J, Barradas AMC, de Ruiter A, Walsh WR, van Blitterswijk CA, de Bruijn JD (2010) Osteoinductive ceramics as a synthetic alternative to autologous bone grafting. Proc Natl Acad Sci U S A 107(31):13614–13619. https://doi.org/10.1073/pnas.1003600107
56. Davison NL, Luo X, Schoenmaker T, Everts V, Yuan H, Barrère-de Groot F, de Bruijn JD (2014) Submicron-scale surface architecture of tricalcium phosphate directs osteogenesis in vitro and in vivo. Eur Cells Mater 27:281–297
57. Bohner M (2001) Physical and chemical aspects of calcium phosphates used in spinal surgery. Eur Spine J 10:114–121
58. Eliaz N, Metoki N (2017) Calcium phosphate bioceramics: a review of their history, structure, properties, coating technologies and biomedical applications. Materials 10(4):104. https://doi.org/10.3390/ma10040334
59. Daculsi G, Laboux O, Malard O, Weiss P (2003) Current state of the art of biphasic calcium phosphate bioceramics. J Mater Sci Mater Med 14(3):195–200
60. Kannan S, Goetz-Neunhoeffer F, Neubauer J, Ferreira JMF (2008) Ionic substitutions in biphasic hydroxyapatite and beta-tricalcium phosphate mixtures: structural analysis by rietveld refinement. J Am Ceramic Soc 91(1):1–12. doi:https://doi.org/10.1111/j.1551-2916.2007.02117.x
61. Kannan S, Lemos AF, Ferreira JMF (2006) Synthesis and mechanical performance of biological-like hydroxyapatites. Chem Mater 18(8):2181–2186
62. Elliott JC (1994) Structure and chemistry of the apatites and other calcium orthophosphates, vol 18. Studies in inorganic chemistry. Elsevier, London
63. LeGeros RZ, LeGeros JP, Daculsi G, Kijkowska R (1995) Encyclopedia handbook of biomaterials and bioengineering, vol 2. Marcel Dekker, New York
64. Monma H, Goto M (1983) Behavior of the α-β phase transformation in tricalcium phosphate. Yohyo-Kyokai-Shi 91:473–475
65. Yin X, Stott MJ, Rubio A (2003) Phys Rev B - Condens Matter Mater Phys 68:205
66. Ginebra MP, Traykova T, Planell JA (2006) Calcium phosphate cements as bone drug delivery systems: a review. J Control Release 113(2):102–110
67. Takahashi Y, Yamamoto M, Tabata Y (2005) Osteogenic differentiation of mesenchymal stem cells in biodegradable sponges composed of gelatin and beta-tricalcium phosphate. Biomater 26:3587–3596
68. Metzger DS, Driskell TD, Paulsrud JR (1982) Tricalcium phosphate ceramic: a resorbable bone implant: review and current status. J Am Dent Assoc 105:1035–1048
69. Kanazawa Te (1989) Inorganic phosphate materials. In: Materials science monographs. Tokyo.
70. Brown PW, Martin R (1999) An analysis of hydroxyapatite surface layer formation. J Phys Chem B 103:1671–1675
71. Habraken W, Habibovic P, Epple M, Bohner M (2016) Calcium phosphates in biomedical applications: materials for the future? Mater Today 19(2):69–87. doi:https://doi.org/10.1016/j.mattod.2015.10.008
72. Brown WE, Chow LC (1983) A new calcium-phosphate setting cement. J Dent Res 62:672–672

73. Ginebra MP, Traykova T, Planell JA (2006) Calcium phosphate cements: competitive drug carriers for the musculoskeletal system? Biomaterials 27(10):2171–2177
74. Dorozhkin SV (2008) Calcium orthophosphates cements for biomedical application. J Mater Sci: Mater in Med 43:3028–3057
75. Bohner M (2007) Reactivity of calcium phosphate cements. J Mater Chem 17(38):3980–3986. https://doi.org/10.1039/B706411j
76. Bohner M, Gbureck U (2008) Thermal reactions of brushite cements. J Biomed Mater Res Part B Appl Biomater 84B(2):375–385. https://doi.org/10.1002/Jbm.B.30881
77. Barralet JE, Lilley KJ, Grover LM, Farrar DF, Ansell C, Gbureck U (2004) Cements from nanocrystalline hydroxyapatite. J Mater Sci Mater Med 15(4):407–411
78. Bauer TW, Muschler GF (2000) Bone graft materials. An overview of the basic science. Clin Orthop Relat Res 371:10–27
79. Pina S, Vieira SI, Torres PMC, Goetz-Neunhoeffer F, Neubauer J, Silva OABdCe, Silva EFdCe, Ferreira JMF (2010) In vitro performance assessment of new brushite-forming Zn- and ZnSr-substituted β-TCP bone cements. J Biomed Mater Res B 94B:414–420
80. http://biomaterials-org.securec7.ezhostingserver.com/week/bio20.cfm
81. Dorozhkin SV (2013) Self-setting calcium orthophosphate formulations. J Funct Biomater 4(4):209–311. https://doi.org/10.3390/jfb4040209
82. Arcos D, Vallet-Regí M (2013) Bioceramics for drug delivery. Acta Mater 61(3):890–911. https://doi.org/10.1016/j.actamat.2012.10.039
83. Mostaghaci B, Loretz B, Lehr CM (2016) Calcium phosphate system for gene delivery: historical background and emerging opportunities. Curr Pharm Design 22(11):1529–1533. https://doi.org/10.2174/1381612822666151210123859
84. Shekhar S, Roy A, Hong D, Kumta PN (2016) Nanostructured silicate substituted calcium phosphate (NanoSiCaPs) nanoparticles - efficient calcium phosphate based non-viral gene delivery systems. Mater Sci Eng C Mater Biol Appl 69:486–495. https://doi.org/10.1016/j.msec.2016.06.076
85. Sherman SL, Thyssen E, Nuelle CW (2017) Osteochondral autologous transplantation. Clin Sports Med 36(3):489–500. https://doi.org/10.1016/J.CSM.2017.02.006
86. Panseri S, Russo A, Cunha C, Bondi A, Di A, bullet M, Patella S, Kon E (2012) Osteochondral tissue engineering approaches for articular cartilage and subchondral bone regeneration. Knee Surg Sports Traumatol Arthrosc 20:1182–1191. https://doi.org/10.1007/s00167-011-1655-1
87. Ng A, Bernhard K (2017) Osteochondral autograft and allograft transplantation in the talus. Clin Podiatr Med Surg 34(4):461–469. https://doi.org/10.1016/j.cpm.2017.05.004
88. Begama H, Nandi SK, Kundu B, Chanda A (2017) Strategies for delivering bone morphogenetic protein for bone healing. Mater Sci Eng C 70:856–869. https://doi.org/10.1016/J.MSEC.2016.09.074
89. Ogawa K, Miyaji H, Kato A, Kosen Y, Momose T, Yoshida T, Nishida E, Miyata S, Murakami S, Takita H, Fugetsu B, Sugaya T, Kawanami M (2016) Periodontal tissue engineering by nano beta-tricalcium phosphate scaffold and fibroblast growth factor-2 in one-wall infrabony defects of dogs. J Periodont Res 51(6):758–767. https://doi.org/10.1111/jre.12352
90. Lv YM, Yu QS (2015) Repair of articular osteochondral defects of the knee joint using a composite lamellar scaffold. Bone Jt Res 4(4):56–64. https://doi.org/10.1302/2046-3758.44
91. Xue D, Zheng Q, Zong C, Li Q, Li H, Qian S, Zhang B, Yu L, Pan Z (2010) Osteochondral repair using porous poly(lactide-co-glycolide)/nano-hydroxyapatite hybrid scaffolds with undifferentiated mesenchymal stem cells in a rat model. J Biomed Mater Res A 94(1):259–270. https://doi.org/10.1002/jbm.a.32691
92. Oliveira JM, Silva SS, Malafaya PB, Rodrigues MT, Kotobuki N, Hirose M, Gomes ME, Mano JF, Ohgushi H, Reis RL (2009) Macroporous hydroxyapatite scaffolds for bone tissue engineering applications: physicochemical characterization and assessment of rat bone marrow stromal cell viability. J Biomed Mater Res A 91A(1):175–186. https://doi.org/10.1002/jbm.a.32213

93. Żylińska B, Stodolak-Zych E, Sobczyńska-Rak A, Szponder T, Silmanowicz P, Łańcut M, Jarosz Ł, Różański P, Polkowska I (2017) Osteochondral repair using porous three-dimensional Nanocomposite scaffolds in a rabbit model. In Vivo 31(5):895–903
94. Li J, Kim K, Roohani-Esfahani S, Guo J, Kaplan D, Zreiqat H (2015) A biphasic scaffold based on silk and bioactive ceramic with stratified properties for osteochondral tissue regeneration. J Mater Chem B Mater Biol Med. 3(26): 5361–5376
95. Guo X, Wang C, Duan C, Descamps M, Zhao Q, Dong L, Lu S, Anselme K, Lu J, Song YQ, Lü S, Anselme K, Lu J, Song YQ (2004) Repair of osteochondral defects with autologous chondrocytes seeded onto bioceramic scaffold in sheep. Tissue Eng 10(11–12):1830–1840. https://doi.org/10.1089/ten.2004.10.1830
96. Bian W, Li D, Lian Q, Li X, Zhang W, Wang K, Jin Z (2012) Fabrication of a bio-inspired beta-Tricalcium phosphate/collagen scaffold based on ceramic stereolithography and gel casting for osteochondral tissue engineering. Rapid Prototyp J 18(1):68–80. https://doi.org/10.1108/13552541211193511
97. Unther Heimke G, Leyen S, Willmann G (2002) Knee arthoplasty: recently developed ceramics offer new solutions. Biomaterials 23:1539–1551
98. Pina S, Vieira SI, Rego P, Torres PMC, da Cruz e Silva OAB, da Cruz e Silva EF, ferreira JMF (2010) Biological responses of brushite-forming Zn- and ZnSr substituted β-tricalcium phosphate bone cements. Eur Cells Mater 20:162–177. doi:https://doi.org/10.22203/eCM.v020a14
99. Flautre B, Maynou C, Lemaitre J, Van Landuyt P, Hardouin P (2002) Bone colonization of B-TCP granules incorporated in brushite cements. J Biomed Mater Res 63(4):413–417. https://doi.org/10.1002/jbm.10262
100. Gotterbarm T, Richter W, Jung M, Berardi Vilei S, Mainil-Varlet P, Yamashita T, Breusch SJ (2006) An in vivo study of a growth-factor enhanced, cell free, two-layered collagen-tricalcium phosphate in deep osteochondral defects. Biomaterials 27(18):3387–3395. https://doi.org/10.1016/j.biomaterials.2006.01.041
101. Malafaya PB, Reis RL (2009) Bilayered chitosan-based scaffolds for osteochondral tissue engineering: influence of hydroxyapatite on in vitro cytotoxicity and dynamic bioactivity studies in a specific double-chamber bioreactor. Acta Biomater 5(2):644–660. https://doi.org/10.1016/j.actbio.2008.09.017
102. Fiedler BA, Ferguson M (2017) Overview of medical device clinical trials. In. Elsevier, pp 17–32. doi:https://doi.org/10.1016/B978-0-12-804179-6.00002-2

# Part II
# Nanotechnology Approaches for Osteochondral Tissue Engineering

# Chapter 4
# Nanomaterials/Nanocomposites for Osteochondral Tissue

**Ohan S. Manoukian, Connor Dieck, Taylor Milne, Caroline N. Dealy, Swetha Rudraiah, and Sangamesh G. Kumbar**

**Abstract** For many years, the avascular nature of cartilage tissue has posed a clinical challenge for replacement, repair, and reconstruction of damaged cartilage within the human body. Injuries to cartilage and osteochondral tissues can be due to osteoarthritis, sports, aggressive cancers, and repetitive stresses and inflammation on wearing tissue. Due to its limited capacity for regeneration or repair, there is a need for suitable material systems which can recapitulate the function of the native osteochondral tissue physically, mechanically, histologically, and biologically. Tissue engineering (TE) approaches take advantage of principles of biomedical engineering, clinical medicine, and cell biology to formulate, functionalize, and apply biomaterial scaffolds to aid in the regeneration and repair of tissues. Nanomaterial science has introduced new methods for improving and fortifying TE scaffolds, and lies on the forefront of cutting-edge TE strategies. These nanomaterials enable unique properties directly correlated to their sub-micron dimensionality including structural and cellular advantages. Examples include electrospun nanofibers and emulsion nanoparticles which provide nanoscale features for biomaterials, more closely replicating the 3D extracellular matrix, providing better

---

O. S. Manoukian · S. G. Kumbar (✉)
Department of Orthopaedic Surgery, University of Connecticut Health, Farmington, CT, USA

Department of Biomedical Engineering, University of Connecticut, Storrs, CT, USA
e-mail: kumbar@uchc.edu

C. Dieck · T. Milne
Department of Biomedical Engineering, University of Connecticut, Storrs, CT, USA

C. N. Dealy
Department of Reconstructive Sciences, University of Connecticut Health, Farmington, CT, USA

S. Rudraiah
Department of Orthopaedic Surgery, University of Connecticut Health, Farmington, CT, USA

Deptartment of Pharmaceutical Sciences, University of Saint Joseph, Hartford, CT, USA

© Springer International Publishing AG, part of Springer Nature 2018
J. M. Oliveira et al. (eds.), *Osteochondral Tissue Engineering*,
Advances in Experimental Medicine and Biology 1058,
https://doi.org/10.1007/978-3-319-76711-6_4

cell adhesion, integration, interaction, and signaling. This chapter aims to provide a detailed overview of osteochondral regeneration and repair using TE strategies with a focus on nanomaterials and nanocomposites.

**Keywords** Osteochondral · Nanomaterials · Regenerative medicine · Tissue engineering · Stem cells

## 4.1 Introduction

Theterm "osteochondral" is derived from the roots "osteo" meaning bone, and "chondro," which refers to cartilage. Thus, osteochondral tissue is tissue composed of or related to bone and cartilage. Osteochondral tissue is found primarily in joints throughout the body, specifically at the smooth end of bones and the articular cartilage that cover them [1]. Injuries to osteochondral tissue are common in the field of orthopedics, and can result in pain and swelling, as well as instability of the joint. Current treatments include tissue transplantation, allografts, as well as the delivery of bioactive agents. However, each treatment carries a number of drawbacks. For example, "autografts and allografts are often associated with limited availability and risks of immunogenicity, respectively" [2]. The aim of tissue engineering is to repair and regenerate tissue, as well as to provide a viable tissue substitute. In practice, the discipline uses one or more of three key components: three-dimensional (3D) scaffolds, healthy harvested cells, and biologically active factors [2]. Any biomaterial used for tissue engineering must meet a number of requirements, including but not limited to biocompatibility, biodegradability, and bioactivity. Composites are often used in order to synthesize the beneficial properties of multiple constituents, and recent advances in nanotechnology have demonstrated the importance of nanoscale structural properties in signaling for cellular regeneration [3]. This chapter outlines the background and clinical relevance of nanomaterials for osteochondral regeneration, as well current tissue engineering techniques and the challenges the discipline faces.

## 4.2 Background and Clinical Relevance

### 4.2.1 Cartilage Tissue Biology

Cartilage is a smooth, elastic tissue found throughout the body. In addition to providing support to various structures in the body such as the rib cage, ear and nose, cartilage acts as a rubber-like padding between bones to minimize friction and provide protection at the joints. Cartilage is classified into three different types: fibrocartilage, elastic cartilage, and hyaline cartilage. Each type of cartilage differs

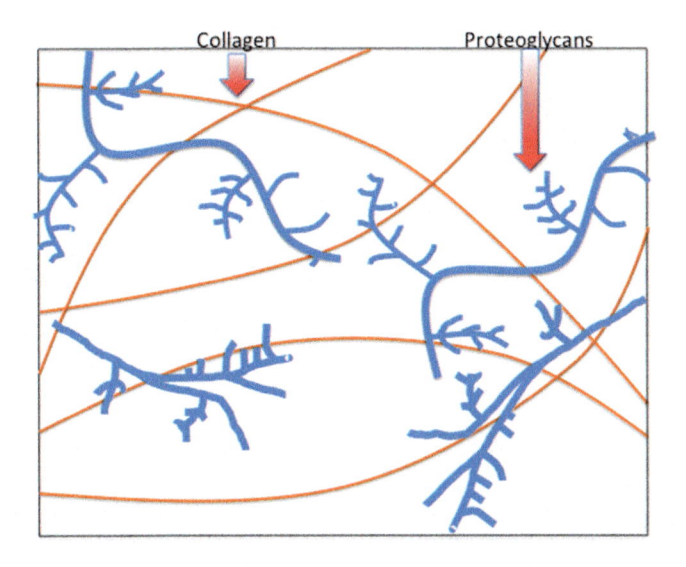

**Fig. 4.1** Ultrastructural level ($10^{-6}$ m–$10^{-8}$ m) of articular cartilage, displaying collagen fibrils and the proteoglycan matrix

based on the amount of collagen and proteoglycans, two proteins that make up much of the structure of cartilage.

There are three types of joints in the body: fibrous, cartilaginous and synovial. Out of these three types, only synovial joints allow for a large degree of motion. This type of joint is covered by the thin, dense, translucent connective tissue known as hyaline cartilage. As it covers the articulating surfaces of bone, this type of connective tissue is also referred to as articular cartilage. The articular cartilage can be considered a "soft tissue composed primarily of a large extracellular matrix with a population of chondrocytes distributed throughout the tissue" [4]. Extracellular matrix (ECM) is composed mainly of tightly wound collagen fibers, which lend the matrix a high tensile strength. A proteoglycan–water gel is also distributed throughout the collagen framework, which allows the hyaline cartilage to withstand compressive forces by attracting and trapping large amounts of water. This structure allows it to perform well as a load-bearing material to support joint movement, showing a low coefficient of friction as well as high wear strength [5]. Figure 4.1 displays the physical structure of cartilage tissue.

It should be noted that the composition and cellular organization of human adult articular cartilage vary depending on the regions of the matrix investigated, with different matrix proteins found in superficial and deep layers. These differences are both qualitative and quantitative. The interterritorial region of the matrix contains a collagen network composed of collagens II, IX, and XI while the pericellular matrix contains collagen VI, fibromodulin, and matrilin 3, but is deficient in or completely lacks type II collagen. The morphology of chondrocytes, the cells responsible for secreting the matrix or cartilage, differs as well, from more flattened nearer to the surface and rounder at the deeper zones [4].

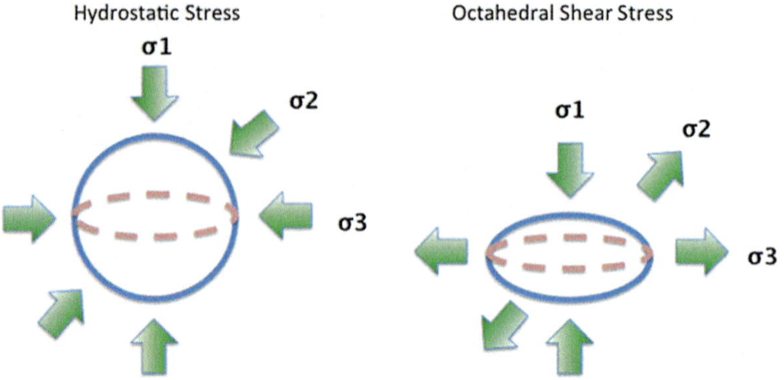

**Fig. 4.2** Diagram displaying hydrostatic and octahedral shear stresses. In hydrostatic stress, σ1 = σ2 = σ2. In octahedral shear stress, σ1 + σ2 + σ3 = 0

## 4.2.2 Cartilage Development

The formation of cartilaginous tissue occurs through a process known as chondrogenesis, and takes place as early as during fetal development. Here, it is a precursor for the process known as endochondral ossification in which "hypertrophic cartilage is replaced by bone," thus giving way to the early skeleton [4]. Chondrogenesis depends "upon signals initiated by cell–cell and cell–matrix interactions and is associated with increased cell adhesion and formation of gap junctions and changes in the cytoskeletal architecture" [6].

The process of chondrogenesis begins with the recruitment, proliferation and condensation of mesenchymal cells. In craniofacial bones, mesenchymal stem cells are recruited from neural crest cells of the neural ectoderm, whereas they are recruited from the sclerotome of the paraxial mesoderm and the somatopleure of the lateral plate mesoderm in axial and appendicular skeleton, respectively [6]. After condensation, these cells differentiate into chondroblasts, which synthesize the cartilage ECM and fibers. As the matrix grows, the chondroblasts mature into chondrocytes. In limb development, the chondrocytes either produce cartilage at the ends of bones or proliferate and undergo terminal differentiation "to hypertrophy, and apoptosis to permit endochondral ossification" [4]. Whether or not a chondrocyte goes down a certain path is determined by positive and negative signaling factors, such as Sox9 and Runx2 [6].

It has been noted that mechanical factors influence the development, maintenance and degradation. Carter et al. report that different types of stresses can either inhibit or promote bone growth, or ossification. Specifically, intermittent hydrostatic compression stress is shown to inhibit ossification, and as a result maintains the cartilage phenotype. On the other hand, intermittent nondestructive octahedral shear stress, resulting in mild tensile stress, promotes ossification and bone growth (The differences in hydrostatic compression stress and octahedral shear stress are illustrated in Fig. 4.2). This trend is supported by the findings that "tensile strain of chondrocytes increases cell proliferation, maturation, and hypertrophy" while "intermittent hydrostatic pressure has been shown to up-regulate aggrecan and collagen II, while inhibiting proinflammatory mediators in chondrocytes" [7].

### 4.2.3 Cartilage Disease and Injury

Articular cartilage is mainly loaded in compression. Therefore, its defects are often related to trauma-induced injuries, but problems can also arise from pediatric growth plate disorders and congenital defects. Injuries to the tissue are fairly common; it has been reported that over 900,000 Americans suffer from articular cartilage injuries each year [8]. Once injured, self-recovery is generally poor due to the lack of blood flow to the area.

Cartilage diseases include such disorders as osteoarthritis, costochondritis, herniation of intervertebral discs, achondroplasia, and relapsing polychondritis. These diseases can be a result of a number of factors, such as the failure of chondrocytes within the cartilage failing to proliferate or the inflammation of cartilage in key areas of the body, but the focus of tissue engineering is to alleviate conditions that result specifically from cartilage degeneration.

Osteoarthritis is an example of a very common disorder resulting from the deterioration of cartilage, with knee arthritis affecting an estimated 6% of adults over the age of 30, and hip arthritis affecting around 3% of the same demographic [9]. As increased functional loading in healthy joints by moderate exercise leads to an increase in articular cartilage thickness, it makes sense that the disease would be more prominent in those with a more sedentary lifestyle, such as the elderly [7]. In fact, the disease is the most common chronic condition affecting patients over the age of 70. Wood et al. characterize the disease as "damage to hyaline articular cartilage [which]... involves the whole joint and has subsequent changes to the subchondral surface involving bone remodeling" [9]. Refer to Fig. 4.3 for a visual representation of the breakdown in hyaline cartilage that takes place in osteoarthritis.

### 4.2.4 Current Treatments

Most current treatments for osteochondral diseases such as osteoarthritis are symptomatic, and attempt to regulate pain and improve mobility. Such treatments include self-care to prevent or reduce risks of disease and anti-inflammatory medication to reduce pain, as well as physical therapy. In some cases, the entire joint may be replaced by surgery. However, none of these measures address the root cause of cartilage degeneration in the joint. Tissue engineering offers the exciting prospect to repair or regenerate tissues, as well as providing alternative substitutes for the lost tissue. This chapter provides an outline for some of the tissue engineering approaches and objectives for osteochondral tissue being researched today.

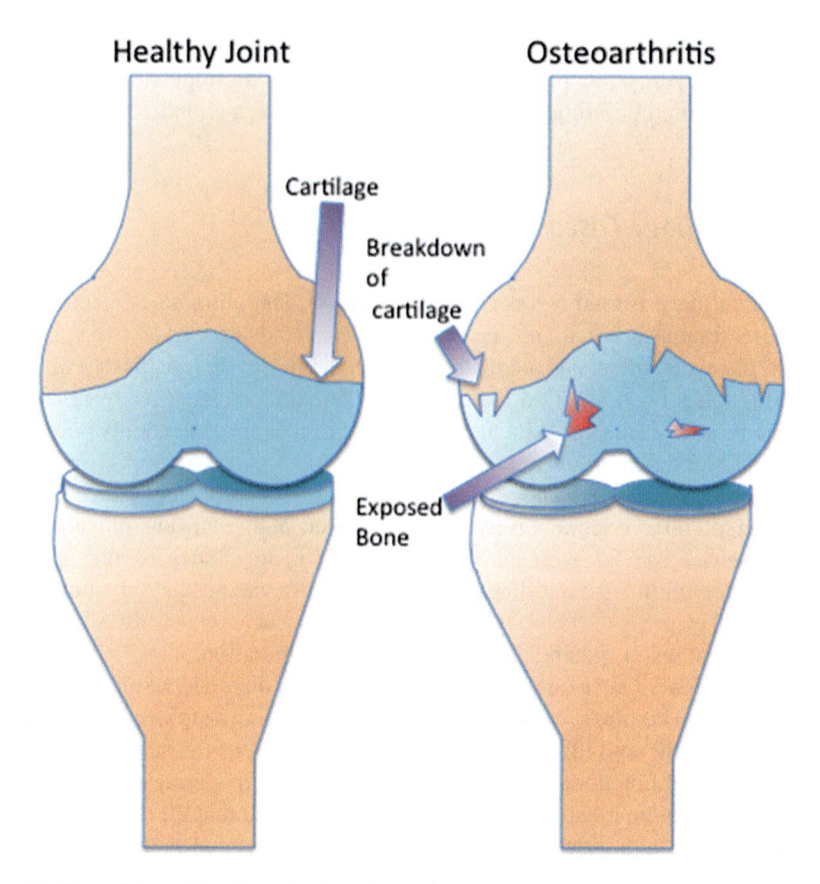

**Fig. 4.3** Degradation of hyaline articular collagen in the knee as a result of osteoarthritis

## 4.3 Tissue Engineering and Regenerative Medicine

### 4.3.1 Tissue Engineering Approaches and Objectives

The approaches to osteochondral tissue regeneration follow from the general tissue engineering objective to repair and restore the function of defect tissues caused by disease or injury. Tissue engineering approaches accomplish successful tissue repair when the resulting material is fully integrated with surrounding tissues and replicates the functionality of the native tissue, while showing no adverse effects. Biomaterial scaffold designs are a central feature in tissue engineering. Specifically, polymeric biomaterial scaffolds are used extensively. The approach to polymeric scaffold design is dependent upon the intended function of the scaffold. Scaffolds are typically used either as structural space fillers that promote tissue development or as delivery vehicles of therapeutic cell treatments. Biomaterials, cells, and bioreactors are the three components that are considered for use in osteochondral tissue engineering design. Combinations of biomaterials with cells and/or bioreactors constitute a strategy that has been extensively studied for tissue engineering applications. Biomaterial scaffolds can provide three-dimensional structural support or morphology and help transport and control delivery or cellular treatments or bioreactor molecules. The biomaterial scaffold, cells, and bioreactors can be used in tissue engineering designs in different combinations to promote tissue regeneration and integration to repair tissue defects with viable tissue substitutes. Understanding the architectural and molecular composition of cartilage as well as the cellular and biochemical interactions characteristic to both the development and function of the native tissue is imperative to engineer a material that matches the physiological, biomechanical, and biochemical signaling properties of the native tissue [10]. Biomaterial selection for scaffold design considers the biocompatibility, mechanical properties, biodegradability, three-dimensional architecture, and bioactivity of the material related to the native tissue [10]. Cartilage has a hierarchical structure with features that can be seen starting at its microstructure and continues down to the nanoscale. This hierarchical architecture inspires scaffold design that incorporates features starting at the nanoscale. Nanomaterials are used to replicate nanolevel features of native cartilage. Cellular and bioreactor components are subsequently added to scaffold materials to enhance their bioactivity in ways relevant to the capacity to regenerate cartilage tissue by facilitating the replication of developmental processes of cartilage formation. Another tissue engineering approach to mimic the native hierarchical structure of cartilage tissue is to create gradient scaffolds with layers representing the layered physiology of articular cartilage at the interface between cartilage and bone (Fig. 4.3).

### 4.3.2 Biomaterials

The three main classes of biomaterials are ceramics, metals, and polymers. Materials within these three classes can be classified as either natural biomaterials or synthetic biomaterials. Natural biomaterials are those derived from either an animal or plant source while synthetic biomaterials are synthesized in a lab [11].

Biomaterials can be further classified as biodegradable or nonbiodegradable. Biodegradability is an attractive feature for tissue engineering and regenerative purposes because it gives a material the capacity to initially function as a structural support then gradually degrade away as the new tissue moves in. A degradation-regeneration approach to tissue repair represents an ideal of tissue engineering to regenerate a fully integrated replication of native tissue. Tissue engineering methods that use biodegradable scaffold material eliminate the problem of long-term durability through the lifetime of the implant that must be considered for nondegradable implants. For biodegradable scaffolds, the implant material should ideally have a tunable degradation rate to ensure that the degradation and resorption of the implant material are compatible with the rate of new tissue generation [11].

Replicating the mechanical properties is particularly important in osteochondral applications in order for the engineered tissue to mimic and restore the functionality of the native tissue. Osteochondral tissues act as support structures in the body and, therefore, scaffold materials intended to function as structural space fillers for osteochondral repairs must support mechanical loads while the new tissue development occurs. Polymer biomaterials are used in osteochondral tissue engineering applications to create both rigid scaffold structures as well as hydrogel scaffolds. The rigid polymers are used in applications where a three-dimensional structural support is a priority while hydrogel scaffolds are more ideal as cell carrier systems. Although polymeric scaffolds are the foundation of most tissue engineering methods, they do not always exhibit entirely ideal properties for their intended applications when used independently. Composite biomaterials are created for osteochondral tissue engineering applications to improve the scaffold properties to more closely match its intended function and to promote effective tissue repair. Composite biomaterials are also used in efforts to mimic the heterogeneous, hierarchical composition of native cartilage. Creating multilayered scaffolds is an approach incorporating composite biomaterials intended to regenerate both cartilage and subchondral bone tissue at the osteochondral interface.

### 4.3.2.1 Natural Biomaterials

Some advantages of natural biomaterials for tissue engineering applications are their bioactivity, biocompatibility, and biodegradability. Natural biomaterials are typically used in the form of polymer hydrogels. Immunogenic incompatibility is a concern when using scaffolds made from natural material, however. Common natural biomaterials used for osteochondral tissue engineering include alginate, chitosan, collagen, fibrin, and hyaluronan [12].

Collagen and hyaluronic acid are both essential components of the ECMError! Bookmark not defined. of native cartilage. Their natural derivation from mammalian tissue allows for the polymers to be recognized by cells, facilitating cell attachment and triggering ECM production unlike plant derived polymers [11]. Experimental results support collagen's potential as a biomaterial useful in tissue engineering applications. An in vivo study using stem cell-seeded type II collagen scaffolds implanted into rabbits for articular cartilage repair showed results of chondrocyte-like cells and extracellular molecules found in the newly formed tissue and with no signs of inflammation after 8 weeks [13]. Chondrocytes and extracellular synthesis are characteristic to cartilage formation and markers of cartilage regeneration potential. Collagen scaffolds the most extensively used material in clinical applications [14]. Natural biomaterials, in general, have nonideal mechanical properties for the load-bearing functions that are characteristic to osteochondral tissues and therefore are used in combinations with other biomaterials. Collagen is an example of a material that is mechanically weaker than native cartilage but shows improved mechanical properties with the addition of other materials to create a composite. Chitosan is a biodegradable natural polymer that shows potential for cartilage tissue engineering. A chitosan-pluronic hydrogel injected for cartilage regeneration yielded a proliferation of chondrocytes and synthesis of GAGs [15]. Another cell-seeded chitosan hydrogel was tested in vivo and was found to fill cartilage defects completely 24 weeks after transplantation [16]. In addition to collagen, hyaluronan and fibrin have also been used clinically for cartilage reconstruction [17, 18]. A list of natural biomaterials and their applications can be found in Table 4.1.

### 4.3.2.2 Synthetic Biomaterials

Synthetic biomaterials allow for a higher degree of variability due to the opportunity to control some of their properties via processing methods. Controlling the composition and structure is a method used to obtain particular mechanical

Table 4.1 Natural biomaterials and applications in tissue engineering

| Material | Applications |
| --- | --- |
| Alginate | • Most common hydrogel scaffold<br>• Microencapsulation with hydrogel beads for cell delivery (growth factors, stem cells)<br>• Nanoparticle coatings (electrostatic interactions with oppositely charged materials)<br>• Electrospun nanofibers<br>• Encapsulate and culture chondrocytes<br>• Hydrogel coating to improve mechanical properties of ceramics |
| Chitosan | • Cell encapsulation/entanglement<br>• Bioactive molecule delivery, controlled drug release (i.e., growth factor microspheres)<br>• Scaffold for chondrocyte culture (hydrogels, fibers, sponges)<br>• Nano/microstructure surface patterning<br>• Support cell growth and adhesion of both bone and cartilage cell types<br>  - Enhances bioactivity, stability, and biocompatibility |
| Collagen | • Cell encapsulation<br>• Nanoparticles for sustained drug release<br>• Bioactive molecule delivery<br>• Hydrogels, sponge, composite nanofiber scaffolds<br>  - Cell adhesion, migration, proliferation<br>• Supports bone and cartilage cell types<br>• Used in combo with other materials because poor mechanical properties on its own |
| Fibrin | • Chondrocyte encapsulation hydrogel microsphere<br>• Deliver autologous chondrocytes to treat full thickness articular cartilage defects [18]<br>• Graft glue |
| Hyaluronan | • Hydrogels<br>• Interacts with stem cells/enhances cell attachment and supports differentiation into chondrocytes<br>• Chemically modified to create nanoparticles/nanofibers |

properties. Synthetic materials used in cartilage tissue engineering are primarily polymers. Synthetic polymer hydrogels are used for their high potential to entrap cells and provide biological stimuli for their migration, proliferation, and differentiation by providing a hydrated environment that facilitates diffusion [19]. Although synthetic polymers lack the bioactive capacity to integrate with surrounding host tissue, they can be functionalized with bioactive molecules. Functionalization of synthetic polymer scaffolds gives the material the ability of cellular interaction to facilitate cell attachment and stimulate matrix production and, therefore, greater potential to modulate cartilage regeneration.

The most common synthetic polymers used in osteochondral tissue engineering applications include poly(ethylene glycol) (PEG), polylactide (PLA) and its derivatives poly(L-lactide) (PLLA) and poly(lactic-co-glycolic acid) (PLGA), poly(ε-caprolactone) (PCL), poly(vinyl alcohol) (PVA), and polyglycolide (PGA). PEG in the form of both a hydrogel and rigid scaffold has been seeded with chondrocytes and proved to support their attachment, viability, proliferation and production of ECM [20]. PLGA scaffolds seeded with MSCs yielded hyaline-like smooth tissue after 12 weeks of implantation into the defect site within rabbit knees [21]. PGA seeded scaffolds show instances of higher expression of the cartilage specific protein, aggrecan, and collagen II when compared with PLGA seeded scaffolds [22]. Polymer scaffolds alone may still lack ideal mechanical strength and therefore composite material strategies are also used for synthetic biomaterials to give the scaffold its required characteristics. Synthetic biomaterials used in osteochondral tissue engineering application are listed in Table 4.2.

**Table 4.2** Synthetic biomaterials and applications in tissue engineering

| Material | Applications |
| --- | --- |
| PEG | • Hydrogel cell encapsulation with tunable hydrolytic degradation<br>• Stem cell nanoencapsulation<br>• Microspheres |
| PLA, PLLA, and PLGA | • Biodegradable porous scaffold<br>• Composite nanofiber scaffolds (hydroxyapatite nanoparticles)<br>• Hydrogel microspheres encapsulation of nanoparticles for bioactive molecule delivery<br>• Modified to create nanosurface<br>• Maintains 3D structure |
| PU | • Scaffolds with nanosurface modifications |
| PVA | • Hydrogel reinforced with nanohydroxyapatite<br>• Binder in scaffold-free method<br>• Promotes cell adhesion |
| PCL | • Electrospun nanofiber scaffolds<br>• Enhance bioactivity<br>• Porous scaffolds with surface modification for nanoscale roughness |
| PGA | • Fibrous scaffolds<br>• Suitable mechanical properties, supports cell growth |

## 4.3.3 Nanomaterials and Nanocomposites

Nanobiomaterials can be introduced to create nanocomposite materials with a nanostructure engineered to mimic the nanoscale level of the hierarchical composition of native cartilage. Adding nanoscale elements into the design improves the functional ability of the material to more closely resemble native tissue behavior with respect to both mechanical properties and biochemical activity.

#### 4.3.3.1 Nanomaterial Strategies

Adding nanomaterials such as nanoparticles (NPs) or nanotubes create nanostructured composition of biomaterial scaffolds. Ceramic NPs help improve the mechanical strength of polymer biomaterials [23, 24]. Adding hydroxyapatite nanoparticles to PVA hydrogels improves their mechanical properties and also creates a bioactive nanocomposite from the synthetic polymer that is not bioactive on its own [25]. NPs are also added for nanosurface modifications. Both metal and ceramic NPs can be used in combination with polymer scaffolds in osteochondral tissue engineering. The NPs are added to scaffolds through chemical treatments. The type and quantity of NPs added to the scaffold can control specific nanosurface properties such as surface area, roughness, and electrical charge.

Another nanomaterial used in tissue engineering is carbon nanotubes (CNTs). Similar to NPs, adding CNTs can alter mechanical and electrical properties of a material as well as influence cellular behavior by increasing the surface area within the scaffold giving it a higher affinity for cellular attachment. Nanoscale surface modifications can be made with physical and chemical treatment methods. Controlling the nanoscale porosity and roughness at the surface can influence scaffolds to promote cellular activity associated with cartilage regeneration. Nanoembossing of both polyurethane (PU) and PCL scaffolds creates highly porous surface with nanoscale surface roughness. The modified surfaces yield increases in chondrocyte numbers, intracellular protein production, and collagen secretion by chondrocytes when compared with the smooth surface scaffolds [26].

Nanostructured fibrous scaffolds are another method of introducing a nanoscale dimension to a material. Nanofiber scaffolds are created by electrospinning a polymer solution or through thermally induced polymer separation (TIPS) techniques. Nanofibers resemble the collagen fibrils of native cartilage. An additional osteochondral tissue engineering strategy using nanostructured scaffold designs incorporates nanocomposite materials in a layered orientation to create a multiphase construct. A layered nanocomposite design includes biomimetic nanoscale properties while also replicating the heterogeneous architecture of articular cartilage that is found as you move up from the subchondral bone to the articulating surface. Multilayered scaffolds with a gradient of nanoscale features create the most structurally similar 3D scaffold replication of native articular cartilage tissue. The layered composition of cartilage at the osteochondral junction is displayed in Fig. 4.4.

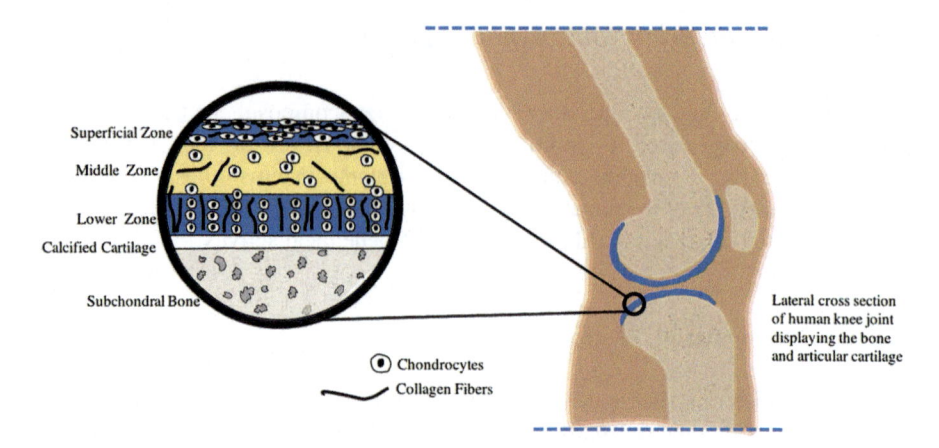

**Fig. 4.4** Layered organization of cartilage tissue at the articulating surface in the knee joint

Multilayer scaffolds show potential for tissue repair at the osteochondral junction where both tissue types must be repaired. In a recent study, by Castro et al. [10] a biphasic, layered, nanocomposite scaffold including both nanocomponents and a microstructure yielded results supporting this tissue engineering design approach and its feasibility for cartilage and bone repair. The high impact polystyrene(HIPS) mold of the scaffold is composed of an osseous layer characterized by a 40% in-fill density and a cartilage layer with a 0% in-fill density. The in-fill density controls the pore density of the respective layers that is representative of their natural composition. A cross-linked poly(ethylene glycol)-diacrylate (PEG-DA): PEG hydrogel is used as the bulk matrix material. Nanostructured hydroxyapatite nanoparticles (nHAs) are added to the osseous layer of the scaffolds and growth factors are added to the cartilage layer. The scaffold is seeded with stem cells and cultured in stem cell media. Scaffolds treated with nHA and the growth factor display higher levels of GAG (a biochemical marker for stem-cell chromogenic differentiation), the presence of proteins indicating type II collagen synthesis, and higher levels of calcium deposition.

The findings support that the addition of nHAs in physiologically relevant concentrations promotes cell adhesion and proliferation. The results demonstrate that the scaffold design provides 3D structural support for cellular attachment and effectively facilitates osteochondral tissue regeneration by incorporating interconnected microchannels, a controlled porosity, nHA nanoparticles, and controlled bioactive factor delivery. This study clearly illustrates how cellular activity that influences tissue formation can be influenced by scaffold geometry and optimizes tissue regeneration and integration.

### 4.3.3.2 Advantages of Nanomaterials and Nanocomposites

Scaffolds must provide support for cellular activity and suitable mechanical properties. Nanocomposites enhance the structural and mechanical properties and influence cellular activity. A more biomimetic structure improves mechanical characteristics of engineered scaffold materials. Mechanical properties of nanocomposite scaffolds match the properties of native cartilage more closely than scaffolds without nanofeatures. The addition of nanoparticles to hydrogels has resulted in native-like mechanical properties for the scaffold. The interconnection between structure and function is especially relevant for the ECM of cartilage tissue. Adding nanoscale features gives scaffolds the ability to stimulate cellular interaction that induces and promotes tissue regeneration. Controlled porosity can facilitate cellular infiltration and migration as well as nutrient flow within the scaffold. Cellular migration promotes tissue integration. Surface roughness and nanofeatures increase the surface area within the scaffold and give it a higher probability for cell attachments. Cell adhesion leads to increased cell proliferation and differentiation and thus, tissue generation. Therefore, nanostructured scaffolds also allow for a more controlled release of bioreactor elements to more closely replicate the dynamic kinematics of biochemical activity in native cartilage development. Nanomaterials and nanocomposites give engineered tissues a more biomimetic structural composition that promotes the restoration of native tissue functionality.

## 4.4 Stem Cell Strategies

Stem cells are capable of differentiating into various other types of cell types found throughout the body. As this multilineage, differential potential is what provides the means to recreate and rebuild tissue, stem cells are a fundamental pillar in tissue engineering.

There are various types of stem cells found in the adult body. The first type of cells is mesenchymal stem cells (MSCs). These types of stem cells are most commonly found in bone marrow, and are among the most popular type of cell used in tissue engineering due to their multipotency. The other types of stem cells are hematopoietic stem cells (HSCs) and skin or epidermal stem cells, found in bone marrow and the epidermis respectively. In regards to osteochondral tissue, mesenchymal stem cells are the stem cells of choice, considering that they are precursors to the chondrocytes.

Mesenchymal stem cells are capable of giving rise to chondrocytes when maintained in a 3D structure and treated with growth factors of the transforming growth factors-$\beta$ (TGF-$\beta$) family. Studies have shown that TGF-$\beta$ can induce in vitro chondrogenesis of mesenchymal stem cells when maintained in aggregates and pellets, as well as when seeded onto nanofibrous scaffolds when treated with proper growth factors. It should be noted that the material used to create the nanofibrous scaffolds has an effect on the tendency of MSCs to initiate chondrogenesis. In the previously discussed study, the synthetic biodegradable polymer poly(ε-caprolactone) (PCL) was used [27]. Therefore, when utilizing stem cells to achieve tissue regeneration, one must consider the extracellular environment, growth factor interaction, and the material and structure of the scaffold onto which the cells are seeded.

## 4.5 Growth Factors

Growth factors, mentioned several times throughout this chapter, are naturally occurring substances, such as proteins or hormones, which are capable of stimulating cellular growth. In terms of osteochondral tissue engineering, growth factors are capable of providing more suitable culture conditions to tissue constructs by supporting chondrogenesis. There are many different types of growth factors available for use by researchers and engineers for a variety of different purposes. In addition to inducing differentiation in stem cells, certain growth factors have been shown to influence the physical properties of engineered cartilage. Factors such as bone morphogenetic protein-2 (BMP-2), IGF-1, and the previously mentioned TGF-β have been shown capable of increasing compressive and tensile properties of engineered cartilage tissues [Elder]. Proliferation of chondrocytes has been increased through the addition of growth factors such as TGF-β, fibroblast growth factor 2 (FGF-2) and platelet–derived growth factor BB (PDGF-BB) [28]. Other studies have shown that when insulin or IGF-1 was added, rabbit auricular chondrocytes showed "increased deposition of cartilaginous ECM, improved mechanical properties, and thicknesses comparable to native auricular cartilage after 4 weeks of growth" [29]. These and other studies show that growth factors play a vital role in influencing the effectiveness of stem cells in regenerating tissue.

## 4.6 Clinical Relevance

While the medical implications for successful tissue engineering are extensive throughout the body, the potential for improvement in osteochondral diseases alone merits a separate discourse. As mentioned earlier in the chapter, damages to articular cartilage alone affect hundreds of thousands of Americans every year. Osteoarthritis alone affects a significant number of the population. Not only is the osteochondral disease the most frequently cited cause of difficulty in walking, but the condition has a significant impact on the economy as well: absence from work and early retirement relating to the disease exceed 2% of the gross domestic product [9].

Current treatments for osteochondral disease such as osteoarthritis are palliative, and "on the basis of medical evidence … do not change the course of the disease". Surgical treatments aim to completely replace the entire joint, and while they provide long-term relief for pain, they do not promote regeneration of tissue and are risky to implement in some elderly patients [30].

Successfully incorporating effective tissue engineering solutions into a clinical setting could provide a greater degree of recovery to a wider pool of patients suffering from osteochondral disease than currently available solutions.

## 4.7 Challenges for OC Tissue Engineering

Several problems face researchers and engineers working to advance the field of osteochondral tissue engineering. These problems include but are not limited to biocompatibility regarding immune response, ethical challenges, and current scientific limitations.

### *4.7.1 Biocompatibility and Immune Response*

In any field dealing with the body, biocompatibility is a primary concern. When a foreign body is introduced to an organism, that organism's immune system will identify it and attempt to protect the surrounding tissue and organs. This can result in inflammation as well as breakdown of the implant. Such a response is associated with allografts, the transplantation of tissue, usually bone, from one person to another. If the body does not recognize the transplant as its own, it will attempt to reject it, resulting in an unsuccessful transplant. Therefore, it is of utmost importance to consider biocompatibility when designing implantable materials. The types of materials chosen to develop the implant and the inclusion of certain bioactive factors play an important role in this aspect.

### *4.7.2 Ethical Issues*

When discussing the use of new technologies on living organisms, it is important to address the ethics involved that come along with it. For instance, stem cells are an important factor in any field of tissue engineering. Nanocomposites may be seeded with stem cells and growth factors in order to induce differentiation into cells that will promote tissue regeneration. However, the use of certain types of stem cells can be a controversial issue depending on the source. Although embryonic stem cells can easily differentiate into many different types of cells, some question the ethics of cell retrieval from undeveloped embryonic tissue. As such, most researchers use stem cells derived from different sources, including adipose-derived stem cells and human mesenchymal stem cells. These cells are also capable of differentiating into various other types of cells, and since they can be retrieved from adult tissue, their use avoids ethical scrutiny.

## 4.7.3 Current Scientific Challenges

Perhaps the most obvious challenge is successfully integrating multidisciplinary techniques to accomplish a wide range of problems. There are currently issues that researchers and engineers do not have answers for yet. For instance, in just osteochondral tissue engineering, there is a major challenge to overcome the inability of "resident chondrocytes to lay down a new matrix with the same properties as it had when it was formed during development" [4]. This problem is seen at the more macro level when considering larger tissue engineering endeavors. Something as grand as complete limb regeneration, an ultimate goal of musculoskeletal tissue engineering, requires the "simultaneous formation of multiple types of tissues and the functional assembly of these tissues into complex organ systems" [2]. However, such multiscale organization is rarely reestablished after surgery, and it is even more difficult to restore functionality similar to the original tissue or organ to affect long term clinical outcome.

Despite these challenges, advances in tissue engineering are made every day, and it remains one of the most promising approaches for tissue and organ recovery. It is possible to envision a future where regenerative tissue engineering will continue to improve with new strategies and technologies that will ultimately push tissue engineering beyond individual tissue repair and be capable to address more complex tissue systems, organs, and limbs.

**Acknowledgements** Authors acknowledge funding support from the National Institute of Biomedical Imaging and Bioengineering of the National Institutes of Health (award number: R01EB020640), the Connecticut Regenerative Medicine Research Fund (grant Number: 15-RMB-UCHC-08) in support of this work. Ohan S. Manoukian acknowledges the National Science Foundation Graduate Research Fellowship under Grant No. (DGE-1747453).

## References

1. Liu H, Liu H (2016) Nanocomposites for musculoskeletal tissue regeneration (Woodhead Publishing Series in Biomaterials). Elsevier Science & Technology, p. 406
2. Roshan J, Cato LT (2014) Musculoskeletal regenerative engineering: biomaterials, structures, and small molecules. Hindawi Publishing Corporation. p. 12
3. Zhang L, Webster TJ (2009) Nanotechnology and nanomaterials: promises for improved tissue regeneration. Nano Today 4:66–80
4. Goldring MB (2012) Chondrogenesis, chondrocyte differentiation, and articular cartilage metabolism in health and osteoarthritis. Therap Adv Musculoskeletal Dis 4:269–285
5. Gloria A, De Santis R, Ambrosio L (2010) Polymer-based composite scaffolds for tissue engineering. J Appl Biomater Biomech 8:57–67
6. Goldring MB, Tsuchimochi K, Ijiri K (2006) The control of Chondrogenesis. J Cell Biochem 97:11
7. Carter DR, BeauprÈ GS, Wong M, Smith RL, Andriacchi TP, Schurman DJ. The mechanobiology of articular cartilage development and degeneration. Clin Orthop Relat Res 427, pp. 69–77, 2004

8. Cheung H-Y, Lau K-T, Lu T-P, Hui D (2006) A critical review on polymer-based bio-engineered materials for scaffold development. 38: 291–300
9. Wood AM, Brock TM, Heil K, Holmes R, Weusten A (2013) A review on the management of hip and knee osteoarthritis. Int J Chronic Dis. 2013(845015): 10
10. Castro NJ, Patel R, Zhang LG (2015) Design of a novel 3D printed bioactive nanocomposite scaffold for improved osteochondral regeneration. Cell Mol Bioeng 8(3):416–432
11. Bernhard JC, Vunjak-Novakovic G. Should we use cells, biomaterials, or tissue engineering for cartilage regeneration? Stem Cell Res Ther. 2016; 7(1): 56
12. Duarte Campos DF, Drescher W, Rath B, Tingart M, Fischer H (Jul 2012) Supporting biomaterials for articular cartilage repair. Cartilage 3(3):205–221
13. Chen WC, Yao CL, Wei YH, Chu IM (2011) Evaluating osteochondral defect repair potential of autologous rabbit bone marrow cells on type II collagen scaffold. Cytotechnology 63(1):13–23
14. Steinwachs M (2009) New technique for cell-seeded collagen-matrix-supported autologous chondrocyte transplantation. Arthroscopy 25(2):208–211
15. Park KM, Lee SY, Joung YK, Na JS, Lee MC, Park KD (2009) Thermosensitive chitosan-Pluronic hydrogel as an injectable cell delivery carrier for cartilage regeneration. Acta Biomater 5(6):1956–1965
16. Hao T et al (2010) The support of matrix accumulation and the promotion of sheep articular cartilage defects repair in vivo by chitosan hydrogels. Osteoarthr Cartil 18(2):257–265
17. Manfredini M, Zerbinati F, Gildone A, Faccini R (2007) Autologous chondrocyte implantation: a comparison between an open periosteal-covered and an arthroscopic matrix-guided technique. Acta Orthop Belg 73(2):207–218
18. Kim MK, Choi SW, Kim SR, Oh IS, Won MH (2010) autologous chondrocyte implantation in the knee using fibrin. Knee Surg Sports Traumatol Arthrosc 18(4):528–534
19. Fedorovich NE, Alblas J, de Wijn JR, Hennink WE, Verbout AJ, Dhert WJ (Aug 2007) Hydrogels as extracellular matrices for skeletal tissue engineering: state-of-the-art and novel application in organ printing. Tissue Eng 13(8):1905–1925
20. Woodfield TB, Van Blitterswijk CA, De Wijn J, Sims TJ, Hollander AP, Riesle J (2005) Polymer scaffolds fabricated with pore-size gradients as a model for studying the zonal organization within tissue-engineered cartilage constructs. Tissue Engl. 11(9–10):1297–1231
21. Uematsu K et al (Jul 2005) Cartilage regeneration using mesenchymal stem cells and a three-dimensional poly-lactic-glycolic acid (PLGA) scaffold. Biomaterials 26(20):4273–4279
22. Zwingmann J, Mehlhorn AT, Südkamp N, Stark B, Dauner M, Schmal H (Sep 2007) Chondrogenic differentiation of human articular chondrocytes differs in biodegradable PGA/PLA scaffolds. Tissue Eng 13(9):2335–2343
23. Hule RA, Pochan DJ (2007) Polymer nanocomposites for biomedical applications. MRS Bulletin 32(4):354–358
24. Winey KI, Vaia RA (2007) Polymer nanocomposites. MRS Bulletin 32(4):314–322
25. Pan Y, Xiong D, Gao F (2008) Viscoelastic behavior of nano-hydroxyapatite reinforced poly(vinyl alcohol) gel biocomposites as an articular cartilage. J Mater Sci Mater Med 19(5):1963–1969
26. Balasundaram G, Storey DM, Webster TJ (2014) Novel nano-rough polymers for cartilage tissue engineering. Int J Nanomedicine 9:1845–1853
27. Lia W-J et al (2005) A three-dimensional nanofibrous scaffold for cartilage tissue engineering using human mesenchymal stem cells. Biomaterials 26(6):10
28. Francioli SE et al (2007) Tissue Engine 13:7
29. Rosa RG, Joazeiro PP, Bianco J, Kunz M, Weber JF, Waldman SD (2014) Growth factor stimulation improves the structure and properties of scaffold-free engineered auricular cartilage constructs. PLoS One 9:e105170
30. Schurman DJ, Smith RL (2004) Osteoarthritis: current treatment and future prospects for surgical, medical, and biologic intervention. PHD Clin Orthop Relat Res. 247:6

# Chapter 5
# Nanofibers and Microfibers for Osteochondral Tissue Engineering

### Zaida Ortega, María Elena Alemán, and Ricardo Donate

**Abstract** The use of fibers into scaffolds is a way to mimic natural tissues, in which fibrils are embedded in a matrix. The use of fibers can improve the mechanical properties of the scaffolds and may act as structural support for cell growth. Also, as the morphology of fibrous scaffolds is similar to the natural extracellular matrix, cells cultured on these scaffolds tend to maintain their phenotypic shape. Different materials and techniques can be used to produce micrfibers- and nanofibers for scaffolds manufacturing; cells, in general, adhere and proliferate very well on PCL, chitosan, silk fibroin, and other nanofibers. One of the most important techniques to produce microfibers/nanofibers is electrospinning. Nanofibrous scaffolds are receiving increasing attention in bone tissue engineering, because they are able to offer a favorable microenvironment for cell attachment and growth. Different polymers can be electrospun, i.e., polyester, polyurethane, PLA, PCL, collagen, and silk. Other materials such as bioglass fibers, nanocellulose, and even carbon fiber and fabrics have been used to help increase bioactivity, mechanical properties of the scaffold, and cell proliferation. A compilation of mechanical properties and most common biological tests performed on fibrous scaffolds is included in this chapter.

#### Highlights

- The use of microfibers and nanofibers allows for tailoring the scaffold properties.
- Electrospinning is one of the most important techniques nowadays to produce fibrous scaffolds.
- Microfibers and nanofibers use in scaffolds is a promising field to improve the behavior of scaffolds in osteochondral applications.

**Keywords** Microfibers · Nanofibers · Fibrous scaffolds · Electrospinning

---

Z. Ortega (✉) · M. E. Alemán · R. Donate
Grupo de investigación en Fabricación Integrada y Avanzada,
Universidad de Las Palmas de Gran Canaria, Las Palmas, Spain
e-mail: zaida.ortega@ulpgc.es

© Springer International Publishing AG, part of Springer Nature 2018
J. M. Oliveira et al. (eds.), *Osteochondral Tissue Engineering*,
Advances in Experimental Medicine and Biology 1058,
https://doi.org/10.1007/978-3-319-76711-6_5

## Abbreviations

| | |
|---|---|
| BDG | Butylene diglycolate |
| bFGF | Basic fibroblast growth factor |
| BMP | Bone morphogenetic protein |
| BMSCs | Bone marrow mesenchymal stem cells |
| BTDG | Butylene thiodiglycolate |
| CPP | Calcium pyrophosphate |
| CPP | Casein phosphopeptide |
| GAG | Glycosaminoglycan |
| HA | Hydroxyapatite |
| hESC | Human embryonic stem cells |
| hMSCs | Human mesenchymal stem cells |
| PA | Polyamide |
| PCL | Polycaprolactone |
| PDLA | Poly D,L-lactic acid |
| PEEK | Poly(ether-ether-ketone) |
| PEG | Poly(ethylene glycol) |
| PEO | Poly(ethylene oxide) |
| PET | Polyethylene terephthalate |
| PGA | Poly glycolic acid |
| PLA | Polylactic acid |
| PLGA | Poly(lactic-co-glycolic acid) |
| PLLA | Poly L-lactic acid |
| PVA | Polyvinyl alcohol |
| PVA-MA | Poly(vinyl alcohol)-methacrylate |
| PVP | Polyvinylpyrrolidone |
| rhBMP | Recombinant human morphogenetic protein |
| SBF | Simulated body fluid |
| TCP | Tricalcium phosphate |
| TFG-$\beta$1 | Transforming growth factor-$\beta$1 |
| TIPS | Thermally induced phase separation |

## 5.1 Introduction

The use of fibers into scaffolds is a way to mimic natural tissues, in which fibrils are embedded in a matrix. Their use can also improve the mechanical properties of the scaffolds and may act as structural support for cell growth. Furthermore, due to their large surface, microfibers and nanofibers can be functionalized by the addition of antibiotics, peptides, RNA or other substances in order to increase their bioactivity or prevent infections, among other possibilities. There are different materials used as fibers within the tissue engineering field, depending on the intended objective, manufacturing process and scaffold material. The materials used as matrix also show a wide range of possibilities, from natural polymers (gelatin or collagen) to bioglass or even carbon fibers. Electrospinning appears to be the most used technique in literature for microfiber and nanofiber production, although novel techniques are also being employed widely.

In the last years, fibers have been produced in a gradually increased materials range, from synthetic polymers (PCL, PLA, polyester, polyurethanes, etc.) to natural ones (silk, fibroin, chitosan, cellulose, etc.), from metals (titanium alloys) to ceramic materials (bioglass or calcium phosphates, even carbon fibers have been used for reinforcement of hyaluronic acid matrices. The main advantage in introducing microfibers or nanofibers within osteochondral tissue engineering is the possibility of tailoring the properties of scaffolds; porosity, pore size, mechanical properties, resilience, flexibility, bioactivity, and hydrophilicity constitute just a short list of potential adaptations. What is also of high interest is the combination of different materials to obtain a wider range of properties, both from the biological and mechanical sides.

As a summary, this is a very promising field, which has suffered a huge development in the last years, although further investigations on materials and manufacturing techniques need still to be performed. The ability of fibrous scaffolds to mimic extracellular matrix makes them definitely suitable for osteochondral applications.

## 5.2 Types of Fibers

### 5.2.1 Synthetic Polymeric Fibers

Different materials have been used to obtain microfibrous scaffolds by electrospinning, as this process is able to produce polymeric fibers from a molten or dissolved polymer at the micrometric and nanometric scale [1]. The benefit of using electrospinning in tissue engineering is that electrospun scaffolds show a similar morphology to the fibrous components of the natural extracellular matrix (ECM) [2], and so cells cultured on them tend to maintain their phenotypic shape [3]. Even though, electrospinning is not yet so widely implemented due to its slow production.

Electrospinning has been used as an efficient processing method to manufacture nanofibrous structures, enhancing cell proliferation and osteogenic differentiation [4]. Moreover, the small scale pores of electrospun nanofibrous scaffolds prevent cell migration, guiding tissue regeneration along the surface of the nanofibrous membrane [5], while porous hierarchical structures enable cell penetration, increasing the surface area for cell adhesion [6]. Furthermore, nanofibers, due to their vast surface, can be functionalized with drugs, antibiotics, bioactive peptides, proteins, RNA, and DNA [7].

Electrospun synthetic polymeric fibers have been widely explored for tissue engineering applications. Biodegradable materials like polylactic acid (PLA) or polycaprolactone (PCL) have suitable mechanical properties for the regeneration of cartilage and bone tissues and they degrade into nontoxic products. The use of polymeric micro – and nano – fibers allows obtaining wide wide range of proper-ties, as summarized in Tables 5.1 and 5.2 (also showing fibrous scaffolds in non – polymeric materials).

**Table 5.1** Mechanical properties under compressive tests for fibrous scaffolds

| Scaffold materials | Method of fabrication | Mechanical property | Value | Ref. |
|---|---|---|---|---|
| PLA nanofibers/alginate-hyaluronic acid hydrogel | Electrospinning and aminolysis, esterification and cross-linking reactions | Young's modulus for a 1:1 hydrogel to fibers weight ratio | 5.40 ± 0.90 kPa | [54] |
| PLLA nanofibers/collagen | Freeze-drying and electrospinning | Young's modulus (week 12 after surgery) | ~ 0.57 MPa | [5] |
| PLLA microfibrous sheets treated with 1-ethyl-3-(3-dimethylaminopropyl) carbodiimide/gelatin–nanoHA | Electrospinning and freeze-drying | Compressive strength analysis (wet state) of a six PLLA layered scaffold | ~ 6.0 MPa | [2] |
| P(LLA-CL) and collagen type I yarn mesh/hyaluronate/TCP | Electrospinning and freeze-drying | Compressive strength of the yarn-collagen type I/hyaluronate hybrid scaffold | ~ 0.25 MPa | [20] |
| PCL microfibrous discs/PLGA | Thermally induced phase separation and electrospinning | Compressive modulus | 125 ± 22 kPa for 90% porosity 75 ± 25 kPa for 95% porosity | [4] |
| | | Increase in the elastic modulus between the first and last cycles of the test (%) | 149 ± 45 for 90% porosity 135 ± 35 for 95% porosity | |
| | | Increase in the strain at peak during fatigue (%) | 204 ± 72 for 90% porosity 152 ± 15 for 95% porosity | |
| Oriented PCL fibrous membrane/collagen type I and hyaluronic acid/TCP | Electrospinning and freeze-drying | Compressive modulus of the chondral phase | 0.205 ± 0.029 MPa | [73] |
| | | Compressive modulus of cylindrical TCP specimens | 216.04 ± 48.08 MPa | |
| PVA nanofibers/hyaluronate/type I collagen/fibrin | Sol-gel processing | Young's modulus at 20% strain of the scaffold enriched with liposomes, basic fibroblast growth factor and insulin | ~ 2.0 MPa | [24] |

(continued)

**Table 5.1** (continued)

| Scaffold materials | Method of fabrication | Mechanical property | Value | Ref. |
|---|---|---|---|---|
| Multiphasic calcium phosphate fibers/chitosan | Freeze-drying | Compressive yield strength for a 1:1 chitosan to fibers weight ratio | ~ 420 kPa | [59] |
| | | Elastic modulus for a 1:1 chitosan to fibers weight ratio | ~ 3.87 MPa | |
| Collagen fibers/hydroxyapatite | Freeze-drying | Young's modulus of a 50HA–50COL scaffold | ~ 7 kPa | [72] |
| Fibrous collagen/PEG hydrogels | Lyophilization and photopolymerization processes | Tangent modulus at 15–20% strain | ~ 400 kPa | [38] |
| Collagen-PCL nanofibers/PCL-coated 45S5 bioactive glass | Foam replication process and electrospinning | Compressive strength of PCL dip-coated 45S5 BG scaffolds | 0.24 ± 0.06 MPa | [7] |
| Collagen fibrils/alginate/hyaluronic acid | Sol-gel processing | Compressive stress at 30% strain | ~ 65 kPa | [11] |
| Alginate/hydroxyapatite/bacterial nanocellulose | | | ~ 80 kPa | |
| Knitted silk-collagen sponge with hESC-MSCs | Knitting technique and freeze-drying | Young's modulus | 34.91 ± 5.08 MPa | [17] |
| Silk fibers/regenerated fibroin | Freeze-drying | Ultimate compressive strength for scaffolds seeded with autologous chondrocytes after 9 months | 0.258 ± 0.158 MPa | [14] |
| | | Young's modulus for scaffolds seeded with autologous chondrocytes after 9 months | 2.661 ± 1.79 MPa | |
| Silk fibroin yarns/polyethylene terephthalate | Knitting technique | Elastic modulus | 41.9 ± 17.1 kPa | [53] |

(continued)

**Table 5.1** (continued)

| Scaffold materials | Method of fabrication | Mechanical property | Value | Ref. |
|---|---|---|---|---|
| Pullulan/cellulose acetate | Electrospinning, cross-linking and freeze-drying | Young's modulus of a P50/CA50 scaffold | 4.13 ± 0.68 MPa | [50] |
| | | Compressive strength of a P50/CA50 scaffold | 0.43 ± 0.01 MPa | |
| | | Strain of a P50/CA50 scaffold (%) | 27.64 ± 2.89 | |

**Table 5.2** Mechanical properties under tensile and flexural tests for fibrous scaffolds

| Scaffold materials | Method of fabrication | Test | Mechanical property | Value | Ref. |
|---|---|---|---|---|---|
| P(LLA-CL) and collagen type I yarn mesh/hyaluronate/TCP | Electrospinning and freeze-drying | Tensile | Tensile strength of the yarn-collagen type I/hyaluronate hybrid scaffold | 3.43 ± 0.15 MPa | [20] |
| PCL microfibrous discs/PLGA | Thermally induced phase separation (TIPS) and electrospinning | Tensile | Elastic modulus | ~ 7 MPa for 90% porosity ~ 5 MPa for 95% porosity | [4] |
| | | | Ultimate stress | ~ 1.6 MPa for 90% porosity ~ 1.1 MPa for 95% porosity | |
| | | | Ultimate strain | 400% for 90% porosity 250% for 95% porosity | |
| | | | Increase in the elastic modulus between the first and last cycles of the test | ~ 120% | |
| | | | Increase in the strain at peak during fatigue | ~ 220% | |

(continued)

5 Nanofibers and Microfibers for Osteochondral Tissue Engineering

**Table 5.2** (continued)

| Scaffold materials | Method of fabrication | Test | Mechanical property | Value | Ref. |
|---|---|---|---|---|---|
| Oriented PCL fibrous membrane/ collagen type I and hyaluronic acid /TCP | Electrospinning and freeze-drying | Tensile | Tensile strength for PCL fibrous membranes | 4.07 ± 0.37 MPa | [73] |
| | | | Tensile modulus for PCL fibrous membranes | 36.14 ± 3.58 MPa | |
| Poly(butylene succinate) mesh | Electrospinning | Tensile | Elastic modulus of polymeric films | ~ 500 MPa | [13] |
| Cellulose acetate nanofibers/ polyethylene terephthalate | XanoMatrix™ (commercial product) | Tensile | Modulus of elasticity | ~ 0.509 GPa | [45] |
| 70S bioactive glass/silk fibroin | Electrospinning | Tensile | Young's modulus | 27.48 ± 3.96 MPa | [52] |
| | | | Elongation at break (%) | 8.52 ± 1.43 | |
| Hydroxyapatite nanofibers/ cellulose | Electrospinning | Tensile | Tensile strength of 5% nano-HA scaffold | ~ 70.6 MPa | [44] |
| | | | Elastic modulus of 5% nano-HA scaffold | ~ 3.12 GPa | |
| | | | Elongation at break of 5% nano-HA scaffold | ~ 5.56% | |
| Collagen-PVA nanofibers/ collagen sponge | Freeze-drying and electrospinning | Tensile | Young's modulus | ~ 0.25 MPa | [12] |
| | | | Ultimate tensile strength | ~ 0.07 MPa | |

(continued)

**Table 5.2** (continued)

| Scaffold materials | Method of fabrication | Test | Mechanical property | Value | Ref. |
|---|---|---|---|---|---|
| Collagen-PCL nanofibers/ PCL-coated 45S5 bioactive glass | Foam replication process and electrospinning | Tensile | Young's modulus of collagen-PCL fibrous meshes | 23 ± 10 MPa | [7] |
| Gelatin mesh | Electrospinning | Tensile | Tensile modulus | 426 ± 39 MPa | [62] |
| Collagen mesh | | | | 262 ± 18 MPa | |
| Elastin mesh | | | | 184 ± 98 MPa | |
| Tropoelastin mesh | | | | ~ 289 MPa | |
| Pullulan/cellulose acetate | Electrospinning, cross-linking and freeze-drying | Tensile | Young's modulus of a P50/CA50 scaffold | 1.54 ± 0.13 MPa | [50] |
| | | | Ultimate tensile strength of a P50/CA50 scaffold | 0.11 ± 0.02 MPa | |
| | | | Strain of a P50/CA50 scaffold (%) | 33.93 ± 2.18 | |
| Titanium fibers/13–93 bioactive glass | Freeform extrusion fabrication | Flexural | Flexural strength of scaffolds made with 0.4 vol% Ti fibers | 14.9 ± 1.3 MPa | [70, 71] |
| | | | Modulus of elasticity of scaffolds made with 0.4 vol% Ti fibers | 15.2 ± 4.1 GPa | |
| | | | Fracture toughness of scaffolds | 0.79 ± 0.07 MPa·m1/2 | |

### 5.2.1.1 Polylactic Acid (PLA)

The use of polymers such as PLA in the fibrous form offers structural support to the cells and is more similar to gelatin or collagen naturally present in terms of resilience, fracture toughness, elasticity and flexibility [4].

Biodegradable microfibrous PLLA/PVA sheets were incorporated into a gelatin–nanoHA matrix, achieving better cellular migration towards the center of the scaffold [4] and reducing the brittleness of the gelatin–nanoHA scaffolds.

The combination of collagen and electrospun PLLA nanofibers has been reported to synergistically promote osteochondral regeneration [5]. These tests have shown a more important osteogenic differentiation in cells seeded on collagen/PLLA scaffolds than in pure collagen ones, also leading to better cartilage formation and, in consequence, to better functional repairing of the osteochondral defects. This can be explained by the lower mechanical properties of collagen sponge to the subchondral bone, thus providing lower mechanical support for cartilage formation [5].

Apart from collagen, other natural polymers have been explored in combination with PLA fibers. For example, Mohabatpour et al. [8] proposed a hydrogel consisting of alginate-graft-hyaluronate. The presence of the nanofibers improved the mechanical properties of the hydrogel alone: the compressive modulus increased around 81%.

PDLLA nanofibers have also been used to coat bioglass scaffolds [9], with a decreased HA mineralization by increased the PDLLA thickness, thus ensuring a strong bond between the glass substrate and the PDLLA nanofibers and a smooth transition of the HA content; in vitro studies with chondrocyte cells shown good cell attachment and proliferation, leading to cell migration into the fibrous network. Hydroxyapatite and PLLA electrospun scaffolds have also been reported, showing differentiation of hMSCs, achieving chondrocyte-like phenotype with generation of a proteoglycan based matrix [10]. Copolymers derived from PLA, such as PLG (poly-D,L-lactide-co-glycolide) have been also explored for the treatment of osteochondral defects. For example, Toyokawa et al. [11] tested this type of material on the femoral condyles of rabbits.

### 5.2.1.2 Polycaprolactone (PCL)

Polycaprolactone is a biocompatible aliphatic polyester widely used in tissue engineering. Vaquette and Cooper-White [1] have combined electrospinning of PCL with a thermally induced phase separation (TIPS), also using PLGA. With this combination, they have been able to produce scaffolds made of electrospun membranes achieving better mechanical properties than scaffolds made by TIPS at shorter times and with no limits in the scaffold thickness. PLGA/PCL electrospun fibrous scaffolds also showed that rat bone marrow cells were infiltrated into the scaffold; GAG assays showed an abundant cartilage matrix after in vitro chondrogenic priming, leading to new bone formation in in vivo analysis [12].

Alginate hydrogels have been also combined with polycaprolactone fibrous matrices [13–15]. For example, the scaffolds proposed by Kook et al. [13] consisted of a nanofiber PCL matrix with infiltrated hydrogel and a second compartment of pure alginate hydrogel. The matrix was treated with oxygen plasma to improve the affinity with the alginate hydrogel. This structure allowed the coculture of different types of cells within the scaffold.

Bioactive glass scaffolds have been covered with PCL to enhance mechanical properties and collagen/PCL were electrospun over the coated scaffold [16]. This structure is justified by the high bioactivity of the PCL-coated bioglass scaffold, acting as support in the bone side, while the microfibers are intended for the cartilage side. Results from in vitro SBF tests show that HA crystals have grown along the surface of the collagen-PCL fibers, confirming their viability for osteochondral tissue engineering.

### 5.2.1.3 Other Synthetic Fibers

Anisotropic scaffolds have been obtained using 45S5 bioactive glass foam as substrate, gelatin as adhesive and short polyamide (PA) fibers, placed on the top surface of the scaffold by electroflocking [17]. This technique allows tailoring the surface porosity of the scaffold by varying the flocking time. After submersing these scaffolds on SBF for 21 days, the surface was entirely covered by HA, thus meaning that mineralization also occurs over the PA fibrils.

Another application of fibrous scaffolds is related to the tailoring of scaffolds properties, not only referred to mechanical ones but also to degradation rates. Chen and collaborators [18] have fabricated electrospun scaffolds from a block poly(butylene succinate)-based copolyesters containing either butylene thiodiglycolate (BTDG) or butylene diglycolate (BDG) sequences. The molecular architecture of the polyesters (and the heteroatom they contained, O or S) made it possible to change the mechanical properties and the hydrolysis rate of produced scaffolds. As a conclusion, they have demonstrated that copolyesters containing thioether links were more favorable for chondrogenesis, while those with ether linkages enhanced scaffold mineralization.

### 5.2.1.4 Fibers Including Additives

To improve the bioactivity of the synthetic fibers, several additives have been proposed, especially natural polymers and biological substances. For example, PVA has been electrospun with liposomes, bFGF, and insulin to obtain nanofibrous scaffolds [7], which, even without been cell seeded prior to implantation, showed enhanced osteochondral regeneration towards hyaline cartilage and/or fibrocartilage.

The incorporation of nanoapatitic particles to a PLGA-based nanofibrous scaffolds [19] significantly improved the tissue response of a subcutaneous implantation, thus demonstrating that the electrospun fibrous scaffolds made of PLGA/PCL at 3/1 rates with up to 30% of nanoapatitic particles allow controlling the in vivo adverse reactions of PLGA materials, leading to optimized clinical application of these materials in biomedical devices. Liu has proved that fibrous scaffolds made of electrospun hydroxyapatite/chitosan fibers show higher proliferation of BMSCs than the membranous compound [20], meaning fibrous scaffolds provide superior ability of bone reconstruction. Similarly, PLGA/PCL scaffold combined with electrospun PCL, hyaluronic acid and chondroitin sulfate nanofibers, also demonstrated that this combination stimulates the different regions of osteochondral tissue regeneration: collagen type II and aggrecan expression in the cartilage region and BMP-2 in the bone area [21]. Oriented poly(L-lactic acid)-copoly(ε-caprolactone) P(LLA-CL)/collagen type I(Col-I) nanofiber mesh made by electrospinning over a collagen I/hyaluronate sponge was fabricated to enhance the mechanical properties of the scaffold, also getting better infiltration [22]. These yarns were also produced over a TCP porous structure, obtaining improved repairing times and good compressive modulus.

Fibrous scaffolds made of PVA-MA and chondroitin sulfate–MA were obtained by electrospinning, obtaining fiber dimensions on the nanoscale for application to articular cartilage repair [23]. The low density of obtained nanofiber scaffolds allows immediate cell infiltration and optimal tissue repair, as shown in the in vitro tests. Furthermore, scaffolds containing chondroitin sulfate nanofibers lead to an increase in the deposition of type II collagen, specific to hyaline cartilage, enhancing the endogenous repair process without exogenous cells. Table 5.3 shows a summary of most usual in vitro tests in fibrous scaffolds and measured parameters.

Electrospinning has also been applied for the production of biphasic nanofiber scaffolds made of poly(lactide co-caprolactone, PLCL) and its mineralized form (obtained after activation in a NaOH solution, and then dipped alternatively in a $CaCl_2$ and $Na_2HPO_4$ solutions) [24]. In vitro studies shown that PLCL favored ECM secretion of cartilage, while mineralized PLCL favored bone secretion; in vivo tests in small animal model (nude mice) revealed that new cartilage and bone tissues were formed in the implanted area. This polymers combination was also used by Cui [25], but impregnating the scaffold into a chitosan-AHP solution, although reported results were similar to those without chitosan. In this case, as scaffolds did not incorporate cells neither bioactive compounds, only bone was formed.

Nanofibers in scaffolds also allow encapsulating active principles. Drugs can be encapsulated in electrospun fibers [26] to achieve a controlled release of the actives during the scaffold degradation; several authors have reported the release of various compounds, such as TFG-β1 from PCL microfibers and nanofibers, BMP-2 from PEG/PCL core/shell nanofibers, and fenbufen from PLGA [26]. Fibrous scaffolds have also been applied to the control of fibrous capsule formation, which leads to tissue fibrosis [27]. Small interfering RNA (siRNA) has been used to virtually make disappear any gene of interest; Rujitanaroj and team have used this approach to modulate fibrous capsule formation by RNAi is collagen type I. siRNA–poly(caprolactone-co-ethylethylene phosphate) nanofibers have been investigated for this purpose [27], leading to a significant decrease in fibrous capsule thickness;

Table 5.3 Most common biological assays for fibrous scaffolds

| Scaffold materials | Method of fabrication | Evaluation | Cell culture/animal model | Biological assay | Ref. |
|---|---|---|---|---|---|
| PLLA nanofibers/collagen | Freeze-drying and electrospinning | In vitro | Rabbit bone marrow mesenchymal stem cells | Methylthiazolyl-diphenyl-tetrazolium-bromide (MTT) | [5] |
| | | | | Von Kossa staining | |
| | | | Human bone marrow mesenchymal stem cells | RNA isolation and semiquantitative reverse transcription polymerase chain reaction | |
| | | In vivo | Adult male New Zealand white rabbits | Hematoxylin and Eosin (H&E) and Safranin O staining | |
| | | | | µ-Ct | |
| Poly(D,L-Lactide) nanofibers/45S5 bioglass derived foams | Foam replica technique and electrospinning | In vitro | ATDC5 chondrocyte cells | Alamar Blue | [39] |
| Mineralized poly(lactide-co-caprolactone) nanofibers | Electrospinning | In vitro | Primary rat articular chondrocytes and bone marrow-derived mesenchymal stem cells | Safranin O and Alcian Blue staining | [74] |
| | | | | Alkaline phosphatase activity | |
| | | | | Gene expression | |
| | | In vivo | Male athymic mice | H&E, Safranin O, and Alizarin Red S staining | |
| | | | | Immunostaining for type II collagen | |
| PCL microfibrous discs/PLGA | Thermally induced phase separation and electrospinning | In vitro | 3T3 fibroblasts | Alamar Blue | [4] |

| | | | | |
|---|---|---|---|---|
| Polyvinyl alcohol-methacrylate/chondroitin sulfate | Electrospinning and UV light cross-linking | In vitro | Goat bone marrow-derived mesenchymal stem cells | DNA quantification | [10] |
| | | | | Sulfated glycosaminoglycan quantification |
| | | | | Total collagen quantification |
| | | | | Immunohistochemical staining for type II collagen |
| | | | | Gene expression |
| | | In vivo | Rat osteochondral defects | Safranin O staining for proteoglycans |
| | | | | Immunohistochemical staining for collagen deposition |
| Carboxymethyl cellulose nanofibers/hydroxyapatite | Electrospinning | In vitro | Mouse osteoblastic MC3T3-E1 cells | Methylthiazolyl-diphenyl-tetrazolium-bromide (MTT) | [49] |
| 70S bioactive glass/silk fibroin | Electrospinning | In vitro | MG63 osteoblasts and primary porcine chondrocytes | Alamar Blue | [52] |
| | | | | Alkaline phosphatase activity |
| | | | | Sirius Red dye based colorimetric assay for total collagen estimation |
| | | | | 1,9-Dimethylmethylene Blue (DMMB) assay for sulfated glycosaminoglycan estimation |
| | | | | Gene expression |
| | | | Murine macrophage cells | Immune response using an ELISA kit |
| Multiphasic calcium phosphate fibers/chitosan | Freeze-drying | In vitro | Immersion in simulated body fluid | FE-SEM/EDX to evaluate the formation of a calcium phosphate-rich layer | [59] |

(continued)

Table 5.3 (continued)

| Scaffold materials | Method of fabrication | Evaluation | Cell culture/animal model | Biological assay | Ref. |
|---|---|---|---|---|---|
| 3D nonwoven silica gel fabric | Electrospinning | In vitro | Mouse calvaria-derived preosteoblastic MC3T3-E1 cells | MTT | [64] |
| | | | | Alkaline phosphatase activity | |
| | | | | Total RNA extraction and reverse transcription-polymerase chain reaction | |
| | | In vivo | New Zealand white male rabbits | Osteoconductivity test | |
| Collagen-PCL nanofibers/PCL-coated 45S5 bioactive glass | Foam replication process and electrospinning | In vitro | Immersion in simulated body fluid | SEM and XRD to evaluate the formation of hydroxyapatite | [7] |
| Silk fibroin yarns/polyethylene terephthalate | Knitting technique | In vitro | Human adipose-derived stem cells | DNA quantification | [53] |
| | | | | Alkaline phosphatase activity | |
| | | | | Alizarin Red staining | |
| | | | | Immunodetection of bone-specific proteins | |
| | | | | RNA isolation and real-time reverse transcriptase-polymerase chain reaction | |
| | | In ovo | White fertilized chicken eggs | Chick chorioallantoic membrane | |
| | | | | Analysis of blood vessel convergence | |
| | | | | Hematoxylin and Eosin staining | |
| | | | | Immunohistochemical detection | |
| | | In vivo | CD-1 mice | H&E staining | |
| Gelatin mesh | Electrospinning | In vitro | Human embryonic palatal mesenchymal cells | Alamar Blue | [62] |
| Collagen mesh | | | | | |
| Elastin mesh | | | | | |
| Tropoelastin mesh | | | | | |

| Carbon nonwoven fabrics/ hyaluronic acid | Thermal treatment and carbonization | In vitro | Human lung adenocarcinoma A549 cells | Trypan Blue staining | [1] |
|---|---|---|---|---|---|
| | | | | Cellular cytokine production to assess inflammatory potential | |
| | | | Human osteoblast-like MG-63 cells | Cell-material interaction evaluated with epifluorescence microscopy | |
| | | In vivo | Knee cartilage of rabbits | Masson-Goldner method for histological analysis | |
| Titanium fibers/13–93 bioactive glass | Freeform extrusion fabrication | In vitro | Immersion in simulated body fluid | SEM and XRD to evaluate the formation of hydroxyapatite | [70, 71] |

the in vitro silencing of collagen I was sustained for at least 4 weeks, in contrast to conventional bolus delivery of siRNA. In this research, scaffolds were obtained by electrospinning PCL and PCLEEP nanofibers, in which siRNA was encapsulated together with cell penetrating peptides.

### 5.2.2 Cellulosic and Cellulosic Derivative Fibers

Cellulose fibers are mainly found to be used in scaffolds for the osseous part, as fibers can act as reinforcement, improving the scaffold stiffness. Bacterial nanocellulose has also been investigated as scaffold for cartilage tissue engineering [24, 28]. Regarding the treatment of osteochondral defects, Iamaguti et al. [29] used cellulose membranes in experimental trochleopasty in dogs. They found that this type of implant could support the migration of chondrogenic cells.

Cellulose fibers have been also tested in combination with other biocompatible materials, such as alginate [11], hydroxyapatite [30, 31] or gelatin [31, 32]. Channel-like pores can be obtained when using nanocellulose in an alginate based scaffold. The use of bacterial nanocellulose fibers lead to an increase in the stiffness of alginate scaffolds under compression tests; this is also observed for the introduction of collagen fibers. No toxic effects have been found for scaffolds containing cellulose, as cell culturing is not influence by the presence of nanocellulose in alginate scaffolds [24]. Chenghong et al. [30] obtained an electrospun scaffold of nanofibrous cellulose and nanohydroxyapatite. The addition of the hydroxyapatite strengthens the matrix: a content of 5% of hydroxyapatite is able to provide a scaffold with a Young's modulus of 3.12 GPa. Moreover, the presence of nanohydroxyapatite also implies an improvement on the bioaffinity of the hybrid scaffolds compared to pure cellulose ones.

Besides, derivatives from cellulose have been also proposed as suitable materials to be used in osteochondral regeneration. For example, XanoMatrix™ is a hybrid material of polyethylene terephthalate and cellulose acetate that was studied by Bhardwaj and Webster [33]. These authors report suitable adhesion and proliferation of chondrocytes for in vitro testing. Furthermore, the cells aligned along the fibers resembling the structure of the natural cartilage. The strategy proposed by Atila et al. [34] was the electrospinning of pullulan and cellulose acetate and their subsequent cross-linking with trisodium trimetaphosphate. The cross-linking is a useful tool to maintain the characteristics of the scaffolds after soaking the samples in PBS because the pullulan component is not dissolved.

Hydroxyapatite coated carboxymethylcellulose nanofiber mat was analyzed by Yamaguchi et al. [35]. This nonwoven mat has potential applications for bone tissue regeneration, owing to its ability to support the growth of osteoblastic cells as shown by the authors.

## 5.2.3 Mineral Fibers

The use of ceramic fibers in scaffolds is mainly justified by the mechanical properties achieved, as these fibers act as reinforcement of hydrogels or polymer matrices. Calcium phosphate salts, like hydroxyapatite, have been used for this purpose. In the last years, other compounds such as bioactive glasses and silicate based ceramics have been investigated [39]. Also, bioactive mesoporous particles have been found to shown hemostatic properties, and so healing materials also tend to be in the form of fibers [36].

Fibers from different materials have been used in calcium phosphate cements to increase the similarity in mechanical properties to the natural bone, mainly in terms of toughness, ductility and fatigue resistance. Chitosan, PA, PCL, PLLA, PGA, carbon and glass fibers have been used to this purpose [40, 41]. The addition of fibers with higher resorption rate than the calcium phosphate matrix would allow creating macropores, thus favoring cell colonization and angiogenesis.

### 5.2.3.1 Glass Fibers

Bioglass nanofibers can be produced in several ways [36, 37]. Concentrating a laser on a bioglass monolith nanofibers can be produced [36]. These fibers, due to their small diameter and their bioactivity, are rapidly dissolved in SBF, leading to hydroxycarbonate apatite tubes. Electrospinning technique has also been recently used to produce nanofibrous scaffolds of bioactive glass [37–39]. Due to their high surface area, bioactive glass nanofibers degrade quickly, converting to HA. The bioactivity of these glass nanofibers is maintained over a larger $SiO_2$ compositional range when compared to melt-derived glasses. Electrospinning can take place from organic or inorganic solutions, being after heated to 600–700 °C to decompose any residual group; fibers prepared in this way exhibit a diameter in the micro to submicron range and are commercially available. Because of their rapid degradation rate, they have a huge potential in the regeneration of non-loaded bone and in the healing of soft tissue.

Submicron 45S5 bioglass fibers (with and without copper) were used in gelatin/collagen scaffolds at a 70/30 ratio (30% of submicron bioglass fibers) [39]. Those scaffolds doped with copper have provided better behavior in terms of cell proliferation and distribution, demonstrating that copper-doped bioglass fibers are non-cytotoxic and that their surface is ideal for osteoblast attachment, growth, viability, and bone regeneration.

In some cases, small amounts of polymer (polyvinyl butyral, PEO or PVA) were firstly introduced to the sol to obtain optimal viscosity for the process; a burning stage was later needed to decompose the polymer and obtain the glass fibers. Hybrid scaffolds (silica and PCL, PLGA, or PLLA) have been successfully electrospun.

Bioactive glass particles have demonstrated to be useful in bone defects regeneration, although approved compositions are not suitable for making fibers. Scaffolds with 50% porosity made from these materials, with 75 μm thick, were completely degraded in 6 months after implantation in rabbit tibia.

### 5.2.3.2 Calcium Phosphate Fibers

Calcium phosphate compounds are widely used in bone regeneration because of their osteoconductive properties [42, 43]. Zhang and collaborators [3] developed a woven-bone-like beta-tricalcium phosphate (β-TCP)/collagen scaffolds by sol-gel electrospinning, preparing pure β-TCP fibers with dimensions close to mineralized collagen fibrils in woven bone. They have observed that osteoblasts showed 3D morphologies and multicellular layers, shortening to time to produce new bone.

Polycrystalline CaP fibers can be obtained by electrospinning an aqueous solution of $CaCl_2$ and $H_3PO_4$, using PEO as spinning aid [44]. Fibers from 10 to 25 μm of diameter were obtained after pyrolysis and sintering to remove the polymer. The so prepared fibers show no cytotoxicity under in vitro tests.

Multiphasic calcium phosphate fibers (HA, β-TCP and CPP) have been used as reinforcement of chitosan matrices, finding an increase in compressive properties, pore size and density and a decrease of porosity and swelling ratio [45]. Calcium phosphate was formed on the scaffold surface after immersion of the scaffolds in a PBS solution, demonstrating their in vitro bioactivity. Fibers were obtained by dissolving $Ca(NO_3)_2 \cdot 4H_2O$ and $(NH_4)_2HPO_4$ in distilled water at pH 3 and with small amounts of urea; the precipitated formed was treated with ethanol and subjected to 800 °C for 2 h. Chitosan scaffolds were obtained by freeze-drying of a chitosan solution containing up to 50% of fibers. Urea has demonstrated to modify the structure of precipitated calcium phosphate fibers [46], depending on the urea concentration and reaction time. Low concentration of urea leads to the production of whisker-like monetite/HA fibers, while higher concentrations tend to produce a combination of whisker-like fibers and spherulites, made of HA and octacalcium phosphate. Reaction times of 10 days allow producing HA monophasic whiskers.

HA fibers can be prepared by treating a block of $\beta\text{-}Ca(PO_3)_2$ fibers with $Ca(OH)_2$ particles heating it at 1000 °C and then treating it with a HCl solution [47]. Also, hydrolysis of TCP in a water-aliphatic alcohol solution at 80 °C and growing the HA fibers in an agar gel system, using $Ca(NO_3)_2 \cdot 4H_2O$ and $(NH_4)_2HPO_4$ solution have been reported to be used for HA fibers obtaining. Wu and collaborators have obtained them by electrospinning a mixture of its precursors (a mixture of $Ca(NO_3)_2 \cdot 4H_2O$ and $(C_2H_5O)_3PO$ with a polymer additive), and then annealing the electrospun fibers (containing the polymer) at 600 °C for 1 h [48]. By this procedure, HA fibers about 25 μm were obtained. Other researchers have used $P_2O_5$ and $Ca(NO_3)_2 \cdot 4H_2O$ as precursors to also obtain submicron fibers by electrospinning, but in this case from a mixture of the gel formed from the mentioned salts with PVP in water and ethanol/water [47]; post-heating is also required to obtain the HA fibers. Diameters of 567 ± 70 nm and 122 ± 32 nm for fibers were obtained starting with a PVP concentration and 50 and 100% in water, respectively. This is due to the higher conductivity and lower viscosity of the water solution, in comparison with the 50% ethanol/water one. Also, composition studies show that after sintering the fibers were made of carbonated hydroxyapatite, as the human bone.

Similarly, the use of PLGA dissolved in hexafluoroisopropanol has been used to produce electrospun fibers of nanohydroxyapatite [49]; thermal processing at 1100 °C is required to evaporate the polymer. These authors also propose the use of PVA or PVP as sacrifice polymers, indicating that polymers with low melting point, such as PCL, are not an option, due to their incapacity to maintain the fibrous structure during the thermal treatment stage. On the other hand, considerations about the low mechanical properties of the so obtained fibers are made in the paper, fact which would need to be solved prior to their use as scaffolds reinforcement.

#### 5.2.3.3 Silica Fibers

Silica fibers, coming from natural sponges skeletons, with an average diameter of 10 μm, have also been used to produce composite scaffolds based on PEEK; to increase mechanical properties, both materials were pretreated by atomization and using citric acid [50]. The composite was prepared by compression molding at 350 °C. The use of silica fibers has led to an increase of over 50% in elastic modulus at flexural testing and of 26.7% in microhardness. High cytocompatibility of the composite was found, as the metabolic activity of fibroblasts was also increased.

Silica fibers have also been produced from tetraethyl orthosilicate, which is hydrolyzed and condensed by water, ethanol and HCl [51]; the solution produced is then electrospun to obtain a nonwoven mesh, which is thermally treated at 300 °C for 3 h. Wide diameter distribution of the fibers is found: from 0.7 to 6 μm. The produced mesh show good cellular behavior, allowing preosteoblastic differentiation and osteoconductivity.

### 5.2.4 Fibers from Animals: Collagen, Silk, and Fibrin

#### 5.2.4.1 Collagen

Collagen is mainly used in scaffolds for chondral applications, as it is naturally found in cartilage regions. Collagen fibrils have been added to the chondral part of biphasic alginate-based hydrogels to improve their mechanical properties, as well as to act as a binding site for living cells [24]. As for cellulose fibers, an increase in the collagen amount of an alginate scaffold reduces the pores density, while the pores diameter is not affected. Collagen in the alginate scaffolds shows no cytotoxicity and does not affect the cells growth.

Collagen–PVA nanofibers have been electrospun onto a collagen sponge to make aligned and random composites [52]. Average diameters for the random nanofibers was 203 ± 74.91 nm and 301.05 ± 96.53 nm after glutaraldehyde cross-linking, while for aligned fibers diameters were significantly smaller: 94.82 ± 25.57 nm before and 198.20 ± 33.61 nm after cross-linking. The swelling ratio was higher for the random nanofibers composite, due to the capillary effect. The aligned composites scaffolds showed higher mechanical properties, making them more suitable for articular cartilage repair, while both scaffolds showed similar cell proliferation and secretion of cartilage II.

Multiphasic composite scaffold made of an upper collagen I fiber layer and a lower part made of PLA, β-TCP and HA, seeded with hMSCs, showed chondrogenic differentiation and a homogeneous cell distribution when cultured in a TGF-β1 medium. Cells were also surrounded by a proteoglycan and collagen type II; also a high deposition of GAGs was measured [53].

Fibrous collagen has also been used as reinforcement of PEG hydrogels [54], obtaining increased modulus and toughness, and decreasing lateral expansion under compressive loading.

In other medical fields, these materials are also showing promising results; for instance, PLLA meshes have been filled in with collagen fibers for the reconstruction of abdominal wall with good results [55].

### 5.2.4.2 Silk

Silk fibroin, extracted from silk fibers, shows low immunogenicity, cell affinity, tunable degradation rates [56], and impressive mechanical properties [57], which provide exceptional advantages over other polymers [58]. The use of embedded silk fibers into a regenerated silk matrix led to the obtaining of scaffolds with great mechanical properties and high porosity levels; it was also found that silk fibers boost the degradation rates, due to an increased number of immigrated cells into the silk matrix. If chondrocytes are seeded in the scaffold, results are better [58]. Similar studies also reported that transplantation of mesenchymal stem cells grown in a silk fibroin/hydroxyapatite scaffold can enhance tissue repairing [59]; also, scaffolds made of silk fibroin containing mesenchymal stem cells and chondroitin provided improved behavior [60].

Silk fibers have also been used by Chen and collaborators to produce a knitted structure in which openings collagen microsponges were placed [27]. Again, seeding the scaffold with hESC-MSCs in in vivo tests provided good tendon healing, with cells differentiation into the tenocyte-lineage morphology. Ribeiro et al. [61] also proposed silk-based biotextiles for bone regeneration. They produced a silk fibroin-PET fabric and they tested the osteogenic differentiation on its surface of human adipose-derived stem cells. The alkaline phosphatase activity quantification showed a higher differentiation on the silk fibroin-PET samples than on PET fabrics taken as reference.

Christakiran et al. [62] developed scaffolds consisting on an osteogenic matrix of 70S bioactive glass and a chondrogenic matrix of silk fibroin. They evaluated the suitability of these scaffolds for the treatment of osteochondral defects by culturing chondrocytes from pigs on the silk membrane and MG63 (osteosarcoma cell line) on the bioactive glass side. They tested two types of silk: non-mulberry and mulberry based ones. The authors concluded that the non-mulberry silk based membrane performs better both from the mechanical and the biological points of view.

### 5.2.4.3 Other Proteins: Fibrin, Elastin

PLGA sponges have also been used in combination with fibrin fiber, BMSC, plasmid DNA TGF-β1. After culturing for 4 weeks under in vitro conditions and implantation for 12 weeks, cartilage defects were completely repaired in rabbits, being the new cartilage well integrated with the surrounding tissue and subchondral bone. GAGs confirmed similar amount and distribution of collagen type II in the new cartilage and in the hyaline one [63, 64].

Apart from collagen and gelatin, α-elastin [65] has also been electrospun to obtain 0.6–3.6 μm width and from 1.4 to 7.4 μm for tropoelastin, depending on the electrospinning parameters (concentration of the solution and delivery rate). Elastin fibers have also found to be more brittle than the other three, although cell viability is higher for elastin, followed by collagen.

## 5.2.5 Carbon Fibers

Carbon fiber is a not biodegradable material that can be obtained both at the nano [66] and micro levels. This feature has attracted the interest of the researchers to include this material as scaffolding in tissue engineering [67]. For the treatment of osteochondral injures, carbon fibers are potentially interesting because they enable the restoration of damaged cartilage [68, 69]. Besides, Aouri et al. [66] demonstrated they are also an effective support for the delivery of recombinant human morphogenetic protein 2 (rhBMP-2). This characteristic was useful to promote bone regeneration. In fact, in this study, SEM observation of samples implanted in mice showed that the carbon fibers and the bone matrix were fully integrated.

Bencano et al. [69] carried out the in vivo assessment of using carbon fibers to treat osteochondral defects. They evaluated the histological progression of the osteochondral defects created on the articular surface of the patella of a population of rabbits and treated with carbon fiber implants. They found that a year after the treatment, the defects had been covered with hyaline cartilage tissue.

Carbon nonwoven fabrics have a higher surface area and an interconnected pore structure, providing increased surface area for cell attachment as well as convenient channels for nutrients transportation, diffusion of gases and cell migration [68].

However, even though carbon fibers are biocompatible they do not have enough biological activity to stimulate the cells proliferation. To overcome this limitation, different modifications have been proposed, such as coatings with hyaluronic acid [68]. This has proofed to provide good cellular attachment and viability and higher speed of tissue regeneration regarding the non-modified carbon nonwovens at in vitro and in vivo studies. Several authors have explored the possibility of obtaining carbon fibers doped with osteoinductive components by previously mixing this component with polyacrylonitrile, precursor of the carbon fibers [67, 70–72]. Following this strategy, Fraczek-Szczypta et al. [67] obtained carbon nonwovens with different ceramic nanoparticles (bioglass and wollastonite). The improvement of the bioactivity of the fabrics was evaluated by the assessment of the apatite forming ability of the material when immersed in SBF solution for 21 days. All the fibers tested promote the apatite precipitation. However, the apatite layer was more uniformly distributed on the nonwoven samples containing wollastonite. On the other hand, Zhang et al. [70] demonstrated that the presence of bioglass in a carbon nanofiber matrix accelerates the proliferation rate of BMSCs when compared to pure carbon nanofiber and, besides, it improves the differentiation ability of the cells.

Another approach is the utilization of composite materials. However, the main limitation for the manufacturing of composite materials containing carbon fibers is their poor dispersion and chemical inertness in the common matrix used for tissue engineering applications [73]. For example, Chlopek et al. [74] proposed a composite of carbon fibers ($d = 7$ μm) in a PGLA matrix. They followed the degradation profile of these implants and pure PGLA ones in vivo on a population of New Zealand rabbits for 48 weeks. In this study, they conclude that the presence of the carbon fibers accelerates bone regeneration and the overall process of resorption of the implant. On the other hand, Shi et al. [75] activated carbon fiber via a high temperature process and subsequent air plasma treatment. With these activated carbon fibers, a composite material with PLGA was obtained. This composite exhibited an improvement on the porosity when compared to the pure PLGA scaffolds.

### 5.2.6 Titanium Fibers

Thomas et al. [76, 77] have produced printed glass scaffolds reinforced with titanium fibers to increase the mechanical properties of the bioactive glass. They started from a composite paste made of bioactive glass and titanium microfibers (16 μm diameter, up to 0.4% in volume fraction) and extruded it at 0–90° orientation. The use of titanium fibers led to an increase in the fracture toughness of about 70%, with an increase of flexural strength near 40%. It has also been demonstrated that the introduction of titanium fibers do not affect bioactivity, as HA is precipitated after 2 weeks of immersion in SBF solution in the same extent and morphology that in bioglass scaffolds. Biodegradation tests on these scaffolds have also been performed by these researchers [76, 77], showing that compressive strength in bioglass is reduced by 30% after 4 weeks in SBF, while this reduction is near 40% for titanium fiber/glass scaffolds (67 MPa and 88 MPa for glass and Ti/glass scaffolds, respectively, after 4 weeks test).

## 5.3 Conclusions

As we observe, the most important technique to obtain fibrous scaffolds is electrospinning. The combination of different materials allows obtaining a wide range of properties, both from the mechanical and biological points of view; what makes fibrous scaffolds especially interesting for osteochondral applications as they are able to mimic extracellular cartilage matrix.

Even if the background in this field is quite large, more effort is needed in order to continue evaluating other material alternatives and their combinations. Furthermore, the adhesion between materials in the scaffold needs to be studied in more detail. Also, other manufacturing techniques, less common, should be further investigated.

## References

1. Vaquette C, Cooper-White J (2013) A simple method for fabricating 3-D multilayered composite scaffolds. Acta Biomater 9:4599–4608
2. Agarwal S, Wendorff JH, Greiner A (2008) Use of electrospinning technique for biomedical applications. Polymer 49:5603–5621
3. Zhang S, Zhang X, Cai Q et al (2010) Microfibrous β-TCP/collagen scaffolds mimic woven bone in structure and composition. Biomed Mater 5(6):065005
4. Shamaz BH, Anitha A, Vijayamohan M et al (2015) Relevance of fiber integrated gelatin-nanohydroxyapatite composite scaffold for bone tissue regeneration. Nanotechnology 26(40):405101
5. Zhang S, Chen L, Jiang Y et al (2013) Bi-layer collagen/microporous electrospun nanofiber scaffold improves the osteochondral regeneration. Acta Biomater 9:7236–7247
6. Yousefi AM, Hoque ME, Prasad R et al (2015) Current strategies in multiphasic scaffold design for osteochondral tissue engineering: a review. J Biomed Mater Res A 103:2460–2481
7. Filová E, Rampichová M, Litvineca A et al (2013) A cell-free nanofiber composite scaffold regenerated osteochondral defects in miniature pigs. Int J Pharm 447:139–149
8. Mohabatpour F, Karkhaneh A, Sharifi AM (2016) A hydrogel/fiber composite scaffold for chondrocyte encapsulation in cartilage tissue regeneration. RSC Adv 6:83135
9. Yunos DM, Ahmad Z, Salih V et al (2011) Stratified scaffolds for osteochondral tissue engineering applications: electrospun PDLLA nanofibre coated bioglass®-derived foams. J Biomater Appl 27(5):537–551
10. Spadaccio C, Rainer A, Trombetta M et al (2009) Poly-L-lactic acid/hydroxyapatite electrospun Nanocomposites induce Chondrogenic differentiation of human MSC. Ann Biomed Eng 37(7):1376–1389

11. Toyokawa N, Fujioka H, Kokubu T et al (2010) Electrospun synthetic polymer scaffold for cartilage repair without cultured cells in animal model. Arthroscopy 26:375–383
12. Yang W, Yang F, Wang Y et al (2013) In vivo bone generation via the endochondral pathway on three-dimensional electrospun fibers. Acta Biomater 9:4505–4512
13. Kook YM, Kang YM, Moon SH et al (2016) Bi-compartmental 3D scaffolds for the co-culture of intervertebral disk cells and mesenchymal stem cells. J Ind Eng Chem 38:113–122
14. Formica FA, Öztürk E, Hess SC et al (2016) A bioinspired ultraporous nanofiber-hydrogel mimic of the cartilage extracellular matrix. Adv Healthc Mater 5:3129–3138
15. Liao IC, Moutos FT, Estes BT et al (2013) Composite three-dimensional woven scaffolds with interpenetrating network hydrogels to create functional synthetic articular cartilage. Adv Funct Mater 23:5833–5839
16. Balasubramanian P, Roether JA, Schubert DW (2015) Bi-layered porous constructs of PCL-coated 45S5 bioactive glass and electrospun collagen-PCL fibers. J Porous Mat 22:1215–1226
17. Balasubramanian P, Boccaccini AR (2015) Bilayered bioactive glass scaffolds incorporating fibrous morphology by flock technology. Mater Lett 158:313–316
18. Chen H, Gigli M, Gualandi C et al (2016) Tailoring chemical and physical properties of fibrous scaffolds from block copolyesters containing ether and thio-ether linkages for skeletal differentiation of human mesenchymal stromal cells. Biomaterials 76:261–272
19. Ji W, Yang F, Seyednejad H et al (2012) Biocompatibility and degradation characteristics of PLGA-based electrospun nanofibrous scaffolds with nanoapatite incorporation. Biomaterials 33:6604–6614
20. Liu H, Peng H, Wua Y et al (2013) The promotion of bone regeneration by nanofibrous hydroxyapatite/chitosan scaffolds by effects on integrin-BMP/Smad signaling pathway in BMSCs. Biomaterials 34:4404–4417
21. Lee P, Manoukian OS, Zhou G et al (2016) Osteochondral scaffold combined with aligned nanofibrous scaffolds for cartilage regeneration. RSC Adv 6:72246–72255
22. Liu S, Wu J, Liu X et al (2014) Osteochondral regeneration using an oriented nanofiber yarn-collagen type I/hyaluronate hybrid/TCP biphasic scaffold. J Biomed Mater Res A 103(2):581–592
23. Coburn JM, Gibson M, Monagle S et al (2012) Bioinspired nanofibers support chondrogenesis for articular cartilage repair. Proc Natl Acad Sci U S A 109(25):10012–10017
24. Despang F, Halm C, Schütz K et al (2016) Bio-inspired regenerative medicine: materials, processes, and clinical applications. Chapter 6: fibre-reinforced, biphasic composite scaffolds with pore channels and embedded stem cells based on alginate for the treatment of Osteochondral defects. Editors: Anna Tampieri and Simone Sprio. 2016. ISBN 978-981-4669-14-6
25. Cui Z, Wright LD, Guzzo R et al (2013) Poly(d-lactide)/poly(caprolactone) nanofiber thermogelling chitosan gel composite scaffolds for osteochondral tissue regeneration in a rat model. J Bioact Compat Polym 28(2):115–125
26. Amler E, Filová E, Buzgo M et al (2014) Functionalized nanofibers as drug-delivery systems for osteochondral regeneration. Nanomedicine 9(7):1083–1094
27. Chen JL, Yin Z, Shen WL et al (2010) Efficacy of hESC-MSCs in knitted silk-collagen scaffold for tendon tissue engineering and their roles. Biomaterials 31:9438–9451
28. Nandgoankar AG, Krause WE, Lucia LA (2016) Fabrication of cellulosic composite scaffolds for cartilage tissue engineering. In: Nanocomposites for musculoskeletal tissue regeneration. 187–212
29. Iamaguti LS, Brandao CVS, Pellizzon CH et al (2008) Histological and morphometric analysis for the use of a biosynthetic cellulose membrane in experimental trochleopasty. Pesqui Vet Bras 28(4):195–200
30. Ao C, Niu Y, Zhang X et al (2017) Fabrication and characterization of electrospun cellulose/nano-hydroxyapatite nanofibers for bone tissue engineering. Int J Biol Macromolec 97:568–573

31. Ran J, Jiang P, Liu S et al (2017) Constructing multi-component organic/inorganic composite bacterial cellulose-gelatin/hydroxyapatite double-network scaffold platform for stem cell-mediated bone tissue engineering. Mater Sci Eng C 78:130–140
32. Wang J, Wan Y, Han J et al (2011) Nanocomposite prepared by immobilising gelatin and hydroxyapatite on bacterial cellulose nanofibres. Micro Nano Lett 6(3):133–136
33. Bhardwaj G, Webster TJ (2016) Enhanced chondrocyte culture and growth on biologically inspired nanofibrous cell culture dishes. Int J Nanomedicine 11:479–483
34. Atila D, Keskin D, Tezcaner A (2016) Crosslinked pullulan/cellulose acetate fibrous scaffolds for bone tissue engineering. Mat Sci Eng C 69:1103–1115
35. Yamaguchi K, Prabakaran M, Ke M et al (2016) Highly dispersed nanoscale hydroxyapatite on cellulose nanofibers for bone regeneration. Mater Lett 168:56–61
36. Jones JR (2013) Review of bioactive glass: from Hench to hybrids. Acta Biomater 9:4457–4486
37. Sharifi E, Ebrahimi-Barough S, Panahi M et al (2016) In vitro evaluation of human endometrial stem cell-derived osteoblast-like cells behavior on gelatin/collagen/bioglass nanofibers scaffolds. J Biomed Mater Res A 104(A):2210–2219
38. Rahaman MN, Day DE, Sonny Bal B et al Bioactive glass in tissue engineering. Acta Biomater 7:2355–2373
39. Sharifi E, Azami M, Kajbafzadeh AM et al (2016) Preparation of a biomimetic composite scaffold from gelatin/collagen and bioactive glass fibers for bone tissue engineering. Mater Sci Eng C 59:533–541
40. Canal C, Ginebra MP (2011) Fibre-reinforced calcium phosphate cements: a review. J Mech Behav Biomed Mater 4:1658–1671
41. Maenz S, Kunisch E, Mühlstädt M et al (2014) Enhanced mechanical properties of a novel, injectable, fiber-reinforced brushite cement. J Mech Behav Biomed Mater 39:328–339
42. Samavedi S, Whittington AR, Goldstein AS (2013) Calcium phosphate ceramics in bone tissue engineering: a review of properties and their influence on cell behavior. Acta Biomater 9:8037–8045
43. Chai YC, Carlier A, Bolander J et al (2012) Current views on calcium phosphate osteogenicity and the translation into effective bone regeneration strategies. Acta Biomater 8:3876–3887
44. Müller A, Schleid T, Doser M et al (2014) Calcium cl/OH-apatite, cl/OH-apatite/Al2O3 and Ca3(PO4)2 fibre non wovens: potential ceramic components for osteosynthesis. J Eur Ceram Soc 34:3993–4000
45. Mohammadi Z, Mesgar ASM, Rasouli-Disfani F (2016) Reinforcement of freeze-dried chitosan scaffolds with multiphasic calcium phosphate short fibers. J Mech Behav Biomed Mater 61:590–599
46. Mohammadi Z, Mesgar ASM, Rasouli-Disfani F (2016) Preparation and characterization of single phase, biphasic and triphasic calcium phosphate whisker-like fibers by homogenous precipitation using urea. Ceram Int 42:6955–6961
47. Franco PQ, João CFC, Silva JC et al (2012) Electrospun hydroxyapatite fibers from a simple sol–gel system. Mater Lett 67:233–236
48. Wu Y, Hench LL, Du J et al (2004) Preparation of hydroxyapatite fibers by electrospinning technique. J Am Ceram Soc 87(10):1988–1991
49. Mouthuy PA, Crossley A, Ye H (2013) Fabrication of calcium phosphate fibres through electrospinning and sintering of hydroxyapatite nanoparticles. Mater Lett 106:145–150
50. Monich PR, Berti FV, Porto LM et al (2017) Physicochemical and biological assessment of PEEK composites embedding natural amorphous silica fibers for biomedical applications. Mater Sci Eng C 79:354–362
51. Kang YM, Kim KH, Seol YJ et al (2009) Evaluations of osteogenic and osteoconductive properties of a non-woven silica gel fabric made by the electrospinning method. Acta Biomater 5:462–469
52. Lin HY, Tsai WC, Chang SH (2017) Collagen-PVA aligned nanofiber on collagen sponge as bi-layered scaffold for surface cartilage repair. J Biomat Sci Polym 28(7):664–678

53. Heymer A, Bradica G, Eulert J et al (2009) Multiphasic collagen fibre–PLA composites seeded with human mesenchymal stem cells for osteochondral defect repair: an in vitro study. J Tissue Eng Regen Med 3:389–397
54. Kinneberg KRC, Nelson A, Stender ME et al (2015) Reinforcement of mono- and bilayer poly(ethylene glycol) hydrogels with a fibrous collagen scaffold. Ann Biomed Eng 43(11):2618–2629
55. Pu F, Rhodes NP, Bayon Y et al (2010) The use of flow perfusion culture and subcutaneous implantation with fibroblast-seeded PLLA-collagen 3D scaffolds for abdominal wall repair. Biomaterials 31:4330–4340
56. Melke J, Midha S, Ghosh S et al (2016) Silk fibroin as biomaterial for bone tissue engineering. Acta Biomater 31:1–16
57. Qi Y, Wang H, Wei K et al (2017) A review of structure construction of silk fibroin biomaterials from single structures to multi-level structures. Int J Molec Sci 18(3):237
58. Kazemnejad S, Khanmohammadi M, Mobini S et al (2016) Comparative repair capacity of knee osteochondral defects using regenerated silk fiber scaffolds and fibrin glue with/without autologous chondrocytes during 36 weeks in rabbit model. Cell Tissue Res 364:559–572
59. Jin J, Wang J, Huang J et al (2014) Transplantation of human placenta-derived mesenchymal stem cells in a silk fibroin/hydroxyapatite scaffold improves bone repair in rabbits. J Biosci Bioeng 118(5):593–598
60. Deng J, She R, Huang W, Dong Z, Mo G, Liu B (2013) A silk fibroin/chitosan scaffold in combination with bone marrow-derived mesenchymal stem cells to repair cartilage defects in the rabbit knee. J Mater Sci Mater Med 24(8):2037–2046
61. Ribeiro VP, Silva-Correia J, Nascimento AI et al (2017) Silk-based anisotropical 3D biotextiles for bone regeneration. Biomaterials 123:92–106
62. Christakiran J, Reardon PJT, Konwarh R et al (2017) Mimicking hierarchical complexity of the osteochondral interface using electrospun silk-bioactive glass composites. App Mater Interf 9:8000–8013
63. Wang W, Li B, Li Y et al (2010) In vivo restoration of full-thickness cartilage defects by poly(lactide-co-glycolide) sponges filled with fibrin gel, bone marrow mesenchymal stem cells and DNA complexes. Biomaterials 31:5953–5965
64. Wang W, Li B, Yang J et al (2010) The restoration of full-thickness cartilage defects with BMSCs and TGF-beta 1 loaded PLGA/fibrin gel constructs. Biomaterials 31:8964–8973
65. Li M, Mondrinos MJ, Gandhi MR et al (2005) Electrospun protein fibers as matrices for tissue engineering. Biomaterials 26:5999–6008
66. Aoki K, Usui Y, Narita N et al (2009) A thin carbon-fiber web as a scaffold for bone-tissue regeneration. Small 5(13):1540–1546
67. Fraczek-Szczypta A, Rabiej S, Szparaga G et al (2015) The structure and properties of the carbon non-wovens modified with bioactive nanoceramics for medical applications. Mater Sci Eng C 51:336–345
68. Rajzer I, Menaszek E, Bacakova L et al (2013) Hyaluronic acid-coated carbon nonwoven fabrics as potential material for repair of osteochondral defects. Fibres text east Eur 3(99):102–107
69. Carranza-Bencano A, Armas-Padrón JR, Gili-Miner M (2000) Carbon fiber implants in osteochondral defects of the rabbit patella. Biomaterials 21:2171–2176
70. Zhang XR, Hu XQ, Jia XL et al (2016) Cell studies of hybridized carbon nanofibers containing bioactive glass nanoparticles using bone mesenchymal stromal cells. Sci Rep 6:38685
71. Rajzer I, Kwiatkowski R, Piekarczyk W et al (2012) Carbon nanofibers produced from modified electrospun PAN/hydroxyapatite precursors as scaffolds for bone tissue engineering. Mater Sci Eng C 32:2562–2569
72. Yang Q, Sui G, Shi YZ et al (2013) Osteocompatibility characterization of polyacrylonitrile carbon nanofibers containing bioactive glass nanoparticles. Carbon 56:288–295
73. Li X, Cui R, Liu W et al (2013) The use of nanoscaled fibers or tubes to improve biocompatibility and bioactivity for biomedical materials. J Nanomater 1:728130

74. Chłopek J, Morawska-Chochół A, Paluszkiewicz C (2008) FTIR evaluation of PGLA-carbon fibers composite behaviour under in vivo conditions. J Mol Struct 875:101–107
75. Shi Y, Han H, Quan H (2014) Activated carbon fibers/poly(lactic-co-glycolic) acid composite scaffolds: preparation and characterizations. Mater Sci Eng C 43:102–108
76. Thomas A, Kolan KCR, Leu MC et al (2017) Freeform extrusion fabrication of titanium fiber reinforced 13-93 bioactive glass scaffolds. J Mech Behav Biomed Mater 70:43–52
77. Thomas A, Kolan KCR, Leu MC (2017) Freeform extrusion fabrication of titanium fiber reinforced 13-93 bioactive glass scaffolds. J Mech Behav Biomed Mater 69:153–162

# Chapter 6
# Micro/Nano Scaffolds for Osteochondral Tissue Engineering

**Albino Martins, Rui L. Reis, and Nuno M. Neves**

**Abstract** To develop an osteochondral tissue regeneration strategy it is extremely important to take into account the multiscale organization of the natural extracellular matrix. The structure and gradients of organic and inorganic components present in the cartilage and bone tissues must be considered together. Another critical aspect is an efficient interface between both tissues. So far, most of the approaches were focused on the development of multilayer or stratified scaffolds which resemble the structural composition of bone and cartilage, not considering in detail a transitional interface layer. Typically, those scaffolds have been produced by the combined use of two or more processing techniques (microtechnologies and nanotechnologies) and materials (organic and inorganic). A significant number of works was focused on either cartilage or bone, but there is a growing interest in the development of the osteochondral interface and in tissue engineering models of composite constructs that can mimic the cartilage/bone tissues. The few works that give attention to the interface between cartilage and bone, as well as to the biochemical gradients observed at the osteochondral unit, are also herein described.

**Keywords** Multiscale organization · Multilayer or stratified scaffolds · Biochemical gradients · Osteochondral interface

---

A. Martins (✉) · R. L. Reis · N. M. Neves
3B's Research Group—Biomaterials, Biodegradables and Biomimetics, University of Minho, Headquarters of the European Institute of Excellence on Tissue Engineering and Regenerative Medicine, Barco, Portugal

ICVS/3B's—PT Government Associate Laboratory, Barco, Guimarães, Portugal
e-mail: amartins@dep.uminho.pt

## 6.1 Introduction

The osteochondral (bone to cartilage) interface plays a critical role is the physiology of joints, since it is the anchorage site of hyaline articular cartilage and subchondral bone. In addition, it provides the mechanical structure to support the energy transfer of biomechanical movements from the joint to the skeleton. Unfortunately, damaged osteochondral tissue is difficult to treat due to the poor regenerative capacity of hyaline cartilage. The presence of complex biological and chemical gradients from the cartilage surface to the underlying subchondral bone is also difficult to recover from injury. As a result, interfacial tissue engineering (TE) has focused on overcoming challenges of connecting various dissimilar tissue types in an effort to better match physiological, biomechanical, and biochemical signaling properties [1].

In the development of scaffolds for osteochondral tissue engineering, efforts have been made to develop or improve new or combined processing strategies to obtain repeatable porous constructs with controlled porous morphology, preferably at different scale levels (e.g.. combination of nanoelements and microelements or pores), comprising different materials that are spatially organized and having the capability to deliver relevant molecules such as growth factors in a controlled way [2]. Such scaffolds have been designed to address particular aspects of the osteochondral tissue, namely the vascularization, the deposition of calcium phosphates in predefined regions, the guidance of regeneration certain directions (through gradient delivery of factors or anisotropic porous architecture), the development of different tissues (i.e., osteochondral defects), or the inhibition of calcification and cell adhesion. Recent achievements on the development of osteochondral scaffolds, facing the abovementioned aspects, are described in this book chapter.

## 6.2 The Multiscale Organization of the Osteochondral Extracellular Matrix

The extracellular matrix (ECM) throughout osteochondral tissue, which is itself secreted and modulated by the encapsulated chondrocytes, presents complex gradients of biochemical cues, such as varying concentrations of glycosaminoglycans and glycoproteins within each region of the tissue, or biophysical (topographical and mechanical) cues, such as nanosized, spatially patterned interactions (with a periodicity of 67 nm) provided by mechanically robust collagen fibers (Fig. 6.1) [3].

Hyaline cartilage is a stratified, multilayered tissue that is anchored to the subchondral bone. Both the phenotype and orientation of the cells (chondrocytes), and the composition and architecture of the extra cellular matrix (ECM) varies substantially along the depth of this complex tissue. The complex architecture throughout the articular cartilage to the subchondral bone interface that constitutes osteochondral tissue spans millimeter (macro) through to nanometer length-scales (Fig. 6.1) [4].

# 6 Micro/Nano Scaffolds for Osteochondral Tissue Engineering

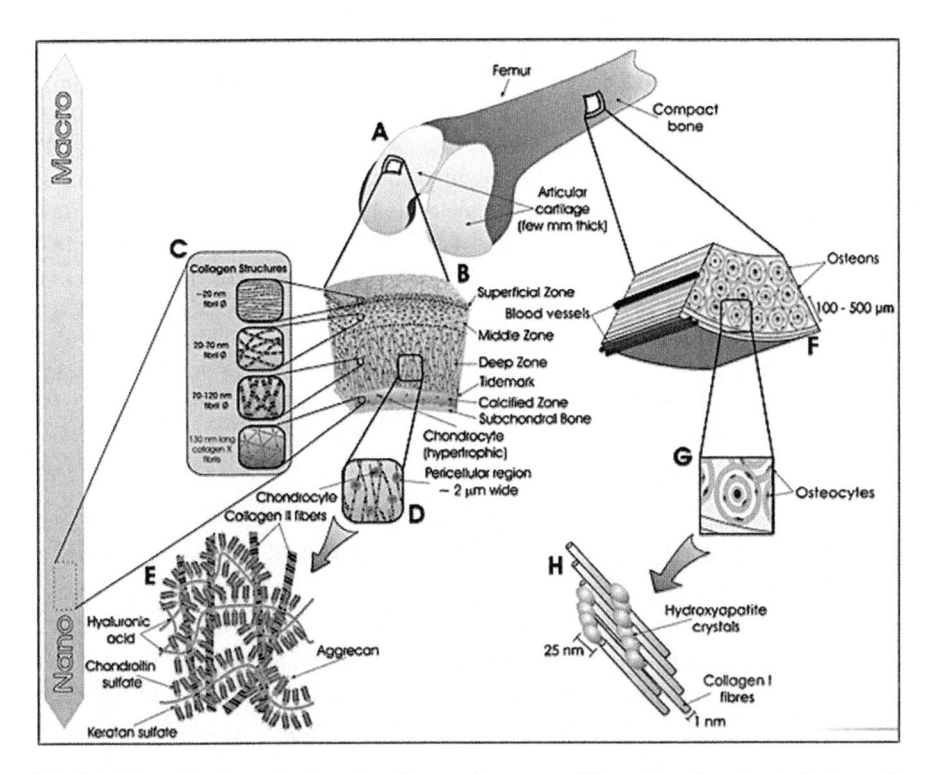

**Fig. 6.1** Hierarchical organization of cartilage and bone over different length scales. Articular cartilage forms a wear-resistant, load-bearing surface that covers bone in diarthrodial joints (**a**). It is organized into distinct zones (**b**) where the organization of the collagen structures varies considerably (**c**). Resident chondrocytes (**d**) are surrounded by super-aggregates of aggrecan/hyaluronic acid and macrofibrillar collagen networks (**e**). Bone mineralizes to form a calcified outer compact layer, which comprises many cylindrical Haversian systems or osteons (**f**). The osteocytes within these systems (**g**) are surrounded by the well-defined nanoarchitecture of the ECM—a dense network of aligned collagen I fibers, which provide templates for the self-assembly of hydroxyapatite crystals (**h**) [5]

At the macroscale, adult articular cartilage is a multizonal material in which three layers (i.e., superficial, middle, and deep zones), accounting for different ECM composition, orientation and cell phenotypes, can be distinguished. The uppermost superficial zone of cartilage is characterized by squamous chondrocytes surrounded by collagen fibrils aligning parallel to the articular surface. In the middle/intermediate zone, rounded chondrocytes are embedded in collagen fibrils less organized relative to the surface. In the deep zone, vertical columns of chondrocytes and collagen fibrils are organized perpendicular to the articular surface. The highest concentration of proteoglycans is found in the deep zone [6]. The base of the deep zone displays the *tidemark* that represents the onset of the calcified area, serving as a transitional zone between the soft cartilaginous tissue and underlying hard bone. The calcified area is rich in hydroxyapatite and alkaline phosphatase and poor in chondrocyte number, serves as an interface between the soft cartilage and the subchondral bone and defines a gradient in mechanical properties between these two tissues [3]. Below the deep zone is the subchondral bone plate. Subchondral bone is a nanocomposite material composed of glycoproteins, such as collagen, laminin, and fibronectin, and hydroxyapatite (HA). Underneath the subchondral bone plate, the subchondral trabecular bone, accounts for a spongy-like structure that is highly vascularized. Trabecular bone is a cellular solid with an interconnected porous structure. The pores have ~1 mm of diameter and the walls (trabeculae) have few micrometres in thickness. The pores appear aligned in the direction of the applied load and the structure is formed by various cell types (i.e., osteocytes, osteoblasts, and osteoclasts), ECM and vasculature (the bone marrow) [3]. Bone is vascularized as well as innervated, and others cells such as neurons and endothelial cells are also present and may play a relevant role in bone biology. In fact, it is generally considered that bone vascularization itself is one of the main reasons for the active self-repair capacity of bone [6].

## 6.3 Scaffold Properties for Osteochondral Tissue Regeneration

In tissue engineering approaches, to restore function or regenerate tissues, one needs a template—a scaffold—that will act as a temporary matrix for cell proliferation and ECM deposition [7]. Moreover, the scaffold also acts as a template for the neo-tissue vascularization and can actively participate in the regenerative process through the release of growth/differentiation factors [8]. In this sense, a 3D scaffold can influence the structure and development of the engineered tissue [9].

The selection of the most appropriate biomaterial to produce a scaffold for bone, cartilage or osteochondral tissue engineering applications is a very important step. The physicochemical properties of the biomaterial will determine, to a great extent, its choice aiming to target a defined tissue composition. After selecting the adequate biodegradable polymer, the next step is to develop or choose an adequate processing method [10]. The selected processing method should not affect the biomaterials properties and characteristics, namely their biocompatibility or chemical properties. The processing method should be accurate and reproducible, regarding pore size, distribution, and interconnectivity. That means that different scaffold batches should exhibit minimal variations in their properties, when processed from the same set of processing parameters and conditions.

Besides the choice of adequate biomaterials and processing method, the architecture of the processed scaffold is an important factor to take into consideration. Indeed, the macrostructural, microstructural, and nanostructural properties of the biomaterials can modulate biological response and the clinical success of the scaffold. Such properties affect cell adhesion, expansion, and also their gene expression and the preservation of their phenotype [7]. In general, all fabrication technologies aim to incorporate a hierarchical element, often in the form of controlled porosity or aligned structures, to imitate the native tissue spatial architecture [11].

Scaffolds should be biocompatible, well integrated in the host's tissue without eliciting an immune response [12]. Scaffolds must possess an open pore, fully interconnected geometry in a highly porous structure with large surface area. This will allow cell in-growth, an accurate cell distribution throughout the porous structure, and will allow the neovascularization of the construct [13]. Porosity and interconnectivity are also important for an accurate diffusion of nutrients, gases and to remove metabolic waste resulting from the cell metabolism. This is of particular importance regarding bone tissue engineering, particularly due to the high rates of mass transfer, even under in vitro culture conditions [14]. The porosity always influences other properties of the scaffolds such as the mechanical stability. This property should always be balanced with the mechanical needs of the particular tissue that is going to be regenerated. Adequate pore size is also important since if the pores employed are too small, pore occlusion by the cells may happen. This will allow cellular penetration, ECM production and neovascularization of the inner areas of the scaffold [15].

The surface properties, both chemical and topographical, can control and affect cellular adhesion and proliferation [16]. Chemical properties are related with the ability of cells to adhere to the biomaterial, as well as with the protein adsorption. Topographical properties are of particular interest when the topic is osteoconduction. The scaffold should also be osteoinductivity that is able to support formation of bone within and/or upon the scaffold [17].

The mechanical properties and biodegradability also have an important role. In vitro, the scaffolds should have adequate mechanical strength to withstand the hydrostatic pressures and to maintain the spaces required for cell in-growth and matrix production. In vivo, and because cartilage and bone tissue are always under continuous stress, the mechanical properties of the implanted construct should ideally be compatible with those of living cartilage and bone, so that an early mobilization of the injured site can be possible. Furthermore, the scaffolds degradation rate must be appropriate to the growth rate of the neotissue, in such a way that by the time the injury site is totally regenerated the scaffold is totally degraded [18].

## 6.4 Micro/Nano Scaffolds for Bone Tissue, Envisioning Osteochondral Regeneration

Recently, additive manufacturing techniques have been employed to develop hierarchical and functionally graded scaffolds. These technologies are characterized by reproducible and highly organized microarchitecture with patient-specific geometry through precise control over scaffold design and structure (porosity, pore size, and interconnectivity), while allowing for the incorporation of bioactive factors rendering the fabricated scaffolds more biomimetic [19].

The incorporation of microscale and nanoscale features on a same scaffold can improve both the mechanical properties and tissue regeneration, through toughening mechanics and better cell adhesion, respectively. Particularly, the multiscale network

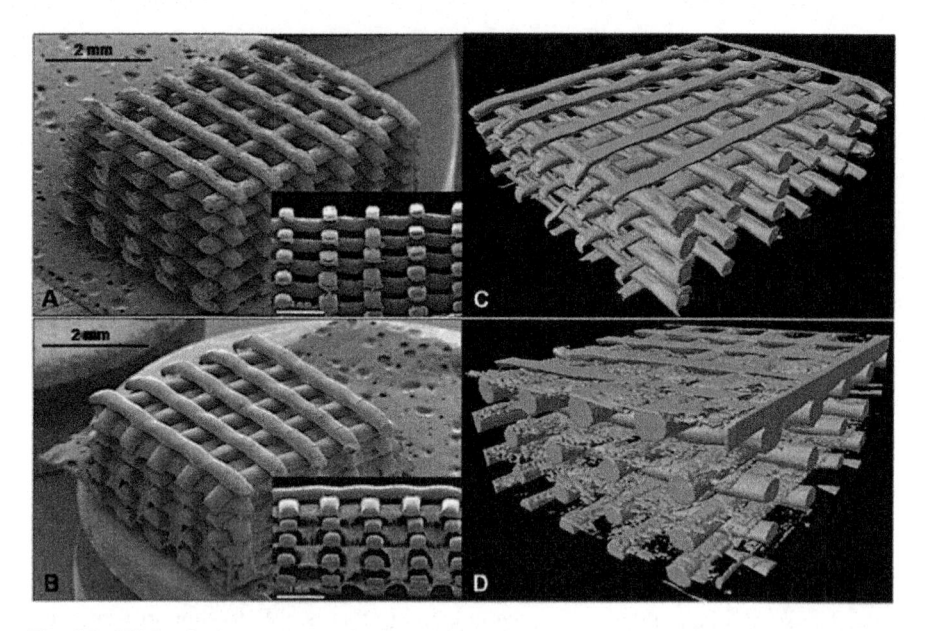

**Fig. 6.2** SEM and micron computed tomography analysis of the starch-based rapid prototyped (**a, c**) and hierarchical fibrous scaffolds (**b, d**). Reproduced with permission from [21]

observed in natural ECMs can be fabricated by the combination of additive manufacturing (AM) and electrospinning (ES) techniques to produce bimodal scaffolds [20]. The resultant multiscale scaffold contained large pore size essential for cell and mass transportation, while the fibrous component provided suitable structures for cell attachment. Moreover, while the 3D rapid prototype scaffold provides structural integrity and mechanical properties, the micro–nano scale of the electrospun fibers mimic the biophysical structure of natural ECM (Fig. 6.2) [21]. Biological results, when human osteobastic cells were dynamically seeded on these hierarchical fibrous scaffolds, showed significantly higher proliferation and maturation. Particularly, scanning electron microscopy (SEM) observation demonstrated that the osteoblastic cells preferentially adhered and spread on the electrospun NFMs, constituting an innovative strategy to enhance cell seeding efficiency/cell adhesion into the microfibrous scaffolds. In a complementary approach, the same hierarchical fibrous scaffolds were able to provide a favorable environment for the proliferation and osteogenic differentiation of human Wharton's jelly derived stem cells [22]. Biochemical data demonstrated that these constructs were in an early mineralization process, because of a significant higher fold change of osteogenic genes typically expressed in the mineralization phase, as well as the identification of calcium and phosphorous elements.

The combination of electrospun nanofibers with microscale to macroscale fibers, processed by other polymer processing techniques (i.e., wet-spinning and fiber extrusion), was been explored by our research group. In a first and simplest approach, electrospun nanofibers were directly deposited over a prefabricated wet-spun microfibrous scaffold [23, 24]. This combined structure was obtained by a two-step methodology and structurally consist of a nano-network incorporated on a macro-fibrous support. Its biological functionality was demonstrated by the culturing of human osteoblast-like cells, bone marrow stromal cells, and endothelial cells (i.e., human umbilical vein endothelial cells and microvascular endothelial cells) [24–26]. This micro/nano structure was developed to mimic the highly organized fibrous structure of bone tissue, not forgetting the vascular network that is identified as the main pitfall in bone tissue engineering and the major hurdle for the clinical application of engineered constructs.

With the intent to reproduce not only the multiscale organization of osteochondral tissue, but also its organic–inorganic composition, it was proposed the development of biphasic scaffolds comprising a polycaprolactone (PCL) cartilage phase and a PCL-tricalcium phosphate (TCP) matrix that served as the bone component. The scaffolds were built using the fused deposition modeling (FDM) process, seeded with mesenchymal stem cells (MSCs) via fibrin encapsulation, and patched with a 20% PCL-collagen electrospun mesh to prevent cell loss and facilitate the diffusion of nutrients from the synovial space [27]. Implantation of such scaffold in a critical size defect, which was created in the medial condyle of the rabbit model, indicated favorable outcomes in the cartilage region, with a reduced incidence of fibrocartilage and improved GAG content when compared to cell-free and mesh-free scaffolds. Furthermore, besides the implant structure and composition, the implantation site appeared to affect the in vivo outcomes (medial condyle vs. patellar groove).

In a similar attempt to create a biphasic scaffold, with a bone and periodontal compartment, FDM was used in addition to an in-house developed melt electrospinning device [28]. Medical grade PCL-TCP membrane scaffolds, acting as the bone compartment, were fabricated using FDM and then coated with calcium phosphate (CaP), while the periodontal compartment was electrospun through a melt electrospinning device. A biphasic scaffold was then assembled by compressing a partially fused CaP-coated bone compartment (FDM scaffold) onto a periodontal compartment (melt electrospun mesh). Subcutaneous implantation of the biphasic scaffold in rats confirmed tissue integration between both compartments, forming a tissue structurally resembling native periodontal tissues, establishing high levels of vascularization and tissue orientation in both bone and periodontal compartments. Despite the dissimilarity between potential tissue engineering applications, this work presents a relevant approach on the development of complex tissues adjacent to bone, such as the osteochondral tissue.

## 6.5 Micro/Nano Scaffolds for Cartilage Tissue, Envisioning Osteochondral Regeneration

So far, most part of these stratified scaffolds were developed envisioning their application in the regeneration of bone tissue. However, the hierarchical organization of collagen fibers is cartilage has been also addressed by the combination of AM with ES. The first approaches on producing 3D micro–nano fibrous scaffolds involve the intercalation of electrospun nanofibers, prefabricated or deposited on time, in between rapid prototyped microfibers [29–32]. Briefly, PCL or PCL/collagen nanofiber meshes were directly electrospun over rapid prototyped (i.e., by 3D plotting or by direct polymer melt deposition) microfibers, during shorter deposition periods. Cell culture experiments demonstrated the preferential adhesion of bovine or porcine primary chondrocytes to the electrospun nanofiber matrices, as well as a statistically significant increment of cell proliferation on the micro–nano fibrous scaffolds.

In another attempt, nanostructured porous polycaprolactone (NSP-PCL) scaffold were produced by the combination of rapid prototyping and thermally induced phase separation methods [33]. The NSP-PCL scaffold expresses macro, micro, and nanopores to benefit mechanical strength, chondrocyte adherence, viability, and differentiation. When implanted in an osteochondral rabbit model, the NSP-PCL scaffold design promotes cartilage ingrowth, but not bone ingrowth.

Keeping the combined use of FDM and ES, a multiphasic scaffold was developed comprising a biphasic PCL scaffold which pores were filled with a 2% alginate hydrogel [34]. To integrate the alginate and PCL components, the alginate hydrogel was partially decrosslinked and press-fitted on top of the biphasic scaffold, which enabled alginate to partially infiltrate the pores of the PCL-FDM scaffolds, and then recrosslinked. Histological analysis of the constructs implanted subcutaneously in rats showed that some alginate constructs had been separated from the PCL scaffolds possibly due to gradual weakening of the interface region.

Another example is the alternation of electrospun PCL fibers with 3D inkjet printing of rabbit chondrocytes in a fibrin–collagen hydrogel, which resulted in 1 mm thick five-layer tissue constructs [35]. The hybrid scaffold demonstrated enhanced mechanical properties compared to conventional hydrogel constructs generated using inkjet printing alone. Furthermore, these tissue constructs produced cartilage-specific ECM both in vitro and in vivo (subcutaneous implantation in immunodeficient mice), as evidenced by the deposition of type II collagen and glycosaminoglycans. This work demonstrated that the combination of controllable scaffold properties with a cell delivery printing process would enable the production of highly functional tissue constructs.

## 6.6 Micro/Nano Scaffolds for Osteochondral Regeneration

Despite the large amount of works reporting the development of micro–nano scaffolds for bone or cartilage tissue engineering approaches, few works addressed the repair of osteochondral defects. Scaffolds targeting the repair of full-thickness osteochondral tissue have combined diverse types of materials such as hydrogels or porous sponges, mimicking the "articular cartilage region," with porous or fibrous rigid scaffolds (made from polymeric or inorganic ceramic-type materials (or combinations of both)) to mimic the "bone region" [36]. As an example, porous bilayered scaffolds produced by freeze-drying and salt leaching techniques, and built up by fully integrating a silk fibroin (SF) layer and a silk-nanoCaP layer, were

6 Micro/Nano Scaffolds for Osteochondral Tissue Engineering 133

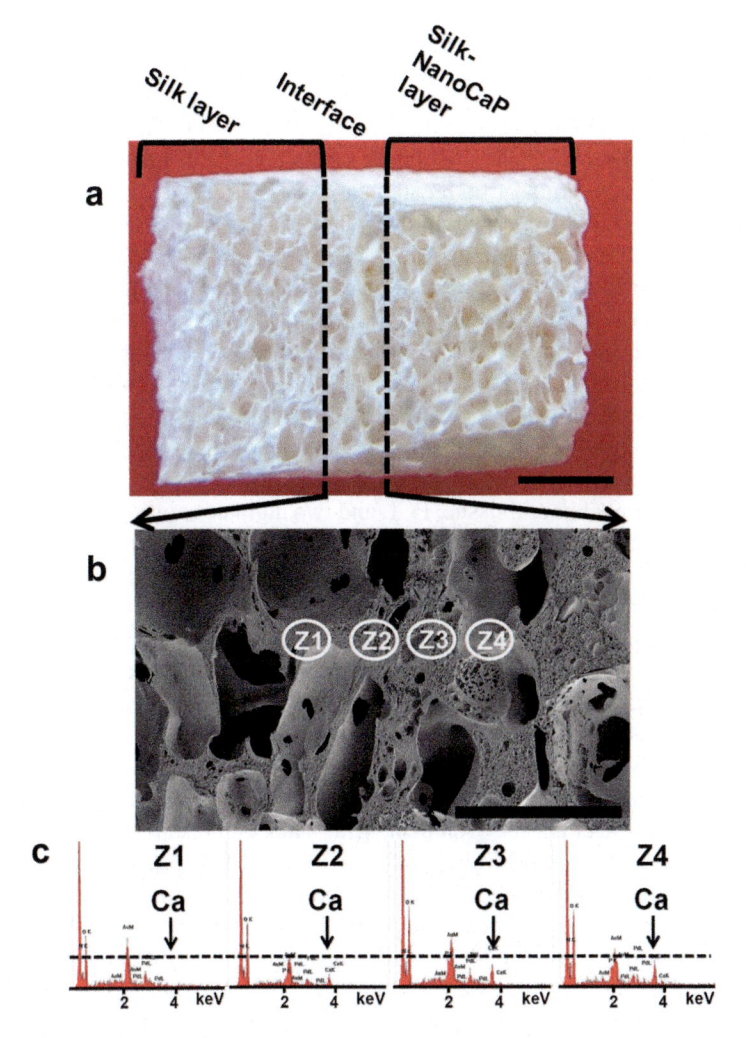

**Fig. 6.3** The interface of the bilayered scaffolds. (**a**) Macroscopic image of the bilayered scaffolds (scale bar: 3 mm). (**b**) SEM image of the interface region in the bilayered scaffold (scale bar: 500 μm). Z1, Z2, Z3, and Z4 indicate different regions from the silk layer to the silk-nanoCaP layer, around the interface area. (**c**) EDX elemental analysis of calcium ions in Z1, Z2, Z3, and Z4 regions [37]

developed for osteochondral tissue regeneration (Fig. 6.3) [37]. The silk-nanoCaP layer of the bilayered scaffolds promoted better osteogenesis differentiation of rabbit bone marrow mesenchymal stromal cells under osteogenic conditions as compared with the SF layer. Furthermore, these scaffolds allowed tissue ingrowth and induced only a very weak foreign body reaction when subcutaneously implanted in rabbit. When implanted in a rabbit knee critical defect, the bilayered scaffolds supported cartilage regeneration in the top silk layer, and encouraged large amounts of subchondral bone ingrowth and angiogenesis in the bottom silk-nanoCaP layer. Using a different processing approach, taking advantage of the sol-derived 70S bioactive glass and of silk fibroin (Indian non-mulberry *Antheraea assama*), a bilayer electrospun mats were proposed to the repair of osteochondral defects [38]. In vitro biological studies revealed that the biphasic mats presented spatial confinement for the growth and maturation of both osteoblasts (MG63 cell line) and chondrocytes (primary porcine ear-derived chondrocytes).

Despite these particular studies based on the use of silk fibroin as biomaterial scaffold, most of the reports relies on the use of additive manufacturing techniques for the production of 3D osteochondral scaffolds [39]. The addition of multiple printing techniques and novel scaffold designs may give rise to advanced 3D printing technologies capable of fabricating higher quality scaffolds for cartilage tissue engineering. Specifically, a multi-head tissue/organ building system (MtoBS) that enabled dispensing of biologically relevant biomaterials, such as PCL and alginate hydrogel, was developed to manufacture 3D tissue and organs [40]. Envisioning the building of an osteochondral tissue, PCL and two alginate solutions with osteoblastic and chondrocytic cell lines were sequentially dispensed, keeping their viability up to 7 days. Considering these promising results, the MtoBS, which overcomes the drawbacks of current cell printing technology, constitutes an interesting method for dispensing multiple cells and biomaterials for heterogeneous tissue regeneration.

Another approach to generate osteochondral scaffolds also relies on the combination of novel nano-inks, composed of organic (i.e., chondrogenic transforming growth-factor beta 1) and inorganic (i.e., nanocrystalline hydroxyapatite) bioactive factors, with advanced tabletop stereolithography 3D printing technology [41]. A series of hierarchical constructs were successfully fabricated which closely mimic the native 3D extracellular environment, with nanocomponents, microarchitecture, and spatiotemporal controlled release of bioactive cues [1]. Experimental data demonstrated that these osteochondral scaffolds promote human bone marrow-derived MSCs adhesion, proliferation, and osteo and chondral differentiation. Also with the attempt to control the release kinetics of transforming growth factor-$\beta$3 (TGF-$\beta$3) and/or insulin-like growth factor-1 (IGF-1), biodegradable bilayered oligo (poly(ethylene glycol) fumarate) composite hydrogels were developed aiming to mimics the distinctive hierarchical structure of native osteochondral tissue [42]. It was achieved higher amounts of active TGF-$\beta$3 released when it was incorporated with gelatin microparticles, as compared to gel phase loading. Single delivery of IGF-1 showed higher scores in subchondral bone morphology, as well as chondrocyte and glycosaminoglycan amount in adjacent cartilage tissue of a rabbit full-thickness osteochondral defect model after 12 weeks, when compared to a dual delivery of IGF-1 and TGF-$\beta$3. The lack of synergy between IGF-1 and TGF-$\beta$3, regardless of TGF-$\beta$3 release kinetics, demonstrates that the dual delivery of GFs does not necessarily confer an improved healing response over the single delivery of GFs in vivo.

In another work that combines 3D bioprinting with multi-nozzle electrospinning, osteochondral scaffolds with multiscale structures are capable of controlling release of multiple biomolecules, namely gentamycin sulfate (GS) and desferoxamine (DFO) [43]. Blend electrospun GS/polyvinyl alcohol (PVA) and coaxial electrospun core PVA-DFO/shell PCL fibers were deposited in between gelatin/sodium alginate struts. The composite scaffold showed its potential to delivery multiple biomolecules with various release profiles over space and time, achieving functional gradient osteochondral scaffolds. In another dual-release approach, a hybrid twin-screw extrusion and electrospinning process was developed for generating osteochondral tissue engineering scaffolds with controlled gradations of concentrations of insulin and β-GP) [44]. In this demonstrative study, the concentration of insulin increased from one side of scaffold to the other, whereas β-GP phosphate concentration decreased. The use of both insulin and β-GP at graded concentrations led to the differentiation of human adipose-derived stromal cells (ADSCs) in a location-dependent manner: higher chondrocytic cell counts and increasing total collagen deposition with increasing concentration of insulin, and different extents of mineralization generated by the β-GP concentration distribution.

Very recently, a biomimetic osteochondral scaffold with continuous multilayer architecture and gradient composition, made of PCL and hydroxyapatite (HA)/PCL microspheres, was also produced via selective laser sintering technique [45]. In vitro and a rabbit osteochondral defect model demonstrated that the multilayer scaffold could successfully induce the formation of multiple tissue types, including articular cartilage and subchondral bone. Due to its controllable forming process, flexible structural design, tailored composition and tunable biomechanical properties, this multilayer scaffold provides a successful platform for the enhanced repair of osteochondral defects.

## 6.7 Concluding Remarks and Future Trends

The repair of osteochondral defects requires a tissue engineering approach that aims at mimicking the physiological properties and structure of two different tissues (i.e., cartilage and bone) using specifically designed scaffold–cell constructs [39]. Furthermore, the transitional zone between these two tissues, i.e., the tidemark, should also be considered in this approach. While polymeric (or even composite) materials offer many possibilities to the field of tissue engineering, they inherently lack the plethora of biological cues provided by the native tissue microenvironment, through cell–ECM and cell–cell communication that facilitate tissue remodeling and repair.

Engineering interactions between culturing cells and biomaterial scaffolds (the "interactome") through mimicry of the hierarchical nature of the native ECM is potentially of great relevance to eliciting control over the molecular and structural cues capable of determining cell fate decisions (e.g., cell migration, proliferation, differentiation, and apoptosis) and neo-tissue formation to achieve functional osteochondral tissue repair [3]. Therefore, in the recent past, multilayered or stratified scaffolds targeting osteochondral repair consisted of only two distinct zones resembling the bone–cartilage interface either chemically, mechanically, or structurally [36]. There are multiple ways to achieve stratification and gradient-based composition. One simple approach is to build composite scaffolds through multilayered scaffold design, to generate structural templates for the cartilaginous layer, the tidemark and calcified cartilage, and the subchondral bone [46]. Such complex but necessary structure is usually accomplished by using two or more different materials. However, these approaches lacked the ability to mimic the architecture of articular cartilage, leading to isotropic cartilaginous tissues that fail to resemble the structure and depth-dependent characteristic of native tissue, and consequently, its mechanical properties.

The scientific literature has shown that the structural stratification alone is not sufficient for establishing effective transition between two tissues as different as cartilage and bone, prompting the need to also establish biochemical gradients, particularly in the interface region [47]. Such biochemical gradients can be achieved by embedding the growth factors, non-growth factor inductive agents (e.g., hydroxyapatite) and other signaling molecules (therapeutic drugs, genes) into the scaffold. Indeed, osteochondral scaffolds are being designed to facilitate tissue-specific growth-factor delivery, mimic connective tissue ECM, be chondroinductive or osteoinductive, and recapitulate the stratified nature of the osteochondral tissue through multiphasic designs [36].

Despite the success of numerous proof-of-concept studies, it is not clear from an accumulation of successes and failure paradigms that the herein described approaches recapitulate the intricate hierarchical organizations of physical structures found in native ECM environments [11]. From our perspective, the successful development of an osteochondral tissue engineering strategy is dependent on the specific biological mechanisms under investigation, which determine the level of complexity. This may vary from the simplest coculture systems to complex bioreactors to generate close-to-native osteochondral constructs, which may have the capability of incorporating other joint tissues, such as the vasculature. Therefore, future attempts to replicate the biological organization of cartilage interfaced with bone may be achieved by recapitulating in vitro key aspects of the in vivo developmental biology.

## References

1. Castro NJ, Patel R, Zhang LJG (2015) Design of a Novel 3D printed bioactive nanocomposite scaffold for improved osteochondral regeneration. Cell Mol Bioeng 8(3):416–432
2. Mano JF, Silva GA, Azevedo HS, Malafaya PB, Sousa RA, Silva SS, Boesel LF, Oliveira JM, Santos TC, Marques AP, Neves NM, Reis RL (2007) Natural origin biodegradable systems in tissue engineering and regenerative medicine: present status and some moving trends. J R Soc Interface 4(17):999–1030
3. Camarero-Espinosa S, Cooper-White J (2017) Tailoring biomaterial scaffolds for osteochondral repair. Int J Pharm 523(2):476–489
4. Stevens MM, George JH (2005) Exploring and engineering the cell surface interface. Science 310(5751):1135–1138. https://doi.org/10.1126/science.1106587
5. Mwenifumbo S, Shaffer MS, Stevens MM (2007) Exploring cellular behaviour with multi-walled carbon nanotube constructs. J Mater Chem 17(19):1894–1902
6. Alexander PG, Gottardi R, Lin H, Lozito TP, Tuan RS (2014) Three-dimensional osteogenic and chondrogenic systems to model osteochondral physiology and degenerative joint diseases. Exp Biol Med 239(9):1080–1095

7. Ingber DE, Mow VC, Butler D, Niklason L, Huard J, Mao J, Yannas I, Kaplan D, Vunjak-Novakovic G (2006) Tissue engineering and developmental biology: going biomimetic. Tissue Eng 12(12):3265–3283
8. Bessa PC, Casal M, Reis RL (2008) Bone morphogenetic proteins in tissue engineering: the road from laboratory to clinic, part II (BMP delivery). J Tissue Eng Regen M 2(2–3):81–96
9. Hutmacher DW, Schantz JT, Lam CXF, Tan KC, Lim TC (2007) State of the art and future directions of scaffold-based bone engineering from a biomaterials perspective. J Tissue Eng Regen M 1(4):245–260
10. Petite H, Viateau V, Bensaid W, Meunier A, de Pollak C, Bourguignon M, Oudina K, Sedel L, Guillemin G (2000) Tissue-engineered bone regeneration. Nat Biotechnol 18(9):959–963
11. Abbah SA, Delgado LM, Azeem A, Fuller K, Shologu N, Keeney M, Biggs MJ, Pandit A, Zeugolis DI (2015) Harnessing hierarchical Nano- and Micro-fabrication Technologies for Musculoskeletal Tissue Engineering. Adv Healthc Mater 4(16):2488–2499
12. Hutmacher DW (2000) Scaffolds in tissue engineering bone and cartilage. Biomaterials 21(24):2529–2543
13. Karageorgiou V, Kaplan D (2005) Porosity of 3D biomaterial scaffolds and osteogenesis. Biomaterials 26(27):5474–5491
14. Stevens B, Yang YZ, Mohanda SA, Stucker B, Nguyen KT (2008) A review of materials, fabrication to enhance bone regeneration in methods, and strategies used engineered bone tissues. J Biomed Mater Res B 85b(2):573–582
15. Fang TD, Salim A, Xia W, Nacamuli RP, Guccione S, Song HM, Carano RA, Filvaroff EH, Bednarski MD, Giaccia AJ, Longaker MT (2005) Angiogenesis is required for successful bone induction during distraction osteogenesis. J Bone Miner Res 20(7):1114–1124
16. Boyan BD, Hummert TW, Dean DD, Schwartz Z (1996) Role of material surfaces in regulating bone and cartilage cell response. Biomaterials 17(2):137–146
17. Riminucci M, Bianco P (2003) Building bone tissue: matrices and scaffolds in physiology and biotechnology. Braz J Med Biol Res 36(8):1027–1036
18. Gomes ME, Godinho JS, Tchalamov D, Cunha AM, Reis RL (2002) Alternative tissue engineering scaffolds based on starch: processing methodologies, morphology, degradation and mechanical properties. Mat Sci Eng C-Bio S 20(1–2):19–26
19. Costa PF, Martins A, Neves NM, Gomes ME, Reis RL (2014) Automating the processing steps for obtaining bone tissue-engineered substitutes: from imaging tools to bioreactors. Tissue Eng Part B-Re 20(6):567–577. https://doi.org/10.1089/ten.teb.2013.0751
20. Dalton PD, Vaquette C, Farrugia BL, Dargaville TR, Brown TD, Hutmacher DW (2013) Electrospinning and additive manufacturing: converging technologies. Biomater Sci 1(2):171–185
21. Martins A, Chung S, Pedro AJ, Sousa RA, Marques AP, Reis RL, Neves NM (2009) Hierarchical starch-based fibrous scaffold for bone tissue engineering applications. J Tissue Eng Regen M 3(1):37–42
22. Canha-Gouveia A, Costa-Pinto AR, Martins AM, Silva NA, Faria S, Sousa RA, Salgado AJ, Sousa N, Reis RL, Neves NM (2015) Hierarchical scaffolds enhance osteogenic differentiation of human Wharton's jelly derived stem cells. Biofabrication 7(3). doi:10.1088/1758-5090/7/3/035009
23. Tuzlakoglu K, Bolgen N, Salgado AJ, Gomes ME, Piskin E, Reis RL (2005) Nano- and microfiber combined scaffolds: a new architecture for bone tissue engineering. J Mater Sci Mater Med 16(12):1099–1104. https://doi.org/10.1007/s10856-005-4713-8
24. Tuzlakoglu K, Santos MI, Neves N, Reis RL (2011) Design of Nano- and Microfiber Combined Scaffolds by electrospinning of collagen onto starch-based fiber meshes: a man-made equivalent of natural extracellular matrix. Tissue Eng Part A 17(3–4):463–473
25. Santos MI, Fuchs S, Gomes ME, Unger RE, Reis RL, Kirkpatrick CJ (2007) Response of micro- and macrovascular endothelial cells to starch-based fiber meshes for bone tissue engineering. Biomaterials 28(2):240–248. doi:S0142-9612(06)00693-4 [pii]. 10.1016/j.biomaterials.2006.08.006

26. Santos MI, Tuzlakoglu K, Fuchs S, Gomes ME, Peters K, Unger RE, Piskin E, Reis RL, Kirkpatrick CJ (2008) Endothelial cell colonization and angiogenic potential of combined nano- and micro-fibrous scaffolds for bone tissue engineering. Biomaterials 29(32):4306–4313
27. Swieszkowski W, Tuan BHS, Kurzydlowski KJ, Hutmacher DW (2007) Repair and regeneration of osteochondral defects in the articular joints. Biomol Eng 24(5):489–495
28. Costa PF, Vaquette C, Zhang QY, Reis RL, Ivanovski S, Hutmacher DW (2014) Advanced tissue engineering scaffold design for regeneration of the complex hierarchical periodontal structure. J Clin Periodontol 41(3):283–294
29. Kim G, Son J, Park S, Kim W (2008) Hybrid process for fabricating 3D hierarchical scaffolds combining rapid prototyping and electrospinning. Macromol Rapid Commun 29(19):1577–1581. https://doi.org/10.1002/marc.200800277
30. Moroni L, Schotel R, Hamann D, de Wijn JR, van Blitterswijk CA (2008) 3D fiber-deposited electrospun integrated scaffolds enhance cartilage tissue formation. Adv Func Mater 18:53–60. https://doi.org/10.1002/adfm.200601158
31. Park SH, Kim TG, Kim HC, Yang DY, Park TG (2008) Development of dual scale scaffolds via direct polymer melt deposition and electrospinning for applications in tissue regeneration. Acta Biomater 4(5):1198–1207
32. Thorvaldsson A, Stenhamre H, Gatenholm P, Walkenstrom P (2008) Electrospinning of highly porous scaffolds for cartilage regeneration. Biomacromolecules 9(3):1044–1049
33. Christensen BB, Foldager CB, Hansen OM, Kristiansen AA, Dang QSL, Nielsen AD, Nygaard JV, Bunger CE, Lind M (2012) A novel nano-structured porous polycaprolactone scaffold improves hyaline cartilage repair in a rabbit model compared to a collagen type I/III scaffold: in vitro and in vivo studies. Knee Surg Sport Tr A 20(6):1192–1204
34. Jeon JE, Vaquette C, Theodoropoulos C, Klein TJ, Hutmacher DW (2014) Multiphasic construct studied in an ectopic osteochondral defect model. J R Soc Interface 11(95):20140184
35. Xu T, Binder KW, Albanna MZ, Dice D, Zhao WX, Yoo JJ, Atala A (2013) Hybrid printing of mechanically and biologically improved constructs for cartilage tissue engineering applications. Biofabrication 5(1):015001
36. Jeon JE, Vaquette C, Klein TJ, Hutmacher DW (2014) Perspectives in multiphasic osteochondral tissue engineering. Anatomical Record-Advances in Integrative Anatomy and Evolutionary Biology 297(1):26–35
37. Yan LP, Silva-Correia J, Oliveira MB, Vilela C, Pereira H, Sousa RA, Mano JF, Oliveira AL, Oliveira JM, Reis RL (2015) Bilayered silk/silk-nanoCaP scaffolds for osteochondral tissue engineering: in vitro and in vivo assessment of biological performance. Acta Biomater 12:227–241
38. Christakiran MJ, Reardon PJT, Konwarh R, Knowles JC, Mandal BB (2017) Mimicking hierarchical complexity of the osteochondral Interface using electrospun silk bioactive glass composites. Acs Appl Mater Inter 9(9):8000–8013
39. Yousefi AM, Hoque ME, Prasad RGSV, Uth N (2015) Current strategies in multiphasic scaffold design for osteochondral tissue engineering: a review. J Biomed Mater Res A 103(7):2460–2481
40. Shim JH, Lee JS, Kim JY, Cho DW (2012) Bioprinting of a mechanically enhanced three-dimensional dual cell-laden construct for osteochondral tissue engineering using a multi-head tissue/organ building system. J Micromech Microeng 22(8):085014
41. Castro NJ, O'Brien J, Zhang LG (2015) Integrating biologically inspired nanomaterials and table-top stereolithography for 3D printed biomimetic osteochondral scaffolds. Nanoscale 7(33):14010–14022
42. Kim K, Lam J, Lu S, Spicer PP, Lueckgen A, Tabata Y, Wong ME, Jansen JA, Mikos AG, Kasper FK (2013) Osteochondral tissue regeneration using a bilayered composite hydrogel with modulating dual growth factor release kinetics in a rabbit model. J Control Release 168(2):166–178

43. Liu YY, Yu HC, Liu Y, Liang G, Zhang T, Hu QX (2016) Dual drug spatiotemporal release from functional gradient scaffolds prepared using 3D bioprinting and electrospinning. Polym Eng Sci 56(2):170–177. https://doi.org/10.1002/pen.24239
44. Erisken C, Kalyon DM, Wang HJ, Ornek-Ballanco C, Xu JH (2011) Osteochondral tissue formation through adipose-derived stromal cell differentiation on biomimetic Polycaprolactone Nanofibrous scaffolds with graded insulin and Beta-Glycerophosphate concentrations. Tissue Eng Pt A 17(9–10):1239–1252
45. Du YY, Liu HM, Yang Q, Wang S, Wang JL, Ma J, Noh I, Mikos AG, Zhang SM (2017) Selective laser sintering scaffold with hierarchical architecture and gradient composition for osteochondral repair in rabbits. Biomaterials 137:37–48
46. Gadjanski I, Vunjak-Novakovic G (2015) Challenges in engineering osteochondral tissue grafts with hierarchical structures. Expert Opin Biol Ther 15(11):1583–1599
47. Di Luca A, Van Blitterswijk C, Moroni L (2015) The osteochondral Interface as a gradient tissue: from development to the fabrication of gradient scaffolds for regenerative medicine. Birth Defects Research Part C-Embryo Today-Reviews 105(1):34–52

# Part III
# Osteochondral Tissue Scaffolding

# Chapter 7
# Mimetic Hierarchical Approaches for Osteochondral Tissue Engineering

**Ivana Gadjanski**

**Abstract** In order to engineer biomimetic osteochondral (OC) construct, it is necessary to address both the cartilage and bone phase of the construct, as well as the interface between them, in effect mimicking the developmental processes when generating hierarchical scaffolds that show gradual changes of physical and mechanical properties, ideally complemented with the biochemical gradients. There are several components whose characteristics need to be taken into account in such biomimetic approach, including cells, scaffolds, bioreactors as well as various developmental processes such as mesenchymal condensation and vascularization, that need to be stimulated through the use of growth factors, mechanical stimulation, purinergic signaling, low oxygen conditioning, and immunomodulation. This chapter gives overview of these biomimetic OC system components, including the OC interface, as well as various methods of fabrication utilized in OC biomimetic tissue engineering (TE) of gradient scaffolds. Special attention is given to addressing the issue of achieving clinical size, anatomically shaped constructs. Besides such neotissue engineering for potential clinical use, other applications of biomimetic OC TE including formation of the OC tissues to be used as high-fidelity disease/healing models and as in vitro models for drug toxicity/efficacy evaluation are covered.

**Highlights**

- Biomimetic OC TE uses "smart" scaffolds able to locally regulate cell phenotypes and dual-flow bioreactors for two sets of conditions for cartilage/bone
- Protocols for hierarchical OC grafts engineering should entail mesenchymal condensation for cartilage and vascular component for bone
- Immunomodulation, low oxygen tension, purinergic signaling, time dependence of stimuli application are important aspects to consider in biomimetic OC TE

**Keywords** Hierarchical scaffold · Bioreactor · Osteochondral · Cartilage, Bone

I. Gadjanski (✉)
BioSense Institute, University of Novi Sad, Dr Zorana Djindjica, Novi Sad, Serbia

Belgrade Metropolitan University, Tadeusa Koscuska 63, Belgrade, Serbia

© Springer International Publishing AG, part of Springer Nature 2018
J. M. Oliveira et al. (eds.), *Osteochondral Tissue Engineering*,
Advances in Experimental Medicine and Biology 1058,
https://doi.org/10.1007/978-3-319-76711-6_7

## 7.1 Introduction

In an osteochondral defect, the osteochondral (OC) unit is disturbed. Native, healthy OC unit is organized in a stratified, hierarchical way, with avascular/aneural cartilaginous zonal layer composed of chondrocytes embedded in the organic extracellular matrix (ECM), situated above the osseous, i.e., bone, part. Bone component comprises subchondral trabecular (cancellous) bone, highly vascular, enervated, with three different cell types (osteoblasts, osteocytes and osteoclasts) in the ECM composed of organic matrix and inorganic hydroxyapatite crystals. Osteochondral tissues are closely connected through the OC interface and function as one unit due to various mechanisms formed during development by the process of endochondral ossification.

In order to attempt any kind of reconstructing such a complex stratified structure comprised of vastly different components, a multifaceted approach needs to be implemented, which addresses both tissues as well as the connections between them, in effect mimicking the developmental processes.

Such approach is termed **biomimetic osteochondral tissue engineering (OC TE)** which aims to recapitulate in vitro the main elements of the in vivo development, i.e., of the endochondral ossification. In practice, this means fabrication of the stratified hierarchical constructs that should, ideally, achieve the structure of the native OC unit. This aim is proving to be very difficult, due to the complexity of the OC unit both from the developmental and structural aspects, particularly when the goal is to engineer living, clinically sized, physiologically stiff neotissue grafts, customized to the patient and to the defect requiring treatment.

Besides such neotissue engineering for potential clinical use, other applications of biomimetic OC TE include formation of the OC tissues to be used as high fidelity disease/healing models and as in vitro models for drug toxicity/efficacy evaluation [1, 2].

**There are several components to take into account in biomimetic OC TE:**

(a) **Cells**: type/source, differentiation protocols

- Cell-free techniques

(b) **Scaffolds**: biomaterials, architecture/design and microstructure, fabrication methods

- Scaffold-less techniques

(c) **Bioreactors:** design, parameters
(d) **Other components:**

- Growth factors
- Mechanical stimulation
- Purinergic signaling
- Low oxygen conditioning
- Immunomodulation
- OC interface engineering methods

Majority of these components are the same whether the goal is to engineer neotissue for potential clinical use or a model system for drug evaluation or disease modeling. However, for the former—the formation of large, clinically sized OC grafts composed of the neotissue, an additional aspect is preferred: **anatomical shape**. To this aim, the use of additive manufacturing and custom-tailored bioreactors is of particular importance. Conceptually, there are several ways these components can be implemented to achieve **stratified hierarchical structural organization in the engineered OC construct**, [3–6]:

1. Scaffold-free cartilage layer and scaffold for the bone layer
2. A different scaffold for each layer, including the OC interface: biphasic, triphasic, multilayered (particularly when mimicking the zonal structure of the cartilage layer)
3. A single heterogenous scaffold for the whole OC construct = scaffolds with morphological/physical gradients
4. A single homogenous scaffold for the whole OC construct = scaffolds with biochemical gradients

However, up to now, there was no defined scaffold structure and biomaterial that was able to meet all the necessary requirements for the formation of a native-like OC-tissue [1] which is why **the current state-of-the-art approach in OC TE is to use multilayered hybrid scaffolds, with biochemical, structural and mechanical gradients**.

The OC constructs can be either cell-free or loaded with the primary OC cells (chondrocytes/osteoprogenitors) or with cells with both chondrogenic and osteogenic capacity, i.e., stem and stromal cells.

## 7.1.1 Cells as Biomimetic System Component

For the small OC lesions, a cell-free approach might be implemented, where only scaffold and the growth factors are used to initiate localized repair and endogenous cell recruitment [7]. However, for the larger, unconfined OC defects, with lesions in the wound bed, the use of cells as part of the OC graft is necessary [3].

There are two main types of cells used in cellular therapy of osteochondral defects: **primary cells** (chondrocytes and osteoblasts-like, i.e., osteoprogenitor cells), preferably autologous and **stem/stromal cells**, autologous or allogeneic, isolated from various tissues. In this chapter, only the tissue engineering methods are covered, while different methods of cellular therapy of OC defects such as ACI (autologous chondrocyte implantation), MACI (Matrix-induced autologous chondrocyte implantation), and mosaicplasty are covered in detail elsewhere [6].

### 7.1.1.1 Chondrocytes

Even though these are the native cells of the cartilage, there are several drawbacks to the use of mature chondrocytes in OC TE: (1) harvesting is not very efficient due to following factors: i) very low number of chondrocytes in the native cartilage tissue—only 5% of total cartilage volume ~ 1 million cells/cm$^3$ [8]; (2) aggressive enzymatic procedure with collagenase needs to be performed to decompose the collagen from the extracellular matrix (ECM), which can also harm the cells (3) phenotype instability of chondrocytes in 2D (monolayer) cell culture [9] that is usually used in order to achieve high cell numbers: chondrocytes in the monolayer undergo dedifferentiation, stop expressing the chondrogenic markers (e.g., collagen II and aggrecan) and lose their distinctive spherical shape while attaining fibroblast-like morphology [10].

Various methods have been utilized in order to achieve maximum harvest yield with optimal cell viability while preserving the chondrocyte phenotype [11–13]. Majority of these are implemented on animal chondrocytes that are usually used as control cells in the experiments with engineered constructs.

### 7.1.1.2 Mesenchymal Stromal Cells (MSCs)

In spite of described drawbacks, chondrocyte-based cartilage tissue engineering remains a useful source of information, particularly when performed in combination with (human) **mesenchymal stromal cells (MSCs)**-based engineering methods [14].

Important to mention is that physiologic mechanical properties can be achieved when engineering cartilage from primary chondrocytes, while the highest compressive moduli reported for cartilage engineered from human MSCs (without enabling mesenchymal condensation—*see below*) was only ~ 50% of the normal values [15].

The use of human chondrocytes in OC TE is still largely prevented by additional challenges: donor-site morbidity and low ECM production in culture.

These challenges can be potentially overcome through the use of **human mesenchymal stromal cells (MSCs)** that possess a number of characteristics advantageous to the OC TE: (i) can be isolated from various sources with very low donor-site morbidity (e.g., from the adipose tissue); (ii) can maintain multipotency even after multiple passages and (iii) can be induced to both chondrogenesis and osteogenesis [16–18].

In addition, MSCs represent a very natural choice of cells for OC TE, since they originate from mesenchymal connective tissues of mesodermal nature that, in the course of development, give rise to all osteochondral components.

Here it is important to note the common confusion regarding the name of this type of cells. Minimal classification criteria for "mesenchymal stem cells" were established by the International Society for Cell Therapy (ISCT): A) plastic adherence B) osteogenic, chondrogenic, and adipogenic differentiation in the bulk culture (not on a single-cell clone) C) cell surface expression of CD73, CD90, and CD105 concurrent with absent expression of CD11b or CD14, CD45, CD34, CD79a, or CD19, and human leukocyte antigen (HLA)-DR [19].

The problem is that only a minority of the cells in the bulk culture (less than half of total cell number) that fulfill these criteria (A-C) also exhibit: (1) high proliferative capacity (colony-forming ability—CFU-F) [20] and (2) multipotency (when appropriately tested on the basis of a single cell clone) [21].

This is why it is not accurate to use the term "mesenchymal stem cells" for the bulk cell population, which is exactly what happens in majority of the tissue engineering studies: the term is non-critically extended to all fibroblast-like cells obtained after one or more culture passages starting from primary bone marrow (and later adipose tissue, cord blood, umbilical cord) mononuclear cells [21].

The only way to detect multipotent stem cells in the bulk population is to assay their colony-forming capacity (CFU-F) according to the initial Friedenstein's functional definition [18, 22]. Only the cells that are able to give rise in vitro to fibroblast colonies (i.e., possess CFU-F ability) can be called stem cells, provided they also exhibit another property: multipotency. If the cells of one single colony are capable of giving rise to at least three cell types (adipocytes, osteoblasts and chondrocytes) then the initial cell that gave rise to the colony was multipotent.

Interestingly, when individual clones were analyzed for their proliferative and differentiation capacities, data showed that only ~34% of CFU-F cells exhibit trilineage potential, ~60% osteogenic and chondrogenic, while 6% can differentiate into only one line (these are termed "committed progenitors") [21, 23].

Different methods were used to select for the "real mesenchymal stem cells", by concentrating the CFU-F in some phenotypically defined populations, but they only allowed enriching of the "real MSC" population to a limited extent.

In conclusion, in order to term cells as multipotent mesenchymal stem cells they need to fulfill two basic conditions: be able to form clonogenic colonies (CFU-F) and differentiate into osteogenic, chondrogenic, and adipogenic lineages. If the CFU capacity has not been evaluated, the most accurate is to term the cells as **Mesenchymal Stromal Cells.**

As mentioned, majority of the tissue engineering studies performed up to now did not pay attention to these aspects. This renders most of the results difficult to transfer to a clinical setting because the conclusions drawn from such studies do not reflect the behavior of the "real stem cells". In fact, the use of such heterogenous populations of mesenchymal cells, without preselection for CFU-F, led to various results such as generation of fibrocartilage and hypertrophic chondrocytes [24] and even non-articular cartilage formation within the defect, after implantation [25, 26].

Based on the above, the abbreviation MSCs in this chapter refers to the **mesenchymal stromal cells**. It is worth noting that there are initiatives (spearheaded by Dr. Arnold Caplan) to change the name of exogenously supplied MSCs (in clinical setting) to Medicinal Signaling Cells to more accurately reflect the fact that these cells home in on sites of injury or disease and secrete bioactive factors that are immunomodulatory and trophic (regenerative). These cells do not differentiate into neotissue, but stimulate via various biofactors the patient's own site-specific and tissue-specific resident stem cells and progenitors that construct the new tissue [27].

Concerning the use of MSCs for engineering the osseous component of the OC construct, the osteogenic capacity has been confirmed for MSCs derived from various sources, where the most used are bone marrow-MSCs (BMSCs) and adipose-derived stromal cells (ASCs). The other MSC types are covered in detail in an excellent review by Vonk et al. [18].

- BMSCs, isolated from bone marrow stroma are the most studied source for bone regeneration. One of the challenges associated with BMSCs use is high inter-patient variability in cell numbers within specific bone marrow aspirate (0.001–0.01% of the nucleated marrow cells) [28] which makes it necessary to expand them in culture to reach clinically relevant numbers for therapeutic purposes [29]. As described above, during expansion, one needs to keep in mind the heterogeneity of the cell population.
- ASCs came to use more recently, but are becoming a solution-of-choice due to the high cell numbers present in lipoaspirates harvested through liposuction techniques [30], that are less invasive than bone marrow aspiration. On average, several liters of lipoaspirate with a relatively high frequency of ASCs (1–5% of isolated nucleated cells) can be obtained [28]. In fact, the stromal vascular fraction of adipose tissue contains more MSCs compared with bone marrow (as measured in a colony-forming unit fibroblast (CFU-F) assay) [18]. Isolation protocols involve density gradient centrifugation of collagenase-digested tissue (lipoaspirate or minced adipose) followed by selection and culture of adherent cell populations. Various studies report successful cultivation of bone-like tissue using scaffolds seeded with ASCs [31, 32].

However, it should be stated that the transplantation of MSCs into bone defects primarily enhances bone repair via immunomodulatory effects, as opposed to their direct differentiation into bone-forming cells [28].

### 7.1.1.3 Osteoblast-Like Cells

Cells with osteoprogenitor characteristics can be harvested from adult bone tissue and periosteum, via preparation of explant cultures from dissected tissues, or enzymatic release of progenitor cells from endosteal and periosteal layers [28, 33, 34]. Osteogeneicity of these cells is confirmed when cultured on porous scaffolds yielding bone-like tissue [35, 36]. Importantly, these cells were also confirmed to have mesenchymal multipotency, demonstrated by single-cell lineage analysis [37].

### 7.1.1.4 Pluripotent Stem Cells

Pluripotent stem cells show unlimited self-renewal and can differentiate into all three germ layers (ectoderm, endoderm and mesoderm). The fact they can differentiate into mesodermal derivatives is of most importance for OC engineering, because of the **mesenchymal condensation** phenomenon—*see below*.

## Human Embryonic Stem Cells (hESC)

hESCs have been used in a number of studies for inducing osteogenic and chondrogenic differentiation: through embryoid bodies (EBs) [38]; by coculture/conditioned culture with fully differentiated chondrocytes [39], MSCs [40], ESC-derived MSCs [41]; or by directed differentiation to chondrogenic and osteogenic cells [42, 43].

It is important to note that, in their directed differentiation protocol, Oldershaw et al. demonstrated that hESCs progress through primitive streak or mesendoderm to mesoderm, before differentiating into a chondrocytic cell aggregates [43], confirming the importance of recapitulating the stage of mesenchymal condensation— *explained in detail below*.

## Human Induced Pluripotent Stem Cells (hiPSCs)

When findings by Yamanaka, Takahashi and Gurdon enabled obtaining autologous pluripotent cells from somatic cells (fibroblasts, keratinocytes, blood cells) of a patient, these naturally seemed like a go-to solution for clinical use.

However, now, more than 10 years after publication of the key papers by Yamanaka and Takahashi, our knowledge on human induced pluripotent stem cells is still not sufficient to allow for a straightforward clinical application of hiPSCs [44]. One of the biggest challenges, raising real safety concerns, is the genomic instability of hiPSCs, which became obvious particularly with the advance of high-throughput technologies such as next-generation sequencing [45].

The application of hiPSCs in OC engineering is also somewhat limited by the current protocols for chondrogenic differentiation that are complicated and inefficient primarily due to the need for intermediate embryoid body (EB) formation, required to generate endodermal, ectodermal, and mesodermal cell lineages [1].

Recently, Nejadnik et al. reported a new, straightforward approach for chondrogenic differentiation of hiPSCs, which avoids embryoid body formation and instead is driving hiPSCs directly into mesenchymal stromal cells (MSC) and chondrocytes. hiPSC-MSC-derived chondrocytes showed significantly increased expression of chondrogenic genes compared to hiPSC-MSCs. Following transplantation of hiPSC-MSC and hiPSC-MSC-derived chondrocytes into osteochondral defects of arthritic joints of athymic rats, MRI studies showed engraftment, and histological correlations showed the production of hyaline cartilage matrix [46].

Suchorska et al. compared four methods to generate chondrocyte-like cells from hiPSCs: (1) monolayer culture with addition of defined mesodermal and chondrogenic growth factors (GFs) (DIRECT protocol), (2) EBs differentiated in chondrogenic medium with TGF-$\beta$3 cells (TGF-$\beta$3 protocol), (3) EBs differentiated in chondrogenic medium conditioned with human chondrocytes (HC-402-05a cell line) (COND protocol) and (4) EBs differentiated in chondrogenic medium conditioned with human chondrocytes and supplemented with TGF-$\beta$3 (TGF-$\beta$3 + COND protocol). Two fastest and most cost-effective methods were the monolayer culture with GFs (DIRECT) and the medium conditioned with human chondrocytes (COND) [47]. De Peppo et al. engineered functional bone substitutes by culturing hiPSC-derived mesenchymal progenitors on osteoconductive scaffolds in perfusion bioreactors, and confirmed their phenotype stability in a subcutaneous implantation model [48].

Along these lines, Wu and colleagues state in their recent review that efficient in vitro differentiation of hiPSCs into downstream cells, such as mesenchymal stem/stromal cells (MSCs), osteoblasts or osteocyte-like cells is necessary to limit undesired tumorigenesis associated with the pluripotency of hiPSCs [49]. They also give good comparisons of the current techniques utilized to confer the induction of hiPSCs into the osteogenic lineage, an evaluation of osteogenic potentials of cells derived from each technique and cells derived from different somatic origins and comparisons of hiPSC-derived MSCs and BMSCs [49].

### 7.1.1.5 Mesenchymal Condensation: Necessary Requirement for Chondrogenesis

From the recent studies using both hESC and hiPSCs, a conclusion emerged that in order to achieve proper differentiation into chondrocytic lineage, one needs to enable the mesenchymal condensation (precartilage condensation) to occur (Fig. 7.1).

Mesenchymal condensation is a key event in the chondrogenic commitment, after which tissue-specific transcription factors and structural proteins begin to accumulate [52, 53]. Main coordinators of this process are transforming growth factor-$\beta$ (TGF-$\beta$) family proteins and Wnt/$\beta$-catenin signaling [54, 55].

In vitro, the mesenchymal condensation is mimicked through the self-assembly methods [56, 57] with TGF-$\beta$ supplementation. Ng et al. used TGF-$\beta$ and thyroxine for both cartilage maintenance and chondrocyte terminal (hypertrophic) differentiation, respectively [58]. Through such biomimetic recapitulation of physiological spatiotemporal signals, Ng and colleagues produced and maintained cartilage discs with functional and phenotypically stable hyaline cartilage with accompanying progressive deep-zone mineralization. The discs remained stable and organized following implantation [58]. Such recapitulation of both temporal and structural aspects of native development is the very essence of the biomimetic approach.

Bhumiratana and Vunjak-Novakovic report that clinically sized pieces of human cartilage with physiologic stratification and biomechanics can be grown in vitro by recapitulating some aspects of the developmental process of mesenchymal condensation [57, 59]. By exposure to TGF-$\beta$, MSCs were induced to aggregate into *condensed mesenchymal bodies* (CMBs) which then formed in vitro an outer boundary after 5 days of culture, as indicated by the expression of mesenchymal condensation genes and deposition of tenascin. Before setting of boundaries, the CMBs could be further fused into homogenous cellular aggregates, without using a scaffolding material, giving rise to well-differentiated and mechanically functional cartilage. The formation of cartilage was initiated by press-molding the CMBs onto the surface of a bone substrate. By image-guided fabrication of the bone substrate and the

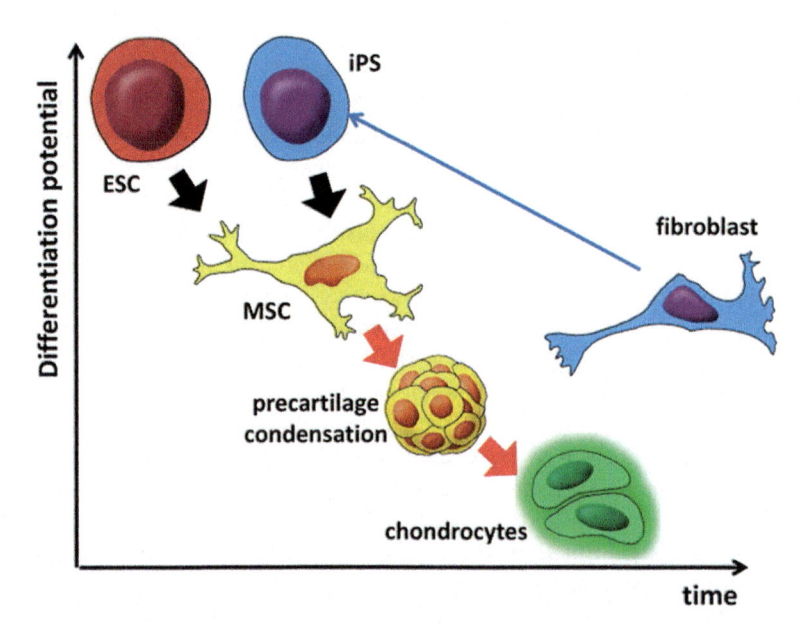

**Fig. 7.1** Importance of mesenchymal (precartilage) condensation in chondrogenesis. Pluripotent cell types (ESC and iPS) have to differentiate into multipotent MSCs in order to form precartilage condensation required for efficient further differentiation into chondrocytes. Chondrocytes, as fully differentiated cells, have lower differentiation potential compared to fibroblasts, which can still be induced to direct differentiation [50] as well as to conversion to iPS cells [51]. Adapted with permission of Springer from Gadjanski I, Spiller K and Vunjak-Novakovic G. *Stem Cell Reviews and Reports* [52]

molds, the OC constructs were engineered in anatomically precise shapes and sizes. Importantly, the cartilage engineered in this way possessed physiologic compressive modulus and lubricative property (Young moduli >0.8 MPa, and friction coefficients <0.3). This method could be highly effective for generating human osteochondral tissue constructs, and for repairing focal cartilage defects to replace currently used dissociated chondrogenic cells [1].

## 7.1.2 Scaffolds as Biomimetic Systems Component

Even though the scaffold-less techniques are gaining impetus, particularly for generating self-assembling tissues [60], scaffolds are still one of the key components for OC tissue engineering. The biggest challenge is how to achieve similar degree of complex hierarchical structure as in the native OC unit, task particularly daunting for the zonal cartilage layer and the complex OC interface, for which many characteristics are still unknown.

Because of this, the prevalent approach is to use multicomponent systems and hybrid scaffolds combining the concepts mentioned earlier.

Scaffolds for the cartilaginous part are frequently hydrogel-based, fostering spherical morphology of the chondrocytes/chondrogenic cells due to hydrogel high water content [61]. Importantly, cell-laden hydrogels, or cell-hydrogel hybrid constructs, can be manufactured in patient-specific anatomical shapes [62]. Injectable hydrogels are particularly convenient materials for in vivo applications. An emerging class of bioinspired polymers for cartilage and bone tissue engineering are glycopolypeptides that mimic naturally occurring glycoproteins, that have been processed into injectable hydrogels, by enzymatic cross-linking of glycopeptides in the presence of horseradish peroxidase (HRP) and hydrogen peroxide ($H_2O_2$) [63]. However, hydrogels, due to their isotropic nature and poor mechanical characteristics, cannot fully mimic the zonal hierarchical structure of the native articular cartilage. This can be improved by adding nanofibers and microfibers, for which electrospinning and melt electrospinning writing techniques are particularly useful [64]. Nanoparticles can be added as well, and loaded with chondrogenic/osteogenic growth factors [65]. In fact, hydrogels with cells and growth factors are proving very useful in engineering OC interface and achieving biochemical gradients [62].

Regarding to above, recent study by Zhu et al. reports a method for rapid formation of tissue-scale gradient hydrogels as a 3D cell niche with tunable biochemical and physical properties. They used photocrosslinkable, multi-arm PEG hydrogel system as a backbone and chondroitin sulfate methacrylate, mixed with two cell-containing precursor solutions (chondrocytes and hMSCs), which, upon exposure to light, quickly formed insoluble cell-laden gradient hydrogels mimicking zonal structure of the native cartilage. The method enabled rapid (~2 min) formation of tissue-scale hydrogels (3 cm × 1 cm × 3 mm) with stiffness and/or ECM molecule gradient cues, while enabling homogeneous cell encapsulation in 3D [66].

Still, multilayered scaffolds can mimic stratified structure to a higher degree, especially important for treatments of full-thickness OC defects. Cartilaginous layer mechanical properties are obtained through the use of hydrogels or porous sponges, while more rigid, porous and fibrous scaffolds are implemented for the bone region [8, 67]. Native ECM components (proteins, GAGs, cell adhesion molecules) are mimicked via chemical functionalization either by chemical binding of peptides on a polymer scaffold [68] or by fabricating a 3D scaffold from self-assembling peptides [67].

The native biological cues are simulated through attached or encapsulated growth factors. To this aim, decellularized extracellular matrices (ECM) are receiving increasing interest as materials capable to induce cell growth/differentiation and tissue repair by physiological presentation of embedded cues [69]. Such ECM are derived from preexisting tissue (native ECM) after isolation and subsequent decellularization (demineralized bone matrix, Matrigel) [70], and, as recently described by Bourgine et al., through designed human cell lines serving as intrinsic tools to achieve efficient ECM deposition and decellularization, offering added possibility of targeted enrichment in the content and delivery of specific molecules. This interesting study reports engineering of ECM materials with customized properties, based on genetic manipulation of immortalized and death-inducible hMSCs, cultured within 3D porous scaffolds under perfusion flow. The strategy allows for robust ECM deposition and subsequent decellularization by deliberate cell-apoptosis induction. As compared to standard production and freeze/thaw treatment, this grants superior preservation of ECM, leading to enhanced bone formation upon implantation in calvarial defects [69].

Proper OC scaffold design should provide hierarchical structure, desired mechanical and mass transport properties (stiffness, elasticity, permeability, diffusion) and ability for processing into precise anatomical shapes [71]. Adequate porosity needs to be achieved as well. Pores of ≤400 μm are recommended by most groups for enhancing new bone formation and the formation of capillaries, and the minimum pore size of ~100 μm, as smaller pores limit cell migration and mass transport [72, 73].

Hierarchical organization needs to comprise all the levels—from nanoscopic to microscopic to macroscopic, in order to meet frequently conflicting requirements for mechanical function, mass transport, and biological regulation [74].

To this aim, various fabrication methods, particularly computer-aided additive manufacturing (CAM), in combination with finite element modeling (FEM) and computational fluid dynamics (CFD) are being developed and implemented [75, 76].

Probably the most utilized method out of CAM technologies is the 3D printing which enables generation of the architectural details that were previously impossible to fabricate. In addition, 3D printing techniques (stereolithography, fused deposition modeling, and selective laser sintering) allow incorporation of gradients into polymer scaffolds to achieve even higher degree of native-like structural, biochemical and mechanical environment. 3D printing can be combined with other approaches, such as self-assembly of nanoparticles [77]. An excellent recent review by Bracaglia et al. covers various 3D techniques for design and fabrication of polymer-based gradient scaffolds in detail [78], while Guo et al. in their review cover the applications of 3D printing for recapitulation of zonal structure of articular cartilage [79]. Importantly, anatomically shaped scaffolds can be made by CAM, tailored to the patient by using the CT images of the defect for creating the CAD (computer-aided design) model [71].

Regarding the use of CAM in OC TE, an interesting study by Hendrikson et al. analyzed the influence of additive manufactured scaffold architecture on distribution of surface strains and fluid flow shear stresses and expected osteochondral cell differentiation [80]. They compared four scaffold designs that only differed in the pore shape while the fiber diameter, spacing, and layer thickness remained constant. Different architectures were obtained by changing the angle of layer deposition and lateral shifting of the layers. Also, μCT-based models of the scaffolds were prepared, and stress and strain distributions within the scaffolds were predicted using CFD and FEM. The results show a distinct effect of the scaffold architecture on surface strains and fluid flow shear stresses under mechanical compression and imposed fluid flow. This implies that regions of the scaffold could be designed favoring specific cell differentiation stimuli. Coupling with biophysical loading regimes a priori in silico could accelerate the design of scaffolds and optimize the loading regimes [80].

One of the CAM methods is biofabrication or 3D bioprinting [81] that allows for the direct incorporation of the live cells in the scaffold fabrication process. There are three major types of 3D bioprinting techniques that are currently available: (1) inkjet bioprinting [82] (2) microextrusion bioprinting [83], and (3) laser-assisted bioprinting [84]. However, it is still challenging to bioprint clinically sized constructs, mostly due to the poor mechanical properties and limited structural integrity of the printed construct. To overcome this limitation, various modifications are tested, such as FRESH method where the tissue construct is built by embedding the printed cell-laden hydrogel within a secondary hydrogel that serves as a temporary, thermoreversible, and biocompatible support [85]. Other option is to combine multiple processing methods, e.g., electrospinning with 3D bioprinting [86].

Currently, one of the main applications of 3D bioprinting is the fabrication of mini-tissues for disease modeling [87]. Lozito et al. constructed an in vitro system with 3D microtissues designed for biological studies of the osteochondral complex of the articular joint [54]. The model was constructed by seeding hMSCs from bone marrow and adipose tissue aspirates into photostereolithographically fabricated biomaterial scaffolds with defined internal architectures. Concerning OC disease-modeling, hiPSCs are also frequently used in the so-called "disease-in-a dish" models. Diseases to be modeled include of course osteoarthritis, but also the numerous hereditary osteochondral dysplasias which result from genetic disorders causing defective cartilage and bone differentiation, formation, and growth and for many of which the disease-causing mutations are already known [88]. Reprogramming patient-specific cells with a genetic predisposition and engineering disease-specific genetic variations into healthy control hiPSC cell lines promises to recapitulate "diseases in a dish" more realistically than immortalized human cell lines and will be an invaluable complementation for animal models [89]. In addition to reprogramming patient-specific cells, novel gene editing methods, such as zinc-finger nuclease (ZFN), transcription activator-like effector nuclease (TALEN), and clustered regularly interspaced short palindromic repeats (CRISPR)/Cas9 [90] allow introducing genetic defects into well-characterized hiPSC lines [89]. Generating stable hiPSC cell lines enables high-throughput drug screening and positions human disease pathophysiology at the core of preclinical drug discovery [91], potentially leading to personalized regenerative medicine therapies [92].

### 7.1.3 Bioreactors as Biomimetic System Component

Bioreactor is a necessary component for maintenance of differentiated cell phenotypes and promoting the OC construct maturation by providing exchange of nutrients and metabolites, control of environmental factors as well as biophysical signaling and mechanical cues. In general, a bioreactor of OC TE should comprise two different compartments—for cartilage and bone, while enabling the interface formation in between. The compartments should allow for specific culture media perfusion as well as biophysical and mechanical stimulation needed for the tissue in question.

Even though some studies report good results with chondrogenesis of undifferentiated hMSCs in chondrogenic medium even in static culture [93], majority utilizes dynamic loading with physiological frequency (1 Hz) to provide both the mechanism for fluid transport through the tissue and the necessary biophysical stimuli [94]. It has been detected that moderate amplitude strains (5%) applied at 1 Hz stimulate chondrogenesis of hMSCs and enable stable chondrocyte-like phenotype, while higher strains and lower frequencies have a negative effect on chondrogenesis [95].

Bioreactor cultivation of the bone, as a tissue that should be vascularized, requires interstitial flow of culture medium through the tissue space, facilitating exchange of nutrients—particularly oxygen, metabolites, and regulatory factors to and from the cells, over minimal diffusional distances, while providing shear stress [96]. In the ideal scenario, the medium would be perfused through a network of channels with endothelial lining, serving as precursors of the vascular network to connect at a later point to the blood supply of the host. Such bioreactor systems are conceptually biomimetic, since they enable convective-diffusive mass transport similar to that occurring in vivo, between blood and tissue, along with dynamic hydrodynamic shear that is an important regulatory factor for bone development and maintenance [1, 96].

In the exemplary study on the effects of medium perfusion achieved through cultivation in a bioreactor, Grayson et al. showed that perfusion culture of prediffentiated osteoblasts or undifferentiated hMSCs with cocktail medium elicited the best osteogenic responses [93]. Bioreactors can be tailored to fit the specific shape, particularly important when engineering anatomically shaped constructs, to provide direct fluid flow through the tissue and/or gradients of biophysical/mechanical cues needed for spatiotemporal recapitulation of cell differentiation, assembly and ECM production [97, 98].

For the detailed overview of the principles of different bioreactor designs, and important parameters to mimic physiological phenomena in OC TE the work by Vunjak-Novakovic, Bhumiratana et al. and Martin et al. is recommended [3, 96, 99, 100].

## 7.1.4 *Other Components in Biomimetic OC TE*

### 7.1.4.1 Growth Factors

There are several key growth factors used in OC TE. These are members of the transforming growth factor-β (TGF-β) superfamily (including Bone Morphogenetic Proteins—BMPs, Growth and Differentiation Factors—GDFs [101]), fibroblast growth factor (FGF) family, insulin-like growth factor-1 (IGF-1), and

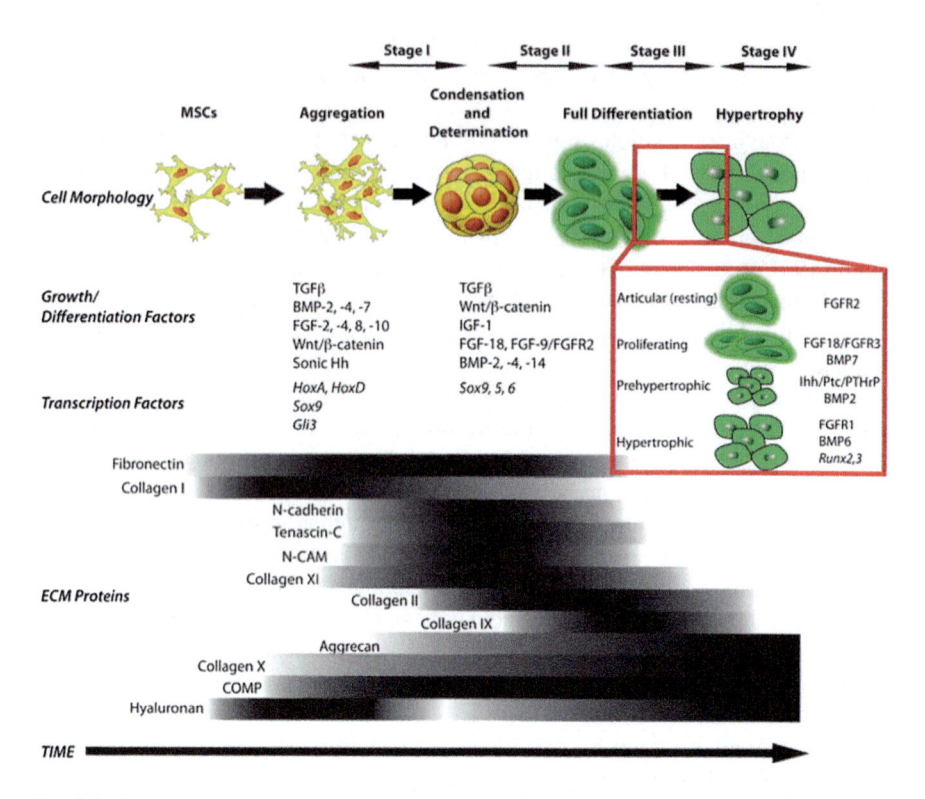

**Fig. 7.2** Sequence of events and time-dependent involvement of growth factors during native chondrogenesis. Adapted with permission of Springer from Gadjanski I, Spiller K and Vunjak-Novakovic G. *Stem Cell Reviews and Reports* [52]

platelet-derived growth factor (PDGF) [8]. Growth factors act through modulation of the local microenvironment (making it chondroinducive or osteoinducive), anabolic cellular effects, and increased matrix production. Additionally, some (e.g., PDFG) are important for vascularization, since they can induce angiogenesis and direct cell migration and support vessel maturation and stabilization [102].

The sequential addition of growth factors (GFs) to cell culture medium has proven useful in stimulating chondrogenesis in vitro [52]. GF addition in a sequence similar to native development, e.g., basic FGF (bFGF) or FGF2 followed by BMP2 or IGF1, TGFβ2 or TGFβ3, increased proliferation and subsequent chondrogenic differentiation [52, 103]. Similarly, exposure of chondrocytes seeded in agarose gels to TGFβ3 for 2 weeks followed by unsupplemented culture medium resulted in enhanced cartilage formation and mechanical properties compared to prolonged exposure to TGFβ3 [104]. The exposure of MSCs in poly(ethylene glycol) (PEG) hydrogels to TGFβ1 for just 7 days resulted in enhanced proteoglycan production compared to prolonged culture, but decreased collagen production [105]. Figure 7.2 shows the sequence of events and time-dependent GF involvement in native chondrogenesis, which should be mimicked in OC TE.

Similarly, sequential GF application proved important for osteogenesis as well. Aksel et al. showed that vascular endothelial growth factor (VEGF) addition in the early phase rather than a continuous presence of both VEGF and BMP-2 enhanced odontogenic/osteogenic differentiation of human dental pulp stem cells (DPSCs) [106]. It was also shown that early delivery of an angiogenic factor (bFGF) combined with sustained exposure to an osteogenic factor (Sonic hedgehog—Shh) can recapitulate the critical aspects of natural bone repair [107]. These data emphasize the importance of controlled duration of GF application.

Generally speaking, in the biomimetic OC construct, the chondrogenic growth factors (e.g., TGF-β family) should be supplied in the cartilage phase (in combination with dynamic loading), while the osteogenic growth factors, e.g., BMPs (combined with medium perfusion) should be applied in the bone phase.

### 7.1.4.2 Vascularization

Native bone tissue is highly vascularized, and its development and function are coordinated by synergistic interactions between the bone cells and vascular cells. The blood vessels supply oxygen and nutrients, as well as calcium and phosphate, the building blocks for mineralization [108]. To a certain degree, the emerging vasculature serves as a template for bone development. Following biomimetic approach, the bone phase in the OC construct should be engineered by synchronizing vascular and bone development in 3D scaffolds [71, 109]. Ideally, the OC construct would provide paracrine signaling between the bone and vascular cells, as well as larger vascular conduits that can help quickly connect the blood to the tissue and establish vascular perfusion following implantation of engineered tissue constructs [1]. However, in practice this is proving very difficult to achieve. Certain advancements have been made through harnessing the proangiogenic effects of immune cells.

### 7.1.4.3 Immunomodulation

Immune response is a major regulator of vascularization and overall functionality of engineered tissues, through the activity of different types of macrophages (proinflammatory M1 and anti-inflammatory M2 phenotype) and the cytokines they secrete [1]. Regarding their contribution to angiogenesis, human macrophages polarized to the M1 or M2 phenotypes behave in different ways. Spiller et al. showed that M1 macrophages express and secrete factors that promote the initiation of angiogenesis, especially VEGF. M2 macrophages secrete factors involved in later stages of angiogenesis, particularly PDGF-BB isoform, which recruits stabilizing pericytes [110]. In addition, M2 macrophages can express high levels of tissue inhibitor of metalloprotease-3 (TIMP3), which inhibits angiogenesis by blocking the actions of metalloproitenase-9 (MMP9) and VEGF [111] and prevents the release of the inflammatory cytokine TNFα [112]. TIMP3 also stabilizes vasculature formation from endothelial cells in vitro [113].

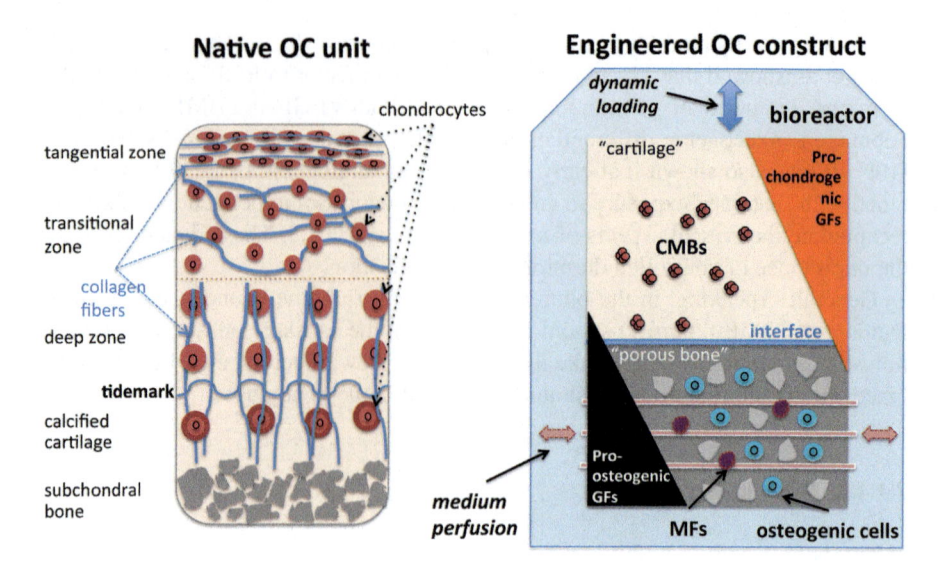

**Fig. 7.3** Paradigm of the biomimetic approach in OC tissue engineering. *CMBs* condensed mesenchymal bodies, *GF* growth factors, *MF* macrophages. Detailed explanations in the text

It is clear that coordinated efforts by both M1 and M2 macrophages are required for angiogenesis and scaffold vascularization [110]. hMSCs have been shown to promote macrophage differentiation toward an M2-like phenotype with a high tissue remodeling potential and anti-inflammatory activity, but also a protumorigenic function [114]. This is in line with previously mentioned hypothesis that many effects of the hMSCs used in regenerative medicine are due to their immunomodulatory effects and not to direct differentiation into specific cell types [115].

To harness immunomodulatory signals to the highest degree, researchers start using "smart" scaffolds that enable sequential release of immunomodulatory factors recruiting the waves of M1 and M2 macrophages [110]. Spiller et al. designed scaffolds for sequential release of pro-M1 (interferon-gamma; IFN-γ) and pro-M2 (interleukin-4—IL4) signals to achieve bone regeneration where IFN-γ was physically adsorbed onto the scaffolds, while IL4 was attached via biotin–streptavidin binding [116] (Fig. 7.3).

### 7.1.4.4 Low Oxygen Tension Conditioning

Oxygen gradients are established early in embryonic development, since the growing tissues of the embryo rapidly deplete local oxygen and nutrient supplies provided via diffusion. During endochondral ossification, the cartilaginous anlage develops into the fetal growth plate, becoming more hypoxic as it grows [117].

Articular cartilage remains hypoxic in adult stage, with spatial oxygen gradient of <1% in the deepest layers, up to <10% at the cartilage surface. Chondrocytes are very adapted to low oxygen tensions present in the avascular environment, but they also promote (by secreting angiogenic stimuli) localized vascularization at the periphery of the cartilage, the key process for the continued development and growth of bone [118]. Oxygen levels and vascularization are connected through the action of hypoxia-inducible factors (HIF) and vascular endothelial growth factor (VEGF). Levels of oxygen in the tissue modulate HIF signaling cascades, the essential mediators of the complex homeostatic responses that enable cells to survive and differentiate in low oxygen environment. VEGF is a downstream target of the HIF pathway and a potent angiogenic factor. Osteoblasts express HIF-1α and HIF-2α, which modulate bone development and homeostasis and angiogenesis. Some of the effects of HIFs on bone and angiogenesis are mediated by VEGF. It is clear that HIFs and VEGF have critical roles in skeletal development and bone homeostasis. Such close spatial and temporal association of osteogenesis and vascularization is now termed as *angiogenic-osteogenic coupling* [118, 119].

It is obvious from abovementioned data that ambient $O_2$ concentration of 21% $O_2$ represents hyperoxic environment for osteochondral cells and should not be used for the in vitro cultivation. Such hyperoxia disturbs the HIF-signaling pathways since when oxygen tension is >5%, the half-life of HIF-1α is very short (<5 min) [120]. This can impair normal anaerobic glycolysis in cartilage and posttranslational modifications of type II collagen [121].

Ambient $O_2$ concentrations are even more detrimental when considering stem/stromal cells that normally reside in "hypoxic" niches in vivo and are well adapted to low $O_2$ tensions. The "hypoxic", or more precisely physioxic (or in situ normoxic) oxygen concentration in adult tissues varies between 1 and 11% [122], while the $O_2$ concentration at the cellular level is estimated to be 1.3–2.5% or even <1% [21, 123]. Atmospheric 21% $O_2$ represents a hyperoxic environment for stem/stromal cells which start losing the phenotypic and molecular markers of stemness [21]. This is the reason why low oxygen conditioning—i.e., cultivation of cells in low $O_2$ tension environment is proving very efficient in stem/stromal based OC TE. If maintenance in low $O_2$ is technically challenging in long-term culture, it is beneficial to perform at least a transient preconditioning, followed by the switch to ambient $O_2$ concentration. Yodmuang et al. showed, using juvenile chondrocytes, that such transient culture of 5% $O_2$ increases expression of cartilaginous genes including COL2A1, ACAN, and SOX9 and increased tissue concentrations of GAG and type II collagen, with accompanying increase in the equilibrium Young's modulus [124]. Henrionnet et al. performed similar study on hMSCs and concluded that better chondrogenic differentiation is achieved when reduced oxygen tension (5% $O_2$) is applied during both expansion and differentiation times, avoiding the in vitro osteogenic commitment of the cells and subsequently the calcification deposition [125].

### 7.1.4.5 Purinergic Signaling

External mechanical stimulation leads to activation of mechanotransduction cascades that promote chemical signaling inside the cell [126]. These intracellular mechanotransduction pathways are still not fully defined [127]. Relatively recently, ATP (adenosine 5'-triphosphate) has been indicated as one of the first molecules to be released from chondrocytes into extracellular space in response to mechanical stimulation [128, 129], subsequently binding to purinergic P2 receptors and activating calcium signaling pathways [130]. Garcia and Knight suggested putative mechanism of ATP release via hemichannels (formed of connexin-43 subunits) in response to cyclic compression [131]. Since then a number of studies has shown that exogenous ATP supplementation, even in the absence of mechanical stimulation, can promote ECM biosynthesis and accumulation providing energy supply to fuel that process [132], increase mechanical properties, particularly through structural organization of the bulk phase and territorial ECM [133]. Exogenous ATP effects are proven to be dose- and time-dependent, where high doses can promote catabolic responses, necessitating the optimization of therapeutic dose range and application timing (e.g., transient vs. continuous) to the cell type and culture system [133–135]. Furthermore, MSCs have been shown to respond to extracellular ATP as well, even more receptively than the chondrocytes. Gadjanski et al. detected that exogenous ATP induced 72% vs. 16% increase in GAG content for human MSCs and chondrocytes, respectively [134], while Steward et al. showed that purinergic signaling regulates the TFG-$\beta$3-induced chondrogenic response of MSC [136]. A mathematical model was defined showing ATP release changes in loaded vs. unloaded cell constructs (chondrocytes and hMSCs) over time [137]. Such model can be of value in determining the potential for pharmacological manipulation of the purinergic mechanotransduction in the engineered osteochondral tissues.

Mechanosensitive purinergic signaling in bone has also been confirmed, where extracellular ATP has been shown to modulate multiple processes including cell proliferation, differentiation, function, and death [138]. Osteoblasts and osteoclasts have been reported to express nearly all the P2Y and P2X receptors to which the extracellular ATP can bind [139].

Additional important aspect to keep in mind is that any mechanical stimulus applied to cells in vitro, even as subtle as fluid movements after a medium change, can increase basal ATP release [140] which was recently again brought to attention in an informative review by Burnstock and Knight [141] who urge the researchers to always include this aspect in the interpretation of their data.

### 7.1.4.6 OC Interface Engineering

In order to engineer a native-like osteochondral interface and complex cell–cell communication between cartilage and bone, it is necessary to fine-tune scaffold properties such as graded molecular composition, structure, and biomechanics, i.e., to specify and precisely implement multiple gradients in scaffolding of the OC construct.

As mentioned several times, one approach is to build composite scaffolds through multilayered scaffold design, to generate structural templates for the cartilaginous layer, the tidemark and calcified cartilage, and the subchondral bone, while allowing the transitional interface layer to efficiently connect cartilage and bone. Usual method for fabrication of composites is by using two or more different materials [1]. Integration between layers (and with native tissue upon implantation) is achieved by suturing [142], cell-mediated ECM formation, use of fibrin and other glues [97], or simply by press fitting [143]. However, such layered composites are susceptible to delamination if the layers are not well connected. To overcome this, the gradient scaffolds are used, which sport gradual changes of physical and mechanical properties, ideally complemented with the biochemical gradients. Such scaffolds can achieve better transition between cartilaginous and osseous components.

Cross et al. present a fabrication method for a scaffold with graded mechanical properties. They used two natural polymers (gelatin methacryloyl (GelMA) and methacrylated kappa carrageenan (MκCA)) reinforced with 2D nanosilicates to mimic the native tissue interface. The addition of nanosilicates results in shear-thinning characteristics of prepolymer solution and increases the mechanical stiffness of cross-linked gradient structure [144]. D'Amora et al. formulated a method for achieving chemical gradients in which CAM and surface modification are combined. They first aminolyzed poly(ε-caprolactone) surface and subsequently covered it with collagen via carbodiimide reaction. These 2D constructs were characterized for their amine and collagen contents, wettability, surface topography and biofunctionality. This functionalization treatment was extended to the 3D printed PCL scaffolds, demonstrating the possibility to manufacture 3D constructs with chemical gradients for OC interface engineering [145]. Dormer et al. achieved biochemical gradients by distributing the microspheres loaded with chondrogenic (TGF-β1) and osteogenic (BMP-2) factors into the two regions of a PLGA scaffold, to produce opposing growth-factor gradients for the formation of cartilage and bone [146]. In addition, therapeutic molecules can be surface-tethered to the microspheres [147]. Using "raw materials," i.e., components like chondroitin sulfate and bioactive glass, in 3D scaffolds was suggested for establishing continuous gradients of both material composition and signaling factors [148].

For now, the best approach seems to be to couple biochemical and structural gradients toward achieving native-like OC interface architecture and integration in large OC constructs intended for implantation [1, 149].

## 7.2 Concluding Remarks

The current state-of-the-art approach in OC TE is to use multilayered hybrid scaffolds, with biochemical, structural and mechanical gradients that eventually lead to functionally graded OC scaffolds. To this aim, the use of additive manufacturing and custom-tailored bioreactors is of particular importance. More research is being directed towards harnessing immunomodulation and low oxygen conditioning in order to improve vascularization and overall functionality of the engineered scaffold, while the complexities of cell-cell communication between osseous and cartilaginous components are being investigated in order to achieve a more native-like osteochondral interface.

**Competing interests** The author participates in a project that has received funding from the European Union's Horizon 2020 research and innovation programme under grant agreement 664387.

**Grant information** This work was supported by the Ministry of Education, Science and Technological Development of the Republic of Serbia (Projects OI174028 and III41007).

The funders had no role in study design, data collection and analysis, decision to publish, or preparation of the manuscript.

# References

1. Gadjanski I, Vunjak-Novakovic G (2015) Challenges in engineering osteochondral tissue grafts with hierarchical structures. Expert Opin Biol Ther 15:1–17. https://doi.org/10.1517/14712598.2015.1070825
2. Vanderburgh J, Sterling JA, Guelcher SA (2017) 3D printing of tissue engineered constructs for in vitro modeling of disease progression and drug screening. Ann Biomed Eng 45(1):164–179
3. Martin I, Miot S, Barbero A, Jakob M, Wendt D (2007) Osteochondral tissue engineering. J Biomech 40(4):750–765. https://doi.org/10.1016/j.jbiomech.2006.03.008
4. Di Luca A, Van Blitterswijk C, Moroni L (2015) The osteochondral interface as a gradient tissue: from development to the fabrication of gradient scaffolds for regenerative medicine. Birth Defects Res C Embryo Today 105(1):34–52. https://doi.org/10.1002/bdrc.21092
5. Yan L, Oliveira JM, Oliveira AL, Reis RL (2015) Current concepts and challenges in osteochondral tissue engineering and regenerative medicine. ACS Biomater Sci Engine
6. Oliveira JM, Reis RL (2016) Regenerative strategies for the treatment of knee joint disabilities. Springer, New York
7. Perdisa F, Sessa A, Filardo G, Marcacci M, Kon E (2017) Cell-free scaffolds for the treatment of chondral and osteochondral lesions. In: Gobbi A, Espregueira-Mendes J, Lane JG, Karahan M (eds) Bio-orthopaedics: a new approach. Springer, Berlin, pp 139–149. https://doi.org/10.1007/978-3-662-54181-4_11
8. Yousefi AM, Hoque ME, Prasad RG, Uth N (2014) Current strategies in multiphasic scaffold design for osteochondral tissue engineering: a review. J Biomed Mater Res A 103:2460. https://doi.org/10.1002/jbm.a.35356
9. von der Mark K, Gauss V, von der Mark H, Müller P (1977) Relationship between cell shape and type of collagen synthesised as chondrocytes lose their cartilage phenotype in culture. Nature 267(5611):531–532
10. Schnabel M, Marlovits S, Eckhoff G, Fichtel I, Gotzen L, Vecsei V, Schlegel J (2002) Dedifferentiation-associated changes in morphology and gene expression in primary human articular chondrocytes in cell culture. Osteoarthritis Res Soc 10(1):62–70. https://doi.org/10.1053/joca.2001.0482
11. Lau TT, Peck Y, Huang W, Wang D-A (2014) Optimization of chondrocyte isolation and phenotype characterization for cartilage tissue engineering. Tissue Eng Part C Methods 21(2):105–111
12. Yonenaga K, Nishizawa S, Fujihara Y, Asawa Y, Sanshiro K, Nagata S, Takato T, Hoshi K (2010) The optimal conditions of chondrocyte isolation and its seeding in the preparation for cartilage tissue engineering. Tissue Eng Part C Methods 16(6):1461–1469

13. Zhou M, Yuan X, Yin H, Gough JE (2015) Restoration of chondrocytic phenotype on a two-dimensional micropatterned surface. Biointerphases 10(1):011003. https://doi.org/10.1116/1.4913565
14. Tan AR, Hung CT (2017) Concise review: mesenchymal stem cells for functional cartilage tissue engineering: taking cues from chondrocyte-based constructs. Stem Cells Transl Med 6(4):1295–1303. https://doi.org/10.1002/sctm.16-0271
15. Erickson IE, Huang AH, Chung C, Li RT, Burdick JA, Mauck RL (2009) Differential maturation and structure-function relationships in mesenchymal stem cell- and chondrocyte-seeded hydrogels. Tissue Eng A 15(5):1041–1052. https://doi.org/10.1089/ten.tea.2008.0099
16. Rodrigues MT, Gomes ME, Reis RL (2011) Current strategies for osteochondral regeneration: from stem cells to pre-clinical approaches. Curr Opin Biotechnol 22(5):726–733. https://doi.org/10.1016/j.copbio.2011.04.006
17. Rodriguez-Fontan F, Piuzzi NS, Chahla J, Payne KA, LaPrade RF, Muschler GF, Pascual-Garrido C (2017) Stem and progenitor cells for cartilage repair: source, safety, evidence, and efficacy. Oper Tech Sports Med 25(1):25–33
18. Vonk LA, De Windt TS, Slaper-Cortenbach IC, Saris DB (2015) Autologous, allogeneic, induced pluripotent stem cell or a combination stem cell therapy? Where are we headed in cartilage repair and why: a concise review. Stem Cell Res Ther 6(1):94
19. Dominici M, Le Blanc K, Mueller I, Slaper-Cortenbach I, Marini F, Krause D, Deans R, Keating A, Prockop D, Horwitz E (2006) Minimal criteria for defining multipotent mesenchymal stromal cells. Int Soc Cell Ther 8(4):315–317
20. Castro-Malaspina H, Ebell W, Wang S (1984) Human bone marrow fibroblast colony-forming units (CFU-F). Prog Clin Biol Res 154:209–236
21. Ivanovic Z, Vlaski-Lafarge M (2015) Anaerobiosis and Stemness. Academic Press
22. Friedenstein A, Chailakhjan R, Lalykina K (1970) The development of fibroblast colonies in monolayer cultures of guinea-pig bone marrow and spleen cells. Cell Prolif 3(4):393–403
23. Muraglia A, Cancedda R, Quarto R (2000) Clonal mesenchymal progenitors from human bone marrow differentiate in vitro according to a hierarchical model. J Cell Sci 113(7):1161–1166
24. Bernhard JC, Vunjak-Novakovic G (2016) Should we use cells, biomaterials, or tissue engineering for cartilage regeneration? Stem Cell Res Ther 7(1):56. https://doi.org/10.1186/s13287-016-0314-3
25. Wakitani S, Nawata M, Tensho K, Okabe T, Machida H, Ohgushi H (2007) Repair of articular cartilage defects in the patello-femoral joint with autologous bone marrow mesenchymal cell transplantation: three case reports involving nine defects in five knees. J Tissue Eng Regen Med 1(1):74–79. https://doi.org/10.1002/term.8
26. Huey DJ, Hu JC, Athanasiou KA (2012) Unlike bone, cartilage regeneration remains elusive. Science 338(6109):917–921. https://doi.org/10.1126/science.1222454
27. Caplan AI (2017) Mesenchymal stem cells: time to change the name! Stem Cells Transl Med 6(6):1445–1451
28. Ng J, Spiller K, Bernhard J, Vunjak-Novakovic G (2017) Biomimetic approaches for bone tissue engineering. Tissue Eng Pt B: Rev
29. Friedenstein A, Chailakhyan R, Gerasimov U (1987) Bone marrow osteogenic stem cells: in vitro cultivation and transplantation in diffusion chambers. Cell Prolif 20(3):263–272
30. Zuk PA, Zhu M, Mizuno H, Huang J, Futrell JW, Katz AJ, Benhaim P, Lorenz HP, Hedrick MH (2001) Multilineage cells from human adipose tissue: implications for cell-based therapies. Tissue Eng 7(2):211–228. https://doi.org/10.1089/107632701300062859
31. Frohlich M, Grayson WL, Marolt D, Gimble JM, Kregar-Velikonja N, Vunjak-Novakovic G (2010) Bone grafts engineered from human adipose-derived stem cells in perfusion bioreactor culture. Tissue Eng A 16(1):179–189. https://doi.org/10.1089/ten.TEA.2009.0164
32. Correia C, Bhumiratana S, Yan L-P, Oliveira AL, Gimble JM, Rockwood D, Kaplan DL, Sousa RA, Reis RL, Vunjak-Novakovic G (2012) Development of silk-based scaffolds for tissue engineering of bone from human adipose-derived stem cells. Acta Biomater 8(7):2483–2492

33. Voegele TJ, Voegele-Kadletz M, Esposito V, Macfelda K, Oberndorfer U, Vecsei V, Schabus R (2000) The effect of different isolation techniques on human osteoblast-like cell growth. Anticancer Res 20(5B):3575–3581
34. Jonsson KB, Frost A, Nilsson O, Ljunghall S, Ljunggren O (1999) Three isolation techniques for primary culture of human osteoblast-like cells: a comparison. Acta Orthop Scand 70(4):365–373
35. Hutmacher DW, Sittinger M (2003) Periosteal cells in bone tissue engineering. Tissue Eng 9(Suppl 1):S45–S64. https://doi.org/10.1089/10763270360696978
36. Puwanun S, Smith RMD, Colley HE, Yates JM, MacNeil S, Reilly GC (2017) A simple rocker-induced mechanical stimulus upregulates mineralization by human osteoprogenitor cells in fibrous scaffolds. J Tiss Eng Regen Med
37. De Bari C, Dell'Accio F, Vanlauwe J, Eyckmans J, Khan IM, Archer CW, Jones EA, McGonagle D, Mitsiadis TA, Pitzalis C, Luyten FP (2006) Mesenchymal multipotency of adult human periosteal cells demonstrated by single-cell lineage analysis. Arthritis Rheum 54(4):1209–1221. https://doi.org/10.1002/art.21753
38. Ko JY, Park S, Im GI (2014) Osteogenesis from human induced pluripotent stem cells: an in vitro and in vivo comparison with mesenchymal stem cells. Stem Cells Dev 23(15):1788–1797. https://doi.org/10.1089/scd.2014.0043
39. Bigdeli N, Karlsson C, Strehl R, Concaro S, Hyllner J, Lindahl A (2009) Coculture of human embryonic stem cells and human articular chondrocytes results in significantly altered phenotype and improved chondrogenic differentiation. Stem Cells 27(8):1812–1821. https://doi.org/10.1002/stem.114
40. Lee TJ, Jang J, Kang S, Bhang SH, Jeong GJ, Shin H, Kim DW, Kim BS (2014) Mesenchymal stem cell-conditioned medium enhances osteogenic and chondrogenic differentiation of human embryonic stem cells and human induced pluripotent stem cells by mesodermal lineage induction. Tissue Eng A 20(7–8):1306–1313. https://doi.org/10.1089/ten.TEA.2013.0265
41. Marolt D, Campos IM, Bhumiratana S, Koren A, Petridis P, Zhang G, Spitalnik PF, Grayson WL, Vunjak-Novakovic G (2012) Engineering bone tissue from human embryonic stem cells. Proc Natl Acad Sci U S A 109(22):8705–8709. https://doi.org/10.1073/pnas.1201830109
42. Cheng A, Hardingham TE, Kimber SJ (2014) Generating cartilage repair from pluripotent stem cells. Tissue Eng Part B Rev 20(4):257–266. https://doi.org/10.1089/ten.TEB.2012.0757
43. Oldershaw RA, Baxter MA, Lowe ET, Bates N, Grady LM, Soncin F, Brison DR, Hardingham TE, Kimber SJ (2010) Directed differentiation of human embryonic stem cells toward chondrocytes. Nat Biotechnol 28(11):1187–1194. https://doi.org/10.1038/nbt.1683
44. Chari S, Mao S (2016) Timeline: iPSCs--the first decade. Cell Stem Cell 18(2):294. https://doi.org/10.1016/j.stem.2016.01.005
45. Yoshihara M, Hayashizaki Y, Murakawa Y (2017) Genomic instability of iPSCs: challenges towards their clinical applications. Stem Cell Rev Rep 13(1):7–16
46. Nejadnik H, Diecke S, Lenkov OD, Chapelin F, Donig J, Tong X, Derugin N, Chan RC, Gaur A, Yang F, Wu JC, Daldrup-Link HE (2015) Improved approach for Chondrogenic differentiation of human induced pluripotent stem cells. Stem Cell Rev 11:242. https://doi.org/10.1007/s12015-014-9581-5
47. Suchorska WM, Augustyniak E, Richter M, Trzeciak T (2017) Comparison of four protocols to generate chondrocyte-like cells from human induced pluripotent stem cells (hiPSCs). Stem Cell Rev Rep 13(2):299–308
48. de Peppo GM, Vunjak-Novakovic G, Marolt D (2014) Cultivation of human bone-like tissue from pluripotent stem cell-derived osteogenic progenitors in perfusion bioreactors. Methods Mol Biol 1202:173–184. https://doi.org/10.1007/7651_2013_52
49. Wu Q, Yang B, Hu K, Cao C, Man Y, Wang P (2017) Deriving osteogenic cells from induced pluripotent stem cells for bone tissue engineering. Tissue Eng Part B Rev 23(1):1–8
50. Outani H, Okada M, Yamashita A, Nakagawa K, Yoshikawa H, Tsumaki N (2013) Direct induction of chondrogenic cells from human dermal fibroblast culture by defined factors. PLoS One 8(10):e77365. https://doi.org/10.1371/journal.pone.0077365

51. Tsumaki N, Okada M, Yamashita A (2015) iPS cell technologies and cartilage regeneration. Bone 70:48–54. https://doi.org/10.1016/j.bone.2014.07.011
52. Gadjanski I, Spiller K, Vunjak-Novakovic G (2012) Time-dependent processes in stem cell-based tissue engineering of articular cartilage. Stem Cell Rev 8(3):863–881. https://doi.org/10.1007/s12015-011-9328-5
53. Shum L, Nuckolls G (2002) The life cycle of chondrocytes in the developing skeleton. Arthritis Res 4(2):94–106. https://doi.org/10.1186/ar396
54. Tuli R, Tuli S, Nandi S, Huang X, Manner PA, Hozack WJ, Danielson KG, Hall DJ, Tuan RS (2003) Transforming growth factor-beta-mediated chondrogenesis of human mesenchymal progenitor cells involves N-cadherin and mitogen-activated protein kinase and Wnt signaling cross-talk. J Biol Chem 278(42):41227–41236. https://doi.org/10.1074/jbc.M305312200
55. Grafe I, Alexander S, Peterson JR, Snider TN, Levi B, Lee B, Mishina Y (2017) TGF-β family signaling in mesenchymal differentiation. Cold Spring Harbor Perspect Biol. a022202
56. Hu JC, Athanasiou KA (2006) A self-assembling process in articular cartilage tissue engineering. Tissue Eng 12(4):969–979. https://doi.org/10.1089/ten.2006.12.969
57. Bhumiratana S, Eton RE, Oungoulian SR, Wan LQ, Ateshian GA, Vunjak-Novakovic G (2014) Large, stratified, and mechanically functional human cartilage grown in vitro by mesenchymal condensation. Proc Natl Acad Sci U S A 111(19):6940–6945. https://doi.org/10.1073/pnas.1324050111
58. Ng JJ, Wei Y, Zhou B, Bernhard J, Robinson S, Burapachaisri A, Guo XE, Vunjak-Novakovic G (2017) Recapitulation of physiological spatiotemporal signals promotes in vitro formation of phenotypically stable human articular cartilage. Proc Natl Acad Sci U S A 114(10):2556–2561. https://doi.org/10.1073/pnas.1611771114
59. Bhumiratana S, Vunjak-Novakovic G (2015) Engineering physiologically stiff and stratified human cartilage by fusing condensed mesenchymal stem cells. Methods 84:109. https://doi.org/10.1016/j.ymeth.2015.03.016
60. Athanasiou KA, Eswaramoorthy R, Hadidi P, Hu JC (2013) Self-organization and the self-assembling process in tissue engineering. Annu Rev Biomed Eng 15:115–136. https://doi.org/10.1146/annurev-bioeng-071812-152423
61. Spiller KL, Maher SA, Lowman AM (2011) Hydrogels for the repair of articular cartilage defects. Tissue Eng Part B Rev 17(4):281–299. https://doi.org/10.1089/ten.TEB.2011.0077
62. Yang J, Zhang YS, Yue K, Khademhosseini A (2017) Cell-laden hydrogels for osteochondral and cartilage tissue engineering. Acta Biomaterialia
63. Ren K, He C, Xiao C, Li G, Chen X (2015) Injectable glycopolypeptide hydrogels as biomimetic scaffolds for cartilage tissue engineering. Biomaterials 51:238–249. https://doi.org/10.1016/j.biomaterials.2015.02.026
64. Visser J, Melchels FP, Jeon JE, van Bussel EM, Kimpton LS, Byrne HM, Dhert WJ, Dalton PD, Hutmacher DW, Malda J (2015) Reinforcement of hydrogels using three-dimensionally printed microfibres. Nat Commun 6:6933. https://doi.org/10.1038/ncomms7933
65. Wang C, Hou W, Guo X, Li J, Hu T, Qiu M, Liu S, Mo X, Liu X (2017) Two-phase electrospinning to incorporate growth factors loaded chitosan nanoparticles into electrospun fibrous scaffolds for bioactivity retention and cartilage regeneration. Mater Sci Eng C 79:507–515
66. Zhu D, Tong X, Trinh P, Yang F (2017) Mimicking cartilage tissue zonal organization by engineering tissue-scale gradient hydrogels as 3D cell niche. Tissue Eng Part A 24:1. https://doi.org/10.1089/ten.TEA.2016.0453
67. Camarero-Espinosa S, Cooper-White J (2017) Tailoring biomaterial scaffolds for osteochondral repair. Int J Pharm 523(2):476–489. https://doi.org/10.1016/j.ijpharm.2016.10.035
68. Sreejalekshmi KG, Nair PD (2011) Biomimeticity in tissue engineering scaffolds through synthetic peptide modifications-altering chemistry for enhanced biological response. J Biomed Mater Res A 96(2):477–491. https://doi.org/10.1002/jbm.a.32980
69. Bourgine PE, Gaudiello E, Pippenger B, Jaquiery C, Klein T, Pigeot S, Todorov A, Feliciano S, Banfi A, Martin I (2017) Engineered extracellular matrices as biomaterials of tunable composition and function. Adv Funct Mater

70. Song JJ, Ott HC (2011) Organ engineering based on decellularized matrix scaffolds. Trends Mol Med 17(8):424–432. https://doi.org/10.1016/j.molmed.2011.03.005
71. Grayson WL, Frohlich M, Yeager K, Bhumiratana S, Chan ME, Cannizzaro C, Wan LQ, Liu XS, Guo XE, Vunjak-Novakovic G (2010) Engineering anatomically shaped human bone grafts. Proc Natl Acad Sci U S A 107(8):3299–3304. https://doi.org/10.1073/pnas.0905439106
72. Karageorgiou V, Kaplan D (2005) Porosity of 3D biomaterial scaffolds and osteogenesis. Biomaterials 26(27):5474–5491. https://doi.org/10.1016/j.biomaterials.2005.02.002
73. Loh QL, Choong C (2013) Three-dimensional scaffolds for tissue engineering applications: role of porosity and pore size. Tissue Eng Part B Rev 19(6):485–502. https://doi.org/10.1089/ten.TEB.2012.0437
74. Hollister SJ (2005) Porous scaffold design for tissue engineering. Nat Mater 4(7):518–524. https://doi.org/10.1038/nmat1421
75. Hendrikson WJ, van Blitterswijk CA, Rouwkema J, Moroni L (2017) The use of finite element analyses to design and fabricate three-dimensional scaffolds for skeletal tissue engineering. Front Bioeng Biotechnol 5:30. https://doi.org/10.3389/fbioe.2017.00030
76. Marrella A, Aiello M, Quarto R, Scaglione S (2016) Chemical and morphological gradient scaffolds to mimic hierarchically complex tissues: from theoretical modeling to their fabrication. Biotechnol Bioeng 113(10):2286–2297. https://doi.org/10.1002/bit.25994
77. Rajasekharan AK, Bordes R, Sandstrom C, Ekh M, Andersson M (2017) Hierarchical and heterogeneous bioinspired composites-merging molecular self-assembly with additive manufacturing. Small 13. https://doi.org/10.1002/smll.201700550
78. Bracaglia LG, Smith BT, Watson E, Arumugasaamy N, Mikos AG, Fisher JP (2017) 3D printing for the design and fabrication of polymer-based gradient scaffolds. Acta Biomater 56:3. https://doi.org/10.1016/j.actbio.2017.03.030
79. Guo T, Lembong J, Zhang LG, Fisher JP (2017) Three-dimensional printing articular cartilage: recapitulating the complexity of native tissue. Tissue Eng Part B Rev 23(3):225–236. https://doi.org/10.1089/ten.TEB.2016.0316
80. Hendrikson WJ, Deegan AJ, Yang Y, van Blitterswijk CA, Verdonschot N, Moroni L, Rouwkema J (2017) Influence of additive manufactured scaffold architecture on the distribution of surface strains and fluid flow shear stresses and expected osteochondral cell differentiation. Front Bioeng Biotech 5
81. Murphy SV, Atala A (2014) 3D bioprinting of tissues and organs. Nat Biotechnol 32(8):773–785. https://doi.org/10.1038/nbt.2958
82. Cui X, Breitenkamp K, Finn MG, Lotz M, D'Lima DD (2012) Direct human cartilage repair using three-dimensional bioprinting technology. Tissue Eng A 18(11–12):1304–1312. https://doi.org/10.1089/ten.TEA.2011.0543
83. Panwar A, Tan LP (2016) Current status of bioinks for micro-extrusion-based 3D bioprinting. Molecules 21(6). https://doi.org/10.3390/molecules21060685
84. Keriquel V, Oliveira H, Remy M, Ziane S, Delmond S, Rousseau B, Rey S, Catros S, Amedee J, Guillemot F, Fricain JC (2017) In situ printing of mesenchymal stromal cells, by laser-assisted bioprinting, for in vivo bone regeneration applications. Sci Rep 7(1):1778. https://doi.org/10.1038/s41598-017-01914-x
85. Hinton TJ, Jallerat Q, Palchesko RN, Park JH, Grodzicki MS, Shue H-J, Ramadan MH, Hudson AR, Feinberg AW (2015) Three-dimensional printing of complex biological structures by freeform reversible embedding of suspended hydrogels. Sci Adv 1(9):e1500758
86. Mellor LF, Huebner P, Cai S, Mohiti-Asli M, Taylor MA, Spang J, Shirwaiker RA (2017) Loboa EG (2017) fabrication and evaluation of electrospun, 3D-bioplotted, and combination of electrospun/3D-bioplotted scaffolds for tissue engineering applications. Biomed Res Int 2017:1
87. Alexander PG, Gottardi R, Lin H, Lozito TP, Tuan RS (2014) Three-dimensional osteogenic and chondrogenic systems to model osteochondral physiology and degenerative joint diseases. Exp Biol Med 239(9):1080–1095. https://doi.org/10.1177/1535370214539232

88. Ikegawa S (2006) Genetic analysis of skeletal dysplasia: recent advances and perspectives in the post-genome-sequence era. J Hum Genet 51(7):581–586. https://doi.org/10.1007/s10038-006-0401-x
89. Diederichs S, Richter W (2017) Induced pluripotent stem cells and cartilage regeneration. Cartilage. Springer, In, pp 73–93
90. Brookhouser N, Raman S, Potts C, Brafman DA (2017) May I cut in? Gene editing approaches in human induced pluripotent stem cells. Cells 6(1):5
91. Grskovic M, Javaherian A, Strulovici B, Daley GQ (2011) Induced pluripotent stem cells—opportunities for disease modelling and drug discovery. Nat Rev Drug Discov 10(12):915–929
92. Wu SM, Hochedlinger K (2011) Harnessing the potential of induced pluripotent stem cells for regenerative medicine. Nat Cell Biol 13(5):497–505
93. Grayson WL, Bhumiratana S, Grace Chao PH, Hung CT, Vunjak-Novakovic G (2010) Spatial regulation of human mesenchymal stem cell differentiation in engineered osteochondral constructs: effects of pre-differentiation, soluble factors and medium perfusion. Osteoarthritis Res Soc 18(5):714–723. https://doi.org/10.1016/j.joca.2010.01.008
94. Steinmetz NJ, Aisenbrey EA, Westbrook KK, Qi HJ, Bryant SJ (2015) Mechanical loading regulates human MSC differentiation in a multi-layer hydrogel for osteochondral tissue engineering. Acta Biomater 21:142–153. https://doi.org/10.1016/j.actbio.2015.04.015
95. Aisenbrey E, Bryant S (2016) Mechanical loading inhibits hypertrophy in chondrogenically differentiating hMSCs within a biomimetic hydrogel. J Mater Chem B 4(20):3562–3574
96. Vunjak-Novakovic G, Meinel L, Altman G, Kaplan D (2005) Bioreactor cultivation of osteochondral grafts. Orthod Craniofac Res 8(3):209–218. https://doi.org/10.1111/j.1601-6343.2005.00334.x
97. Grayson WL, Chao PH, Marolt D, Kaplan DL, Vunjak-Novakovic G (2008) Engineering custom-designed osteochondral tissue grafts. Trends Biotechnol 26(4):181–189. https://doi.org/10.1016/j.tibtech.2007.12.009
98. Temple JP, Yeager K, Bhumiratana S, Vunjak-Novakovic G, Grayson WL (2014) Bioreactor cultivation of anatomically shaped human bone grafts. Methods Mol Biol 1202:57–78. https://doi.org/10.1007/7651_2013_33
99. Bhumiratana S, Bernhard J, Cimetta E, Vunjak-Novakovic G (2013) Principles of bioreactor design for tissue engineering. Prin Tiss Eng. 261–278
100. Petrenko Y, Petrenko A, Martin I, Wendt D (2017) Perfusion bioreactor-based cryopreservation of 3D human mesenchymal stromal cell tissue grafts. Cryobiology 76:150–153
101. Murphy MK, Huey DJ, Hu JC, Athanasiou KA (2015) TGF-beta1, GDF-5, and BMP-2 stimulation induces chondrogenesis in expanded human articular chondrocytes and marrow-derived stromal cells. Stem Cells 33(3):762–773. https://doi.org/10.1002/stem.1890
102. Fortier LA, Barker JU, Strauss EJ, McCarrel TM, Cole BJ (2011) The role of growth factors in cartilage repair. Clin Orthop Relat Res 469(10):2706–2715
103. Martin I, Suetterlin R, Baschong W, Heberer M, Vunjak-Novakovic G, Freed LE (2001) Enhanced cartilage tissue engineering by sequential exposure of chondrocytes to FGF-2 during 2D expansion and BMP-2 during 3D cultivation. J Cell Biochem 83(1):121–128
104. Byers BA, Mauck RL, Chiang IE, Tuan RS (2008) Transient exposure to transforming growth factor beta 3 under serum-free conditions enhances the biomechanical and biochemical maturation of tissue-engineered cartilage. Tissue Eng A 14(11):1821–1834. https://doi.org/10.1089/ten.tea.2007.0222
105. Buxton AN, Bahney CS, Yoo JU, Johnstone B (2011) Temporal exposure to chondrogenic factors modulates human mesenchymal stem cell chondrogenesis in hydrogels. Tissue Eng A 17(3–4):371–380. https://doi.org/10.1089/ten.TEA.2009.0839
106. Aksel H, Huang GT (2017) Combined effects of vascular endothelial growth factor and bone morphogenetic protein 2 on Odonto/osteogenic differentiation of human dental pulp stem cells in vitro. J Endod 43(6):930–935. https://doi.org/10.1016/j.joen.2017.01.036
107. Song K, Rao NJ, Chen ML, Huang ZJ, Cao YG (2011) Enhanced bone regeneration with sequential delivery of basic fibroblast growth factor and sonic hedgehog. Injury 42(8):796–802. https://doi.org/10.1016/j.injury.2011.02.003

108. Stegen S, van Gastel N, Carmeliet G (2015) Bringing new life to damaged bone: the importance of angiogenesis in bone repair and regeneration. Bone 70:19–27
109. Correia C, Grayson WL, Park M, Hutton D, Zhou B, Guo XE, Niklason L, Sousa RA, Reis RL, Vunjak-Novakovic G (2011) In vitro model of vascularized bone: synergizing vascular development and osteogenesis. PLoS One 6(12):e28352. https://doi.org/10.1371/journal.pone.0028352
110. Spiller KL, Anfang RR, Spiller KJ, Ng J, Nakazawa KR, Daulton JW, Vunjak-Novakovic G (2014) The role of macrophage phenotype in vascularization of tissue engineering scaffolds. Biomaterials 35(15):4477–4488. https://doi.org/10.1016/j.biomaterials.2014.02.012
111. Qi JH, Ebrahem Q, Moore N, Murphy G, Claesson-Welsh L, Bond M, Baker A, Anand-Apte B (2003) A novel function for tissue inhibitor of metalloproteinases-3 (TIMP3): inhibition of angiogenesis by blockage of VEGF binding to VEGF receptor-2. Nat Med 9(4):407–415. https://doi.org/10.1038/nm846
112. Mohammed FF, Smookler DS, Taylor SE, Fingleton B, Kassiri Z, Sanchez OH, English JL, Matrisian LM, Au B, Yeh WC, Khokha R (2004) Abnormal TNF activity in Timp3−/− mice leads to chronic hepatic inflammation and failure of liver regeneration. Nat Genet 36(9):969–977. https://doi.org/10.1038/ng1413
113. Saunders WB, Bohnsack BL, Faske JB, Anthis NJ, Bayless KJ, Hirschi KK, Davis GE (2006) Coregulation of vascular tube stabilization by endothelial cell TIMP-2 and pericyte TIMP-3. J Cell Biol 175(1):179–191. https://doi.org/10.1083/jcb.200603176
114. Luz-Crawford P, Jorgensen C, Djouad F (2017) Mesenchymal stem cells direct the immunological fate of macrophages. In: Macrophages. Springer, pp 61–72
115. Hoogduijn MJ (2017) Immunomodulation by mesenchymal stem cells: lessons from vascularized composite Allotransplantation. Transplantation 101(1):30–31
116. Spiller KL, Nassiri S, Witherel CE, Anfang RR, Ng J, Nakazawa KR, Yu T, Vunjak-Novakovic G (2015) Sequential delivery of immunomodulatory cytokines to facilitate the M1-to-M2 transition of macrophages and enhance vascularization of bone scaffolds. Biomaterials 37:194–207
117. Schipani E, Ryan HE, Didrickson S, Kobayashi T, Knight M, Johnson RS (2001) Hypoxia in cartilage: HIF-1alpha is essential for chondrocyte growth arrest and survival. Genes Dev 15(21):2865–2876. https://doi.org/10.1101/gad.934301
118. Maes C, Carmeliet G, Schipani E (2012) Hypoxia-driven pathways in bone development, regeneration and disease. Nat Rev Rheumatol 8(6):358–366
119. Schipani E, Maes C, Carmeliet G, Semenza GL (2009) Regulation of osteogenesis-angiogenesis coupling by HIFs and VEGF. J Bone Miner Res 24(8):1347–1353
120. Wang GL, Jiang BH, Rue EA, Semenza GL (1995) Hypoxia-inducible factor 1 is a basic-helix-loop-helix-PAS heterodimer regulated by cellular O2 tension. Proc Natl Acad Sci U S A 92(12):5510–5514
121. Myllyharju J, Schipani E (2010) Extracellular matrix genes as hypoxia-inducible targets. Cell Tissue Res 339(1):19–29. https://doi.org/10.1007/s00441-009-0841-7
122. Carreau A, Hafny-Rahbi BE, Matejuk A, Grillon C, Kieda C (2011) Why is the partial oxygen pressure of human tissues a crucial parameter? Small molecules and hypoxia. J Cell Mol Med 15(6):1239–1253
123. Gnaiger E (2003) Oxygen conformance of cellular respiration. A perspective of mitochondrial physiology. Adv Exp Med Biol 543:39–55
124. Yodmuang S, Gadjanski I, Chao PH, Vunjak-Novakovic G (2013) Transient hypoxia improves matrix properties in tissue engineered cartilage. J Orthop Res 31(4):544–553. https://doi.org/10.1002/jor.22275
125. Henrionnet C, Liang G, Roeder E, Dossot M, Wang H, Magdalou J, Gillet P, Pinzano A (2017) Hypoxia for mesenchymal stem cell expansion and differentiation: the best way for enhancing TGFß-induced Chondrogenesis and preventing calcifications in alginate beads. Tissue Eng A
126. Ramage L, Nuki G, Salter DM (2009) Signalling cascades in mechanotransduction: cell-matrix interactions and mechanical loading. Scand J Med Sci Sports 19(4):457–469. https://doi.org/10.1111/j.1600-0838.2009.00912.x

127. Bougault C, Paumier A, Aubert-Foucher E, Mallein-Gerin F (2009) Investigating conversion of mechanical force into biochemical signaling in three-dimensional chondrocyte cultures. Nat Protoc 4(6):928–938. https://doi.org/10.1038/nprot.2009.63
128. Pingguan-Murphy B, El-Azzeh M, Bader D, Knight M (2006) Cyclic compression of chondrocytes modulates a purinergic calcium signalling pathway in a strain rate-and frequency-dependent manner. J Cell Physiol 209(2):389–397
129. Graff RD, Lazarowski ER, Banes AJ, Lee GM (2000) ATP release by mechanically loaded porcine chondrons in pellet culture. Arthritis Rheum 43(7):1571–1579. https://doi.org/10.1002/1529-0131(200007)43:7<1571::AID-ANR22>3.0.CO;2-L
130. Knight MM, McGlashan SR, Garcia M, Jensen CG, Poole CA (2009) Articular chondrocytes express connexin 43 hemichannels and P2 receptors - a putative mechanoreceptor complex involving the primary cilium? J Anat 214(2):275–283. https://doi.org/10.1111/j.1469-7580.2008.01021.x
131. Garcia M, Knight MM (2010) Cyclic loading opens hemichannels to release ATP as part of a chondrocyte mechanotransduction pathway. J Orthop Res 28(4):510–515. https://doi.org/10.1002/jor.21025
132. Gonzales S, Wang C, Levene H, Cheung HS, Huang CY (2015) ATP promotes extracellular matrix biosynthesis of intervertebral disc cells. Cell Tissue Res 359(2):635–642. https://doi.org/10.1007/s00441-014-2042-2
133. Waldman SD, Usprech J, Flynn LE, Khan AA (2010) Harnessing the purinergic receptor pathway to develop functional engineered cartilage constructs. Osteoarthritis Res Soc 18(6):864–872. https://doi.org/10.1016/j.joca.2010.03.003
134. Gadjanski I, Yodmuang S, Spiller K, Bhumiratana S, Vunjak-Novakovic G (2013) Supplementation of exogenous adenosine 5′-triphosphate enhances mechanical properties of 3D cell-agarose constructs for cartilage tissue engineering. Tissue Eng A 19(19–20):2188–2200. https://doi.org/10.1089/ten.TEA.2012.0352
135. Brady MA, Waldman SD, Ethier CR (2014) The application of multiple biophysical cues to engineer functional neocartilage for treatment of osteoarthritis. Part II: signal transduction. Tissue Eng Part B Rev 21(1):20–33
136. Steward AJ, Kelly DJ, Wagner DR (2016) Purinergic signaling regulates the transforming growth factor-β3-induced Chondrogenic response of mesenchymal stem cells to hydrostatic pressure. Tissue Eng A 22(11–12):831–839
137. Gadjanski I, Filipovic N Mathematical modeling of ATP release in response to mechanical stimulation of chondrogenic cells. In: Bioinformatics and Bioengineering (BIBE), 2015 IEEE 15th International Conference on, 2015. IEEE, pp 1–5
138. Rumney RM, Wang N, Agrawal A, Gartland A (2012) Purinergic signalling in bone. Front Endocrinol 3
139. Dixon SJ, Sims SM (2000) P2 purinergic receptors on osteoblasts and osteoclasts: potential targets for drug development. Drug Dev Res 49(3):187–200
140. Lazarowski ER, Boucher RC, Harden TK (2000) Constitutive release of ATP and evidence for major contribution of ecto-nucleotide pyrophosphatase and nucleoside diphosphokinase to extracellular nucleotide concentrations. J Biol Chem 275(40):31061–31068. https://doi.org/10.1074/jbc.M003255200
141. Burnstock G, Knight GE (2017) Cell culture: complications due to mechanical release of ATP and activation of purinoceptors. Cell Tissue Res 370:1–11
142. Schaefer D, Martin I, Shastri P, Padera RF, Langer R, Freed LE, Vunjak-Novakovic G (2000) In vitro generation of osteochondral composites. Biomaterials 21(24):2599–2606
143. Jeon JE, Vaquette C, Theodoropoulos C, Klein TJ, Hutmacher DW (2014) Multiphasic construct studied in an ectopic osteochondral defect model. J Royal Soc Interf/Royal Soc 11(95):20140184. https://doi.org/10.1098/rsif.2014.0184
144. Cross LM, Shah K, Palani S, Peak CW, Gaharwar AK (2017) Gradient nanocomposite hydrogels for Interface tissue engineering. Nanomedicine: Nanotechnol Biol Med
145. D'Amora U, D'Este M, Eglin D, Safari F, Sprecher CM, Gloria A, De Santis R, Alini M, Ambrosio L (2017) Collagen density gradient on 3D printed poly (ε-Caprolactone) scaffolds for Interface tissue engineering. J Tissue Eng Regen Med

146. Dormer NH, Singh M, Wang L, Berkland CJ, Detamore MS (2010) Osteochondral interface tissue engineering using macroscopic gradients of bioactive signals. Ann Biomed Eng 38(6):2167–2182. https://doi.org/10.1007/s10439-010-0028-0
147. Perez RA, Won JE, Knowles JC, Kim HW (2013) Naturally and synthetic smart composite biomaterials for tissue regeneration. Adv Drug Deliv Rev 65(4):471–496. https://doi.org/10.1016/j.addr.2012.03.009
148. Mohan N, Gupta V, Sridharan B, Sutherland A, Detamore MS (2014) The potential of encapsulating "raw materials" in 3D osteochondral gradient scaffolds. Biotechnol Bioeng 111(4):829–841. https://doi.org/10.1002/bit.25145
149. Qu D, Mosher CZ, Boushell MK, Lu HH (2015) Engineering complex orthopaedic tissues via strategic biomimicry. Ann Biomed Eng 43(3):697–717. https://doi.org/10.1007/s10439-014-1190-6

# Chapter 8
# Porous Scaffolds for Regeneration of Cartilage, Bone and Osteochondral Tissue

**Guoping Chen and Naoki Kawazoe**

**Abstract** Porous scaffolds play an important role as a temporary support for accommodation of seeded cells to control their functions and guide regeneration of functional tissues and organs. Various scaffolds have been prepared from biodegradable polymers and calcium phosphate. They have also been hybridized with bioactive factors to control differentiation of stem cells. Except the composition, porous structures of scaffolds are also extremely important for cell adhesion, spatial distribution and tissue regeneration. The method using preprepared ice particulates has been developed to precisely control surface and bulk pore structures of porous scaffolds. This chapter summarizes the design and preparation of porous scaffolds of biodegradable polymers and their hybrid scaffolds with calcium phosphate nanoparticles and bioactive factors. Their applications for regeneration of cartilage, bone and osteochondral tissue will be highlighted.

**Highlights**

- Porous scaffolds of naturally derived polymers and their hybrid scaffolds with biodegradable synthetic polymers have been prepared for cartilage tissue engineering. The surface and bulk pore structures of the scaffolds are controlled by using preprepared ice particulates. The scaffolds facilitate cartilage tissue engineering when they are used for three-dimension culture of chondrocytes.
- PLGA–collagen–BMP4 and collagen–CaP nanoparticles–dexamethasone hybrid scaffolds have been prepared and used for culture of mesenchymal stem cells. The hybrid scaffolds facilitate osteogenic differentiation of mesenchymal stem cells and ectopic bone tissue regeneration during in vitro culture and in vivo implantation.

G. Chen (✉) · N. Kawazoe
Research Center for Functional Materials, National Institute for Materials Science, Tsukuba, Japan
e-mail: guoping.chen@nims.go.jp

- Osteochondral tissue engineering has been realized by laminating two different layers of cartilage and subchondral bone or by using stratified scaffolds for simultaneous regeneration of cartilage and subchondral bone.

**Keywords** Porous scaffold · Hybrid scaffold · Biodegradable synthetic polymer · Calcium phosphate · Bioactive factor · Cartilage · Bone · Osteochondral

## 8.1 Scaffolds for Cartilage Tissue Engineering

### 8.1.1 Porous Scaffolds with Funnel-Like Surface Pore Structures

Scaffolds with interconnected pore structures are extremely important for homogeneous cell adhesion and distribution in the porous scaffolds, thus facilitating cartilage tissue engineering. Although a number of three-dimensional porous scaffolds have been developed from various types of biodegradable polymers, not all the scaffolds can ensure homogenous tissue formation in the scaffolds. In general, cells are easily allocated and distributed in the peripheral areas, resulting in tissue regeneration only in the outermost peripheral layers of the scaffolds. Uniform cell delivery and distribution throughout the entire scaffold are desirable for functional tissue engineering.

Some methods have been developed for controlling pore structure of scaffolds, such as pore size, porosity and interconnectivity [1–4]. Porogen-leaching method offers many advantages for the easy manipulation and control of pore size and porosity. However, the porogen materials cannot initiate the formation of surrounding pores. As a result, isolated pores are often formed in the scaffold, a situation which is not desirable for tissue engineering scaffolds.

A method by using preprepared ice particulates as a porogen material has been developed to efficiently control the pore structure of scaffolds [5]. To make the surface pores open, embossed ice particulates have been used. To increase the pore interconnectivity, free ice particulates have been used to initiate the formation of new ice crystals during the preparation of porous scaffolds [6].

Embossed ice particulates have been used to prepare collagen porous scaffolds with funnel-like open surface pore structures for cartilage tissue engineering [7]. At first, embossing ice particulate templates are prepared by freezing microsized water droplets on a perfluoroalkoxy (PFA)-film wrapped copper plate at −30 °C. The water droplets are formed by spraying pure water onto the PFA film with a trigger sprayer. Templates with ice particulates having three different diameters of about 180 μm, 400 μm and 720 μm are prepared and used for preparation of collagen porous scaffolds. Subsequently, the ice particulate templates are moved and kept in a low-temperature chamber for 1 h to maintain their temperatures at −1 °C, −3 °C, −5 °C or −10 °C. An aqueous solution of porcine type I collagen is poured onto

8 Porous Scaffolds for Regeneration of Cartilage, Bone, and Osteochondral Tissue 173

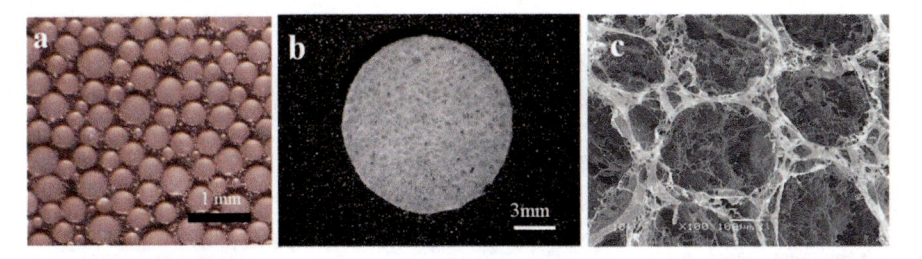

**Fig. 8.1** Photomicrograph of ice particulates with a mean diameter of about 400 mm (**a**), gross appearance (**b**) and top surface SEM image (**c**) of 400 μm_3 °C funnel-like collagen sponge. Reproduced with permission [7]

each of the embossing ice template with a 1-mm thick silicone frame to maintain the collagen thickness. The collagen solution is stored at −1 °C for 24 h for cooling before being poured onto the templates. The collagen solution does not freeze during cooling at −1 °C in the absence of a nucleating agent. After being poured on the ice particulate templates, the top of the collagen solution is covered with a glass plate wrapped with polyvinylidene chloride film and the whole set is placed in a low temperature chamber. The temperature of the chamber is kept at −1 °C, −3 °C, −5 °C, or −10 °C to initiate the growth of ice crystals. The chamber is kept at this temperature for 1 h to allow the formation of ice crystals. Subsequently, the frozen collagen solution and the template are moved to a freezer with a temperature of −80 °C for 5 h. Finally, the frozen construct is freeze-dried for 24 h and the freeze-dried collagen sponges are cross-linked for 4 h by glutaraldehyde vapor that is saturated with 25% aqueous glutaraldehyde solution at 37 °C. The unreacted aldehyde group is blocked by treatment with 0.1 M aqueous glycine solution. The collagen sponges are washed with Milli-Q water, freeze-dried and kept at 4 °C before further investigation. Control collagen sponges are prepared by using the same procedure at −3 °C but without the use of the ice particulate template.

The ice particulates have hemispherical morphology (Fig. 8.1a). Large surface pores are visible on the top surface (Fig. 8.1b). SEM observation shows large hemispherical pores are formed on the surface of the collagen sponges (Fig. 8.1c). The collagen sponge has a hierarchical porous structure of two layers: a surface porous layer and a bulk porous layer. The surface porous layer is composed of large open pores and the bulk porous layer is composed of inner bulk pores that are interconnected with the large surface pores and extended into the bulk body of the sponge. Such structure is very similar to a funnel and therefore the collagen sponges prepared by this method are referred as funnel-like collagen sponges.

By using ice particulates of different sizes (180 μm, 400 μm, and 720 μm) and different freezing temperatures (−1 °C, −3 °C, −5 °C, and −10 °C), 6 types of collagen sponges with different pore structures and 1 control collagen sponge are prepared. They were designated in abbreviated form as: 400 μm_-1 °C (400 μm ice particulate template and freezing temperature of −1 °C), 400 μm_−3 °C, 400 μm_−5 °C, 400 μm_−10 °C, 180 μm_−3 °C, 720 μm_−3 °C and control (with no embossing ice particulate template). The size of the large surface pores in the

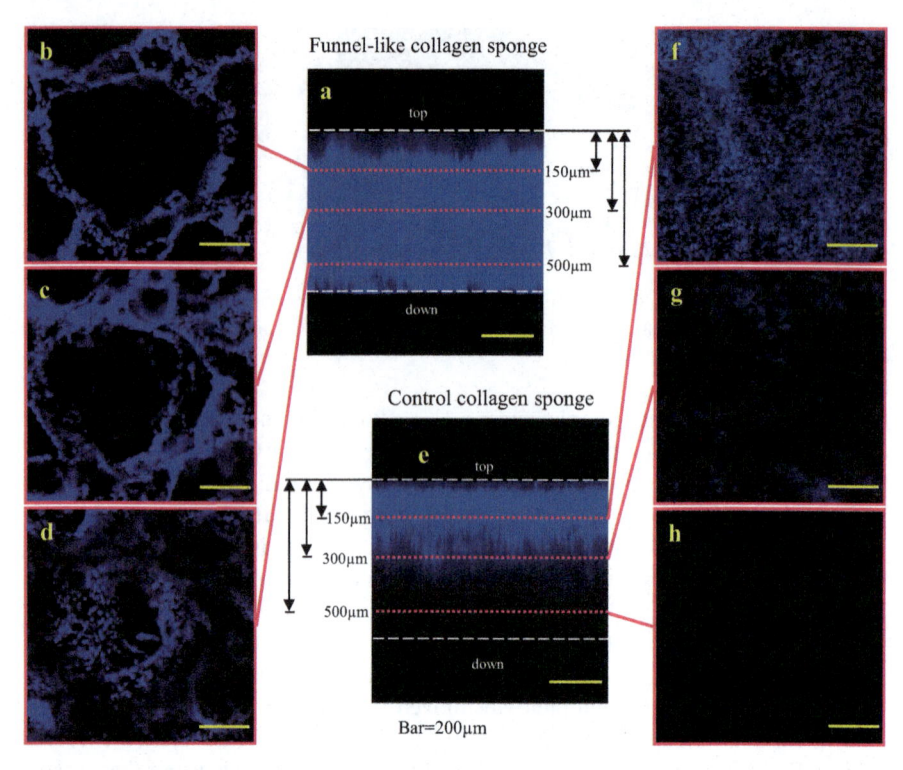

Fig. 8.2 LSCM observation of cell adhesion in the 400 μm −3 °C funnel-like collagen sponge (**a-d**) and control collagen sponge (**e-h**). (**a**) and (**e**) are the gross images of sponges viewing from XZ direction. (**b-d**) and (**f-h**) are images of sections at certain depth viewing from XY direction as shown in the insert. Blue dots indicate cell nuclei stained with Hoechst 33,258. Scale bar = 200 μm. Reproduced with permission [7]

funnel-like collagen sponges is determined by the embossing ice particulates and is very close to the size of the respective ice particulates. The size of the inner bulk pores is determined by the temperature of freezing. Therefore the pore sizes of funnel-like collagen sponges can be controlled by the ice particulate size and freezing temperature.

The funnel-like collagen sponges are used for three-dimensional culture of bovine articular chondrocytes to investigate their pore structure on cartilage tissue engineering. Bovine articular chondrocytes that are isolated from knee articular cartilage of an 8-week-old male calf are subcultured once and the subcultured chondrocytes are seeded in the funnel-like and control collagen sponges. The cell–scaffold constructs are cultured in DMEM serum medium. After in vitro culture for 2 h, the chondrocyte distribution in the 400 μm_−3 °C funnel-like and control collage sponges is estimated by observing cell nuclei that are stained with Hoechst 33,258. A domain of the collagen sponge 600 μm high is scanned using a laser scanning confocal microscope (Fig. 8.2). The cells adhere on both the funnel-like and control collagen sponges. However, the cell distribution is dissimilar. Cell nuclei are observed in the whole funnel-like collagen sponge, indicating that the cells are

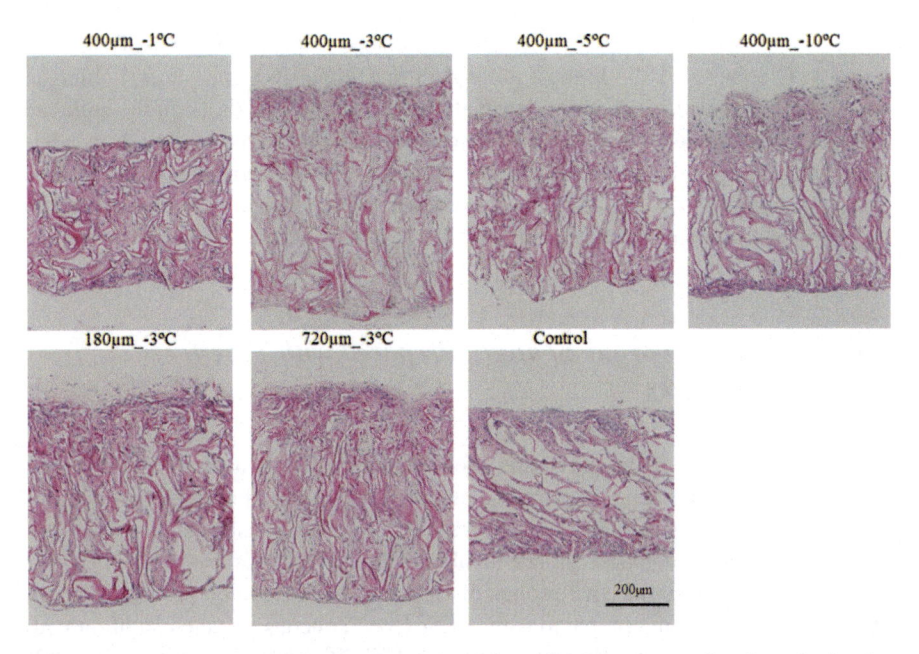

**Fig. 8.3** HE staining photomicrographs of the cell–scaffold constructs after 3 weeks in vivo implantation. Reproduced with permission [7]

evenly distributed in the funnel-like collagen sponge, spanning all parts of the sponge. However, in the control collagen sponge, the cells only accumulate on the top surface (200~300 μm). Few cells are observed in areas beneath the surface layer of the control collagen sponge.

SEM observation shows that chondrocytes are distributed more homogeneously in the funnel-like collagen sponges than in the control collagen sponge. In a comparison of the funnel-like collagen sponges, the chondrocytes adhere and distribute more homogeneously in the 400 μm_−1 °C, 400 μm_−3 °C and 720 μm_−3 °C funnel-like sponges than in the 400 μm_−5 °C, 400 μm_−10 °C and 180 μm_−3 °C funnel-like sponges.

After in vitro culture for 1 week, the cell–scaffold constructs are subcutaneously transplanted into athymic nude mice for 3 weeks. Hematoxylin and eosin staining shows that the tissue regenerated in the control sponges has a middle-hollow structure, which may be due to the limited penetration of cells into the inner parts of the sponge (Fig. 8.3). The 400 μm_−5 °C, 400 μm_−10 °C and 180 μm_−3 °C funnel-like collagen sponges show nonhomogeneous cell distribution: more cells are in the top layer and fewer cells are in the bottom layer of the sponges. The cells and ECM are relatively homogeneously distributed in the 400 μm_−1 °C, 400 μm_−3 °C and 720 μm_−3 °C funnel-like collagen sponges. The tissue formed in the 400 μm_−3 °C sponge is thicker than that formed in the 400 μm_−1 °C and 720 μm_−3 °C sponges.

Expression of genes encoding type I collagen, type II collagen, and aggrecan of the implants is analyzed by real-time RT-PCR. The cells express higher levels of *Col2a1* and *Acan* genes in the funnel-like collagen sponges than they do in the control collagen sponge. The *Col2a1* and *Acan* gene expression level of the 400 μm_−3 °C sponges is the highest among all the funnel-like and control collagen sponges. These results indicate the formation of cartilage-like tissue in the collagen sponges. The 400 μm_−3 °C funnel-like collagen sponge supports chondrogenesis more strongly than do the other collagen sponges. The 400 μm_−3 °C funnel-like collagen sponge provides optimal surface pores and inner bulk pores for cell seeding, cell infiltration and distribution.

## 8.1.2 Collagen Porous Sponges Prepared with Free Ice Particulates

Free ice particulates are prepared and used to prepare collagen sponges with interconnected pore structure [8]. Ice particulates are prepared by spraying Milli Q water into liquid nitrogen using a trigger sprayer. The ice particulates are sieved by sieves with 335 and 425 μm mesh pores to obtain ice particulates having a diameter between 335 and 425 μm. Two groups of collagen sponges are prepared. Group A: the collagen sponges are prepared using a 2% (w/v) aqueous collagen solution with a ratio of ice particulates/collagen solution of 25%, 50% and 75% (w/v). Group B: the collagen sponges are prepared using 1%, 2% and 3% (w/v) aqueous collagen solution with a ratio of ice particulates/collagen solution of 50% (w/v).

The collagen sponges are prepared by using the free ice particulates (Fig. 8.4a). At first, collagen solution is prepared by dissolving freeze-dried porcine type I collagen in a solution (10:90 (v/v)) of ethanol and 0.1 M acetic acid (pH 3.0) at

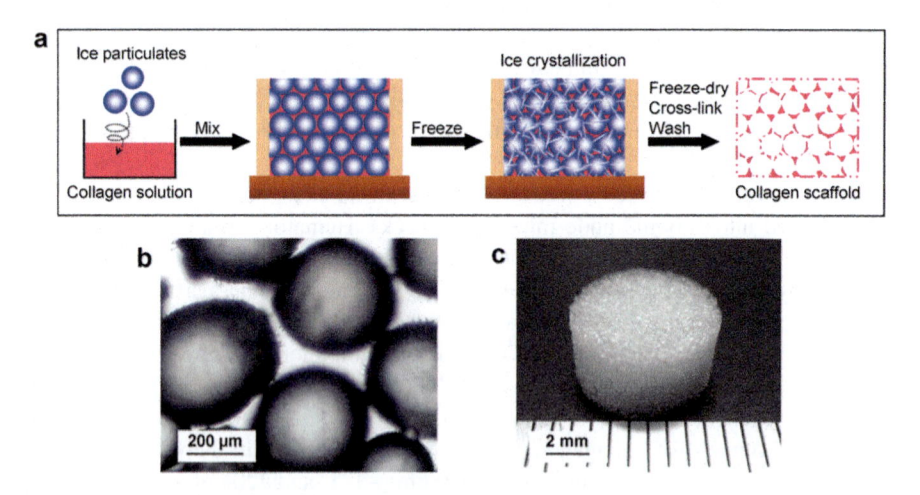

**Fig. 8.4** Preparation scheme for the ice–collagen sponges prepared by using ice particulates as a porogen material (**a**), a phase-contrast photomicrograph of ice particulates (**b**) and a photo of the ice–collagen sponge (**c**). Reproduced with permission [8]

4 °C. The acetic acid/ethanol mixture is used to decrease the freezing temperature of the collagen solution below −4 °C. Before mixing, the ice particulates and collagen solution are kept in a −4 °C chamber for 6 hours to balance the temperature. Subsequently, the ice particulates are added to the collagen solution and mixed thoroughly with a steel spoon to ensure that the ice particulates and collagen solution are homogeneously mixed. The mixture is poured onto a PFA-film wrapped copper plate with a 10 mm thick silicone frame, flattened with a steel spatula and covered with a glass plate wrapped with polyvinylidene chloride film.

The manipulation was conducted in a −4 °C chamber. After mixing, the whole set is kept at −80 °C for 6 h. The frozen structures are freeze-dried for 3 days in a freeze-dryer under a vacuum of 20 Pa to form collagen sponges. Finally, the freeze-dried collagen sponges are cross-linked for 6 h with glutaraldehyde vapor. After cross-linking, the scaffolds are immersed in a 0.1 M aqueous glycine solution to deactivate any unreacted aldehyde groups. The scaffolds are washed 6 times with pure water. The washed scaffolds are freeze-dried again for SEM observation. The control collagen sponge is prepared by the same procedure with a 2% (w/v) concentration of collagen solution without the use of ice particulates.

The preprepared ice particulates are spherical (Fig. 8.4b). The collagen sponges prepared with ice particulates are defined as ice–collagen sponges (Fig. 8.4c). SEM observation shows that there are interconnected large pores and small pores in the ice–collagen sponges (Fig. 8.5). The large pores were spherical and of the same size as the ice particulates. The small pores have a random morphology and different sizes. The small pores surround the large spherical pores. The large pores are negative replicas of the preprepared ice particulates, while the small ice particulates are from the ice crystals that are formed during freezing. The density of the large spherical pores is low when 25% ice particulates are used to prepare the ice–collagen scaffolds and increases with increasing percentages of ice particulates. The ice–collagen sponges prepared with 50% ice particulates have the most homogenous pore structure. When 75% ice particulates are used, some collapsed large pores are observed due to an insufficient amount of collagen matrix and incomplete mixing.

The effect of the collagen concentration on the pore structure is investigated by fixing the ice particulate ratio at 50% (w/v) and changing the collagen concentration from 1 to 3% (w/v). Collapsed large pores are observed in ice–collagen sponges prepared with 1% and 3% aqueous collagen solutions. The collapsed large pores in collagen sponges prepared with the 1% aqueous collagen solution can have occurred because the low concentration results in a less dense collagen matrix surrounding the large pores. When a 3% aqueous collagen solution is used, the imperfect structures may result from incomplete mixing because the 3% collagen solution is too viscous. The ice–collagen sponge prepared with the 2% collagen solution and an ice particulate/collagen solution ratio of 50% shows the most homogeneous pore structure. The control collagen sponge prepared without ice particulates has a heterogeneous lamellar pore structure (Fig. 8.5 k and l).

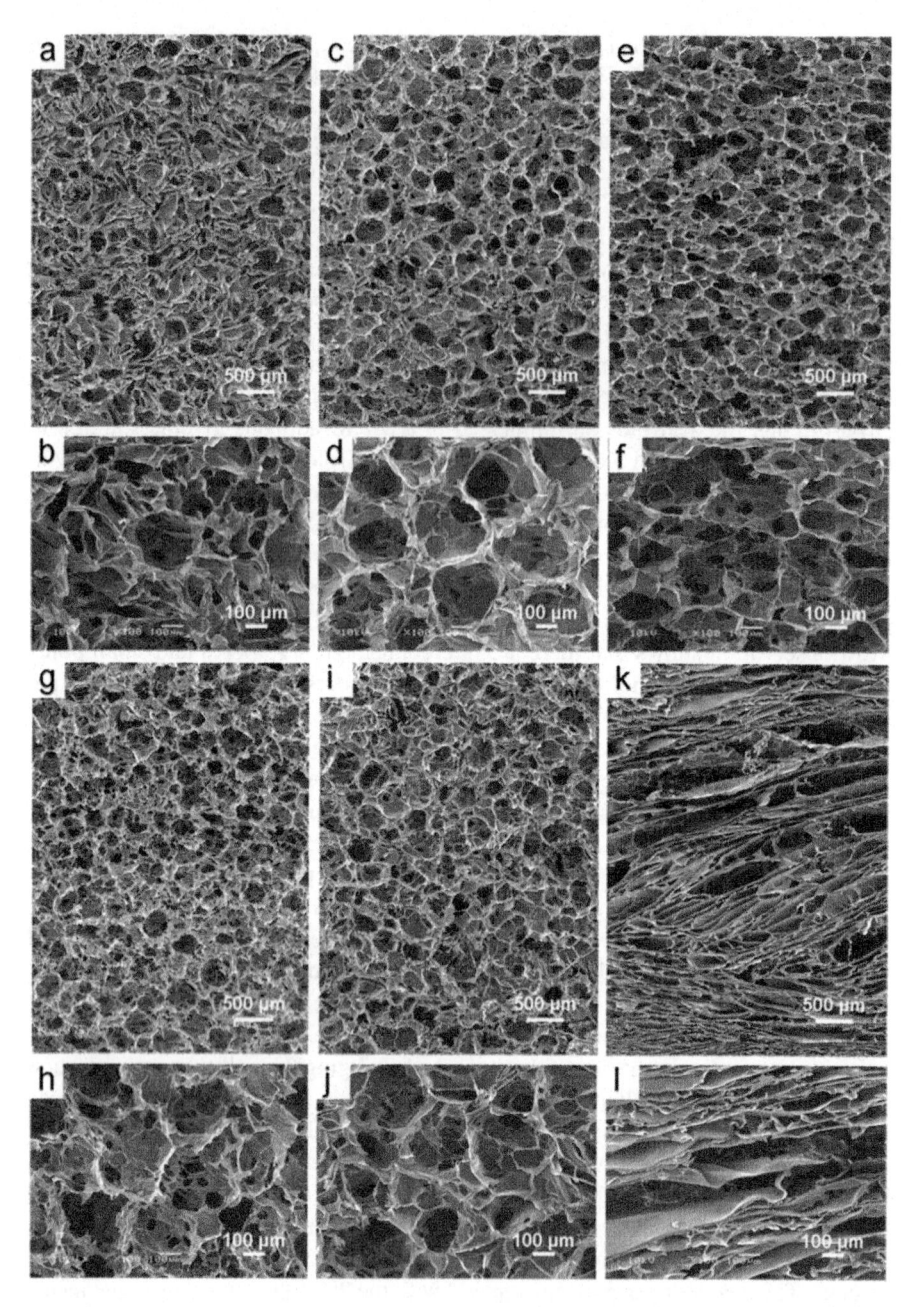

**Fig. 8.5** SEM images of cross sections of the ice–collagen sponges prepared with 2% aqueous collagen solution and ice particulates at a ratio of ice particulates/collagen solution of 25% (**a, b**), 50% (**c, d**) and 75% (**e, f**); ice–collagen sponges prepared with a ratio of ice particulates–collagen solution of 50% and a collagen solution concentration of 1% (**g, h**) and 3% (**i, j**) and control collagen sponge prepared with 2% aqueous collagen solution without the use of ice particulates (**k, l**). The freezing temperature was −80 °C. The images are shown in low (**a, c, e, g, i, k**) and high (**b, d, f, h, j, l**) magnification. Reproduced with permission [8]

Young's modulus of the ice–collagen sponges prepared with 2% collagen and 25% ice particulates, 2% collagen and 50% ice particulates, 3% collagen and 50% ice particulates is significantly higher than that of the control collagen sponge. When collagen concentration is fixed at 2% and the ratio of ice particulates is changed, Young's modulus of the ice–collagen scaffolds increases in the following order: 75%<25%<50%. The ice–collagen sponge prepared with 50% ice particulates has the highest Young's modulus. The differences in the mechanical properties are mainly ascribed to the different pore structures. The spherical pores formed by ice particulates are thought to resist mechanical loading, therefore reinforcing the collagen sponges. The high mechanical strength of the collagen sponge prepared with 50% ice particulates should be due to the most appropriate packing of the large spherical pores and appropriate filling of the collagen matrix between the large spherical pores. The low mechanical strength of the collagen sponge prepared with 75% ice particulates should be due to the partially collapsed large pore structure. Young's modulus increases as the collagen concentration increases, which can be explained by the presence of a dense collagen matrix surrounding the large pores when the collagen concentration increases.

The ice–collagen sponges are used for three-dimensional culturing of bovine articular chondrocytes. Bovine chondrocytes are seeded on one side of the ice–collagen and control collagen scaffolds. After in vitro culture for 1 week, the ice–collagen sponge prepared with 2% collagen and 50% ice particulates shows the most homogeneous cell distribution due to its homogeneous pore structure. The ice–collagen sponges prepared with 25% and 75% ice particulates at a fixed collagen concentration of 2% and different collagen concentrations at a fixed ratio of ice particulates of 50% show less homogenous cell distributions than the scaffolds prepared with 50% ice particulates and 2% collagen solution. However, they show a more homogeneous cell distribution than the control collagen scaffold.

The cell distribution throughout the scaffold is a premise for uniform tissue formation. From the above results, the ice–collagen sponges prepared with 2% collagen and 50% ice particulates shows the most homogeneous cell distribution. Therefore, this scaffold is chosen for cartilage regeneration and compared to the control collagen sponge. The chondrocyte/scaffold constructs after 1 week in vitro culture are transplanted subcutaneously in athymic nude mice for 8 weeks. Gross appearance indicates that the ice–collagen sponge implant maintains its original shape, while the control collagen sponge implant deforms (Fig. 8.6a, b). The high mechanical strength of the ice–collagen scaffold protects it from cell-mediated contraction and suppression by the surrounding tissue during transplantation. Hematoxylin and eosin staining shows a uniform spatial distribution of cells, a uniform ECM distribution and homogeneous tissue formation in the ice–collagen sponge (Fig. 8.6c, d). However, the control collagen sponge shows an uneven cell distribution and some void spaces remain in the control sponge (Fig. 8.6e, f).

The sGAG–DNA ratio increases significantly from 1 week of in vitro culture to 8 weeks of in vivo transplantation, which indicates that the chondrocytes produce ECM continually during in vivo transplantation. The sGAG–DNA ratio in the ice–collagen sponge is significantly higher than that in the control collagen scaffold.

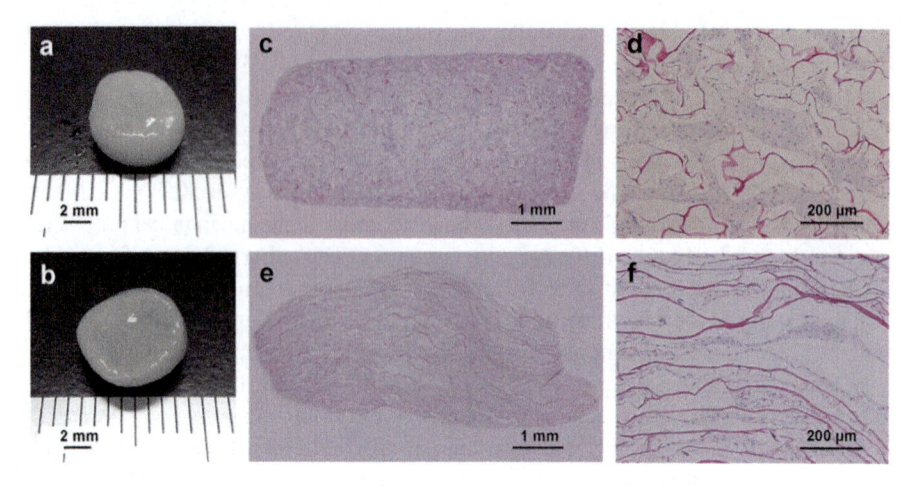

**Fig. 8.6** Gross appearance (**a, b**) and photomicrographs of HE staining (**c-f**) of engineered cartilage after 8 weeks of in vivo implantation of the cells/scaffold of the ice–collagen sponge prepared with 50% ice particulates and 2% aqueous collagen solution (**a, c, d**) and the control collagen sponge (**b, e, f**). Reproduced with permission [8]

The result indicates that the ice–collagen sponge is more favorable to the production of cartilaginous ECM and chondrocyte maturation than the control scaffold.

Immunohistological staining of aggrecan and type II collagen shows that aggrecan and type II collagen in the ice–collagen sponge are much more homogeneously and strongly stained than those in the control sponge. The compression Young's modulus of the regenerated cartilage tissue in the ice–collagen sponge and in the control collagen sponge are 199.0 ± 8.7 and 118.7 ± 7.5 kPa, respectively. Cartilage regenerated in the ice–collagen sponge has significantly higher mechanical property than that regenerated in the control collagen sponge. The ice–collagen sponge promotes cartilage tissue regeneration more strongly than do the control collagen sponge.

## 8.1.3 PLGA–Collagen Hybrid Scaffolds for Cartilage Tissue Engineering

Porous scaffolds prepared from biodegradable synthetic polymers and naturally derived polymers have their respective advantages. Naturally derived polymers such as collagen and hyaluronic acid have hydrophilic surfaces and specific cell interaction peptides, which are excellent for cell growth, but their weak mechanical property makes it very difficult to withstand compression when implanted into the cartilage defect. On the other hand, biodegradable synthetic polymers such as poly(lactic acid) (PLA), poly(glycolic acid) (PGA) and their copolymer poly(lactic-co-glycolic acid) (PLGA) can easily be formed into designed shapes with relatively

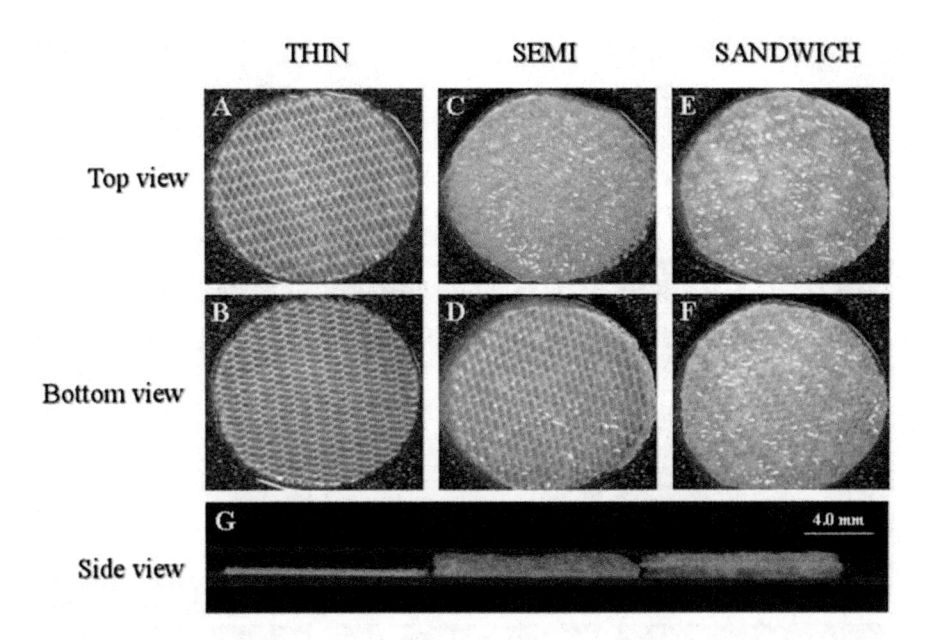

8 Porous Scaffolds for Regeneration of Cartilage, Bone, and Osteochondral Tissue 181

**Fig. 8.7** Gross view of the THIN-, SEMI- and SANDWICH-type PLGA–collagen hybrid scaffolds. (**a, b, c**) top view of the scaffolds, (**d, e, f**) bottom view of the scaffolds and (**g**) side view of the scaffolds. Reproduced with permission [10]

high mechanical strength, but their hydrophobic surface is not favorable for cell seeding. They do not have specific cell interaction motif either. Ideal scaffold for cartilage tissue engineering should have good cell affinity and enough mechanical strength to serve as an initial support.

Therefore, hybrid scaffolds have been prepared by forming microsponges of naturally derived polymers in the openings of a porous synthetic polymer skeleton [9]. The hybrid scaffolds exhibit the properties of high mechanical strength, ease of handling, easy cell seeding and uniform distribution and facilitate the formation of cartilage tissue in in vivo and in vitro experiments. As an example, three types of PLGA–collagen hybrid scaffolds having different thickness are prepared by forming collagen microsponges in the openings of a PLGA knitted mesh and collagen sponge on one or both sides of the PLGA knitted mesh (Fig. 8.7) [10]. In the THIN-type PLGA–collagen hybrid scaffold, web-like collagen microsponges are formed in the openings of the synthetic polymer knitted mesh. The collagen microsponges are connected by the collagen fibers that lay on the PLGA fiber bundles. The SEMI-type PLGA–collagen hybrid scaffolds are prepared by forming a thick collagen sponge layer on one side of the thin PLGA–collagen hybrid scaffold. The SANDWICH-type PLGA–collagen hybrid scaffolds are prepared by forming a thick collagen sponge layer on both sides of the thin PLGA–collagen hybrid scaffold. The thickness of the collagen sponge layer can be adjusted depending on the size of engineered cartilage.

THIN          SEMI          SANDWICH

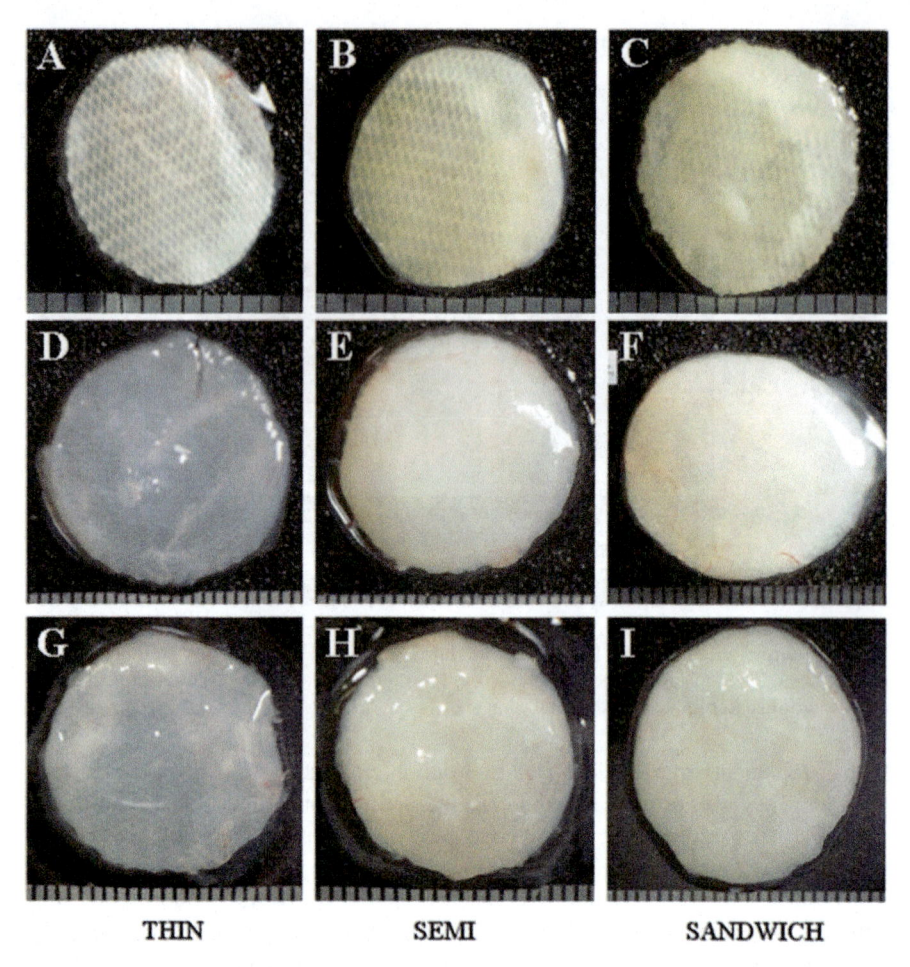

**Fig. 8.8** Gross view of the implants of cell-seeded scaffolds 2, 4 and 8 weeks after implantation. (**a, d, g**) gross view of THIN scaffold 2, 4 and 8 weeks after transplantation, (**b, e, h**) gross view of SEMI scaffold 2, 4 and 8 weeks after transplantation and (**d, f, i**) gross view of SANDWICH scaffold 2, 4 and 8 weeks after transplantation. Bar = 1.0 mm. Reproduced with permission [10]

Bovine chondrocytes after once subculture are seeded in the THIN-, SEMI- and SANDWICH-type PLGA–collagen hybrid scaffolds. After in vitro culture for 1 week, the cells–scaffold constructs are transplanted subcutaneously in the dorsa of athymic nude mice for 2, 4 and 8 weeks. Cartilage-like tissues are regenerated and appear glistening white (Fig. 8.8).

All the three groups of transplants show a spatially even cell distribution, natural chondrocyte morphology and abundant cartilaginous extracellular matrix deposition. The SEMI- and SANDWICH-type PLGA–collagen hybrid scaffolds show larger cell accommodation, greater cell seeding efficiency, higher production of GAGs per DNA and higher expression of type II collagen and aggrecan mRNA and therefore thicker newly formed cartilage is observed when compared to the

THIN-type PLGA–collagen hybrid scaffold. Histological structure and mechanical property of the engineered cartilage using the SEMI- and SANDWICH-type PLGA–collagen hybrid scaffolds match the native bovine articular cartilage better than do that regenerated in the THIN-type PLGA–collagen hybrid scaffolds. These scaffolds with the designed structure can be used for tissue regeneration of articular cartilage with an adjustable thickness.

## 8.2 Scaffolds for Bone Tissue Engineering

### 8.2.1 PLGA–Collagen–BMP4 Hybrid Scaffolds for Bone Tissue Engineering

In bone tissue engineering, porous scaffolds can provide 3D space for cell growth and extracellular matrix formation and structural support for the newly formed bone tissue. Biodegradable polymers have been frequently used to prepare tissue engineering scaffolds due to their versatile properties. Naturally derived polymers such as collagen and synthetic polymers such as PLGA have been extensively studied for tissue engineering. Furthermore, the hybridization of mechanically weak collagen sponges with mechanically strong synthetic polymer skeletons has been shown to combine the advantages of both collagen and synthetic polymers. The hybrid scaffolds show high mechanical strength and good cell interaction and show great potential for tissue engineering applications. Despite these advances, the great challenge involves increasing the osteoinductive capacity because the scaffolds alone are very limited in the regenerative stimulation of large bone defects. One strategy to solve this problem is the incorporation of cell growth factors into the 3D scaffolds.

Growth factors such as bone morphogenetic proteins (BMPs) are often employed to promote bone regeneration. BMPs are osteoinductive growth factors that can induce bone formation both in vivo and in vitro. They have been widely used in tissue engineering approaches for the repair of bone injuries and bone defects. One of the BMP family proteins, BMP4, has been shown to be one of the attractive factors that induces osteogenic differentiation of osteoblasts and osteo-progenitors and promotes bone formation. The BMPs can be delivered by direct injection and release from a carrier. Compared to these conventional administrations, the immobilization method has attracted much attention as a new delivery method. It has been reported that immobilized growth factors can be more efficiently used than free growth factors, and therefore, the amount of growth factors required can be reduced. Furthermore, the immobilized growth factors can be localized and retained in the designated location to maintain the stimulation effect for a long period. Immobilized growth factors can mimic the in vivo microenvironment where growth factors bind to extracellular matrices to regulate their bioactivities. Typically, chemical crosslinking is used to immobilize growth factors into biomaterials and scaffolds. However, crosslinking by covalent bonds may cause structural changes in large protein molecules. Use of fusion proteins of growth factors with a binding domain

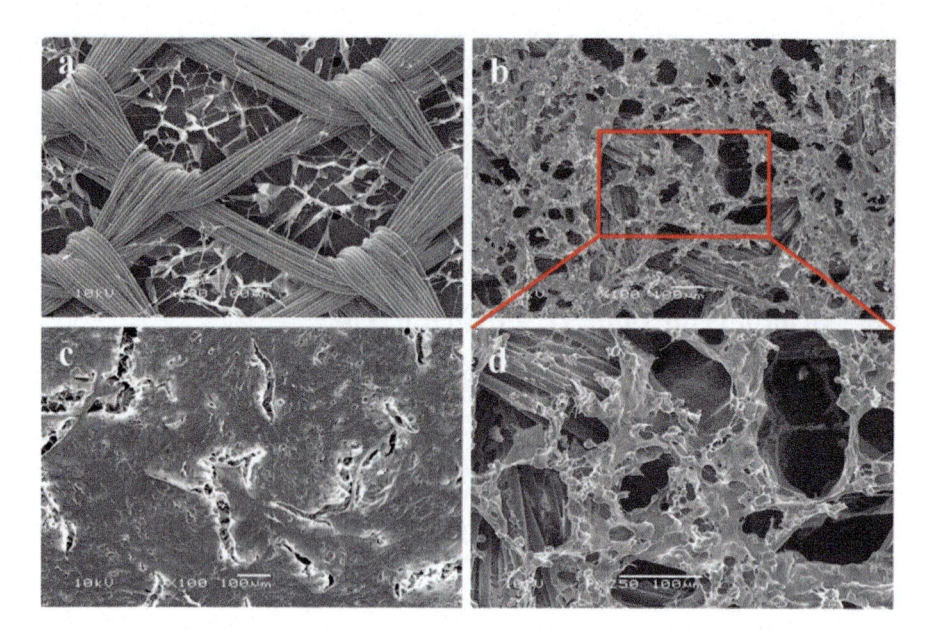

**Fig. 8.9** SEM micrographs of the BMP4-immobilized collagen–PLGA hybrid scaffold before cell culture (**a**) and after MSC culture for 1 h (**b, d**) and 24 h (**c**). (**d**) is a higher magnification of (**b**). Scale bar = 100 μm. Reproduced with permission [11]

derived from the extracellular matrix is another attractive strategy for immobilization of growth factors into the respective extracellular matrix while maintaining their bioactivity. A collagen binding domain (CBD) derived from fibronectin has been shown to prolong the retention of the growth factors at the site of injury, thereby enhancing its activity. Adding CBD to BMP4 may increase the binding capacity of BMP4 to collagen-containing scaffolds for immobilization while maintaining its high bioactivity.

BMP4 has been spatially immobilized into a PLGA–collagen hybrid scaffold using a CBD-BMP4 fusion protein [11]. BMP4 is first fused to a polypeptide sequence that is derived from the fibronectin CBD in a way that does not affect the bioactivity of the fusion protein. The CBD-BMP4 is produced from transgenic silkworms. Subsequently, CBD-BMP4 is immobilized into the PLGA–collagen hybrid scaffold. The BMP4-immobilized PLGA–collagen hybrid scaffold has the same pore structure as that of the PLGA-collagen hybrid scaffold where cobweb-like collagen microsponges are formed in the openings of the synthetic PLGA mesh (Fig. 8.9a). Human bone marrow-derived mesenchymal stem cells (hMSCs) are seeded and cultured in the PLGA–collagen–BMP4 hybrid scaffold. The hMSCs adhere to the hybrid scaffold and show uniform distribution in the hybrid mesh (Fig. 8.9b, d). The cells proliferate and produce ECM that fills the spaces in the mesh after 24 h of culture (Fig. 8.9c).

After in vitro culture for 24 h, the cells/scaffold constructs are subcutaneously transplanted in athymic nude mice for 4 weeks. The hMSCs in the PLGA–collagen–BMP4 hybrid scaffolds show higher ALP activity than the cells cultured in the PLGA–collagen hybrid scaffolds treated with free BMP4, CBD or PBS. The hMSCs cultured in the PLGA–collagen–BMP4 hybrid scaffold show obvious positive staining of deposited calcium.

Gene expression results indicate that the hMSCs cultured in the PLGA–collagen–BMP4 hybrid scaffold show higher expression of genes encoding type I collagen (*COL1A1*), alkaline phosphatase (*ALPL*), osteopontin (*SPP1*) and osteocalcin (*BGLAP*) compared with the hMSCs cultured in the PLGA–collagen hybrid scaffold treated with free BMP4, CBD or PBS after 2 and 4 weeks transplantation. The PLGA–collagen–BMP4 hybrid scaffold shows strong osteogenic induction activity to hMSCs.

## 8.2.2 Collagen–CaP–DEX Hybrid Scaffolds for Bone Tissue Engineering

To mimic the extracellular matrix composition of bone, collagen and calcium phosphate (CaP) are hybridized to prepare collagen–CaP hybrid scaffolds. To further render the scaffolds with good osteoinductivity, dexamethasone (DEX), as a low-molecular-weight osteoinductive cue, is incorporated in scaffolds to compose collagen–CaP–DEX hybrid scaffolds [12]. At first, DEX is added in $Ca(NO_3)_2 \cdot 4H_2O$ aqueous solution. The mixture solution is reacted with $(NH_4)_2HPO_4$ aqueous solution to prepare DEX-loaded BCP nanoparticles (NPs). Subsequently, the DEX-loaded BCP NPs are mixed with collagen aqueous solution and preprepared ice particulates. The mixture is frozen and freeze-dried to form collagen–CaP–DEX hybrid scaffolds. Finally, the freeze-dried hybrid scaffolds are cross-linked with 50 mM 1-ethyl-3-(3-dimethylaminopropyl) carbodiimide and 20 mM N-hydroxysuccimide in an 80% (v/v) ethanol aqueous solution at room temperature under gentle shaking for 8 h. After cross-linking, the hybrid scaffolds are washed three times with Milli-Q water and immersed in a 0.1 M glycine aqueous solution to block unreacted NHS residues.

The ratio of BCP NPs/collagen can affect the pore structure and mechanical property of the hybrid scaffolds. When the mass ratio of BCP NPs–collagen is 1:1, the hybrid scaffolds have the best interconnected pore structure and still very high mechanical property. Therefore, the BCP NPs–collagen mass ratio of 1:1 is used for preparation of the hybrid scaffolds. The Collagen–CaP–DEX scaffolds have well interconnected large pores (Fig. 8.10a). The spherical large pores are distributed homogeneously throughout the scaffolds and their size was almost equal to the size of the preprepared ice particulates used as porogen material. The spherical large pores should be derived from the preprepared ice particulates, while the small pores should be derived from the new ice crystals formed during freezing process. During the pre-freezing process, many small ice crystals should be in situ-formed around the preprepared large ice particulates. The small pores provided good interconnections among the spherical large pores. The surface of spherical large pores in the Collagen–CaP–DEX is much rougher than that of the control Col scaffold. The DEX-loaded BCP NPs are homogeneously embedded into the collagen matrices on the pore walls and therefore increase the pore surface roughness of the hybrid scaffolds (Fig. 8.10b).

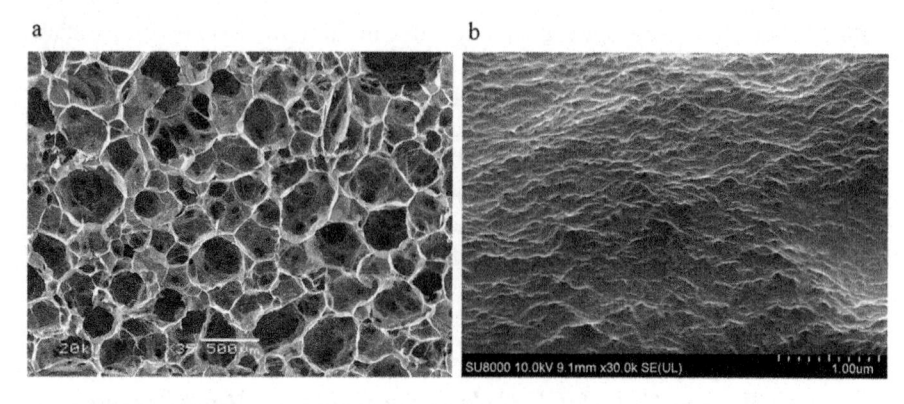

**Fig. 8.10** SEM images of the cross sections of the collagen–CaP–DEX hybrid scaffold at low (**a**) and high (**b**) magnifications

The loading amount and cumulative release amount of DEX from the collagen–CaP–DEX hybrid scaffolds increase with increasing of the DEX feeding amount in the DEX-loaded BCP NPs embedded in the hybrid scaffolds. The release of DEX can last over 35 days. The hMSCs are cultured in the hybrid scaffolds. The hMSCs well adhere on the walls of spherical large pores in the hybrid scaffolds and have high viability. The collagen–CaP–DEX hybrid scaffolds with different DEX loading amount show good biocompatibility and stimulate osteogenic differentiation of hMSCs during in vitro culture. Subcutaneous implantation of the hybrid scaffolds at the dorsa of athymic nude mice demonstrates that they facilitate the ectopic bone tissue regeneration. The collagen–CaP–DEX hybrid scaffold with the highest DEX loading amount has the most promising potential for bone tissue engineering. The collagen–CaP–DEX hybrid scaffolds should provide some useful guidance for bone tissue engineering.

## 8.3 Stratified Scaffolds for Osteochondral Tissue Engineering

### 8.3.1 Osteochondral Tissue Engineering Using PLGA–Collagen Hybrid Mesh

Fixation and integration with surrounding tissues is an important issue for clinical application of tissue engineered tissues. Osteochondral tissue engineering is an attractive strategy to simultaneously stimulate the regeneration of cartilage and subchondral bone, therefore facilitating the fixation of implants and their integration with surrounding cartilage and subchondral bone. Scaffolds with stratified structures should facilitate regeneration of each tissue and promote integration of neighboring tissues. Each phase of the scaffolds should have structures that mimic the in vivo microenvironments of each tissue. Strategies have been considered for osteochondral tissue engineering by laminating two different layers of cartilage and subchondral bone or by using stratified scaffolds for simultaneous regeneration of cartilage and subchondral bone.

One of the strategies in osteochondral tissue engineering is to culture bone marrow stromal cells and chondrocytes in biodegradable scaffolds to construct subchondral bone and cartilage layers, respectively, and then put the two layers together. This method can regenerate a biphasic osteochondral tissue showing a distinct interface between the two layers.

The PLGA–collagen hybrid mesh scaffold has been for culture of canine bone marrow stromal cells and articular chondrocytes to regenerate cartilage tissue and subchondral bone, respectively [13]. The two parts are laminated to construct osteochondral tissue. Canine bone marrow stromal cells are seeded in the PLGA–collagen hybrid mesh and cultured in vitro in osteogenic medium for 1 week. Canine articular chondrocytes are also seeded in the PLGA–collagen hybrid mesh and cultured in vitro in DMEM serum medium for 1 week. Both marrow stromal cells and articular chondrocytes well adhere and spread in the hybrid mesh. They proliferate and produce extracellular matrices to fill the spaces in the hybrid meshes with the increase of culture time. After 1 week in vitro culture, the PLGA–collagen hybrid meshes seeded with chondrocytes are laminated to construct cartilage layer while the PLGA–collagen hybrid meshes seeded with stromal cells are laminated to form subchondral layer. The two layers are further placed together to construct an osteochondral implant (Fig. 8.11). The osteochondral implant is transplanted subcutaneously in the dorsum of athymic nude mouse to confirm ectopic regeneration of osteochondral tissue.

**Fig. 8.11** Lamination of PLGA–collagen hybrid meshes seeded with bone marrow stromal cells and chondrocytes. Reproduced with permission [13]

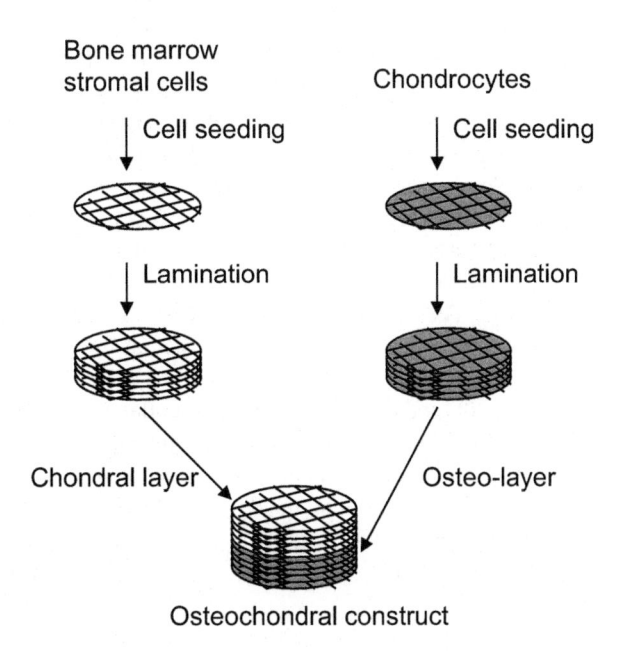

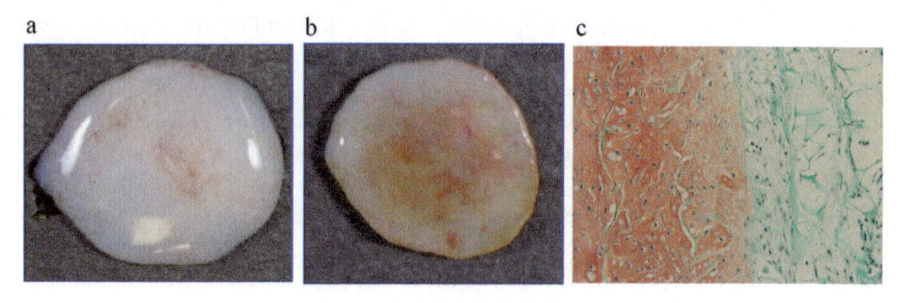

**Fig. 8.12** Gross appearance of cartilage layer (**a**) and subchondral bone layer (**b**) of osteochondral implant and safranin O/fast green staining of the implant (**c**). Reproduced with permission [13]

After 9 weeks following subcutaneous implantation, cartilage layer of the implant appears glistening white while the subchondral bone layer appears red (Fig. 8.12a, b). Capillary blood vessels are evident in the subchondral bone layer while not observed in the cartilage layer. The osteochondral implant preserves its original round disc shape during the transplantation. Safranin-O/fast green staining shows that the stromal cells and chondrocytes are evenly distributed throughout the scaffold (Fig. 8.12c). The cells in the cartilage layer show a round morphology and are surrounded with redly stained extracellular matrices. On the other hand, the cells in the subchondral bone layer show a spindle-like morphology and are surrounded with greenly stained extracellular matrices. Toluidine blue staining demonstrates the typical metachromasia of the articular cartilage matrix in the cartilage layer (purple), whereas the subchondral bone layer is negatively stained as blue.

Many methods have been developed for osteochondral tissue engineering that include binding separately engineered cartilage and bone layers by suture or glue and engineering an osteochondral tissue in a biphasic scaffold. With whichever method, it is important to seed a sufficient number of cells in the scaffold and promote tissue formation. However, cell seeding efficiency decreases as the scaffold thickness increases. Even for scaffolds with porosity greater than 97%, there are still problems concerning cell seeding. In this part, the thin PLGA–collagen hybrid mesh is used for tissue engineering of osteochondral tissue. The web-like collagen microsponges promote cell seeding and cell adhesion. Cell distribution is homogenous after lamination because the cells are seeded in each of the meshes before lamination. The laminated meshes become integrated with extracellular matrices. The spatially even distribution of a sufficient number of cells facilitates the formation of osteochondral tissue.

The results indicate that the PLGA–collagen hybrid mesh scaffold supports the adhesion and proliferation of canine articular chondrocytes and bone marrow stromal cells and promotes the formation of osteochondral-like tissue. The cell-seeded mesh layers are bounded and integrated by the secreted extracellular matrices that function as a glue. In this method, cells are separately seeded in the PLGA–collagen mesh that can guarantee homogeneous cell seeding and distribution. The differentiation induction culture can also be conducted separately before integration of multilayers. The osteochondral tissue has a very clear interface. However, this method need to suture or glue the cell-seeded mesh together to keep the integrity of the implant.

## 8.3.2 PLGA–Collagen/Collagen Biphasic Scaffold

A biphasic scaffold with a stratified two-layer structure has been prepared from PLGA and collagen [14]. One layer of the scaffold is a collagen sponge that is used for regeneration of cartilage layer. The other layer is a PLGA–collagen hybrid sponge that is used for regeneration of subchondral bone. At first, a PLGA sponge cylinder is prepared by porogen leaching using NaCl particulate. Subsequently, a collagen/PLGA–collagen biphasic sponge cylinder is prepared by introducing collagen sponge into the pores of the PLGA sponge and forming collagen sponge at one side of the PLGA sponge (Fig. 8.13a). The bilayer structure of the scaffold is obvious from the gross appearance (Fig. 8.13b). SEM observation shows the

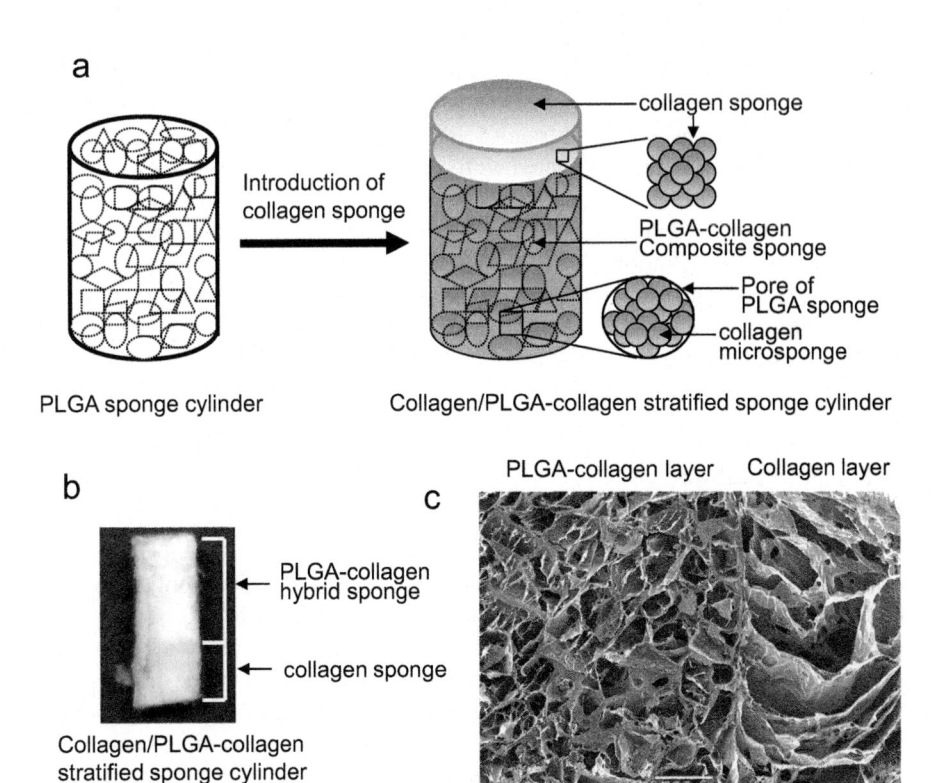

Fig. 8.13 Preparation scheme of the collagen/PLGA–collagen biphasic sponge (a). Gross appearance (b) and SEM photomicrograph (c) of the collagen/PLGA–collagen biphasic sponge. Reproduced with permission [14]

stratified structure of the collagen/PLGA–collagen biphasic scaffold (Fig. 8.13c). One layer of the biphasic scaffold is highly porous collagen sponge while the other layer is a hybrid sponge with collagen sponge formed in the pores of a PLGA sponge. The collagen sponges in the two layers are connected.

The biphasic scaffold is used for the culture of canine bone marrow-derived mesenchymal stem cells, MSCs. Implantation of the cell-seeded scaffolds in the knees of beagles demonstrates that the biphasic scaffold seeded with MSCs facilitates the integration with the surrounding tissue and promotes osteochondral tissue regeneration. The PLGA–collagen hybrid sponge layer provides the biphasic scaffold with high mechanical strength and thus facilitates the fixation and the integration of the implant. The collagen sponge layer facilitates regeneration of articular cartilage. The PLGA–collagen biphasic sponge will be useful for osteochondral tissue engineering.

Canine mesenchymal stem cells are seeded into the biphasic scaffold by dropping cell suspension solution. The cell suspension solution is absorbed and diffused into the scaffold. After 1 week in vitro culture, the cell–scaffold constructs are transplanted in the knee of the same beagle. The implants are harvested after 4 months. Gross appearance shows that the defect treated with cell–scaffold construct presents a smoother surface and better integration with the surrounding tissue than does the scaffold without cells. Histological examination of the implants indicates that cartilage-like and subchondral bone-like tissues are regenerated. The cartilage-like tissue is stained intensively by safranin O and well integrated with the surrounding tissue. However, the defect implanted with scaffold without the cells does not show any evidence of hyaline cartilage regeneration. The biphasic scaffold seeded with MSCs facilitates the integration with the surrounding tissue and promotes osteochondral tissue regeneration. The PLGA–collagen hybrid sponge layer provides the biphasic scaffold with high mechanical strength and thus facilitates the fixation and the integration of the implant. The collagen sponge layer facilitates regeneration of articular cartilage.

## 8.4 Conclusions

Various porous scaffolds have been designed and prepared for osteochondral tissue engineering. Biodegradable synthetic polymers, naturally derived polymers, calcium phosphate and bioactive factors such as BMP4 and DEX have been used to construct porous scaffolds and their hybrid scaffolds. Their hybrid scaffolds can combine the advantages of the respective components. Furthermore, preprepared ice particulates are useful to precisely control the pore structures. These porous scaffolds can be used independently for tissue engineering of cartilage and bone. The different parts can also be constructed in a stratified structure for simultaneous regeneration of cartilage and subchondral bone. These porous scaffolds should be useful for osteochondral tissue engineering applications.

## References

1. Zhang H, Hussain I, Brust M, Butler MF, Rannard SP, Cooper AI (2005) Aligned two- and three-dimensional structures by directional freezing of polymers and nanoparticles. Nature Materials 4:787. https://doi.org/10.1038/nmat1487
2. Yoon JJ, Kim JH, Park TG (2003) Dexamethasone-releasing biodegradable polymer scaffolds fabricated by a gas-foaming/salt-leaching method. Biomaterials 24(13):2323–2329. https://doi.org/10.1016/S0142-9612(03)00024-3
3. Hou Q, Grijpma DW, Feijen J (2003) Porous polymeric structures for tissue engineering prepared by a coagulation, compression moulding and salt leaching technique. Biomaterials 24(11):1937–1947. https://doi.org/10.1016/S0142-9612(02)00562-8
4. Harris LD, Kim B-S, Mooney DJ (1998) Open pore biodegradable matrices formed with gas foaming. Journal of Biomedical Materials Research 42(3):396–402. https://doi.org/10.1002/(SICI)1097-4636(19981205)42:3<396::AID-JBM7>3.0.CO;2-E
5. Chen G, Ushida T, Tateishi T (2001) Preparation of poly(l-lactic acid) and poly(dl-lactic-co-glycolic acid) foams by use of ice microparticulates. Biomaterials 22(18):2563–2567. https://doi.org/10.1016/S0142-9612(00)00447-6
6. Ko Y-G, Kawazoe N, Tateishi T, Chen G (2010) Preparation of Novel Collagen Sponges Using an Ice Particulate Template. Journal of Bioactive and Compatible Polymers 25(4):360–373. https://doi.org/10.1177/0883911510370002
7. Lu H, Ko Y-G, Kawazoe N, Chen G (2010) Cartilage tissue engineering using funnel-like collagen sponges prepared with embossing ice particulate templates. Biomaterials 31(22):5825–5835. https://doi.org/10.1016/j.biomaterials.2010.04.019
8. Zhang Q, Lu H, Kawazoe N, Chen G (2013) Preparation of collagen scaffolds with controlled pore structures and improved mechanical property for cartilage tissue engineering. Journal of Bioactive and Compatible Polymers 28(5):426–438. https://doi.org/10.1177/0883911513494620
9. Chen G, Ushida T, Tateishi T (2000) Hybrid biomaterials for tissue engineering: A preparative method for PLA or PLGA-collagen hybrid sponges. Adv Mater 12(6):455–457. https://doi.org/10.1002/(sici)1521-4095(200003)12:6<455::aid-adma455>3.0.co;2-c
10. Dai W, Kawazoe N, Lin X, Dong J, Chen G (2010) The influence of structural design of PLGA/collagen hybrid scaffolds in cartilage tissue engineering. Biomaterials 31(8):2141–2152. https://doi.org/10.1016/j.biomaterials.2009.11.070
11. Lu H, Kawazoe N, Kitajima T, Myoken Y, Tomita M, Umezawa A, Chen G, Ito Y (2012) Spatial immobilization of bone morphogenetic protein-4 in a collagen-PLGA hybrid scaffold for enhanced osteoinductivity. Biomaterials 33(26):6140–6146. https://doi.org/10.1016/j.biomaterials.2012.05.038
12. Chen Y, Kawazoe N, Chen G (2018) Preparation of dexamethasone-loaded biphasic calcium phosphate nanoparticles/collagen porous composite scaffolds for bone tissue engineering. Acta Biomaterialia 67:341–351. https://doi.org/10.1016/j.actbio.2017.12.004
13. Chen G, Tanaka J, Tateishi T (2006) Osteochondral tissue engineering using a PLGA–collagen hybrid mesh. Mater Sci Eng C 26(1):124–129. https://doi.org/10.1016/j.msec.2005.08.042
14. Chen G, Sato T, Tanaka J, Tateishi T (2006) Preparation of a biphasic scaffold for osteochondral tissue engineering. Mater Sci Eng C 26(1):118–123. https://doi.org/10.1016/j.msec.2005.07.024

# Chapter 9
# Layered Scaffolds for Osteochondral Tissue Engineering

Diana Ribeiro Pereira, Rui L. Reis, and J. Miguel Oliveira

**Abstract** Despite huge efforts, tissue engineers and orthopedic surgeons still face a great challenge to functionally repair osteochondral (OC) defects. Nevertheless, over the past decade great progress has been made to find suitables strategies towards OC regeneration. In the clinics, some osteochondral tissue engineering (OCTE) strategies have already been applied although with some incongruous outcomes as OC tissue is complex in its architecture and function. In this chapter, we have summarized current OCTE strategies that are focused on hierarchical scaffold design, mainly layered scaffolds. Most suitable candidates towards functional regeneration of OC tissues are envisaged from monophasic to layered scaffolds. Herein is documented a variety of strategies with their intrinsic properties for further application as bare scaffolds or in combination with biologics. Both in vitro and in vivo approaches have been thoroughly studied aiming at functional OC regeneration. The most noteworthy studies in OC regeneration developed within the past 5 years are herein documented as well as some current clinical trials.

**Keywords** Osteochondral defects · Hierarchical scaffolds · Multilayered scaffolds · Tissue integration · Functional OC regeneration

---

D. R. Pereira
3B's Research Group – Biomaterials, Biodegradables and Biomimetics, University of Minho, Headquarters of the European Institute of Excellence on Tissue Engineering and Regenerative Medicine, Barco, Portugal

ICVS/3B's - PT Government Associate Laboratory, Barco, Guimarães, Portugal

R. L. Reis · J. M. Oliveira (✉)
3B's Research Group – Biomaterials, Biodegradables and Biomimetics, University of Minho, Headquarters of the European Institute of Excellence on Tissue Engineering and Regenerative Medicine, Barco, Portugal

ICVS/3B's - PT Government Associate Laboratory, Barco, Guimarães, Portugal

The Discoveries Centre for Regenerative and Precision Medicine, Headquarters at University of Minho, Barco, Guimarães, Portugal
e-mail: miguel.oliveira@dep.uminho.pt

© Springer International Publishing AG, part of Springer Nature 2018
J. M. Oliveira et al. (eds.), *Osteochondral Tissue Engineering*,
Advances in Experimental Medicine and Biology 1058,
https://doi.org/10.1007/978-3-319-76711-6_9

## 9.1 Introduction

Osteochondral tissue engineering (OCTE) has its main focus on the to regulation of cell function and further functional tissue formation in respect to both cartilage and bone. Therefore, recapitulating intrinsic features of the native osteochondral (OC) environment demands for successful strategies capable of integration of an engineered long-lasting functional construct. These strategies are seen as potential alternatives to the conservative methods or potential combinatorial approaches. Hierarchical scaffolds have been designed to deliver relevant tissue-specific cues taking into consideration the dissimilar OC tissue structure, composition and nutritional requirements. Novel OCTE strategies under development include not only cartilage and bone regeneration but additionally aim to establish the OC interface formation and tissue integration. Several advances have been made in OCTE; however, the suitable construct has not yet been engineered and thus, widely accepted in the clinics. The combination of biodegradable polymers and bioactive ceramics in a variety of composite constructs is very promising, whereby the fabrication methods, cell sources, and signalling factors determine a strategy's success. The objective of this review is to present and discuss approaches that are currently proposed for OCTE. With focus on hierarchical scaffolds our intent is to encourage the creation of novel engineered constructs able to mimic the native OC tissue.

## 9.2 Complexity of OC Tissue

The intimate contact between articular cartilage and subchondral bone forms a functional unit called OC tissue. The OC tissue is peculiarly complex and divided into two major parts; the upper cartilage and the underlying subchondral bone [1]. Articular cartilage is a highly specialized tissue with pivotal roles in the joint. Hyaline cartilage, commonly name given to articular cartilage in the joints, acts as a low-friction and low-bearing surface that allows smooth motions in diarthrodial joints. Friction is reduced at the edges of long bones and hyaline cartilage acts as a cushion for pressure while protects subchondral bone from high stresses. The cartilage region exhibits a particular structural organization composed of four distinct layers consisting of superficial, middle, deep and calcified cartilage zones (Fig. 9.1) [2]. The three upper layers are mainly composed of type II collagen (Col II) varying the compactness and orientation of the fibers. The calcified cartilage is the layer responsible for anchoring the whole cartilage zone to the subchondral bone. The unique feature of cartilage tissue limits its self-renewal capabilities due to lack of vascularization and innervation [3] and remodelling is hardly seen as a consequence of low cell number and metabolic activity [4, 5]. Therefore, cartilage defects are usually irreparable and when left untreated lead to a resurgence of a variety of

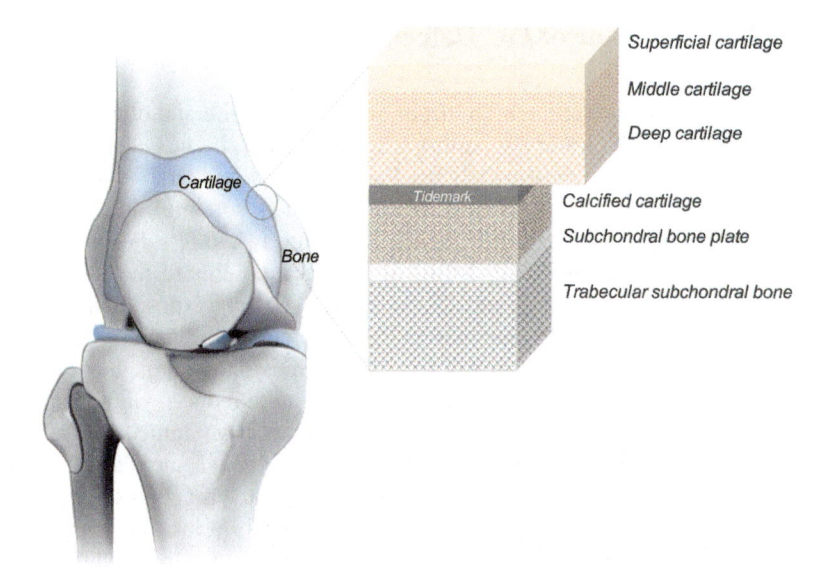

9 Layered Scaffolds for Osteochondral Tissue Engineering

**Fig. 9.1** Osteochondral unit and hierarchical organization. Zonal organization of cartilage, interface region, and underlying subchondral bone

symptoms such as severe pain along joint deformity and loss of motion [6]. The natural cartilage self-healing, when assisted by macrophages and mesenchymal stem cells (MSCs) infiltration from the bone marrow, rather leads to the formation of fibrocartilage with mechanical properties very dissimilar from those of the native hyaline cartilage. As consequence, subchondral bone lesions are developed due to the lack of efficient protection from hyaline cartilage. Anatomically, subchondral bone is composed of two distinct entities: the subchondral bone plate and the trabecular bone. Subchondral bone plate encompasses a thin cortical lamella, lying immediately beneath the calcified cartilage with marked porosity. It exhibits channels populated by vessels and nerves that penetrate the calcified cartilage. The trabecular bone is responsible for shock absorbing and serves as supportive structure while supplying essential nutrients to the cartilage. Its structure exhibits higher porosity than that of subchondral bone plate and the place of bone marrow. Both subchondral layers are composed mainly of type I collagen (Col I) and hydroxyapatite (HAp). A gradual transition between cartilage and bone tissues is seen at the interface region which is composed of calcified cartilage. Vertically orientated Col I fibers bridge the deep layer in the cartilage region to the calcified cartilage forming a wavy tidemark responsible for absorbing and releasing weight-bearing pressures to the subchondral bone. Between calcified cartilage and subchondral bone plate there is an absence of the vertical Col I fibers, but a sharp borderline called the "cement line."

## 9.3 Clinical Repair of OC Defects

Damage to the articular cartilage can occur at any time in life and often affects the subchondral bone with further osteoarthritis development [7]. The Outerbridge classification (Fig. 9.2) is the most widely used and accepted standard for diagnosis of OC defects in which, Grade 0: Normal cartilage; Grade I: softening and swelling of cartilage tissue; Grade II: Partial thickness defect not reaching subchondral bone (<1.5 cm diameter); Grade III: Full thickness defect, fissures to the subchondral bone level (>1.5 cm diameter) and Grade IV: OC defect (exposed subchondral bone). The option for surgical intervention in OC defects is often made on an individual patient basis and on the size defect. Conservative non-surgical approaches are the first-line treatment towards OC defects helping to reduce symptoms and to enhance self-healing. However, non-surgical approaches have limited recovery and often require posterior surgical interventions. Conventional surgical treatments undergo arthroscopic debridement and lavage, fixation procedures, and bone-marrow stimulation such as drilling [8], microfracture [9], abrasion arthroplasty [10], and chondroplasty [11]. The main objective is to alleviate pain while aiming for some self-healing regeneration. Through bone marrow stimulation techniques, such as subchondral drilling, abrasion and microfracture, which attempt to the MSCs recruitment from the bone marrow to the site defect, often result in repair rather than tissue regeneration. Fibrocartilaginous tissue formation is one of the most unsatisfactory outcomes from performing such techniques [12, 13]. The cartilage tissue appears less stiff and more likely to break down than the hyaline cartilage. Often, the repaired cartilage leads to undesirable function and failure [9, 14]. Additionally, large defects or degenerative diseases, such as osteoarthritis, have not proved to benefit from such procedures [15].

Over the last decade, more complex strategies have been developed with autologous or allogeneic OC grafting, i.e., autologous chondrocytes implantation (ACI) [16, 17]. ACI requires the harvesting of chondrocytes from healthy tissues with further in vitro expansion until transplantation into the chondral defect. Carticel® is a commercial product for ACI approved by FDA and currently applied in the clinic.

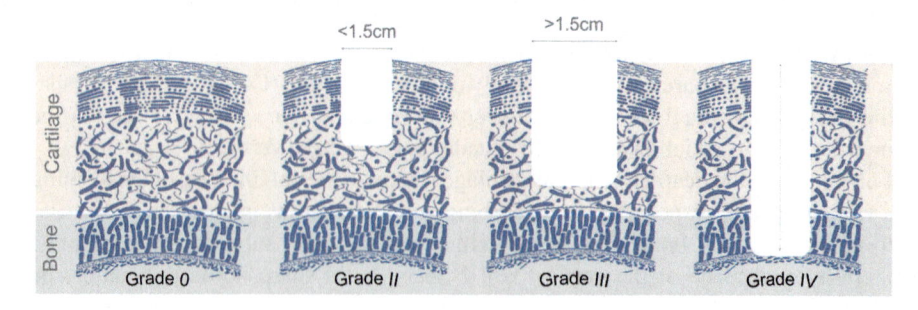

**Fig. 9.2** Osteochondral defect classification. Representative model of Outerbridge classification system. Grade 0: Normal; Grade II: Partial thickness defect; Grade III: Full thickness defect, and Grade IV: Osteochondral defect

Alike, mosaicplasty technique [18] as the most widely accepted technique for articular cartilage reconstructions shows superior outcomes demonstrated for up to 5 years follow-up. Small autologus cylindrical plugs (from 2.7 to 9 mm) are harvested from the femoral condyle and transplanted into the site defect. Nonetheless, several limitations arise from such a procedure including the limited availability of material, the donor site morbidity and the fixation of the graft within the size defect. Hence, methods making use of chondrocytes, i.e. ACI, have been improved and so far show relative success. Autologous chondrocytes, prior cultured in vitro, are incorporated within biomaterials that can be implanted in the OC defect to what is called Matrix-assisted ACI (MACI) [19]. This procedure can be stated as the first TE strategy that makes use of autologous cells to populate biodegradable matrix to guide and stimulate cartilage tissue formation. Hyaluronic acid-based cartilage grafts in combination with autologous chondrocytes, under the name Hyalograft® C, appears as a MACI-based clinical approach. Despite numerous scaffolds available as matrix-assisted few technical reports about MACI are found in the literature. In other respects excellent reviews on cellular techniques such as autologous chondrocyte implantation (ACI) and matrix-induced autologous chondrocyte implantation (MACI) are well documented in the literature [20, 21]. A valid study, 1 year follow-up, by Ventura et al. reported MACI technique performed in 53 patients undergoing OC defects with a completely repair of the defect with slight subchondral bone abnormalities. Subsequently, a 5 year follow-up study proved the complete integration of the construct with the surrounding native cartilage, absence of edema and no detachment at the interface and bone region [22].

The design of tissue engineering (TE) strategies for OC defects is in increasing expansion but so far few approaches already developed were effectively brought to the clinics due to the complexity of the OC tissue and the regulatory hurdles in clinical translation.

## 9.4 Hierarchical Structured Scaffolds

Load-bearing tissues require a comprehensive correlation between structure and function in order to develop TE design criteria. The study of both, cartilage and bone tissue engineering, has come a long way over the years [23–25]. Overall, an OC scaffold must be biocompatible, possess sufficient mechanical integrity to be able to support loading, guide cartilage and bone formation. Additionally, interface region between these two dissimilar tissues must be capable of functional scaffold performance at structural/mechanical and biological levels.

Primarily, regeneration of OC defects was only focused on cartilage repair due to the single cell type, apparent simplicity and also clinical relevance. Thus, the subchondral bone was totally neglected at that time [26]. Taking advantage of the compelling healing capability of bone, cartilage was integrated into full-thickness defects and neocartilage was naturally anchoring the underlying bone. The formation of neocartilage, in direct apposition with bone, created a transitional cartilage–bone region similar to the native tissue interface [27]. Besides the good histological scores, a functional interface was not achieved often presenting delamination and separation [28]. Thus, the design of OC scaffolds proceeded with an underlying subchondral region, also crucial for supporting neocartilage formation. A variety of scaffolds focus on the seeding of chondrogenic cells directly on top of osteoconductive biomaterials (i.e., coated tantalum, bioactive glasses, ceramics, etc.) [29, 30] were of preference. Although some of these strategies showed an ability of bone bonding promoted bone healing and cartilage formation, not all OC tissue requirements were fulfilled. Functional stratification, fixation, interface region, etc. were not achieved and thus failing to serve a long-lasting functional strategy. OCTE strategies must possess an optimal environment enabling the cell-to-cell communication and cell-to-matrix interaction. Moreover, it is paramount to generate OC constructs that can ensure a mechanically stable interface between tissues [31]. The scaffold architecture is fully acknowledged nowadays, as it defines the ultimate shape of the newly formed tissues, for both cartilage and bone [32]. The improved understanding of OC biology and the advance of technology drives OCTE towards the design of dissimilar structures. Thus, structural constructs for the simultaneous regeneration of both cartilage and subchondral bone must also possess a functional and efficient interface region. Hierarchical scaffolds allow the creation of optimized tissue-specific biological environments by variation of the mechanical, structural, and chemical properties. The major challenge that TE faces when engineering OC constructs is to generate a long-term functionally integrated stratified cartilage–bone structure. Thus, one intact scaffold must provide a structural environment to support both chondrogenesis and osteogenesis without neglecting the interface region. Among all the components that a TE strategy requires, biomaterial scaffold properties are fundamental to serve as artificial support while recreating the native environment. Therefore, the types of scaffold must enable proper cell function in guiding cell responses and promoting tissue growth. The first criteria to be considered in an OCTE strategy concern the scaffold biomaterial host response. Thus, the scaffold must not elicit strong inflammatory responses due to toxicity and it must be biocompatible. The regulatory approval for TE products is still elusive but by following some essential standardized requirements may result in biomaterial scaffolds with valid clinical application [33]. Biomaterial scaffolds should be osteoinductive, osteoconductive, chondroinductive, or chondroconductive, and must balance out scaffold degradation with neotissue formation. Additionally, the breakdown products must not elicit any host response.

The interdependency of the different scaffold constituents and their synergistic effects are difficult to determine or be sufficiently controlled. Nevertheless, it is widely accepted that hierarchical scaffolds must be foreseen when considering an OCTE strategy with particular attention being given to their structure and mechanical properties. Until now, no biomaterial is put forward as the optimal that can meet all the requirements for the fabrication of complex, stratified and dissimilar tissues [34]. The prevalent approach uses multicomponent systems and often hybrid scaffolds, indiscriminately from natural and synthetic biomaterials [32]. As cartilage and bone possess unique compositions, architectures and tissue growth mechanisms, the preferred biomaterial source often reflects this variability. Natural materials such as proteins and carbohydrates demonstrated superior performance in biological host interaction as they closely mimic the native environment. Yet the immunogenicity and extremely poor control over reproducibility due to batch-to-batch variability means natural biomaterials have only limited clinical relevance while their use raises enormous regulatory issues [35]. On the other hand, synthetic materials benefits from their highly controllable and accurate reproducibility and are thus easy to scale up at industrial level allowing off-the-shelf clinical products. Synthetic but biodegradable biomaterials are usually suggested for TE cartilage which benefit from easy and consistent processability, controllable degradation rate, and FDA approval with, however, some drawbacks on cell attachment [36]. Product breakdown from biomaterial degradation, with special focus on synthetic materials, must elicit a weak inflammatory response to avoid activation of cell death mechanisms. The perfect balance between scaffold degradation and neotissue formation is hard to determine due to its interdependence on the material-based scaffold and the animal model whereas these strategies are tested. Conversely, tailoring the degradation rate of the scaffold may result in undesirable responses especially those that weaken the mechanical properties. Natural and synthetic biomaterials can be processed into feasible scaffolds. Among others, fibrin [37, 38], hyaluronan (HA) [39, 40], collagen (Col) [41, 42], chitosan (CH) [43, 44], alginate [45, 46], silk fibroin (SF) [47], and gellan gum (GG) [48] have been widely applied as OC biomaterial scaffolds mostly for chondral lesions. Conversely, synthetic biomaterials as polyglycolic acid (PGA) [49], polylactide acid (PLA) [50], poly(lactide-co-glycolide) (PLGA) [51], and poly(ε-caprolactone) (PCL) [52] have been applied in both cartilage and subchondral bone-based scaffolds by exploiting their easy processability and by tailoring degradation rates and mechanical properties. Additionally, inorganic materials such as metals and ceramic materials including hydroxyapatite (HAp) [44], β-tricalcium phosphate (β-TCP) [53], and bioactive glasses (BG) [54] have been widely used in subchondral bone regeneration, alone or in combination with either natural or synthetic biomaterials. The combination of biomaterials allows the enhancement of cell attachment while tailoring physicochemical properties [55]. In addition, it can reinforce the mechanical properties of the scaffolds to serve as a suitable support for subchondral bone while promoting excellent osteoconductivity and osteoinductivity [56]. β-TCP alone or in combination with HA [57], PCL [58, 59], Col [60], or Hap [61] can improve the mechanical strength of the scaffolds for subchondral bone regeneration. BGs are a good option to serve as a subchondral bone layer and, when combined with PLGA, showed the best histological scores although failing in the spongy bone structure.

OCTE aims at the recapitulation of the native structure where scaffold design is paramount. Generally, the scaffolds are categorized into monophasic or hierarchical scaffolds (Fig. 9.3). Initially, OCTE strategies would be focused on monophasic scaffolds with a homogenous structure and inadequate cartilage–subchondral bone interface region. Hierarchical scaffolds are usually reported as layered scaffolds mostly addressing the physical properties or as gradient-based scaffolds which involve the loading of biochemical cues [62]. Monophasic scaffolds are

**Hierarchical structured scaffolds**

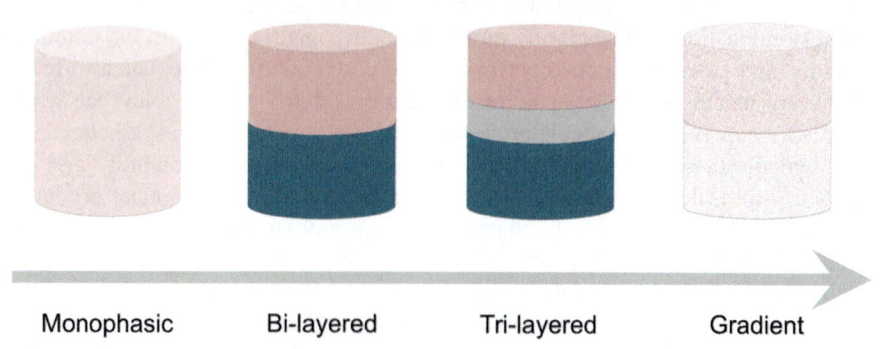

Monophasic   Bi-layered   Tri-layered   Gradient

**Fig. 9.3** Different tissue engineering strategies to regenerate osteochondral defects. From the earlier monophasic constructs to the newly hierarchical constructs, either layered or gradient scaffolds. A particular strategy can be applied taking in account the type of defect/tissue regeneration. Chondral defects can still be resolved making use of monophasic scaffolds, while more exacerbated defects, i.e., Outerbridge IV, demand for more complex approaches. Often, OCTE strategies are applied in combination with current clinical approaches, i.e. MACI and microfracture

composed of one single biomaterial/composite with no spatiotemporal variation, either physical or biological. The very first OC scaffolds were engineered from one single biomaterial with uniform architecture and porosity, a single cell type distributed within the scaffold and no variation in the biological cues (i.e., growth factors) [63]. In contrast, hierarchical layered scaffold can present two (bilayered) or three (trilayered) different compartments, with dissimilar architectures and made of different biomaterials/composites. Hierarchical layered scaffold can also be made of one single biomaterial; however, they must present variation in physical properties (i.e., porosity, interconnectivity, pore size, etc.) with dissimilarities between layers. Gradient scaffolds are usually categorized as monophasic in regard to the single material used for their design, however, they present hierarchical stratification concerning the loading of therapeutic molecules or/ and cells in a gradient fashion.

## 9.4.1 Bilayered Scaffolds

Bilayered scaffolds encompass two dissimilar layers intended for the simultaneous regeneration of cartilage and subchondral bone. Bilayered scaffolds can be made of two or more materials, thus usually named as composites, i.e. bilayered HAp–chitosan scaffolds [44]. Some others present only a single material varying their architecture, i.e. bi-layered collagen scaffold [64].

For most of the approaches involving bilayered scaffolds the combination of individual scaffolds that have already been suggested for regeneration of either cartilage or bone is privileged. The methods to combine two independent layers would make use of fibrin sealant in between the two layers, press fitting both scaffolds, suturing scaffolds between them and to native tissue or use external fixation. All procedure randomly performed either before or after implantation [65, 66]. However, the aforementioned methods by which two layers are integrated in one single unit do not meet the OC tissue requirements. At the interface region, delamination and total layer separation occur due to the scarce ECM deposition, consequence of the lower cell population. Innovative techniques focusing on bioadhesive materials, i.e., chondroitin sulfate (CS) and fibrin [67], have been explored targeting mostly the scaffold integration and thus, a better performance of the bilayered. Several remarkable reviews on the subject were recently published [24, 63, 68]. Often, the bilayered scaffolds are prepared based on their native mechanical properties. Lower strength scaffolds are usually assigned for cartilage and higher strength scaffolds for the underlying subchondral bone [69]. Natural materials, in particular, do not always match the mechanical strength of the tissue to be regenerated yet such property are not totally necessary at the primary stage of scaffolds implantation. Management of weight bearing, post surgery should be envisaged while neotissue is being formed [70].

Table 9.1 presents the most recent work done in engineering bilayered OC scaffolds.

Table 9.1 Bilayered scaffolds for OCTE

| Biomaterial | | | | | |
|---|---|---|---|---|---|
| *Chondral* | *Subchondral* | Cells | Outcome | Study | References |
| Gelatin, CS, HA | Gelatin and ceramic bovine bone | Chondrocytes and BMSCs | Hyaline like cartilage and bone formation with the presence of tidemark at 36 weeks post implantation | In vivo (rabbits) | Deng et al. [71] |
| Collagen | GAG phosphate | – | Hyaline-like cartilage formation and less bone cyst formation | In vivo (goats) | Getgood et al. [41] |
| SF | SF-nano CaP | – | Fully integration of scaffolds within host tissue. Collagen II positive cartilage and GAG deposition for neocartilage formation. Novo bone ingrowth and vessel formation in subchondral layer | In vivo (rabbits) | Yan et al. [47] |
| Col | Col-HAp | hBMSCs | Chondrogenesis and osteogenesis differentiation in each respective layer. | In vitro | Zhou et al. [72] |

(continued)

Table 9.1 (continued)

| Biomaterial | | Cells | Outcome | Study | References |
| --- | --- | --- | --- | --- | --- |
| *Chondral* | *Subchondral* | | | | |
| Gelatin-CH | Cancellous bone | ADSCs | Suitable porous structure. ADSc proliferation, adhesion, and differentiation. Higher cytokine levels secretion. | In vitro | Song et al. [73] |
| PVA/gel/V | HAp/PA6 | BMSCs | Layered structure with intervening nonporous layer. Integration within surrounding tissue. Cartilage and subchondral bone formation. | In vitro, in vivo (rabbits) | Li et al. [74] |
| ACECM | ACECM-HAp | Chondrocytes | Porous structure with gradual interfacial region with no delamination. Chondrocytes in upper layer and some at interface. | In vitro | Wang et al. [75] |
| Cartilage-ECM derived | PLGA-TCP | BMSCs | Cartilage repair with large amounts of lyaloid cartilage | In vivo (goats) | Zhang et al. [76] |
| Methacrylated CS | PCL-PEG-PCL copolymer | Chondrocytes, osteoblasts | Spatially controllable porosity. Weak inflammatory response. Cartilage formation similar to native tissue | In vivo (rabbits) | Liao et al. [77] |
| polyHEMA-HA | polyHEMA-HAp | hMSCs and chondrocytes | Integrated scaffolds with predesigned architecture and pore morphology. Scaffold supported simultaneous growth of chondrocytes and differentiation of hMScs into osteoblasts. Maintenance of chondrocytes phenotype. | In vitro | Galperin et al. [78] |

(continued)

**Table 9.1** (continued)

| Biomaterial | | | | | |
|---|---|---|---|---|---|
| Chondral | Subchondral | Cells | Outcome | Study | References |
| Cartilage ECM110-derived | PLGA-TCP-Col I | BMSCs | Compact layer in a biphasic scaffold improves OC regeneration. Chondro and osteogenic induced BMSCs with independent environments. | In vivo (rabbits) | Da et al. [79] |
| HA | Col I | Chondrocytes and osteoblasts | Col I-based matrix better for bone TE and HA-based matrix better for cartilage TE. Survival and cell functions maintained for up to 14 days I culture. | In vitro | Park et al. [80] |
| OPF-gelatin microparticles | OPF-gelatin microparticles | BMSCs | Cartilage regeneration enhanced by predifferentiated MSCs | In vivo (rabbits) | Lam et al. [81, 82] |
| Col I | Col I Mg-doped HAp | hMSCs | Scaffold supported chondro and osteogenic differentiation. Presence of neo angiogenesis at 4 weeks in subcutaneous implantation. | In vivo (mice) | Sartori et al. [83] |

## 9.4.2 Trilayered Scaffolds

Engineering the OC interface is one of the major challenges in OCTE. A feasible interface is paramount for the formation of the calcified cartilage that lies between the non-calcified cartilage and the underlying subchondral bone. The cell population at the interface is composed of hypertrophic chondrocytes with ECM composed of Col II and X, calcium deposits, and vertical fibers [64]. Trilayered scaffolds aim to regenerate cartilage, calcified cartilage, and subchondral bone simultaneously, with the transitional interface. The vitality of not only trilayered but all hierarchical scaffolds relies on the linkage between cartilage and bone. Thus, a functional interface region is critical to promote fully integration of scaffold and native tissues. In several approaches, limited tissue integration is often seen due to the imbalance of scaffold

Table 9.2 Trilayered scaffolds for OCTE

| Biomaterial | | | Cells | Outcome | Study | References |
|---|---|---|---|---|---|---|
| Chondral | Interface | Subchondral | | | | |
| Agarose | Agarose-PLGA-BG microspheres | PLGA-BG microspheres | Chondro/Osteo | Three distinct integrated regions with chondrocyte mineralized interface region. | In vitro | Jiang et al. [84] |
| Col | Col | Col | BMSCs Chondro/Osteo | Hypertrophic chondrocytes at the interface and presence of Col II and X as well as calcium deposits. | | Cheng et al. [64] |
| PCL | Col I -HAp | TCP | BMSCs | Orientation of BMSCs by electrospun PCL membranes. Cartilage-like tissue formation. | In vivo (rabbits) | Liu et al. [85] |
| PGA | Col I-HAp | PLLA-PCL | L929 cell line | Integrated trilayered plug with stratified architecture. Good biocompatibility. | In vitro | Aydin [86] |
| Silk | Silk-HAp | Silk-HAp | ADSCs | Good biocompatibility, supporting cell growth, proliferation, and infiltration. Chondrogenesis and osteogenesis differentiation in each respective layer. Intermediate layer capable of isolating cells in each layer. | In vitro | Ding et al. [87] |
| PGA | PLLA | PLLA-PCL-Col I-TCP | – | Cartilage formation within 6 months. | In vivo (sheep) | Yucekul et al. [88] |
| Col I-Col II-HA | Col I-Col II-HAp | Col I-HAp | MC3T3 cell line | Integrated scaffold with gradient structure with optimal pore size, porosity, and mechanical properties. Good cell attachment and proliferation. | In vitro | Levingstone et al. [89] |

(continued)

**Table 9.2** (continued)

| Biomaterial | | | Cells | Outcome | Study | References |
|---|---|---|---|---|---|---|
| Chondral | Interface | Subchondral | | | | |
| Col I-Col II-HA | Col I-Col II-HAp | Col I-HAp | – | OC defect repair after 12 weeks implantation. Repair of subchondral bone and formation of cartilaginous layer with an intermediate tidemark. | In vivo (rabbits) | Levingstone et al. [90] |
| Gelatin | Gelatin-HAp (30) | Gelatin-HAP (50) | hBMSCs Chondro/ Osteo | Interconnected porosity and tailored mechanical properties. Maintenance of cells phenotype and ECM secretion | In vitro | Amadori et al. [91] |

degradation. Additionally, collapse of neotissue under the natural mechanical loads results in further interface. In Table 9.2 some relevant studies summarize the latest OCTE strategies using trilayered scaffolds. Kon et al. [92] reported the good outcome at the midterm follow-up using a trilayered scaffold made of Col I (cartilage), Col I-HAp (60–40 wt.%, interface) and Col I-HAp (30–70 wt.%, subchondral bone) [93]. St-Pierre et al. [94] successfully developed a calcified cartilage formed by chondrocytes seeded on substrates of calcium phosphates. The interface also hampered the inorganic calcium phosphates of going into the cartilage layer, thus preventing endochondral ossification. The generation of the interface layer allowed the construct to undergo a 3.3-fold change increase in interfacial shear strength. Another study from Cheng et al. [64] reported the generation of an OC interface of mesenchymal stem cells (MSCs)-derived. MSCs were encapsulated within collagen microspheres with further cell differentiation into both chondrogenic and osteogenic lineages. Afterwards, both types of microspheres aggregates were integrated in a construct with an interfacial layer of MSCs-collagen.

### 9.4.3 Gradient Scaffolds

Several efforts are also been made towards the fabrication of graded scaffolds that allow cartilage–subchondral bone interface, similar to that of the native OC tissue [24, 95].

Gradient scaffolds are capable of OC regeneration by modulating cellular interactions and can be further enhanced by incorporation of therapeutic molecules. These gradient scaffolds, unlike layered scaffolds, consist of gradual or continuous transitional composition [96]. Some studies report the development of scaffolds with compositional gradient granted by the incorporation of bioactive cues and effective methodologies adopted [97, 98]. Therefore, structurally homogenous scaffolds were fabricated presenting spatial variation in stiffness or spatial patterning.

Additionally, numerous gradient scaffold have been developed to control the release of therapeutic molecules either spatial and temporally. Physical properties, such as pore size, porosity, and cross-linking degrees are per se strategies for the controlled release of therapeutic molecules. The employed methods involve the loading of the therapeutic molecules within the biomaterials or the impregnation of those after scaffold fabrication. It is thus acknowledged the development of OCTE strategies that somehow can recapitulate some spatiotemporal aspects of the native development, as they may be successful candidates when it comes to functionally regenerate OC tissues.

Table 9.3 summarizes some strategies of gradient scaffolds developed towards OCTE regeneration.

**Table 9.3** Gradient scaffolds loaded with therapeutic molecules

| Signalling molecule | Gradient scaffold | Outcome | References |
|---|---|---|---|
| TGF-$\beta$1 | SPU (PLGA microsphere) PLGA | Cartilage and subchondral bone formation. | Reyes et al. [46] |
| PRP | Chondromimetic™ (Col I-GAG) | Hyaline-like cartilage formation and reduction of subchondral cyst formation | Getgood et al. [99] |
| BMP-7 | PLGA | Cartilage formation with positive col II and GAG staining | Jung et al. [100] |
| VEGF | PLGA | Cartilage formation and bone ingrowth at early stage. | Sakata et al. [101] |
| IGF-1 | OPF (gelatin microspheres) | IGF-1 improved subchondral bone morphology as well as chondrocytes and GAG amount. | Kim et al. [102] |
| Dex | CH-alginate (CA/PEC) CA/PEC-TCP (60-40) CA/PEC-TCP (30-70) | Good tissue biocompatibility and biodegradability. Blood vessels growth within scaffold. | Algul et al. [103] |
| TGF-$\beta$1 | GS-GMA | Multi-layered graded structure with dissimilar porosity and mechanical properties. Remarkably recovery of OC defects in vivo with formation of neocartilage and bone. | Han et al. [104] |
| TGF-$\beta$1 | CH-gelatin | Bi-layered gene activated scaffold of TGF-$\beta$1 for cartilage layer and BMP-2-activated for subchondral bone layer induce MSCs into chondro and osteogenesis, respectively. Cartilage and subchondral bone formation in rabbit's OC defects 12 weeks after implantation. | Chen et al. [43] |

(continued)

9 Layered Scaffolds for Osteochondral Tissue Engineering

**Table 9.3** (continued)

| Signalling molecule | Gradient scaffold | Outcome | References |
|---|---|---|---|
| TGF-β1, BMP-2, HAp, BMP-2 + HAp | PLGA microspheres | Spatial patterning of different bioactive signals. OC regeneration with bone ingrowth and overlying cartilage layer with high GAG content. Lateral integration and vertical integration | Mohan et al. [95] |
| BMP-4 | Gelatin-FA | Chondrogenic differentiation of BMSCs maintained for 3 weeks. BMP-4 fusion protein with a collagen binding domain (CBD) was retained in the hydrogels. OC defect was repaired using gelatin-FA hydrogels CBD-BMP-4 and BMSCs resulting in articular cartilage-like tissue and regenerated subchondral bone | Mazaki et al. [105] |

*CA/PEC* chitosan and alginate polyelectrolyte complex, *Gelatin-FA* furfurylamine-conjugated gelatin, *CS-GMA* carboxymethyl chitosan and methacrylated gelatin, *OPF* oligo(poly(ethylene glycol)fumarate), *SPU* segmented polyurethane, *BMP* Bone morphogenetic protein, *TGF* Transforming growth factor, *Dex* dexamethasone, *PRP* Platelet-rich plasma, *IGF* Insulin growth factor, *VEGF* Vascular endothelial growth factor

## 9.5 Integration of OC Scaffolds

Tissue regeneration by OCTE strategies relies almost entirely on the scaffold fixation and its integration (Fig. 9.4). Integration is a vital process as it provides stable biologic fixation, proper load distribution and promotes the adequate mechanotransduction for tissue homeostasis [106]. Bone integration (or osseointegration) is widely reported in the literature for a variety of materials such as metallic implants, collagen scaffolds, and porous ceramics due to the high bone cell metabolism. Bone tissue engineering is still favored and thus results in a higher number of studies reported in literature.

Presently, small surgically repaired OC defects do not encompass any fixation procedure. The fibrin clot or the use of biocompatible glues (i.e., fibrin glue) grants the stability of the scaffold at the site defect. Nevertheless, some limitations are found in the use of fibrin glue as it shows poor adhesion between native cartilage and scaffold [107]. The suturing of cartilage may be an option in fractures. The cartilage shift during the surgical procedure favor the use of sutures. Neverthless, cartilage resorption may happen in postoperative period [40]. Many others, such as the use of fixation devices and the transosseous fixation technique are considered as alternatives to conventional procedures, however, not able to be applied in the fixation of matrix [108]. Recently, scaffold fixation via magnetic forces application is being applied to firmly hold the scaffold in place [109].

When considering integration, two important notions need to be addressed: (1) vertical integration which considers the bonding of cartilage to the underlying subchondral bone, and (2) lateral integration which concerns to the integration of the

**Fig. 9.4** Scaffold integration to adjacent tissues (lateral) and scaffold integration (vertical). OCTE strategies must enable lateral and vertical biologic fixation with further tissue integration to functionally regenrate and restore the OC unit

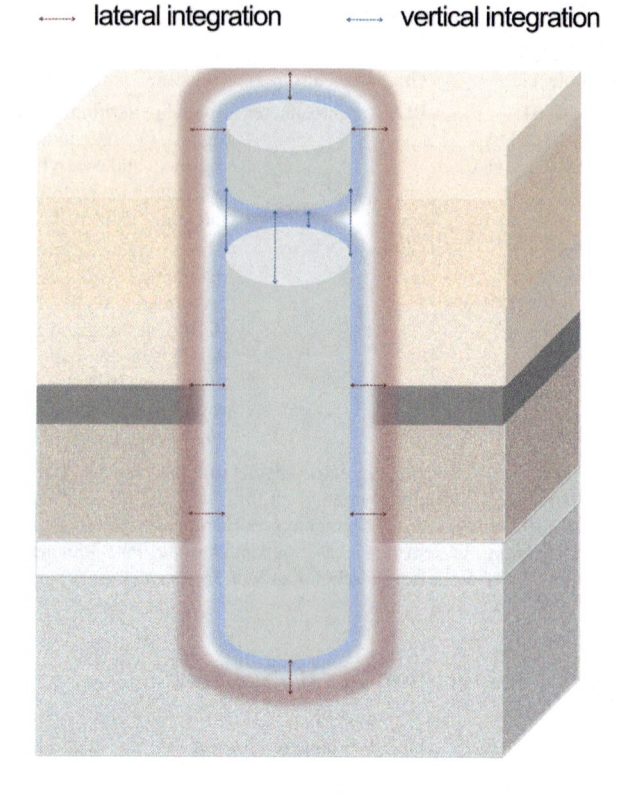

neocartilage to the adjacent native cartilage. Lateral integration is rarely seen and vertical integration is mainly driven by bone and not by cartilage [27]. After implantation at the site of defect, osteoinductive scaffolds, with particular attention to the inorganic materials, must facilitate stable fixation. The low metabolism of chondrocytes accompanied with the existence of dense and anti-adhesive extracellular matrix (ECM) hinders lateral cartilage integration [110]. In a clinical scenario, mismatches between the biomechanical properties of the native cartilage and engineered scaffold can result in stress concentrations that hamper lateral integration. Currently, some in vitro strategies seeking to enhance lateral integration suggest the inclusion of antiapoptotic agents to diminish cell death at the defect edge and the use of matrix-degrading enzymes (MMP) to decrease ECM antiadhesive properties [110, 111]. Recently, the functionalization of the scaffolds with collagen adhesion proteins (CAN) attempts to increase the number of cells at the site defect to promote ECM deposition and thus, reinforcing the boundary between native and neocartilage [112].

Despite the improved outcomes in OCTE strategies, OC grafts are still predominantly considered for the formation of neotissues in the clinics, i.e., ACI and mosaicplasty. Recently, some studies have been done with OC grafts to replicate the native cartilage by a method called scaffoldless. Clinically sized pieces of human cartilage with physiologic stratification and biomechanics are grown in vitro by recapitulating some features of the developmental process of mesenchymal condensation. The use of condensed mesenchymal cells (CMs) exposed to TGF-β induces the formation of cell bodies (CMBs) with superior ability of bonding in vitro and possessing mechanically stable interface. CMBs were fused onto the bone structure surface by press molding the CMBs. With the help of image-guided techniques to precisely match the site and shape defect, it was possible to place the engineered CMBs constructs [113]. Overall, scaffoldless technique in OCTE may have advantages not only in repairing chondral defects but also in OC regeneration for generating self-assembling tissues [114]. The intimate contact of graft native tissue without cell–material interactions may enable cells to better sense the native stimuli. Thus, it is expected a higher secretion of proteins and ECM deposition enhancing the lateral integration [20]. The scaffoldless strategy may greatly benefit cartilage formation not only helping in the lateral integration as the mismatch between native and neocartilage is reduced but also in recapitulating its native multilayered structure. Shimomura et al. [115] developed an implant based on the hybridization of scaffold-free tissue-engineered construct (TEC) derived from MSCs and HAp artificial bone-like structure. The construct was implanted in OC defects in rabbit's knee for 6 months. Results reported good attachment of TEC onto the surface of HAp artificial structure. OC defects exhibited more rapid subchondral bone repair with TEC-HAp implants and development of cartilaginous tissue with good lateral integration than the respective controls. Biomechanically, TEC-HAp restored tissue stiffness similar to that of native OC tissue. Therefore, the strategies showed promising results in earlier restoration of subchondral bone and good tissue integration of cartilage, clinically relevant in patients with OC defects.

In addition, living grafts are currently being used as model systems for comprehensive studies. The main purpose is to focus on the understanding of the biological environment, the study of disease pathologies, the drug screening and identification of therapeutic targets. OCTE would greatly benefit from the currently research being done in developmental biology and thus translational research towards engineering.

## 9.6 Clinical Applications of OC Scaffolds

Recently, bilayered scaffolds were regarded as the most feasible approaches for a more accurate mechanical interface with a marked number of OCTE scaffolds being preclinically evaluated [24, 116]. A wide variety of strategies were tested in rodent models exploiting the low cost and large sample size [66, 117]. Nevertheless, the poor experimental design or those that were well designed but poorly reported, the majority of the studies lack standardization. Despite the numerous preclinical studies in this extensive research area, few scaffolds exist that are clinically relevant [118]. In Table 9.4, the most significant OCTE scaffolds in the clinics are summarized.

MaioRegen® [19, 92, 93, 119–122] and Trufit® [123–125] are two bilayered scaffolds approved in the clinics that have been systematically evaluated in patients. MaioRegen®, a Col I (cartilage-like layer) and magnesium-enriched HAp

**Table 9.4** Clinically relevant OC scaffolds

| Application | Biomaterial | Commercial name | Company | Current status |
|---|---|---|---|---|
| Articular cartilage | HA | Hyalograft C® | Anika therapeutics, USA | Withdrew |
| | CH | BST-CarGel® | Piramal Healthcare, Canada | Long-term results expected |
| OC unit | Fibrin-PLGA, PLA, PCL | Bioseed-C® | BioTissue Technologies GmbH, Germany | Any significant benefit over first-generation ACI |
| | Col-HAp | MaioRegen® | Finceramica, Italy | Promising results |
| | PGA-PLGA | TruFit CB plug® | Smith & Nephew, UK | Poor tissue restoration, pain and swelling and high rate of reoperations |

(subchondral-like layer) showed promising results in repairing OC defects. Unlike, results showed that Trufit® did not show any superiority or equality compared with conservative methods or mosaicplasty/microfracture method. Noteworthy, the number of studies for MaioRegen® implant was higher than Trufit® implantation. Nevertheless, approval of MaioRegen® was done several years earlier. The ambiguous conclusions derived from localized clinical trials ask for more accurate studies with effective standardized procedures.

## 9.7 Conclusions

The challenging regeneration of OC tissue demands for complex but suitable engineered strategies to mimic the native OC tissue environment towards development, regeneration and remodelling. To replicate the natural OC tissue, hierarchical scaffolds should be foreseen aiming at structural and functional integration of cartilage and subchondral bone. To generate functional OC tissues, such as cartilage and subchondral bone which possess very dissimilar tissues, cellular phenotypes are needed to be carefully controlled in a spatiotemporally manner. Each individual tissue requires specific environmental cues capable of modulating the tissue cell fate. Therefore, the surrounding environment is of extreme importance as cells will respond to different physical, chemical, biological or mechanical stimulus and will synergistically regulate the tissue regeneration. Nevertheless, a scaffold with all multiple and interconnected features and properties is unfeasible. Current technologies face the problem of providing the adequate signals (dosages, gradients, and timing) for the regeneration process to occur. Therefore, researchers are focusing on the understanding of natural mechanisms looking at different developmental tissue formation to be able to recapitulate biologically engineered constructs. Therefore, engineering living OC grafts, customized on a patient-basis and on a defect-basis, are being tremendously foreseen. Engineering OC tissues would serve as platforms for the screening of disease or healing processes. Despite all advances, the different healing abilities of cartilage and subchondral bone along the complex interface transitional layer often results in tissue repair than regeneration. New OCTE methodologies of additive manufacturing and 3D precision printing technology may be promising approaches to fabricate complex scaffolds. Hence, multilayered scaffolds are critical to replicate the native OC tissue in order to satisfy long-term clinical outcomes. The latest advances in generation of OCTE pursue the cell–scaffold–bioreactor model for the development and maturation of cells similar to native tissue to yield functional engineered tissue. Therefore, engineering OC tissues will progress in functionally integrating stratified cartilage–subchondral bone structure while incorporating vasculature into the bone with lateral and vertical integration. Moreover, reduction in micromotion at the graft–tissue interface is crucial for functional regeneration. The use of cells and/ or therapeutic molecules may present beneficial outcomes. However, embedding these in commercial scaffolds raises huge concerns regarding the storage and the transportation conditions and thus, challenging their approval as effective clinical OCTE strategies. Likewise, the species, surgical defect, engineered scaffold, biomaterial nature, cell and drug concentration, short follow-up time, etc. accurately reflect the translational limitation from animals to humans; bench side to clinics.

Notwithstanding OCTE limitations, we have to deeply ackowledge the huge effort put forward in different fields and with many different professionals envolved that arise numerous noteworthy studies leading to crucial conclusions in OCTE regeneration. And, from all those conclusions, professionnals from research to clinics are yet focused in pursuing for the optimal long-term functionally integrated stratified cartilage–bone structure.

**Acknowledgments** DR Pereira acknowledges the Foundation for Science and Technology (FCT), Portugal, for an individual grant (SFRH/BD/ 81356/2011). JM Oliveira also thanks the FCT for the funds provided under the program Investigator FCT 2012 and 2015 (IF/00423/2012 and IF/01285/2015). Thanks to Mr. Maciej Doczyk for his help with the graphic illustrations.

# References

1. Lories RJ, Luyten FP (2011) The bone-cartilage unit in osteoarthritis. Nat Rev Rheumatol 7(1):43–49. https://doi.org/10.1038/nrrheum.2010.197
2. Athanasiou KA, Darling EM, Hu JC (2009) Articular cartilage tissue engineering. Syn Lect Tiss Eng 1(1):1–182. https://doi.org/10.2200/S00212ED1V01Y200910TIS003
3. Temenoff JS, Mikos AG (2000) Review: tissue engineering for regeneration of articular cartilage. Biomaterials 21(5):431–440
4. Hunziker EB (2000) Articular cartilage repair: problems and perspectives. Biorheology 37(1–2):163–164
5. Pacifici M, Koyama E, Iwamoto M, Gentili C (2000) Development of articular cartilage: what do we know about it and how may it occur? Connect Tissue Res 41(3):175–184
6. Moisio K, Eckstein F, Chmiel JS, Guermazi A, Prasad P, Almagor O, Song J, Dunlop D, Hudelmaier M, Kothari A, Sharma L (2009) Denuded subchondral bone and knee pain in persons with knee osteoarthritis. Arthritis Rheum 60(12):3703–3710. https://doi.org/10.1002/art.25014

7. Curl WW, Krome J, Gordon ES, Rushing J, Smith BP, Poehling GG (1997) Cartilage injuries: a review of 31,516 knee arthroscopies. Arthroscopy 13(4):456–460
8. Blevins FT, Steadman JR, Rodrigo JJ, Silliman J (1998) Treatment of articular cartilage defects in athletes: an analysis of functional outcome and lesion appearance. Orthopedics 21(7):761–767; discussion 767–768
9. Steadman JR, Rodkey WG, Rodrigo JJ (2001) Microfracture: surgical technique and rehabilitation to treat chondral defects. Clin Orthop Relat Res 391(Suppl):S362–S369
10. Caffey S, McPherson E, Moore B, Hedman T, Vangsness CT Jr (2005) Effects of radiofrequency energy on human articular cartilage: an analysis of 5 systems. Am J Sports Med 33(7):1035–1039. https://doi.org/10.1177/0363546504271965
11. Spahn G, Kahl E, Muckley T, Hofmann GO, Klinger HM (2008) Arthroscopic knee chondroplasty using a bipolar radiofrequency-based device compared to mechanical shaver: results of a prospective, randomized, controlled study. Knee Surg Sports Traumatol Arthrosc 16(6):565–573. https://doi.org/10.1007/s00167-008-0506-1
12. Upmeier H, Bruggenjurgen B, Weiler A, Flamme C, Laprell H, Willich SN (2007) Follow-up costs up to 5 years after conventional treatments in patients with cartilage lesions of the knee. Knee Surg Sports Traumatol Arthrosc 15(3):249–257. https://doi.org/10.1007/s00167-006-0182-y
13. Laupattarakasem W, Laopaiboon M, Laupattarakasem P, Sumanonont C (2008) Arthroscopic debridement for knee osteoarthritis. Cochrane Database Syst Rev 1:CD005118. https://doi.org/10.1002/14651858.CD005118.pub2
14. Mankin HJ (1982) The response of articular cartilage to mechanical injury. J Bone Joint Surg Am 64(3):460–466
15. Falah M, Nierenberg G, Soudry M, Hayden M, Volpin G (2010) Treatment of articular cartilage lesions of the knee. Int Orthop 34(5):621–630. https://doi.org/10.1007/s00264-010-0959-y
16. Vasiliadis HS, Wasiak J (2010) Autologous chondrocyte implantation for full thickness articular cartilage defects of the knee. Cochrane Database Syst Rev 10:CD003323. https://doi.org/10.1002/14651858.CD003323.pub3
17. Brittberg M, Lindahl A, Nilsson A, Ohlsson C, Isaksson O, Peterson L (1994) Treatment of deep cartilage defects in the knee with autologous chondrocyte transplantation. N Engl J Med 331:889. https://doi.org/10.1056/NEJM199410063311401
18. Espregueira-Mendes J, Pereira H, Sevivas N, Varanda P, da Silva MV, Monteiro A, Oliveira JM, Reis RL (2012) Osteochondral transplantation using autografts from the upper tibiofibular joint for the treatment of knee cartilage lesions. Knee Surg Sports Traumatol Arthrosc 20(6):1136–1142. https://doi.org/10.1007/s00167-012-1910-0
19. Filardo G, Kon E, Di Martino A, Busacca M, Altadonna G, Marcacci M (2013) Treatment of knee osteochondritis dissecans with a cell-free biomimetic osteochondral scaffold: clinical and imaging evaluation at 2-year follow-up. Am J Sports Med 41(8):1786–1793. https://doi.org/10.1177/0363546513490658
20. Makris EA, Gomoll AH, Malizos KN, Hu JC, Athanasiou KA (2015) Repair and tissue engineering techniques for articular cartilage. Nat Rev Rheumatol 11(1):21–34. https://doi.org/10.1038/nrrheum.2014.157
21. Dunkin BS, Lattermann C (2013) New and emerging techniques in cartilage repair: MACI. Oper Tech Sports Med 21(2):100–107. https://doi.org/10.1053/j.otsm.2013.03.003
22. Ventura A, Memeo A, Borgo E, Terzaghi C, Legnani C, Albisetti W (2012) Repair of osteochondral lesions in the knee by chondrocyte implantation using the MACI® technique. Knee Surg Sports Traumatol Arthrosc 20(1):121–126. https://doi.org/10.1007/s00167-011-1575-0
23. Amini AR, Adams DJ, Laurencin CT, Nukavarapu SP (2012) Optimally porous and biomechanically compatible scaffolds for large-area bone regeneration. Tissue Eng A 18(13–14):1376–1388. https://doi.org/10.1089/ten.TEA.2011.0076
24. Nukavarapu SP, Dorcemus DL (2013) Osteochondral tissue engineering: current strategies and challenges. Biotechnol Adv 31(5):706–721. https://doi.org/10.1016/j.biotechadv.2012.11.004
25. Danisovic L, Varga I, Zamborsky R, Bohmer D (2012) The tissue engineering of articular cartilage: cells, scaffolds and stimulating factors. Exp Biol Med (Maywood) 237(1):10–17. https://doi.org/10.1258/ebm.2011.011229

26. Huey DJ, Hu JC, Athanasiou KA (2012) Unlike bone, cartilage regeneration remains elusive. Science 338(6109):917–921. https://doi.org/10.1126/science.1222454
27. Khan IM, Gilbert SJ, Singhrao SK, Duance VC, Archer CW (2008) Cartilage integration: evaluation of the reasons for failure of integration during cartilage repair. A review. Eur Cell Mater 16:26–39
28. Niemeyer P, Pestka JM, Kreuz PC, Erggelet C, Schmal H, Suedkamp NP, Steinwachs M (2008) Characteristic complications after autologous chondrocyte implantation for cartilage defects of the knee joint. Am J Sports Med 36(11):2091–2099. https://doi.org/10.1177/0363546508322131
29. van Bergen CJA, Zengerink M, Blankevoort L, van Sterkenburg MN, van Oldenrijk J, van Dijk CN (2010) Novel metallic implantation technique for osteochondral defects of the medial talar dome: a cadaver study. Acta Orthop 81(4):495–502. https://doi.org/10.3109/17453674.2010.492764
30. Mardones RM, Reinholz GG, Fitzsimmons JS, Zobitz ME, An KN, Lewallen DG, Yaszemski MJ, O'Driscoll SW (2005) Development of a biologic prosthetic composite for cartilage repair. Tissue Eng 11(9–10):1368–1378. https://doi.org/10.1089/ten.2005.11.1368
31. Stock UA, Vacanti JP (2001) Tissue engineering: current state and prospects. Annu Rev Med 52:443–451. https://doi.org/10.1146/annurev.med.52.1.443
32. Hutmacher DW (2000) Scaffolds in tissue engineering bone and cartilage. Biomaterials 21(24):2529–2543. https://doi.org/10.1016/S0142-9612(00)00121-6
33. Lee SH, Shin H (2007) Matrices and scaffolds for delivery of bioactive molecules in bone and cartilage tissue engineering. Adv Drug Deliv Rev 59(4–5):339–359. https://doi.org/10.1016/j.addr.2007.03.016
34. Seo SJ, Mahapatra C, Singh RK, Knowles JC, Kim HW (2014) Strategies for osteochondral repair: focus on scaffolds. J Tissue Eng 5:2041731414541850. https://doi.org/10.1177/2041731414541850
35. Malafaya PB, Silva GA, Reis RL (2007) Natural-origin polymers as carriers and scaffolds for biomolecules and cell delivery in tissue engineering applications. Adv Drug Deliv Rev 59(4–5):207–233. https://doi.org/10.1016/j.addr.2007.03.012
36. Cosson S, Otte EA, Hezaveh H, Cooper-White JJ (2015) Concise review: tailoring bioengineered scaffolds for stem cell applications in tissue engineering and regenerative medicine. Stem Cells Transl Med 4(2):156–164. https://doi.org/10.5966/sctm.2014-0203
37. Shao XX, Hutmacher DW, Ho ST, Goh JC, Lee EH (2006) Evaluation of a hybrid scaffold/cell construct in repair of high-load-bearing osteochondral defects in rabbits. Biomaterials 27(7):1071–1080. https://doi.org/10.1016/j.biomaterials.2005.07.040
38. Kazemnejad S, Khanmohammadi M, Mobini S, Taghizadeh-Jahed M, Khanjani S, Arasteh S, Golshahi H, Torkaman G, Ravanbod R, Heidari-Vala H, Moshiri A, Tahmasebi MN, Akhondi MM (2016) Comparative repair capacity of knee osteochondral defects using regenerated silk fiber scaffolds and fibrin glue with/without autologous chondrocytes during 36 weeks in rabbit model. Cell Tissue Res 364(3):559–572. https://doi.org/10.1007/s00441-015-2355-9
39. Gao J, Dennis JE, Solchaga LA, Goldberg VM, Caplan AI (2002) Repair of osteochondral defect with tissue-engineered two-phase composite material of injectable calcium phosphate and hyaluronan sponge. Tissue Eng 8(5):827–837. https://doi.org/10.1089/10763270260424187
40. Frenkel SR, Bradica G, Brekke JH, Goldman SM, Ieska K, Issack P, Bong MR, Tian H, Gokhale J, Coutts RD, Kronengold RT (2005) Regeneration of articular cartilage--evaluation of osteochondral defect repair in the rabbit using multiphasic implants. Osteoarthritis Cartilage 13(9):798–807. https://doi.org/10.1016/j.joca.2005.04.018
41. Getgood AM, Kew SJ, Brooks R, Aberman H, Simon T, Lynn AK, Rushton N (2012) Evaluation of early-stage osteochondral defect repair using a biphasic scaffold based on a collagen-glycosaminoglycan biopolymer in a caprine model. Knee 19(4):422–430. https://doi.org/10.1016/j.knee.2011.03.011
42. Schleicher I, Lips KS, Sommer U, Schappat I, Martin AP, Szalay G, Hartmann S, Schnettler R (2013) Biphasic scaffolds for repair of deep osteochondral defects in a sheep model. J Surg Res 183(1):184–192. https://doi.org/10.1016/j.jss.2012.11.036

43. Chen J, Chen H, Li P, Diao H, Zhu S, Dong L, Wang R, Guo T, Zhao J, Zhang J (2011) Simultaneous regeneration of articular cartilage and subchondral bone in vivo using MSCs induced by a spatially controlled gene delivery system in bilayered integrated scaffolds. Biomaterials 32(21):4793–4805. https://doi.org/10.1016/j.biomaterials.2011.03.041
44. Oliveira JM, Rodrigues MT, Silva SS, Malafaya PB, Gomes ME, Viegas CA, Dias IR, Azevedo JT, Mano JF, Reis RL (2006) Novel hydroxyapatite/chitosan bilayered scaffold for osteochondral tissue-engineering applications: scaffold design and its performance when seeded with goat bone marrow stromal cells. Biomaterials 27(36):6123–6137. https://doi.org/10.1016/j.biomaterials.2006.07.034
45. Re'em T, Witte F, Willbold E, Ruvinov E, Cohen S (2012) Simultaneous regeneration of articular cartilage and subchondral bone induced by spatially presented TGF-beta and BMP-4 in a bilayer affinity binding system. Acta Biomater 8(9):3283–3293. https://doi.org/10.1016/j.actbio.2012.05.014
46. Reyes R, Delgado A, Sanchez E, Fernandez A, Hernandez A, Evora C (2014) Repair of an osteochondral defect by sustained delivery of BMP-2 or TGFbeta1 from a bilayered alginate-PLGA scaffold. J Tissue Eng Regen Med 8(7):521–533. https://doi.org/10.1002/term.1549
47. Yan LP, Silva-Correia J, Oliveira MB, Vilela C, Pereira H, Sousa RA, Mano JF, Oliveira AL, Oliveira JM, Reis RL (2015) Bilayered silk/silk-nanoCaP scaffolds for osteochondral tissue engineering: in vitro and in vivo assessment of biological performance. Acta Biomater 12:227–241. https://doi.org/10.1016/j.actbio.2014.10.021
48. Pereira DR, Canadas RF, Correia JS, Marques AP, Reis RL, Oliveira JM (2014) Gellan gum-based hydrogel bilayered scaffolds for osteochondral tissue engineering. Key Eng Mater 587:255–260. https://doi.org/10.4028/www.scientific.net/KEM.587.255
49. Mahmoudifar N, Doran PM (2013) Osteogenic differentiation and osteochondral tissue engineering using human adipose-derived stem cells. Biotechnol Prog 29(1):176–185. https://doi.org/10.1002/btpr.1663
50. Narayanan LK, Huebner P, Fisher MB, Spang JT, Starly B, Shirwaiker RA (2016) 3D-bioprinting of polylactic acid (PLA) nanofiber–alginate hydrogel bioink containing human adipose-derived stem cells. ACS Biomater Sci Eng 2(10):1732–1742. https://doi.org/10.1021/acsbiomaterials.6b00196
51. Sidney LE, Heathman TRJ, Britchford ER, Abed A, Rahman CV, Buttery LDK (2015) Investigation of localized delivery of diclofenac sodium from poly(D,L-lactic acid-co-glycolic acid)/poly(ethylene glycol) scaffolds using an in vitro osteoblast inflammation model. Tissue Eng A 21(1–2):362–373. https://doi.org/10.1089/ten.tea.2014.0100
52. Koupaei N, Karkhaneh A (2016) Porous crosslinked polycaprolactone hydroxyapatite networks for bone tissue engineering. Tissue Eng Regen Med 13(3):251–260. https://doi.org/10.1007/s13770-016-9061-x
53. Lv YM, Yu QS (2015) Repair of articular osteochondral defects of the knee joint using a composite lamellar scaffold. Bone Joint Res 4(4):56–64. https://doi.org/10.1302/2046-3758.44.2000310
54. Yao Q, Nooeaid P, Detsch R, Roether JA, Dong Y, Goudouri O-M, Schubert DW, Boccaccini AR (2014) Bioglass®/chitosan–polycaprolactone bilayered composite scaffolds intended for osteochondral tissue engineering. J Biomed Mater Res A 102(12):4510–4518. https://doi.org/10.1002/jbm.a.35125
55. Oh SH, Lee JH (2013) Hydrophilization of synthetic biodegradable polymer scaffolds for improved cell/tissue compatibility. Biomed Mater (Bristol, England) 8(1):014101
56. Perez RA, Won JE, Knowles JC, Kim HW (2013) Naturally and synthetic smart composite biomaterials for tissue regeneration. Adv Drug Deliv Rev 65(4):471–496. https://doi.org/10.1016/j.addr.2012.03.009
57. Ahn JH, Lee TH, Oh JS, Kim SY, Kim HJ, Park IK, Choi BS, Im GI (2009) Novel hyaluronate-atelocollagen/beta-TCP-hydroxyapatite biphasic scaffold for the repair of osteochondral defects in rabbits. Tissue Eng A 15(9):2595–2604. https://doi.org/10.1089/ten.TEA.2008.0511

58. Cui W, Wang Q, Chen G, Zhou S, Chang Q, Zuo Q, Ren K, Fan W (2011) Repair of articular cartilage defects with tissue-engineered osteochondral composites in pigs. J Biosci Bioeng 111(4):493–500. https://doi.org/10.1016/j.jbiosc.2010.11.023
59. Ho ST, Hutmacher DW, Ekaputra AK, Hitendra D, Hui JH (2010) The evaluation of a biphasic osteochondral implant coupled with an electrospun membrane in a large animal model. Tissue Eng A 16(4):1123–1141. https://doi.org/10.1089/ten.TEA.2009.0471
60. Pei M, He F, Boyce BM, Kish VL (2009) Repair of full-thickness femoral condyle cartilage defects using allogeneic synovial cell-engineered tissue constructs. Osteoarthritis Cartilage 17(6):714–722. https://doi.org/10.1016/j.joca.2008.11.017
61. Gupta V, Lyne DV, Barragan M, Berkland CJ, Detamore MS (2016) Microsphere-based scaffolds encapsulating Tricalcium phosphate and hydroxyapatite for bone regeneration. J Mater Sci Mater Med 27(7):121–121. https://doi.org/10.1007/s10856-016-5734-1
62. Singh M, Berkland C, Detamore MS (2008) Strategies and applications for incorporating physical and chemical signal gradients in tissue engineering. Tissue Eng Part B Rev 14(4):341–366. https://doi.org/10.1089/ten.teb.2008.0304
63. Jeon JE, Vaquette C, Klein TJ, Hutmacher DW (2014) Perspectives in multiphasic osteochondral tissue engineering. Anat Rec (Hoboken, NJ) 297(1):26–35. https://doi.org/10.1002/ar.22795
64. Cheng HW, Luk KD, Cheung KM, Chan BP (2011) In vitro generation of an osteochondral interface from mesenchymal stem cell-collagen microspheres. Biomaterials 32(6):1526–1535. https://doi.org/10.1016/j.biomaterials.2010.10.021
65. Schaefer D, Martin I, Shastri P, Padera RF, Langer R, Freed LE, Vunjak-Novakovic G (2000) In vitro generation of osteochondral composites. Biomaterials 21(24):2599–2606
66. Jeon JE, Vaquette C, Theodoropoulos C, Klein TJ, Hutmacher DW (2014) Multiphasic construct studied in an ectopic osteochondral defect model. J R Soc Interface 11(95):20140184. https://doi.org/10.1098/rsif.2014.0184
67. Filardo G, Drobnic M, Perdisa F, Kon E, Hribernik M, Marcacci M (2014) Fibrin glue improves osteochondral scaffold fixation: study on the human cadaveric knee exposed to continuous passive motion. Osteoarthritis Cartilage 22(4):557–565. https://doi.org/10.1016/j.joca.2014.01.004
68. Yousefi AM, Hoque ME, Prasad RG, Uth N (2015) Current strategies in multiphasic scaffold design for osteochondral tissue engineering: a review. J Biomed Mater Res A 103(7):2460–2481. https://doi.org/10.1002/jbm.a.35356
69. Yang J, Zhang YS, Yue K, Khademhosseini A (2017) Cell-laden hydrogels for osteochondral and cartilage tissue engineering. Acta Biomater 57:1–25. https://doi.org/10.1016/j.actbio.2017.01.036
70. Brittberg M (2010) Cell carriers as the next generation of cell therapy for cartilage repair: a review of the matrix-induced autologous chondrocyte implantation procedure. Am J Sports Med 38(6):1259–1271. https://doi.org/10.1177/0363546509346395
71. Deng T, Lv J, Pang J, Liu B, Ke J (2014) Construction of tissue-engineered osteochondral composites and repair of large joint defects in rabbit. J Tissue Eng Regen Med 8(7):546–556
72. Zhou J, Xu C, Wu G, Cao X, Zhang L, Zhai Z, Zheng Z, Chen X, Wang Y (2011) In vitro generation of osteochondral differentiation of human marrow mesenchymal stem cells in novel collagen-hydroxyapatite layered scaffolds. Acta Biomater 7(11):3999–4006
73. Song K, Li L, Yan X, Zhang Y, Li R, Wang Y, Wang L, Wang H, Liu T (2016) Fabrication and development of artificial osteochondral constructs based on cancellous bone/hydrogel hybrid scaffold. J Mater Sci Mater Med 27(6):114
74. Li X, Li Y, Zuo Y, Qu D, Liu Y, Chen T, Jiang N, Li H, Li J (2015) Osteogenesis and chondrogenesis of biomimetic integrated porous PVA/gel/V-n-HA/pa6 scaffolds and BMSCs construct in repair of articular osteochondral defect. J Biomed Mater Res A 103(10):3226–3236
75. Wang Y, Meng H, Yuan X, Peng J, Guo Q, Lu S, Wang A (2014) Fabrication and in vitro evaluation of an articular cartilage extracellular matrix-hydroxyapatite bilayered scaffold with low permeability for interface tissue engineering. Biomed Eng Online 13:80

76. Zhang T, Zhang H, Zhang L, Jia S, Liu J, Xiong Z, Sun W (2017) Biomimetic design and fabrication of multilayered osteochondral scaffolds by low-temperature deposition manufacturing and thermal-induced phase-separation techniques. Biofabrication 9(2):025021
77. Liao J, Tian T, Shi S, Xie X, Ma Q, Li G, Lin Y (2017) The fabrication of biomimetic biphasic CAN-PAC hydrogel with a seamless interfacial layer applied in osteochondral defect repair. Bone Res 5:17018
78. Galperin A, Oldinski RA, Florczyk SJ, Bryers JD, Zhang M, Ratner BD (2013) Integrated bi-layered scaffold for osteochondral tissue engineering. Adv Healthc Mater 2(6):872–883
79. Da H, Jia SJ, Meng GL, Cheng JH, Zhou W, Xiong Z, Mu YJ, Liu J (2013) The impact of compact layer in biphasic scaffold on osteochondral tissue engineering. PLoS One 8(1):e54838
80. Park JY, Choi JC, Shim JH, Lee JS, Park H, Kim SW, Doh J, Cho DW (2014) A comparative study on collagen type I and hyaluronic acid dependent cell behavior for osteochondral tissue bioprinting. Biofabrication 6(3):035004
81. Lam J, Lu S, Lee EJ, Trachtenberg JE, Meretoja VV, Dahlin RL, van den Beucken JJ, Tabata Y, Wong ME, Jansen JA, Mikos AG, Kasper FK (2014) Osteochondral defect repair using bilayered hydrogels encapsulating both chondrogenically and osteogenically pre-differentiated mesenchymal stem cells in a rabbit model. Osteoarthritis Cartilage 22(9):1291–1300
82. Lam J, Lu S, Meretoja VV, Tabata Y, Mikos AG, Kasper FK (2014) Generation of osteochondral tissue constructs with chondrogenically and osteogenically predifferentiated mesenchymal stem cells encapsulated in bilayered hydrogels. Acta Biomater 10(3):1112–1123
83. Sartori M, Pagani S, Ferrari A, Costa V, Carina V, Figallo E, Maltarello MC, Martini L, Fini M, Giavaresi G (2017) A new bi-layered scaffold for osteochondral tissue regeneration: In vitro and in vivo preclinical investigations. Mater Sci Eng C Mater Biol Appl 70(Pt 1):101–111
84. Jiang J, Tang A, Ateshian GA, Guo XE, Hung CT, Lu HH (2010) Bioactive stratified polymer ceramic-hydrogel scaffold for integrative osteochondral repair. Ann Biomed Eng 38(6):2183–2196
85. Liu X, Liu S, Liu S, Cui W (2014) Evaluation of oriented electrospun fibers for periosteal flap regeneration in biomimetic triphasic osteochondral implant. J Biomed Mater Res B Appl Biomater 102(7):1407–1414
86. Aydin HM (2011) A three-layered osteochondral plug: structural, mechanical, and in vitro biocompatibility analysis. Adv Eng Mater 13(12):B511–B517
87. Ding X, Zhu M, Xu B, Zhang J, Zhao Y, Ji S, Wang L, Wang L, Li X, Kong D, Ma X, Yang Q (2014) Integrated trilayered silk fibroin scaffold for osteochondral differentiation of adipose-derived stem cells. ACS Appl Mater Interfaces 6(19):16696–16705
88. Yucekul, A., D. Ozdil, N. H. Kutlu, E. Erdemli, H. M. Aydin and M. N. Doral (2017) Tri-layered composite plug for the repair of osteochondral defects: in vivo study in sheep. J Tissue Eng 8:2041731417697500
89. Levingstone TJ, Matsiko A, Dickson GR, O'Brien FJ, Gleeson JP (2014) A biomimetic multi-layered collagen-based scaffold for osteochondral repair. Acta Biomater 10(5):1996–2004
90. Levingstone TJ, Thompson E, Matsiko A, Schepens A, Gleeson JP, O'Brien FJ (2016) Multi-layered collagen-based scaffolds for osteochondral defect repair in rabbits. Acta Biomater 32:149–160
91. Amadori S, Torricelli P, Panzavolta S, Parrilli A, Fini M, Bigi A (2015) Multi-layered scaffolds for osteochondral tissue engineering: in vitro response of co-cultured human mesenchymal stem cells. Macromol Biosci 15(11):1535–1545
92. Kon E, Delcogliano M, Filardo G, Busacca M, Di Martino A, Marcacci M (2011) Novel nano-composite multilayered biomaterial for osteochondral regeneration: a pilot clinical trial. Am J Sports Med 39(6):1180–1190. https://doi.org/10.1177/0363546510392711
93. Kon E, Filardo G, Di Martino A, Busacca M, Moio A, Perdisa F, Marcacci M (2014) Clinical results and MRI evolution of a nano-composite multilayered biomaterial for osteochondral regeneration at 5 years. Am J Sports Med 42(1):158–165. https://doi.org/10.1177/0363546513505434
94. St-Pierre JP, Gan L, Wang J, Pilliar RM, Grynpas MD, Kandel RA (2012) The incorporation of a zone of calcified cartilage improves the interfacial shear strength between in vitro-formed car-

tilage and the underlying substrate. Acta Biomater 8(4):1603–1615. https://doi.org/10.1016/j. actbio.2011.12.022
95. Mohan N, Dormer NH, Caldwell KL, Key VH, Berkland CJ, Detamore MS (2011) Continuous gradients of material composition and growth factors for effective regeneration of the osteochondral interface. Tissue Eng Part A 17(21–22):2845–2855. https://doi.org/10.1089/ten. tea.2011.0135
96. Phillips JE, Burns KL, Le Doux JM, Guldberg RE, García AJ (2008) Engineering graded tissue interfaces. Proc Natl Acad Sci U S A 105(34):12170–12175
97. Singh M, Sandhu B, Scurto A, Berkland C, Detamore MS (2010) Microsphere-based scaffolds for cartilage tissue engineering: using subcritical CO(2) as a sintering agent. Acta Biomater 6(1):137–143. https://doi.org/10.1016/j.actbio.2009.07.042
98. Li X, Xie J, Lipner J, Yuan X, Thomopoulos S, Xia Y (2009) Nanofiber scaffolds with gradations in mineral content for mimicking the tendon-to-bone insertion site. Nano Lett 9(7):2763–2768. https://doi.org/10.1021/nl901582f
99. Getgood A, Henson F, Skelton C, Herrera E, Brooks R, Fortier LA, Rushton N (2012) The augmentation of a collagen/glycosaminoglycan biphasic osteochondral scaffold with platelet-rich plasma and concentrated bone marrow aspirate for osteochondral defect repair in sheep: a pilot study. Cartilage 3(4):351–363. https://doi.org/10.1177/1947603512444597
100. Jung MR, Shim IK, Chung HJ, Lee HR, Park YJ, Lee MC, Yang YI, Do SH, Lee SJ (2012) Local BMP-7 release from a PLGA scaffolding-matrix for the repair of osteochondral defects in rabbits. J Control Release 162(3):485–491. https://doi.org/10.1016/j.jconrel.2012.07.040
101. Sakata R, Kokubu T, Nagura I, Toyokawa N, Inui A, Fujioka H, Kurosaka M (2012) Localization of vascular endothelial growth factor during the early stages of osteochondral regeneration using a bioabsorbable synthetic polymer scaffold. J Orthop Res 30(2):252–259. https://doi.org/10.1002/jor.21502
102. Kim K, Lam J, Lu S, Spicer PP, Lueckgen A, Tabata Y, Wong ME, Jansen JA, Mikos AG, Kasper FK (2013) Osteochondral tissue regeneration using a bilayered composite hydrogel with modulating dual growth factor release kinetics in a rabbit model. J Control Release 168(2):166–178. https://doi.org/10.1016/j.jconrel.2013.03.013
103. Algul D, Gokce A, Onal A, Servet E, Dogan Ekici AI, Yener FG (2016) In vitro release and in vivo biocompatibility studies of biomimetic multilayered alginate-chitosan/β-TCP scaffold for osteochondral tissue. J Biomater Sci Polym Ed 27(5):431–440. https://doi.org/10.10 80/09205063.2016.1140501
104. Han F, Yang X, Zhao J, Zhao Y, Yuan X (2015) Photocrosslinked layered gelatin-chitosan hydrogel with graded compositions for osteochondral defect repair. J Mater Sci Mater Med 26(4):160. https://doi.org/10.1007/s10856-015-5489-0
105. Mazaki T, Shiozaki Y, Yamane K, Yoshida A, Nakamura M, Yoshida Y, Zhou D, Kitajima T, Tanaka M, Ito Y, Ozaki T, Matsukawa A (2014) A novel, visible light-induced, rapidly cross-linkable gelatin scaffold for osteochondral tissue engineering. Sci Rep 4:4457. https://doi. org/10.1038/srep04457
106. Moffat KL, Wang IN, Rodeo SA, Lu HH (2009) Orthopedic interface tissue engineering for the biological fixation of soft tissue grafts. Clin Sports Med 28(1):157–176. https://doi. org/10.1016/j.csm.2008.08.006
107. Efe T, Fuglein A, Heyse TJ, Stein T, Timmesfeld N, Fuchs-Winkelmann S, Schmitt J, Paletta JR, Schofer MD (2012) Fibrin glue does not improve the fixation of press-fitted cell-free collagen gel plugs in an ex vivo cartilage repair model. Knee Surg Sports Traumatol Arthrosc 20(2):210–215. https://doi.org/10.1007/s00167-011-1571-4
108. Siparsky PN, Bailey JR, Dale KM, Klement MR, Taylor DC (2017) Open reduction internal fixation of isolated chondral fragments without osseous attachment in the knee: a case series. Orthop J Sports Med 5(3):232596711769628. https://doi.org/10.1177/2325967117696281
109. Tampieri A, Landi E, Valentini F, Sandri M, D'Alessandro T, Dediu V, Marcacci M (2011) A conceptually new type of bio-hybrid scaffold for bone regeneration. Nanotechnology 22(1):015104. https://doi.org/10.1088/0957-4484/22/1/015104
110. van de Breevaart BJ, In der Maur CD, Bos PK, Feenstra L, Verhaar JAN, Weinans H, van Osch GJVM (2004) Improved cartilage integration and interfacial strength after enzymatic

treatment in a cartilage transplantation model. Arthritis Res Ther 6(5):R469. https://doi.org/10.1186/ar1216
111. Gilbert SJ, Singhrao SK, Khan IM, Gonzalez LG, Thomson BM, Burdon D, Duance VC, Archer CW (2009) Enhanced tissue integration during cartilage repair in vitro can be achieved by inhibiting chondrocyte death at the wound edge. Tissue Eng A 15(7):1739–1749. https://doi.org/10.1089/ten.tea.2008.0361
112. Allon AA, Ng KW, Hammoud S, Russell BH, Jones CM, Rivera JJ, Schwartz J, Hook M, Maher SA (2012) Augmenting the articular cartilage-implant interface: functionalizing with a collagen adhesion protein. J Biomed Mater Res A 100(8):2168–2175. https://doi.org/10.1002/jbm.a.34144
113. Bhumiratana S, Eton RE, Oungoulian SR, Wan LQ, Ateshian GA, Vunjak-Novakovic G (2014) Large, stratified, and mechanically functional human cartilage grown in vitro by mesenchymal condensation. Proc Natl Acad Sci U S A 111(19):6940–6945. https://doi.org/10.1073/pnas.1324050111
114. Athanasiou KA, Eswaramoorthy R, Hadidi P, Hu JC (2013) Self-organization and the self-assembling process in tissue engineering. Annu Rev Biomed Eng 15:115–136. https://doi.org/10.1146/annurev-bioeng-071812-152423
115. Shimomura K, Moriguchi Y, Ando W, Nansai R, Fujie H, Hart DA, Gobbi A, Kita K, Horibe S, Shino K, Yoshikawa H, Nakamura N (2014) Osteochondral repair using a scaffold-free tissue-engineered construct derived from synovial mesenchymal stem cells and a hydroxyapatite-based artificial bone. Tissue Eng A 20(17–18):2291–2304. https://doi.org/10.1089/ten.tea.2013.0414
116. Keeney M, Pandit A (2009) The osteochondral junction and its repair via bi-phasic tissue engineering scaffolds. Tissue Eng Part B Rev 15(1):55–73. https://doi.org/10.1089/ten.teb.2008.0388
117. Yan L-P, Oliveira JM, Oliveira AL, Reis RL (2015) Current concepts and challenges in osteochondral tissue engineering and regenerative medicine. ACS Biomater Sci Eng 1(4):183–200. https://doi.org/10.1021/ab500038y
118. Smith BD, Grande DA (2015) The current state of scaffolds for musculoskeletal regenerative applications. Nat Rev Rheumatol 11(4):213–222. https://doi.org/10.1038/nrrheum.2015.27
119. Filardo G, Kon E, Perdisa F, Di Matteo B, Di Martino A, Iacono F, Zaffagnini S, Balboni F, Vaccari V, Marcacci M (2013) Osteochondral scaffold reconstruction for complex knee lesions: a comparative evaluation. Knee 20(6):570–576. https://doi.org/10.1016/j.knee.2013.05.007
120. Kon E, Filardo G, Venieri G, Perdisa F, Marcacci M (2014) Tibial plateau lesions. Surface reconstruction with a biomimetic osteochondral scaffold: results at 2 years of follow-up. Injury 45(Suppl 6):S121–S125. https://doi.org/10.1016/j.injury.2014.10.035
121. Berruto M, Delcogliano M, de Caro F, Carimati G, Uboldi F, Ferrua P, Ziveri G, De Biase CF (2014) Treatment of large knee osteochondral lesions with a biomimetic scaffold: results of a multicenter study of 49 patients at 2-year follow-up. Am J Sports Med 42(7):1607–1617. https://doi.org/10.1177/0363546514530292
122. Christensen BB, Foldager CB, Jensen J, Jensen NC, Lind M (2016) Poor osteochondral repair by a biomimetic collagen scaffold: 1- to 3-year clinical and radiological follow-up. Knee Surg Sports Traumatol Arthrosc 24(7):2380–2387. https://doi.org/10.1007/s00167-015-3538-3
123. Dhollander AA, Liekens K, Almqvist KF, Verdonk R, Lambrecht S, Elewaut D, Verbruggen G, Verdonk PC (2012) A pilot study of the use of an osteochondral scaffold plug for cartilage repair in the knee and how to deal with early clinical failures. Arthroscopy 28(2):225–233. https://doi.org/10.1016/j.arthro.2011.07.017
124. Bekkers JE, Bartels LW, Vincken KL, Dhert WJ, Creemers LB, Saris DB (2013) Articular cartilage evaluation after TruFit plug implantation analyzed by delayed gadolinium-enhanced MRI of cartilage (dGEMRIC). Am J Sports Med 41(6):1290–1295. https://doi.org/10.1177/0363546513483536
125. Verhaegen J, Clockaerts S, Van Osch GJ, Somville J, Verdonk P, Mertens P (2015) TruFit plug for repair of osteochondral defects-where is the evidence? Systematic review of literature. Cartilage 6(1):12–19. https://doi.org/10.1177/1947603514548890

# Part IV
# Advanced Processing Methodology

# Chapter 10
# Preparation of Polymeric and Composite Scaffolds by 3D Bioprinting

**Ana Mora-Boza and María Luisa Lopez-Donaire**

**Abstract** Over the recent years, the advent of 3D bioprinting technology has marked a milestone in osteochondral tissue engineering (TE) research. Nowadays, the traditional used techniques for osteochondral regeneration remain to be inefficient since they cannot mimic the complexity of joint anatomy and tissue heterogeneity of articular cartilage. These limitations seem to be solved with the use of 3D bioprinting which can reproduce the anisotropic extracellular matrix (ECM) and heterogeneity of this tissue. In this chapter, we present the most commonly used 3D bioprinting approaches and then discuss the main criteria that biomaterials must meet to be used as suitable bioinks, in terms of mechanical and biological properties. Finally, we highlight some of the challenges that this technology must overcome related to osteochondral bioprinting before its clinical implementation.

**Keywords** 3D bioprinting · Cellular bioprinting · Acellular bioprinting · Bioink · Extracellular matrix

## 10.1 Introduction

Bioprinting has emerged over the recent years as a promising technique for osteochondral TE applications. Bioinks (biomaterials and bioactive cues) via 3D bioprinting can be deposited in a spatiotemporally accurate layer-by-layer manner, allowing for high cell seeding density and strong cell–cell interactions [1–4]. This technology can be classified into two different categories based on whether the bioink contains living cells (cellular 3D bioprinting) or not (acellular 3D bioprinting). Although to date, applications of 3D bioprinting have been focused on cardiovascular, skin regeneration, tracheal splints, cartilaginous structures, and hard tissues like bon, among others. The uniquely capacity of this technique to mimic heterogeneous and anisotropic properties of ECM has attracted much attention to osteochondral tissues [2, 5–9]. In this perspective, 3D fabrication techniques have raised as an alternative for grafting methodologies, which remain as the common gold standard treatment for joint degenerative diseases such as osteoarthritis (OA) or trauma [7]. Osteochondral grafts exhibit low integration at the bone–cartilage interface and poor tissue formation de novo [5, 8]. For its part, 3D bioprinting provides the fabrication of scaffolds with interconnected macroporosity and microporosity which improves nutrient diffusion and removal of waste products, and facilitates the ECM deposition and ingrowth of blood vessels [10, 11]. In the case of osteochondral tissue, considerations regarding to heterogeneity and anisotropy are of special importance due to mechanical and composition requirements, which differ from cartilage to bone tissues. Thus, 3D bioprinting can be advantageous.

To obtain effective and biologically relevant tissue constructions that mimic the native microenvironment, several specifications must be considered (Fig. 10.1).

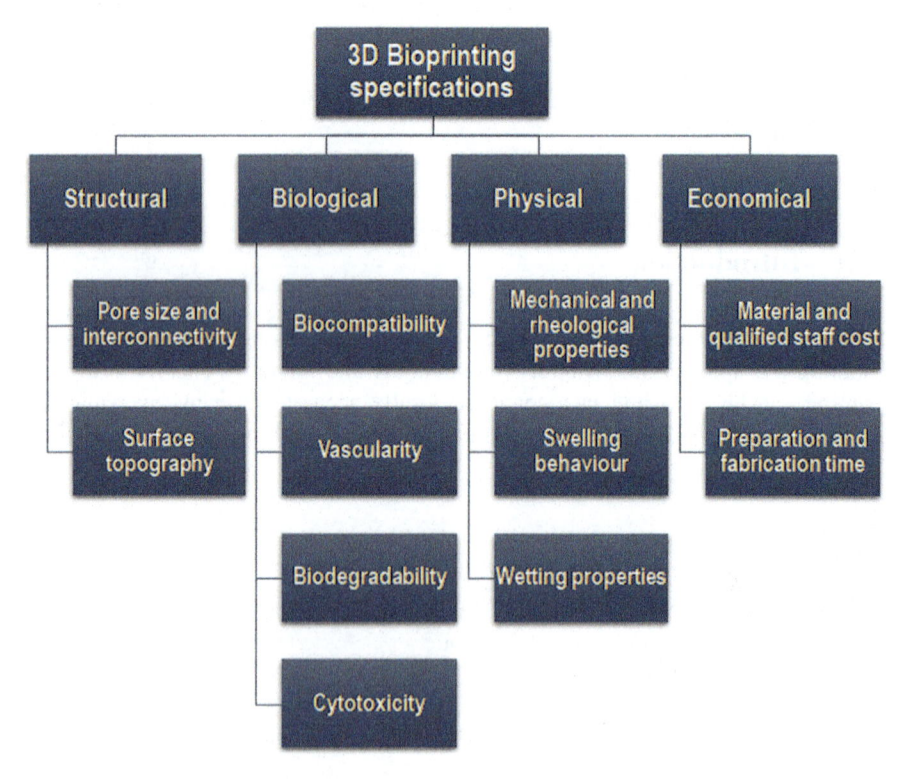

**Fig. 10.1** 3D bioprinting considerations regarding structural, physical, biological, and economical specifications

Among these essential aspects, structural and physical specifications such as bulk properties and surface topography play a key role in the development of bioactive tissue constructs. 3D bioprinting is an appropriate fabrication technique with high spatial resolution by which achieves these aspects. However, appropriate biomaterials (bioinks) should be developed with optimal rheological and biological properties, since this is the main limitation of the technique as it will be exposed in this chapter [12–15]. Material viscosity, gelation method, and speed must be optimized to obtain 3D architectures with enough structural integrity and mechanical properties that allow for not only interactions with the materials but also cell communication [2, 12, 16–19]. In addition, many manufacturing techniques can also be employed to improve the relatively weak mechanical properties of soft hydrogels, such as ultraviolet (UV) curing [20, 21], pre-cross-linking procedures, or the incorporation of additional elements and materials such as poly (ε-caprolactone) (PCL) [22] or graphene oxide (GO) elements [23, 24]. Some of these elements can be sacrificial since they will not form part of the final constructs.

Moreover, the 3D bioprinting processes must ensure compliance with some biological specifications. Certainly, biocompatibility and absence of cytotoxicity are essential requisites, but the considerable efforts made over the recent decades on the TE field have demonstrated that bioactive constructs are indispensable. By this way, the need for vascularization remains one of the most daunting challenges in the development of 3D complexes. Molecular diffusion limitations make necessary a minimun distance ($\approx$ 100 µm) between cells and the nearest capillary to facilitate the exchange of nutrients and oxygen, which would be impossible without an adequate vascular network [12, 25, 26]. Finally, for a succeed integration with surrounding environment, degradation and absorption kinetics of the constructs must be fast to avoid side effects. 3D bioprinting provides some advantages regarding to these biological aspects among other biofabrication techniques, since it can facilitate a controlled deposition of cells, maintaining their viability during the process [12, 27].

Finally, economic issues regarding manufacturing requirements, overall cost of materials and fabrication devices, and necessary production time are crucial aspects for successful clinical translation of 3D bioprinting in osteochondral restoration applications. Although the cost of specialized equipment and experienced personnel could be high, there are progressively more affordable 3D fabrication systems with intuitive interfaces for inexperienced users [5, 7, 12].

In this chapter, we discuss how the advent of 3D bioprinting has provided new opportunities for osteochondral TE, and the current advances and challenges that must be addressed by current 3D bioprinting approaches and bioinks for the preparation of polymeric and composited scaffolds.

## 10.2 3D Bioprinting Fabrication Strategies for Osteochondral TE

As explained in the introduction, 3D bioprinting techniques have attracted much attention to the treatment of osteochondral degeneration and diseases such as osteoarthritis through osteochondral tissue regeneration. Currently, autografts and allografts are being applied to reduce donor site morbidity and matching mainly in young patients. However, the research is moving towards developing de novo tissue constructions to improve integration with host tissue and nutrient diffusion in larger macroporous scaffolds through cell-based repairs such as autologous chondrocyte implantation [5, 7, 12, 28, 29].

Recently, bioprinting has been subclassified into two categories: scaffold-based and scaffold-free bioprinting. While scaffold-based bioprinting implies the generation of scaffolding materials by 3D printing where cells can be seeded during or post-fabrication, scaffold-free bioprinting is based on the self-assembly of cellular components mimicking embryonic development [30–32]. This chapter focuses only on the description of the scaffold-based bioprinting techniques for the development of osteochondral complex tissue. We have made a subdivision between cellular and acellular scaffold-based bioprinting depending on if their bioink formulation contains living cells or not.

### 10.2.1 Cellular Bioprinting

In this first paragraph, the most commonly used cellular 3D bioprinting processes will be presented, namely, extrusion-based bioprinting, droplet-based bioprinting, laser-based bioprinting, and stereolithography, [5, 7, 9, 33, 34]. An illustration of these 3D bioprinting processes with its main components is shown in Fig. 10.2.

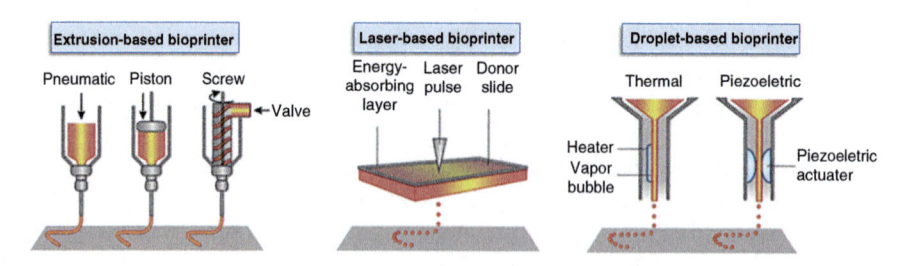

**Fig. 10.2** Schemes of 3D bioprinting process and its main components [35]. Adapted from Biomaterials 83, D. Tang. et al., Biofabrication of bone tissue: approaches, challenges and translation for bone regeneration, pp. 363–382, Copyright 2016, with permission from Elsevier

### 10.2.1.1 Extrusion-Based Bioprinting

The extrusion-based bioprinting (EBB) consists of the dispensation of bioinks using an air-force pump, solenoid or mechanical screw plunger. EBB addresses the challenges of droplet bioprinting process, which cannot deposit very viscous materials or high cell density solutions [1, 12, 33]. As discussed later in the chapter, high viscosity values of the bioinks are desirable to obtain high shape fidelity of the tissue constructs but in some cases, high concentration of the components of the bioinks can result in less cell viability due to cytotoxicity [36]. Nevertheless, EBB is the most used technique for TE applications due to its moderate cost in comparison to the good resolution it provides, as well as, the high cellular concentrated bioinks that can be printed [1]. In addition, good-shaped fidelity can be obtained through a fast phase change from a liquid bioink to a more solid network by different cross-linking procedures, that can be classified into chemical (reversible) and physical (irreversible) cross-linkings. Among all chemical cross-linkings processes photo-initiated free radical polymerization reaction is a commonly used alternative for rapid cross-linking despite its cytotoxicity. This process is widely accepted due to its effectiveness, efficiency, and controllability. Duchi and collaborators developed a coaxial core–shell system for EBB to avoid the cytotoxicity that can trigger UV photocuring due to the generation of free radicals and exposure of the cells to UV. They demonstrated that these problems can be addressed with an accurate control of the deposition parameters [37]. In another work, O'Connell et al. developed an easy-to-handle device for medical surgery, named "biopen," in an attempt of bringing together 3D printing technology and surgical processes. The tool could print gelatin methacrylamide (GelMA) and hyaluronic acid methacrylate (HAMA) hydrogels, which were photocrosslinked. The process was compatible with the deposition of adipose stem cells at chondral wound side protocol [38].

Many authors have used of EBB for the development of multilayered compound scaffolds in the context of cartilage and osteochondral regeneration [10, 37, 39–41]. For example, Bartnikowski and collaborators developed a 3D plotted scaffold composed of alginate and hydroxyapatite (HAp), mixed with GelMA, or GelMA with HAMA for the regeneration of a zone of calcified cartilage, concluding that the incorporation of HAMA in these hydrogels improved chondrogenesis [11]. T. Ahlfeld and coworkers used EBB to obtain 3D constructs by printing alginate and methylcellulose with clinically relevant dimensions thanks to the addition of laponite, a nanosilicate clay that improves mechanical properties of the matrices. The cellular viability was maintained for 21 days, making this approach as a promising alternative for 3D bioprinting materials [39].

Another important aspect to consider in EBB is the geometry of the needle, which can play a crucial role in cellular viability, since the shear stress under the extrusion can affect cellular behavior and well-being. Muller et al. developed an interesting study where they used different needle geometries and sizes to print alginate and nanocellulose bioinks for cartilage applications. The computational fluid dynamic analysis of different needle geometries is shown in Fig. 10.3. In conclusion, they demonstrated that the appropriate selection of the needle geometry is as important as bioink optimization for high printing resolution and cell viability [13].

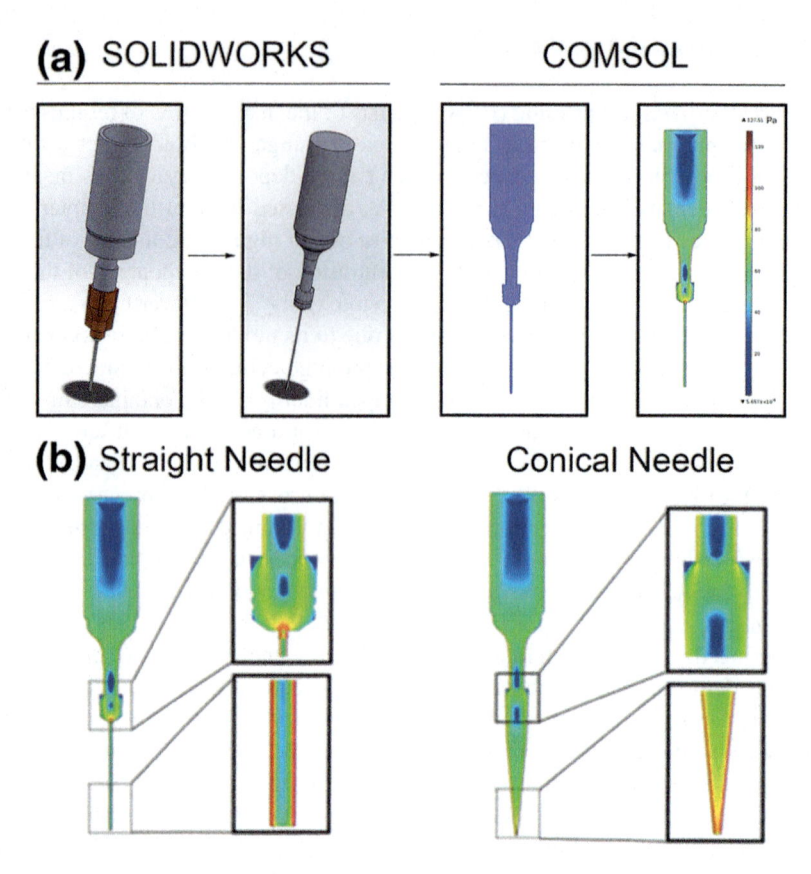

**Fig. 10.3** Computational fluid dynamic analysis for a straight and a conical needle, respectively. Regions of high shear stress are indicated in red/orange colors. Clear differences can be observed between the two geometries [13]. Reproduced from "Alginate Sulfate–Nanocellulose Bioinks for Cartilage Bioprinting Applications", Annals of Biomedical Engineering, Vol. 45, No. 1, January 2017, pp. 210–223, Muller et al. with permission of Springer

### 10.2.1.2 Laser-Based Bioprinting and Stereolithography

Laser-based bioprinting (LBB) is implemented by laser-induced forward transfer (LIFT), which is a method to deposit inorganic materials onto a platform construction through a patterned substrate. Odde and Renn used this technique for the first time in 1999 for the deposition of biological materials and cell patterning into clusters to obtain 2D and 3D structures [12, 34, 42, 43]. LIFT uses very high-powered pulsed laser, and a glass or quartz print ribbon is coated with a thin film of metal or other laser-absorbing material to protect the cells from the laser power. Then, a cell suspension is spread onto the bottom of the ribbon. This suspension is vaporized with a laser pulse focused onto the metal layer, which propels the cell suspension from the ribbon to a receiving platform construction [44]. LBB is very useful for bioinks with very low viscosity, allowing for microscale resolution. However, it is restricted only to very thin structures and presents a high cost and complex manufacturing [12, 45]. Gruene and collaborators demonstrated in their work that LBB is suitable for 3D scaffold-free autologous tissue grafts

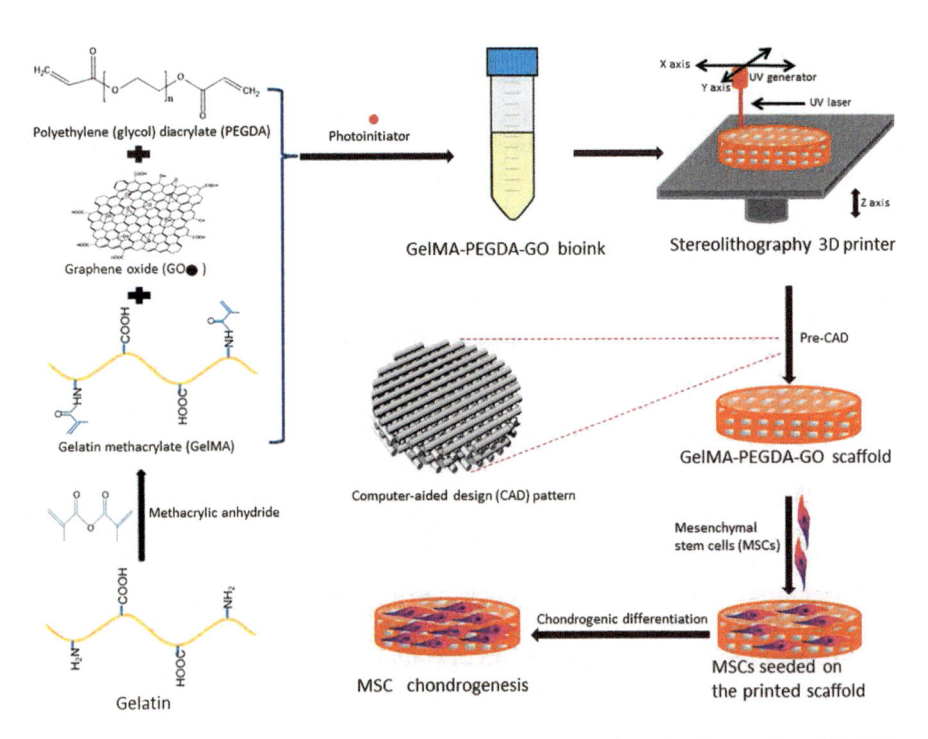

# 10 Preparation of Polymeric and Composite Scaffolds by 3D Bioprinting

**Fig. 10.4** Illustration of 3D printed GO scaffolds for enhancing chondrogenesis of hMSCs through SLA approach [24]. Reprinted from 3D bioprinted graphene oxide-incorporated matrix for promoting chondrogenic differentiation of human bone marrow MSCs, Zhou et al., Carbon, volume 116, pp. 615–624, Copyright 2017, with permission from Elsevier

with high cell density enough to promote chondrogenesis. In addition, the printed mesenchymal stem cells (MSCs) tolerated the complete process maintaining their functionality [46]. Other similar techniques to LIFT are used in LBB approaches. For example, absorbing film-assisted laser-induced forward transfer (AFA-LIFT) uses a 100 nm sacrificial metal layer to interact with the laser. There is also a version of AFA-LIFT, known as biological laser processing (BioLP), which uses motorized receiving stages and a charge-coupled device (CCD) camera to focus the laser. The sacrificial metallic layer allows having a rapid thermal expansion, to reduce the heating of the small cell suspension volume that is propel from the ribbon to the substrate. Finally, matrix-assisted pulsed laser evaporation direct writing (MAPLE DW) is similar to AFA-LIFT, but it uses a low powered laser operating in the UV or near-UV region. In addition, the ribbon is coated with a sacrificial biological layer to allow the initial cell attachment [44].

On the other hand, stereolithography (SLA) consists of the irradiation of a photopolymerizable macromer solution with a laser to cross-link patterns with high resolution in the polymerization plane. This technique allows the fabrication of accurate microstructured scaffolds [1]. Thus, this technique is only valid for photopolymerizable materials, exhibiting high microscale resolution and printing speed [12]. X. Zhou et al. used SLA to produce GelMA, poly(ethylene glycol) diacrylate (PEGDA) and GO scaffolds that induced chondrogenic differentiation of human MSCs (hMSCs) by promotion of glycosaminoglycan (GAG) and collagen levels. A scheme of the scaffold fabrication is showed in Fig. 10.4 [24].

### 10.2.1.3 Droplet-Based Bioprinting

Droplet-based bioprinting (DBB) is a deposition method were prepolymer solution droplets are jetted onto a platform in a predefined pattern. It could be performed by the aid of piezoelectric or thermal actuators (Fig. 10.2). The polymerization takes place after deposition by UV light, ionic, thermal or chemical cross-linking processes. The main advantages of this bioprinting technique are its low cost and the wide range of polymers that can be used. However, the viscosity range of this solution is very limited and cell density cannot be very high [12].

In addition, the bioprinting process can make a negative impact on cellular viability. Regarding to this and in order to understand better the process that can affect them, Hendriks et al. have developed an analytical model with which they can relate the cell survival to the cell membrane elongation and this last one, with the size and speed of the droplet, as well as, substrates characteristics [47]. Another interesting work is the one carried out by Graham et al., where they developed high-resolution 3D geometries by DBB, which consisted of the 3D printing of aqueous droplets containing Human Embryonic Kidney (HEK) cells and ovine MSCs (oMSCs). These platforms included arborized cell junctions and osteochondral interfaces, exhibiting high viability. In addition, oMSCs showed a chondrogenic differentiation to cartilage-like structures after 5 weeks of culture [48].

## 10.2.2 Acellular Bioprinting Techniques

Acellular bioprinting covers the generation of nonliving material constructs based on the pattern and assembly of materials and the successively cell post-printing seeding [3]. This strategy offers several advantages over printed cellular constructs such as higher resolution and greater shape complexity due to the manufacturing conditions in which is avoided the printing of either cells or heat-sensitive biological cues [49]. Acellular tissue scaffolds, alone or in combination with cellular techniques, have shown promising results for bone (BTE) and cartilage (CTE) TE.

### 10.2.2.1 Fused Deposition Modeling (FDM)

FDM, also known as fused filament fabrication (FFF), is based on extrusion, through a computer-guided nozzle, of melting or semimolten thermoplastic filaments which are finally deposited onto a platform where its solidification takes place in a layer-by-layer fashion [23]. Thus, this printing technique, which later helps in the development of other bioprinting techniques, concretely extrusion based bioprinting [3], has been widely applied in the synthesis of acellular porous scaffolds for osteochondral TE due to the fact that the final construct provides a mechanical properties in a closer magnitude to articular cartilage and cancellous bone [50–52]. Strengths such as its rapid printing capability, the ability to obtain large construct with good mechanical integrity, easy scalability, and the no need of solvent and support structure has made this technique widely explored, especially for bone tissue. However several disadvantages should be mentioned such as the reduce number of filament materials that can be used, or the high temperature required to melt the filament which limits the printability with cells or temperature sensitive biological cues. In addition, it is very complicated the fabrication of constructs with small pore size while maintaining the porosity (100 μm) [53–55] .Thermoplastic polymers such as PCL, poly-lactic acid (PLA) and poly(lactic-co-glycolic acid) (PLGA) [52, 56, 57], which are the most common biodegradable synthetic polymers used in this manufacturing process, are the main responsible for this mechanical properties, especially in the case of PCL [55].

The replacement of the hot rollers system of FDM by a pressurized syringe with a thermostatically controlled heating jacket, defined as extrusion printing, has increased the number of synthetic materials used for 3D biofabrication [58]. For example, Woodfield et al. have shown the success bioprinting of an amphiphilic biodegradable poly(ether ester) multiblock copolymers as carrier materials for articular cartilage repair based on hydrophilic poly(ethylene glycol)-terephthalate (PEGT) and hydrophobic poly(butylene terephthalate) (PBT) (PEGT/PBT). Furthermore, constructs with a gradient in pore-size trying to mimic the complex zonal structure of cartilage were designed showing an efficiency inhomogeneous chondrocyte distribution but no differences in cartilage-like tissue formation related to cell density were observed [58]. More recently, Schuurman et al. have demonstrated the production of highly cartilage-like tissue abundance by improving the efficiency of cell seeding by distributing the cells along the PEGT/PBT scaffolds in form of pellets. However, additional options should be explored in order to generate de novo cartilage zonally organized [59].

The presence of nanoscale features in the constructs plays an important role in the generation of TE by affecting cell attachment, proliferation and cytoskeletal assembly. However, FDM, as well as other AM techniques, have not fulfilled this biomimetic nano-resolution. In this sense and in order to overcome this limitation, recent strategies have been proposed for the post-fabrication functionalization with techniques such as layer-by-layer deposition (LbL) [60], plasma deposition [61], and the attachment based on mussel-inspired materials [62]. These strategies include not only the change of topography surface, also the incorporation of some thermal labile biological cues which should be incorporated afterwards. Regarding to this, dexamethasone which is an osteoinductive drug has been incorporated in 3D PCL/poloxamer scaffold during FDM without affecting its properties [63]. However, some labile compounds require their incorporation using the post-fabrication treatment mentioned before or be printed by other bioprinting techniques. Examples of the last one are described in Sect. 10.3.1.

### 10.2.2.2 Melt Electrospinning Writing

Melt-electrospinning writing (MEW) is an emerging manufacturing approach wherein major principles of melt extrusion-based additive manufacturing (AM) and electrospinning are combined. A melt polymer is extruded through a nozzle and beginning electrically charged due to the application of a high voltage between the nozzle tip and the collected platform where fiber are deposited upon each other [64]. The main different in comparison with electrospinning is the lack of organic solvent as in MEW the polymer is melted. This fact allows to improve cell viability and to obtain 3D structures with well-orientated fibers by avoiding both their mechanical and electrical coiling [65]. On the other hand, this fibrous construct can be based on fibers with diameters down to 1 µm [66, 67], far away from the >200 µm provided by FDM manufacture technique [65]. All these aspects provide a really well organized network construct that can be built to millimeters thickness with a convenience pore size for allowing the cell invasion and vascularization of de novo tissue [65, 68]. The potential of this technique for the reinforcement of soft hydrogel matrices has been recently published because it is well known that the actual TE scaffolds based on hydrogels are unable to reach the stiffness and therefore the biological requirements to promote the neotissue. Concretely, electrospun PCL fibers obtained by MEW are infused with GelMA, providing a scaffold with mechanical properties in the range of articular cartilage [66]. More recently, and following the same strategy, constructs based on highly negatively charged star-shaped poly(ethylene glycol)–heparin hydrogel (sPEG/Hep) reinforced with medical grade PCL (mPCL) fibers by MEW were also obtained for articulate CTE. Despite the fact that the fibers provide an outstanding increase in mechanical properties such as anisotropy and viscoelasticity, the system does not meet the expectation under simulated dynamic load-bearing conditions, the necessity to explore different composite material soft fiber-reinforced hydrogels [69]. In this sense, it is interesting to mention the importance of trying to mimic the natural fiber structure in natural soft tissue which is mainly based on collagen. Thus, Bas et al. have compared the behavior of soft network composites reinforced with either stretchable curvy or straight mPCL fibers presenting the curvy fibers the more similar behavior to natural soft tissue [70].

### 10.2.2.3 Selective Lase Sintering (SLS)

SLS, which was developed at the University of Texas [71], is an AM technique where a construct is obtained by sequential deposition of biopolymers, bioceramics, or biocomposites powders which are spread in the bed with a roller following by their fusion via the increase in temperature coming from a computer controlled high-power carbon dioxide laser. Thus, a first thing layer (100–200 nm) is formed and the process is repeated layer-by-layer. Features of the powder such as particle size and shape can affect the SLS process [72]. In comparison with FDM, it might be easy to incorporate composite materials such as polymers-bioceramics as there is no requirement of the materials to be in filament form [73]. Other advantages are the high precision, nonrequirement of solvent or porogens, and the manufacturing of mimetic scaffolds with complicated geometries [74]. Therefore, SLS has found its potential application for BTE and more concretely in bone complex structure and intricate shapes such as maxillofacial and craniofacial [75]. Materials that do not decompose under the laser beam [73, 74] can be used for SLS. Thereby, apart from the metallic devices which are the most common one fabricated by SLS, it has also been explored for BTE using biodegradable polymers such as PCL [74], PLA [76], and polyvinyl alcohol (PVA) [77], polymer–ceramics composites such as nano-HAp-PCL [78], aliphatic polycarbonate-HAp [79], PLA-HAp [80], PLA-carbonated HAp microsphere [24],calcium phosphate (Ca-P)/poly(hydroxybutyrate-co-hydroxyvalerate (PHBV) [75], polyamide-HAp [81], GO reinforced PVA [82], PLA-(Ca-P) [83], and PLGA/HAp and Beta tricalcium phosphate (β-TCP) [84]. Nevertheless, this technique has hardly been applied for CTE but it is worth to mention the modification of SLS defined as microsphere-based SLS technique [85] where the powder used has a spherical shape in the microscale. This version has led to the subsequent application in the manufacturing of scaffolds that mimic the complex multiple tissue structure of osteochondral defects (subchondral bone, intermediate calcified cartilage and the superficial cartilage region) [28, 86]. Pointedly, an approach trying to obtain HAp gradient scaffolds has been built by sintering PCL and PCL-HAp microparticles by SLS. The potential of SLS in the regeneration of osteochondral tissue was showed in vivo" experiments in a rabbit model by forming new tissue with both, articular cartilage and subchondral bone regions [87].

### 10.2.2.4 Cellular/Acellular Bioprinting Techniques

Cellular/acellular bioprinting techniques arise from the need to overcome the actual limitation of both main types of bioinks, natural and synthetic polymers. Hydrogel-based bioprinting constructs are restricted in term of mechanical strength especially when their applications rely on the treatment of load-bearing tissue such as osteochondral tissue. On the other hand, synthetic polymers present limited cell affinity due to the lack of surface cell recognition sites [88, 89]. Furthermore, a common disadvantage for both of them is the inefficiency of in vivo hydrolytic and enzymatic degradation which should match the speed of tissue in-growth. For example, PCL presents a very low degradation rate (1.5–2 years) [90] while natural polymers such as Chitosan shown a variable enzymatic degradation depending of the host response [91].

At this point, both the concept of substrate support and sacrificial templates are introduced due to the important role that they play in these hybrid bioprinting strategies. Sacrificial templates are usually synthetic polymers that are used during the manufacturing process of hydrogel-based bioink to provide to each layer with the requirement mechanical properties during the layer by layer deposition and they are removed in a second step [1]. Alternatively, they have found a great application when trying to obtain vascularized tissue such as BTE because the vascular channels in the scaffolds are printing with these sacrificial materials and subsequently removed [1, 3]. On the other hand, substrate support includes the pre-printing of templates that are not removed after hydrogel addition. Thus, the hybrid system can encompass the advantages of both systems, the good mechanical properties of the thermoplastic polymers and the good cells adhesion of natural polymers [92]. Examples of hybrid techniques for the development of these systems are described below.

MHDS is a solid free-form fabrication which allows for obtaining hybrid constructs with more than one bioinks. Concretely, those bioinks (thermoplastic polymers, natural polymers) are loaded in different thermostatically controlled syringes and parameters such as temperature, pneumatic pressure and motion are stabilized independently for each syringe. Thus, alternant layers of different bioinks either loaded or unloaded with cells, some reinforce additives and biological cues can be co-printed [93, 94]. Several works have been developed based on MHDS for bone and cartilage tissue regeneration based on the hybrid system PCL–alginate [92, 94, 95]. Although, initial thought about MHDS techniques point to a possible reduction of cell viability when alternating thermoplastic polymer-natural polymer loaded with cells layer are deposited. Recent study based on the system PCL-Alginate loaded with primary chondrocytes isolated from chick embryos have demonstrated the high cell vitality after deposition (higher than 80%). The melting PCL cool down faster enough to minimize the effect on cell viability [95]. Furthermore, the mechanical stability conferred by the thermoplastic polymer allows for the printing of hydrogels with lower cross-linking density which could be beneficial for cell viability [92].

Template-Fused Deposition Modelling (t-FDM) has been used to create sacrificial templates. The template is printed by FDM and a cross-linkable material is poured onto the template where the polymerization takes places [96]. The template whether is removed or not should have biocompatible properties. An example of sacrificial template can be found in Guo et al. where a polyurethane construct was obtained with well-defined topological properties (in the range of trabecular bone) after its cross-linking polymerization on a PLA template [97]. An example where the template is not removed is described by Dong et al. where hybrid chitosan–PCL scaffolds have been described for their application in BTE. In vitro results of these systems when encapsulating rabbit bone marrow mesenchymal stem cells (BMMSCs) in the chitosan matrix have shown an improvement of osteogenesis differentiation compared to PCL control scaffolds alone [98]. However, this hybrid system can fail under mechanical stresses due to an inefficient thermoplastic–natural hydrogel interface adhesion. In this sense, a covalent attachment between both materials has been proposed for the hybrid system based on GelMa and poly(hydroxymethylglycolide-co-ε-caprolactone)/poly(ε-caprolactone) (pHMGCL/PCL), showing an increase of mechanical integrity while also keeping their ability to promote ECM formation [99]. Additional strategies to increase the mechanical properties are described in Sect. 10.3.1.

## 10.3 3D Printing Polymeric and Composited Materials for Osteochondral Tissue Engineering

Biomaterials used for 3D printing fabrication must meet different criteria to successful development of the scaffolds. The first requisite is having good rheological properties, which means that bioinks must be mechanically suitable for printing depending on the used bioprinting technique, and provide an appropriate environment to the cells after bioprinting to promote adhesion, proliferation and differentiation. Secondly, it is essential that the material maintains its structural integrity, in other words, high shape fidelity after the deposition process. This is directly related to printability, which refers to the relationship between the substrates and the bioinks. The bioprinting solutions should maintain vertical tension having a high contact angle with the substrate surface, and it normally depends on how fast is the cross-linking process. Finally, the bioinks must provide a biocompatible and not cytotoxic environment for cell encapsulation and deposition. However, many materials usually meet one or two requisites, being necessary the development of bioinks which present all these criteria. Usually, materials that are printable and maintain their structure after bioprinting through a rapid cross-linking, make necessary the use of high temperature for thermal curing or UV light for photopolymerization, which compromise the encapsulation of cells in the bioinks. In addition, the most biocompatible materials do not exhibit good rheological properties for extrusion or bioprinting deposition, like for example hydrogels [1, 21, 100]. Hydrogels are highly appropriate biomaterials for 3D bioprinted scaffolds for osteochondral TE due to its high biocompatibility, which make them suitable for cell encapsulation, and biodegradability properties [101]. Hydrogels are networks of 3D cross-linked polymers that able to uptake huge amount of water due to their inherent hydrophilic properties. This capability can be modulate depending on the biological tissue of interest. In addition, hydrogels pose injectability properties for minimally invasive therapies of cartilaginous-like tissues [8, 11, 101, 102]. One approach to improve mechanical properties of hydrogel bioinks is to increase the concentration of the components, obtaining highly viscous solutions with suitable printability. However, cell viability is usually decreased in high concentrated bioinks due to higher stress must be applied to the solution [6, 13, 36, 103].

Among all biomaterials explored for 3D bioprinting technology, we can distinguish between those derived from natural polymers, such as collagen, gelatin, alginate, chitosan, and hyaluronic acid (HA), or synthetic-derived polymers, such as PCL, PLGA, PLA, PEG, and PEGDA. As it has been explained, synthetic materials exhibit robust mechanical properties, but poor biocompatibility and toxic degradation products. For these reasons, the use of composites is more widespread. Composites are a combination of two or more than three individual materials. They are used for enhancing mechanical strength and fabrication of more intricately designed constructs, as well as improving their long-term stability. Thanks to this combination, the suitable strength and mechanical properties of the scaffold can be suitably modulated depending on the properties of the native tissue [1, 8, 101].

Nanoclays and PEGDA, for example, have been incorporated into some hydrogels solutions to control their viscosity [101]. One interesting example is the work developed by Yang et al., who synthesized a biphasic graft consisting of cartilage and subchondral bone, using synthetic (PLGA) and natural (alginate) polymers and a multi-nozzle deposition system [14]. Over the recent years the use of decellularized extracellular matrix (dECM) has been investigated for osteochondral regeneration. dECM consists of a complex of GAGs, collagen, and elastin that mimics the native tissue environment. In addition, the ECM can lead and mediate the differentiation of stem cells [101].

HA is a naturally derived polysaccharide that has been amply used in osteochondral tissue regenerative therapies. It is an anionic, GAG distributed widely throughout connective, epithelial, and neural tissues. As it is also one of the main components of the ECM, contributing significantly to cell proliferation and migration. All these properties make to HA a suitable material for 3D bioprinting application [11, 21, 38, 40, 104–106]. For example, Shaoquan et al. developed a semi-interpenetrating polymer network (semi-IPN) based on HA and hydroxyethylmethacrylate-derivatized dextran (dex-HEMA), which showed shear thinning rheology and mechanical strength. The scaffolds exhibited high porous structure, supporting the viability of encapsulated chondrocytes [107]. Ju Young and coworkers used HA with alginate and chondrocytes in Dulbecco's Modified Eagle Medium (DMEM) for chondral section, while collagen-I in DMEM constituted the osteo-section. Thus, they fabricated a two-compartment scaffold for osteochondral tissue mimetic structures [105].

Gelatin is a naturally derived polysaccharide widely used in bioprinting techniques due to its thermosensitive properties which eases the development of shaped fidelity structures [20, 21, 38, 104]. Gelatin is the denatured form of collagen, which resembles the ECM environments providing key biological motifs for cell adhesion and proliferation [102]. An example within numerous studies developed with gelatin or its methacrylate form, is the one carried out by Levato et al., who developed novel constructs consisting of GelMA and gellan gum for osteogenic and chondrogenic differentiation of MSCs [100, 108]. Gelatin has been also found to participate in some regulation ways for chondrogenesis. For example, Chameettachal et al. developed tyrosinase cross-linked silk–gelatin bioinks and demonstrated that these bioinks could upregulate the expression of hypoxia markers such as hypoxia inducible factor 1 alpha (HIF1A) which positively regulated also the expression of chondrogenic markers such as aggrecan or cartilage oligomeric matrix protein 1 (COMP1). The gelatin, particularly, showed the induction of matrix metalloproteinase 2 (MMP2) activity, which is known to promote the creation of a pericellular zone for the accumulation of growth factors and de novo matrix [109]. Costantini and collaborators also used GelMA for the development of 3D bioprinted constructs through a coaxial needle system. The bioinks, composed of GelMA, chondroitin sulfate amino ethyl methacrylate (CS-AEMA), and HAMA, showed the upregulated expression of chondrogenic markers, like COL2A1 and aggrecan, as well as osteogenic markers like COL1A1. Thus, the presented approach demonstrated to be a suitable candidate for 3D bioprinted applications for cartilage TE field [40]. In addition, gelatin is usually combined with HA since it is known not only for promoting chondrogenesis in the 3D constructs but also for improving mechanical properties of the constructs during the bioprinting process [59, 106, 110].

Apart from bioink design, cross-linking mechanisms are another aspect to be optimized in order to obtain more complex constructs reducing undesirable side effects. Cross-linking procedures in bioprinting need to be secure for cell encapsulation and fast, promoting a state change from liquid (viscous) to almost solid network. The cross-linking can be physical, chemical, or a combination of both, but it must maintain native cell adhesion properties of the biomaterial [8]. Chemical cross-linking processes are the most accepted due to its effectiveness, efficiency and controllability, being able to synthetize handle scaffolds with good mechanical properties and stiffness. Photopolymerization is one of the most commonly approach used for the development of 3D bioprinted scaffolds. The chemical reaction can be triggered by the irradiation of a photoinitiator (PI) containing-hydrogel at a specific wavelength. However, photocuring also shows some drawbacks due mainly to the cytotoxicity and inflammation reactions that are provoked by the generation of free radicals by UV exposure that can damage DNA and cellular components [20, 111, 112]. For this reason, activated PIs under visible or A-UV light are being extensively used during the last years [20, 113]. However, many authors have demonstrated in their studies that a proper adjustment of the UV irradiation time, intensity, and wavelength could ensure cell viability [37, 38, 104, 106, 110, 114, 115].

### 10.3.1 Incorporation of Additives for Enhancing Mechanical Properties

In order to achieve 3D bioprinted scaffolds with clinical relevant dimensions, there are two main strategies without using a bath as supporting medium. The first one consisted of the improvement of mechanical properties of the solution through a rapid cross-linking process, as it has been discussed in the previous section. The second one is the incorporation of support materials, such as PCL, which can confer space and structural integrity to low viscous bioinks [5, 39, 116, 117]. Daly and coworkers developed a study to compare the printability of different bioinks for 3D bioprinting of hyaline and fibrocartilage, using the most common hydrogels: agarose, GelMA, alginate and BioINK™, which consists of a poly(ethylene glycol) methacrylate (PEGMA) based hydrogel. The tissue staining for type II collagen revealed that alginate and agarose based bioinks supported properly the development of hyaline-like cartilage, while GelMA and BioINK™ supported the growth of fibrocartilage. They used PCL filaments to reinforce the mechanical properties of the hydrogels, being able to synthesized constructs with a compressive moduli similar to articular cartilage [22].

In another interesting work, Ahlfeld et al. used Laponite, a synthetic nanosilicate clay which is known for its drug delivery properties. They combined alginate and methylcellulose with Laponite to develop constructs with high printing fidelity as well as, controlled released of active compounds [39]. Co-printing approaches with PLGA or nanocellulose rather than PCL as supporting materials have also been boarded [118]. Nanocellulose, for example, is able to increase the viscosity of an alginate solution bioink up to sevenfold, improving therefore the bioprintability [5]. In a work by Markstedt and collaborators, a bioink composed of nanofibrillate cellulose and alginate was developed to a patterned meniscus cartilage in a single-step bioprinting process [119, 120]. Müller et al. also developed a sulfate alginate-based bioink in combination with nanocellulose to make it printable. This mix was photocurable, arising as a good alternative for cartilage tissue regeneration applications [13]. In addition, to avoid the limitations of PCL as a reinforcing material, the increasing of porosity of the reinforcing phase can be also an alternative approach [9, 14].

## 10.3.2 Incorporation of Bioactive Compounds

The incorporation of active compounds such as growth factors and inorganic compounds is a very common approach to enhance cell adhesion and proliferation, as well as, differentiation to a specific tissue. Bartnikowski et al. incorporated a paste of HAp to a GelMA, and GelMA-HAMA bioink for the development of a zone of calcified cartilage, as well as improvement of bioinks printability. They concluded that the incorporation of HAMA enhanced chondrogenesis and the bioprinted scaffolds showed good cell culture viability for 28 days [11]. In another interesting work, Wang et al. studied the effect of HAp in an HA-based bioink. They demonstrated that a small amount of HAp enhanced chondrogenesis and hypertrophic differentiation of adipose derived MSCs. In addition, they were able to develop stratified scaffolds with mineralized and nonmineralized layers (HA-HAp based and HA-based) [121]. In another work, Zhou and coworkers incorporated GO to their gelatin-based 3D bioprinted scaffolds to promote chondrogenic differentiation, demonstrating that multifunctional carbon-based nanomaterials can be a suitable additive for osteochondral TE approaches [24].

Traditionally, several growth factors including transforming growth factors (TGFs), insulin-like growth factors (IGFs) and bone morphogenetic protein (BMPs) have been incorporated to osteochondral TE scaffolds to promote chondrogenic or osteogenic stem cells differentiation as it has been reviewed recently [122]. Similar approach has been also used in AM scaffolds. Until now, TGF-β has been incorporated either directly to the cell culture media [41, 123, 124] or by physical encapsulation in the hydrogel [94, 125, 126]. An example of TGF-β physical encapsulation has been reported by Kundu et al. where an alginate–TGF-β–BMMSCs printed scaffold reinforced with PCL has been manufactured by MHDS. Scaffolds loaded with TGF-β produced higher GAGs content after 4 weeks compared to the unloaded ones [94]. However, recent studies focusing on silk–gelatin constructs incubated with TGF-β1 have shown hypertrophy instead of articular cartilage MSC differentiation. This evidence has led to an increasing need to find new strategies which could avoid the hypertrophic differentiation. In this sense, the overexpression of nuclear receptor subfamily 2 group F member 2 (NR2F2) in MSCs was promoted previous to scaffold cell implantation. This overexpression has provided the generation of abundant cartilage matrix [127]. Another strategy focused on the 3D bioprinting encapsulation of bioactive drug Y27632 [(+)-(R)-trans-4-(1-aminoethyl)-N-(4-pyridyl)cyclohexanecarboxamide dihydrochloride] which has been shown to reduce the hypertrophic market collagen X (Col X) in comparison with TGF-$\beta$ when MSCs were seeded on polyurethane (PU)–HA constructs [126]. BMPs are another group of growth factors widely applied for promoting osteogenic differentiation. In a recent work the surface of PLA constructs has been modified by the assembly of multilayer nanocoating based on gelatin (Gel) and poly-lysine (PLL) finally cross-linked with genipin (GnP). An increase cell adhesion of both human umbilical vein endothelial cells (HUVECs) and hMSCs respect to the control (unmodified PLA construct was reported. More interesting, this approach allowed for the smart release of growth factors such as recombinant human bone morphogenetic protein (rhBMP-2) and recombinant human vascular endothelial (rhVEGF) by promoting osteogenic differentiation of hMSCs and .proliferation and differentiation of HUVECs Thus, it has been possible the generation of vascularized bone grafts [60]. In order to avoid undesirable growth factors degradation when adding directly to the cell culture media, Dong et al. have developed hybrid chitosan–PCL scaffolds loaded with BMP-2. A sustained in vitro release of BMP-2 promoted BMMSC osteogenic differentiation [98].

## 10.4 Conclusion and Future Perspectives

In general, the arrival and development of 3D bioprinting has made a huge impact on tissue regeneration field. Its implementation in osteochondral TE field is highly appropriate owing to the particular heterogeneity and anisotropy that the osteochondral tissue exhibits. 3D bioprinting allows for fabricating very intricate heterogeneous 3D constructs by an accurate spatiotemporal positioning of cells and biomolecules, controlling the structure, size, shape, pore, and orientation of each component with micrometer precision. In addition, the porosity and gradient created in the scaffolds by 3D bioprinting ensures a good cell–cell communication and vascularization of the construct, which is essential for an appropriate distribution of oxygen and nutrients, and thus for long-term stability.

However, despite all the advantages that this technology holds in the field, some challenging aspects have to be solved before translation of the technology to the clinic occurs. It cannot be denied that 3D bioprinting will be responsible for a new generation of personalized therapeutic approach, but the materials and technology should be meticulously chosen when aiming for translation to the clinic. Currently, the most daunting challenges that restrict the clinical translation of this technology are the capacity for large-scale fabrication, sterilization process, stringent quality control for 3D scaffolds for human trials, and the affordability of the medical expenditure. Although numerous preclinical studies are being developed, clinical trials are very limited due to regulatory issues, differences in patient responses, as well as implantation constraints. In addition, the necessity of skilled experts and cost efficacy of the fabrication devices are still bottlenecks for the clinical translation of the technology.

In conclusion, the emergence of printing technologies for the construction of mimetic scaffolds for the regeneration of osteochondral tissue seems to be a significant milestone. As all novel technologies, 3D bioprinting should face regulatory hurdles for clinical translation that must be solved in the following years, as these technologies provide real benefits and advantages to really complicated osteochondral diseases and lesions.

## References

1. Mandrycky C, Wang Z, Kim K, Kim DH (2016) 3D bioprinting for engineering complex tissues. Biotechnol Adv 34(4):422–434. https://doi.org/10.1016/j.biotechadv.2015.12.011
2. Murphy SV, Atala A (2014) 3D bioprinting of tissues and organs. Nat Biotechnol 32(8):773–785. https://doi.org/10.1038/nbt.2958
3. Cui H, Nowicki M, Fisher JP, Zhang LG (2017) 3D bioprinting for organ regeneration. Adv Healthc Mater 6(1):1601118. https://doi.org/10.1002/adhm.201601118
4. Arslan-Yildiz A, Assal RE, Chen P, Guven S, Inci F, Demirci U (2016) Towards artificial tissue models: past, present, and future of 3D bioprinting. Biofabrication 8(1):014103. https://doi.org/10.1088/1758-5090/8/1/014103
5. O'Connell G, Garcia J, Amir J (2017) 3D bioprinting: new directions in articular cartilage tissue engineering. ACS Biomater Sci Eng 3:2657. https://doi.org/10.1021/acsbiomaterials.6b00587
6. Muller M, Becher J, Schnabelrauch M, Zenobi-Wong M (2015) Nanostructured Pluronic hydrogels as bioinks for 3D bioprinting. Biofabrication 7(3):035006. https://doi.org/10.1088/1758-5090/7/3/035006
7. Ozbolat IT (2017) Bioprinting of osteochondral tissues: a perspective on current gaps and future trends. Int J Bioprint. 3(2). doi:https://doi.org/10.18063/ijb.2017.02.007
8. Radhakrishnan J, Subramanian A, Krishnan UM, Sethuraman S (2017) Injectable and 3D bioprinted polysaccharide hydrogels: from cartilage to osteochondral tissue engineering. Biomacromolecules 18(1):1–26. https://doi.org/10.1021/acs.biomac.6b01619
9. Daly AC, Freeman FE, Gonzalez-Fernandez T, Critchley SE, Nulty J, Kelly DJ (2017) 3D bioprinting for cartilage and osteochondral tissue engineering. Adv Healthc Mater 6. https://doi.org/10.1002/adhm.201700298
10. Fedorovich NE, Schuurman W, Wijnberg HM, Prins HJ, van Weeren PR, Malda J, Alblas J, Dhert WJ (2012) Biofabrication of osteochondral tissue equivalents by printing topologically defined, cell-laden hydrogel scaffolds. Tissue Eng Part C Methods 18(1):33–44. https://doi.org/10.1089/ten.TEC.2011.0060
11. Bartnikowski M, Akkineni AR, Gelinsky M, Woodruff MA, Klein TJ (2016) A hydrogel model incorporating 3D-plotted hydroxyapatite for osteochondral tissue engineering. Materials (Basel) 9(4). https://doi.org/10.3390/ma9040285
12. Pedde RD, Mirani B, Navaei A, Styan T, Wong S, Mehrali M, Thakur A, Mohtaram NK, Bayati A, Dolatshahi-Pirouz A, Nikkhah M, Willerth SM, Akbari M (2017) Emerging biofab-

rication strategies for engineering complex tissue constructs. Adv Mater 29(19). https://doi.org/10.1002/adma.201606061
13. Muller M, Ozturk E, Arlov O, Gatenholm P, Zenobi-Wong M (2017) Alginate sulfate-Nanocellulose bioinks for cartilage bioprinting applications. Ann Biomed Eng 45(1):210–223. https://doi.org/10.1007/s10439-016-1704-5
14. Yang SS, Choi WH, Song BR, Jin H, Lee SJ, Lee SH, Lee J, Kim YJ, Park SR, Park S-H, Min B-H (2015) Fabrication of an osteochondral graft with using a solid freeform fabrication system. Tiss Eng Regen Med 12(4):239–248. https://doi.org/10.1007/s13770-015-0001-y
15. Kang HW, Lee SJ, Ko IK, Kengla C, Yoo JJ, Atala A (2016) A 3D bioprinting system to produce human-scale tissue constructs with structural integrity. Nat Biotechnol 34(3):312–319. https://doi.org/10.1038/nbt.3413
16. Ahadian S, Yamada S, Ramon-Azcon J, Estili M, Liang X, Nakajima K, Shiku H, Khademhosseini A, Matsue T (2016) Hybrid hydrogel-aligned carbon nanotube scaffolds to enhance cardiac differentiation of embryoid bodies. Acta Biomater 31:134–143. https://doi.org/10.1016/j.actbio.2015.11.047
17. Hölzl K, Lin S, Tytgat L, Van Vlierberghe S, Gu L, Ovsianikov A (2016) Bioink properties before, during and after 3D bioprinting. Biofabrication 8(3):032002. https://doi.org/10.1088/1758-5090/8/3/032002
18. Ji S, Guvendiren M (2017) Recent advances in bioink design for 3D bioprinting of tissues and organs. Front Bioeng Biotechnol 5. https://doi.org/10.3389/fbioe.2017.00023
19. Donderwinkel I, van Hest JCM, Cameron NR (2017) Bio-inks for 3D bioprinting: recent advances and future prospects. Polym Chem 8(31):4451–4471. https://doi.org/10.1039/c7py00826k
20. Klotz BJ, Gawlitta D, Rosenberg AJ, Malda J, Melchels FP (2016) Gelatin-Methacryloyl hydrogels: towards biofabrication-based tissue repair. Trends Biotechnol 34(5):394–407. https://doi.org/10.1016/j.tibtech.2016.01.002
21. Skardal A, Zhang J, McCoard L, Xu X, Oottamasathien S, Prestwich GD (2010) Photocrosslinkable hyaluronan-gelatin hydrogels for two-step bioprinting. Tissue Eng Part A 16(8):2675–2685. https://doi.org/10.1089/ten.TEA.2009.0798
22. Daly AC, Critchley SE, Rencsok EM, Kelly DJ (2016) A comparison of different bioinks for 3D bioprinting of fibrocartilage and hyaline cartilage. Biofabrication 8(4):045002. https://doi.org/10.1088/1758-5090/8/4/045002
23. Brunello G, Sivolella S, Meneghello R, Ferroni L, Gardin C, Piattelli A, Zavan B, Bressan E (2016) Powder-based 3D printing for bone tissue engineering. Biotechnol Adv 34(5):740–753. https://doi.org/10.1016/j.biotechadv.2016.03.009
24. Zhou X, Nowicki M, Cui H, Zhu W, Fang X, Miao S, Lee S-J, Keidar M, Zhang LG (2017) 3D bioprinted graphene oxide-incorporated matrix for promoting chondrogenic differentiation of human bone marrow mesenchymal stem cells. Carbon 116:615–624. https://doi.org/10.1016/j.carbon.2017.02.049
25. Kolesky DB, Homan KA, Skylar-Scott MA, Lewis JA (2016) Three-dimensional bioprinting of thick vascularized tissues. Proc Natl Acad Sci U S A 113(12):3179–3184. https://doi.org/10.1073/pnas.1521342113
26. Dalby MJ, Gadegaard N, Tare R, Andar A, Riehle MO, Herzyk P, Wilkinson CD, Oreffo RO (2007) The control of human mesenchymal cell differentiation using nanoscale symmetry and disorder. Nat Mater 6(12):997–1003. https://doi.org/10.1038/nmat2013
27. Ikada Y (2006) Challenges in tissue engineering. J R Soc Interface 3(10):589–601. https://doi.org/10.1098/rsif.2006.0124
28. Boushell MK, Hung CT, Hunziker EB, Strauss EJ, Lu HH (2016) Current strategies for integrative cartilage repair. Connect Tissue Res 58(5):393–406. https://doi.org/10.1080/03008207.2016.1231180
29. Ahmed TAE, Hincke MT (2009) Strategies for articular cartilage lesion repair and functional restoration. Tissue Eng Part B Rev 16(3):305–329. https://doi.org/10.1089/ten.teb.2009.0590

30. Hospodiuk M, Dey M, Sosnoski D, Ozbolat IT (2017) The bioink: a comprehensive review on bioprintable materials. Biotechnol Adv 35(2):217–239. https://doi.org/10.1016/j.biotechadv.2016.12.006
31. Yu Y, Moncal KK, Li J, Peng W, Rivero I, Martin JA, Ozbolat IT (2016) Three-dimensional bioprinting using self-assembling scalable scaffold-free "tissue strands" as a new bioink. Sci Rep 6:28714. https://doi.org/10.1038/srep28714. https://www.nature.com/articles/srep28714#supplementary-information
32. Ozbolat IT (2015) Scaffold-based or scaffold-free bioprinting: competing or complementing approaches? J Nanotechnol Eng Med 6(2):024701–024706. https://doi.org/10.1115/1.4030414
33. Ozbolat IT, Hospodiuk M (2016) Current advances and future perspectives in extrusion-based bioprinting. Biomaterials 76:321–343. https://doi.org/10.1016/j.biomaterials.2015.10.076
34. Pereira RF, Bártolo PJ (2015) 3D bioprinting of photocrosslinkable hydrogel constructs. J Appl Polym Sci 132(48):n/a-n/a. doi:https://doi.org/10.1002/app.42458
35. Tang D, Tare RS, Yang LY, Williams DF, Ou KL, Oreffo RO (2016) Biofabrication of bone tissue: approaches, challenges and translation for bone regeneration. Biomaterials 83:363–382. https://doi.org/10.1016/j.biomaterials.2016.01.024
36. Ahn G, Min KH, Kim C, Lee JS, Kang D, Won JY, Cho DW, Kim JY, Jin S, Yun WS, Shim JH (2017) Precise stacking of decellularized extracellular matrix based 3D cell-laden constructs by a 3D cell printing system equipped with heating modules. Sci Rep 7(1):8624. https://doi.org/10.1038/s41598-017-09201-5
37. Duchi S, Onofrillo C, O'Connell CD, Blanchard R, Augustine C, Quigley AF, Kapsa RMI, Pivonka P, Wallace G, Di Bella C, Choong PFM (2017) Handheld co-axial bioprinting: application to in situ surgical cartilage repair. Sci Rep 7(1):5837. https://doi.org/10.1038/s41598-017-05699-x
38. O'Connell CD, Di Bella C, Thompson F, Augustine C, Beirne S, Cornock R, Richards CJ, Chung J, Gambhir S, Yue Z, Bourke J, Zhang B, Taylor A, Quigley A, Kapsa R, Choong P, Wallace GG (2016) Development of the biopen: a handheld device for surgical printing of adipose stem cells at a chondral wound site. Biofabrication 8(1):015019. https://doi.org/10.1088/1758-5090/8/1/015019
39. Ahlfeld T, Cidonio G, Kilian D, Duin S, Akkineni AR, Dawson JI, Yang S, Lode A, Oreffo ROC, Gelinsky M (2017) Development of a clay based bioink for 3D cell printing for skeletal application. Biofabrication 9(3):034103. https://doi.org/10.1088/1758-5090/aa7e96
40. Costantini M, Idaszek J, Szoke K, Jaroszewicz J, Dentini M, Barbetta A, Brinchmann JE, Swieszkowski W (2016) 3D bioprinting of BM-MSCs-loaded ECM biomimetic hydrogels for in vitro neocartilage formation. Biofabrication 8(3):035002. https://doi.org/10.1088/1758-5090/8/3/035002
41. Chawla S, Kumar A, Admane P, Bandyopadhyay A, Ghosh S (2017) Elucidating role of silk-gelatin bioink to recapitulate articular cartilage differentiation in 3D bioprinted constructs. Bioprinting 7:1–13. https://doi.org/10.1016/j.bprint.2017.05.001
42. Odde DJ, Renn MJ (1999) Laser-guided direct writing for applications in biotechnology. Trends Biotechnol 17(10):385–389. https://doi.org/10.1016/S0167-7799(99)01355-4
43. Odde DJ, Renn MJ (2000) Laser-guided direct writing of living cells. Biotechnol Bioeng 67(3):312–318. https://doi.org/10.1002/(sici)1097-0290(20000205)67:3<312::aid-bit7>3.0.co;2-f
44. Schiele NR, Corr DT, Huang Y, Raof NA, Xie Y, Chrisey DB (2010) Laser-based direct-write techniques for cell printing. Biofabrication 2(3):032001. https://doi.org/10.1088/1758-5082/2/3/032001
45. Kingsley DM, Dias AD, Chrisey DB, Corr DT (2013) Single-step laser-based fabrication and patterning of cell-encapsulated alginate microbeads. Biofabrication 5(4):045006. https://doi.org/10.1088/1758-5082/5/4/045006
46. Gruene M, Deiwick A, Koch L, Schlie S, Unger C, Hofmann N, Bernemann I, Glasmacher B, Chichkov B (2011) Laser printing of stem cells for biofabrication of scaffold-free autologous grafts. Tissue Eng Part C Methods 17(1):79–87. https://doi.org/10.1089/ten.TEC.2010.0359

47. Hendriks J, Willem Visser C, Henke S, Leijten J, Saris DB, Sun C, Lohse D, Karperien M (2015) Optimizing cell viability in droplet-based cell deposition. Sci Rep 5:11304. https://doi.org/10.1038/srep11304
48. Graham AD, Olof SN, Burke MJ, Armstrong JPK, Mikhailova EA, Nicholson JG, Box SJ, Szele FG, Perriman AW, Bayley H (2017) High-resolution patterned cellular constructs by droplet-based 3D printing. Sci Rep 7(1):7004. https://doi.org/10.1038/s41598-017-06358-x
49. O'Brien CM, Holmes B, Faucett S, Zhang LG (2015) Three-dimensional printing of nanomaterial scaffolds for complex tissue regeneration. Tissue Eng Part B Rev 21(1):103–114. https://doi.org/10.1089/ten.teb.2014.0168
50. Hutmacher DW, Schantz JT, Lam CXF, Tan KC, Lim TC (2007) State of the art and future directions of scaffold-based bone engineering from a biomaterials perspective. J Tissue Eng Regen Med 1(4):245–260. https://doi.org/10.1002/term.24
51. Chen SS, Falcovitz YH, Schneiderman R, Maroudas A, Sah RL (2001) Depth-dependent compressive properties of normal aged human femoral head articular cartilage: relationship to fixed charge density. Osteoarthr Cartil 9(6):561–569. https://doi.org/10.1053/joca.2001.0424
52. Goldstein SA (1987) The mechanical properties of trabecular bone: dependence on anatomic location and function. J Biomech 20(11):1055–1061. https://doi.org/10.1016/0021-9290(87)90023-6
53. Adepu S, Dhiman N, Laha A, Sharma CS, Ramakrishna S, Khandelwal M (2017) Three-dimensional bioprinting for bone tissue regeneration. Cur Opin Biomed Eng? 2(supplement C):22–28. doi:https://doi.org/10.1016/j.cobme.2017.03.005
54. Mota C, Puppi D, Chiellini F, Chiellini E (2015) Additive manufacturing techniques for the production of tissue engineering constructs. J Tissue Eng Regen Med 9(3):174–190. https://doi.org/10.1002/term.1635
55. Woodruff MA, Hutmacher DW (2010) The return of a forgotten polymer—Polycaprolactone in the 21st century. Prog Polym Sci 35(10):1217–1256. https://doi.org/10.1016/j.progpolymsci.2010.04.002
56. Huang Q, JCHG, Hutmacher DW, Lee EH (2004) In Vivo Mesenchymal Cell Recruitment by a Scaffold Loaded with Transforming Growth Factor β1 and the Potential for in Situ Chondrogenesis. Tiss Eng 8(3):469–482. doi:https://doi.org/10.1089/107632702760184727
57. Zein I, Hutmacher DW, Tan KC, Teoh SH (2002) Fused deposition modeling of novel scaffold architectures for tissue engineering applications. Biomaterials 23(4):1169–1185. https://doi.org/10.1016/S0142-9612(01)00232-0
58. Woodfield TBF, Malda J, de Wijn J, Péters F, Riesle J, van Blitterswijk CA (2004) Design of porous scaffolds for cartilage tissue engineering using a three-dimensional fiber-deposition technique. Biomaterials 25(18):4149–4161. https://doi.org/10.1016/j.biomaterials.2003.10.056
59. Schuurman W, Levett PA, Pot MW, van Weeren PR, Dhert WJA, Hutmacher DW, Melchels FPW, Klein TJ, Malda J (2013) Gelatin-Methacrylamide hydrogels as potential biomaterials for fabrication of tissue-engineered cartilage constructs. Macromol Biosci 13(5):551–561. https://doi.org/10.1002/mabi.201200471
60. Cui H, Zhu W, Holmes B, Zhang LG (2016) Biologically inspired smart release system based on 3D bioprinted perfused scaffold for vascularized tissue regeneration. Adv Sci 3(8):1600058. doi:10.1002/advs.201600058
61. Domingos M, Intranuovo F, Gloria A, Gristina R, Ambrosio L, Bártolo PJ, Favia P (2013) Improved osteoblast cell affinity on plasma-modified 3-D extruded PCL scaffolds. Acta Biomater 9(4):5997–6005. https://doi.org/10.1016/j.actbio.2012.12.031
62. Lee SJ, Lee D, Yoon TR, Kim HK, Jo HH, Park JS, Lee JH, Kim WD, Kwon IK, Park SA (2016) Surface modification of 3D-printed porous scaffolds via mussel-inspired polydopamine and effective immobilization of rhBMP-2 to promote osteogenic differentiation for bone tissue engineering. Acta Biomaterialia 40(supplement C):182–191. doi:https://doi.org/10.1016/j.actbio.2016.02.006

63. Costa PF, Puga AM, Díaz-Gomez L, Concheiro A, Busch DH, Alvarez-Lorenzo C (2015) Additive manufacturing of scaffolds with dexamethasone controlled release for enhanced bone regeneration. Int J Pharm 496(2):541–550. https://doi.org/10.1016/j.ijpharm.2015.10.055
64. Brown TD, Dalton PD, Hutmacher DW (2011) Direct writing by way of melt electrospinning. Adv Mater 23(47):5651–5657. https://doi.org/10.1002/adma.201103482
65. Bas O, De-Juan-Pardo EM, Chhaya MP, Wunner FM, Jeon JE, Klein TJ, Hutmacher DW (2015) Enhancing structural integrity of hydrogels by using highly organised melt electrospun fibre constructs. Eur Polym J 72(supplement C):451–463. doi:https://doi.org/10.1016/j.eurpolymj.2015.07.034
66. Visser J, Melchels FPW, Jeon JE, van Bussel EM, Kimpton LS, Byrne HM, Dhert WJA, Dalton PD, Hutmacher DW, Malda J (2015) Reinforcement of hydrogels using three-dimensionally printed microfibres. 6:6933. https://doi.org/10.1038/ncomms7933. https://www.nature.com/articles/ncomms7933#supplementary-information
67. Gernot Hochleitner TJ, Brown TD, Hahn K, Moseke C, Jakob F, Dalton PD, Groll J (2015) Additive manufacturing of scaffolds with sub-micron filaments via melt electrospinning writing. Biofabrication 7(3)
68. Brown TD, Edin F, Detta N, Skelton AD, Hutmacher DW, Dalton PD (2014) Melt electrospinning of poly(ε-caprolactone) scaffolds: phenomenological observations associated with collection and direct writing. Mater Sci Eng C 45(supplement C):698-708. doi:https://doi.org/10.1016/j.msec.2014.07.034
69. Onur B, Elena MD-J-P, Christoph M, Davide DA, Jeremy GB, Laura JB, Wellard RM, Stefan K, Ernst R, Carsten W, Travis JK, Isabelle C, Dietmar WH (2017) Biofabricated soft network composites for cartilage tissue engineering. Biofabrication 9(2):025014
70. Bas O, D'Angella D, Baldwin JG, Castro NJ, Wunner FM, Saidy NT, Kollmannsberger S, Reali A, Rank E, De-Juan-Pardo EM, Hutmacher DW (2017) An integrated design, material, and fabrication platform for engineering biomechanically and biologically functional soft tissues. ACS Appl Mater Interfaces 9(35):29430–29437. https://doi.org/10.1021/acsami.7b08617
71. Deckard CR (1997) Apparatus for producing parts by selective sintering. Google Patents.
72. Schmid M, Amado A, Wegener K (2015) Polymer powders for selective laser sintering (SLS). AIP Conf Proc 1664(1):160009. https://doi.org/10.1063/1.4918516
73. Almoatazbellah Y, Scott JH, Paul DD (2017) Additive manufacturing of polymer melts for implantable medical devices and scaffolds. Biofabrication 9(1):012002
74. Williams JM, Adewunmi A, Schek RM, Flanagan CL, Krebsbach PH, Feinberg SE, Hollister SJ, Das S (2005) Bone tissue engineering using polycaprolactone scaffolds fabricated via selective laser sintering. Biomaterials 26(23):4817–4827. https://doi.org/10.1016/j.biomaterials.2004.11.057
75. Duan B, Wang M, Zhou WY, Cheung WL, Li ZY, Lu WW (2010) Three-dimensional nanocomposite scaffolds fabricated via selective laser sintering for bone tissue engineering. Acta Biomater 6(12):4495–4505. https://doi.org/10.1016/j.actbio.2010.06.024
76. Kanczler JM, Mirmalek-Sani S-H, Hanley NA, Ivanov AL, Barry JJA, Upton C, Shakesheff KM, Howdle SM, Antonov EN, Bagratashvili VN, Popov VK, Oreffo ROC (2009) Biocompatibility and osteogenic potential of human fetal femur-derived cells on surface selective laser sintered scaffolds. Acta Biomater 5(6):2063–2071. https://doi.org/10.1016/j.actbio.2009.03.010
77. Cijun S, Zhongzheng M, Haibo L, Yi N, Huanlong H, Shuping P (2013) Fabrication of porous polyvinyl alcohol scaffold for bone tissue engineering via selective laser sintering. Biofabrication 5(1):015014
78. Xia Y, Zhou P, Cheng X, Xie Y, Liang C, Li C, Xu S (2013) Selective laser sintering fabrication of nano-hydroxyapatite/poly-ε-caprolactone scaffolds for bone tissue engineering applications. Int J Nanomedicine 8:4197–4213. https://doi.org/10.2147/ijn.s50685
79. XiaoHui S, Wei L, PingHui S, QingYong S, QingSong W, YuSheng S, Kai L, WenGuang L (2015) Selective laser sintering of aliphatic-polycarbonate/hydroxyapatite compos-

ite scaffolds for medical applications. Int J Adv Manuf Technol 81(1):15–25. https://doi.org/10.1007/s00170-015-7135-x
80. Kuznetsova D, Prodanets N, Rodimova S, Antonov E, Meleshina A, Timashev P, Zagaynova E (2017) Study of the involvement of allogeneic MSCs in bone formation using the model of transgenic mice. Cell Adhes Migr 11(3):233–244. https://doi.org/10.1080/19336918.2016.1202386
81. Savalani MM, Hao L, Dickens PM, Zhang Y, Tanner KE, Harris RA (2012) The effects and interactions of fabrication parameters on the properties of selective laser sintered hydroxyapatite polyamide composite biomaterials. Rapid Prototyp J 18(1):16–27. https://doi.org/10.1108/13552541211193467
82. Shuai C, Feng P, Gao C, Shuai X, Xiao T, Peng S (2015) Graphene oxide reinforced poly(vinyl alcohol): nanocomposite scaffolds for tissue engineering applications. RSC Adv 5(32):25416–25423. https://doi.org/10.1039/c4ra16702c
83. Chong W, Qilong Z, Min W (2017) Cryogenic 3D printing for producing hierarchical porous and rhBMP-2-loaded ca-P/PLLA nanocomposite scaffolds for bone tissue engineering. Biofabrication 9(2):025031
84. Zhou WY, Lee SH, Wang M, Cheung WL, Ip WY (2008) Selective laser sintering of porous tissue engineering scaffolds from poly(l-lactide)/carbonated hydroxyapatite nanocomposite microspheres. J Mater Sci Mater Med 19(7):2535–2540. https://doi.org/10.1007/s10856-007-3089-3
85. Du Y, Liu H, Shuang J, Wang J, Ma J, Zhang S (2015) Microsphere-based selective laser sintering for building macroporous bone scaffolds with controlled microstructure and excellent biocompatibility. Colloids and Surf B Biointerf 135(supplement C):81–89. doi:https://doi.org/10.1016/j.colsurfb.2015.06.074
86. Di Bella C, Fosang A, Donati DM, Wallace GG, Choong PFM (2015) 3D bioprinting of cartilage for orthopedic surgeons: reading between the lines. Front Surg 2:39. https://doi.org/10.3389/fsurg.2015.00039
87. Du Y, Liu H, Yang Q, Wang S, Wang J, Ma J, Noh I, Mikos AG, Zhang S (2017) Selective laser sintering scaffold with hierarchical architecture and gradient composition for osteochondral repair in rabbits. Biomaterials 137 (Supplement C):37–48. doi:https://doi.org/10.1016/j.biomaterials.2017.05.021
88. Cai Q, Wan Y, Bei J, Wang S (2003) Synthesis and characterization of biodegradable polylactide-grafted dextran and its application as compatilizer. Biomaterials 24(20):3555–3562. https://doi.org/10.1016/S0142-9612(03)00199-6
89. Ciardelli G, Chiono V, Vozzi G, Pracella M, Ahluwalia A, Barbani N, Cristallini C, Giusti P (2005) Blends of poly-(ε-caprolactone) and polysaccharides in tissue engineering applications. Biomacromolecules 6(4):1961–1976. https://doi.org/10.1021/bm0500805
90. Sun H, Mei L, Song C, Cui X, Wang P (2006) The in vivo degradation, absorption and excretion of PCL-based implant. Biomaterials 27(9):1735–1740. https://doi.org/10.1016/j.biomaterials.2005.09.019
91. Chung C, Burdick JA (2008) Engineering cartilage tissue. Adv Drug Deliv Rev 60(2):243–262. https://doi.org/10.1016/j.addr.2007.08.027
92. Schuurman W, Khristov V, Pot MW, PRv W, Dhert WJA, Malda J (2011) Bioprinting of hybrid tissue constructs with tailorable mechanical properties. Biofabrication 3(2):021001
93. Shim J-H, Huh J-B, Park JY, Jeon Y-C, Kang SS, Kim JY, Rhie J-W, Cho D-W (2012) Fabrication of blended Polycaprolactone/poly (lactic-co-glycolic acid)/β-Tricalcium phosphate thin membrane using solid freeform fabrication Technology for Guided Bone Regeneration. Tissue Eng A 19(3–4):317–328. https://doi.org/10.1089/ten.tea.2011.0730
94. Kundu J, Shim J-H, Jang J, Kim S-W, Cho D-W (2015) An additive manufacturing-based PCL–alginate–chondrocyte bioprinted scaffold for cartilage tissue engineering. J Tissue Eng Regen Med 9(11):1286–1297. https://doi.org/10.1002/term.1682

95. Izadifar Z, Chang T, Kulyk W, Chen X, Eames BF (2015) Analyzing biological performance of 3D-printed, cell-impregnated hybrid constructs for cartilage tissue engineering. Tissue Eng Part C Methods 22(3):173–188. https://doi.org/10.1089/ten.tec.2015.0307
96. Margaret AN, Nathan JC, Michael WP, Lijie Grace Z (2016) 3D printing of novel osteochondral scaffolds with graded microstructure. Nanotechnology 27(41):414001
97. Guo R, Lu S, Page JM, Merkel AR, Basu S, Sterling JA, Guelcher SA (2015) Fabrication of 3D scaffolds with precisely controlled substrate modulus and pore size by templated-fused deposition modeling to direct osteogenic differentiation. Adv Healthc Mater 4(12):1826–1832. https://doi.org/10.1002/adhm.201500099
98. Dong L, Wang S-J, Zhao X-R, Zhu Y-F, Yu J-K (2017) 3D- printed poly(ε-caprolactone) scaffold integrated with cell-laden chitosan hydrogels for bone tissue engineering. Sci Rep 7(1):13412. https://doi.org/10.1038/s41598-017-13838-7
99. Boere KWM, Visser J, Seyednejad H, Rahimian S, Gawlitta D, van Steenbergen MJ, Dhert WJA, Hennink WE, Vermonden T, Malda J (2014) Covalent attachment of a three-dimensionally printed thermoplast to a gelatin hydrogel for mechanically enhanced cartilage constructs. Acta Biomater 10(6):2602–2611. https://doi.org/10.1016/j.actbio.2014.02.041
100. Levato R, Webb WR, Otto IA, Mensinga A, Zhang Y, van Rijen M, van Weeren R, Khan IM, Malda J (2017) The bio in the ink: cartilage regeneration with bioprintable hydrogels and articular cartilage-derived progenitor cells. Acta Biomater 61:41–53. https://doi.org/10.1016/j.actbio.2017.08.005
101. Kim JE, Kim SH, Jung Y (2016) Current status of three-dimensional printing inks for soft tissue regeneration. Tiss Eng Regen Med 13(6):636–646. https://doi.org/10.1007/s13770-016-0125-8
102. Chuah YJ, Peck Y, Lau JE, Hee HT, Wang DA (2017) Hydrogel based cartilaginous tissue regeneration: recent insights and technologies. Biomater Sci 5(4):613–631. https://doi.org/10.1039/c6bm00863a
103. Zhai X, Ma Y, Hou C, Gao F, Zhang Y, Ruan C, Pan H, Lu WW, Liu W (2017) 3D-printed high strength bioactive supramolecular polymer/clay nanocomposite hydrogel scaffold for bone regeneration. ACS Biomater Sci Eng 3(6):1109–1118. https://doi.org/10.1021/acsbiomaterials.7b00224
104. Camci-Unal G, Cuttica D, Annabi N, Demarchi D, Khademhosseini A (2013) Synthesis and characterization of hybrid hyaluronic acid-gelatin hydrogels. Biomacromolecules 14(4):1085–1092. https://doi.org/10.1021/bm3019856
105. Ju Young P, Jong-Cheol C, Jin-Hyung S, Jung-Seob L, Hyoungjun P, Sung Won K, Junsang D, Dong-Woo C (2014) A comparative study on collagen type I and hyaluronic acid dependent cell behavior for osteochondral tissue bioprinting. Biofabrication 6(3):035004
106. Levett PA, Melchels FP, Schrobback K, Hutmacher DW, Malda J, Klein TJ (2014) A biomimetic extracellular matrix for cartilage tissue engineering centered on photocurable gelatin, hyaluronic acid and chondroitin sulfate. Acta Biomater 10(1):214–223. https://doi.org/10.1016/j.actbio.2013.10.005
107. Shaoquan B, He M, Junhui S, Cai H, Sun Y, Liang J, Fan Y, Zhang X (2016) The self-crosslinking smart hyaluronic acid hydrogels as injectable three-dimensional scaffolds for cells culture, vol 140. doi:https://doi.org/10.1016/j.colsurfb.2016.01.008
108. Riccardo L, Jetze V, Josep AP, Elisabeth E, Jos M, Miguel AM-T (2014) Biofabrication of tissue constructs by 3D bioprinting of cell-laden microcarriers. Biofabrication 6(3):035020
109. Chameettachal S, Midha S, Ghosh S (2016) Regulation of Chondrogenesis and hypertrophy in silk fibroin-gelatin-based 3D bioprinted constructs. ACS Biomater Sci Eng 2(9):1450–1463. https://doi.org/10.1021/acsbiomaterials.6b00152
110. Levett PA, Melchels FP, Schrobback K, Hutmacher DW, Malda J, Klein TJ (2014) Chondrocyte redifferentiation and construct mechanical property development in single-component photocrosslinkable hydrogels. J Biomed Mater Res A 102(8):2544–2553. https://doi.org/10.1002/jbm.a.34924

111. Annabi N, Tamayol A, Uquillas JA, Akbari M, Bertassoni LE, Cha C, Camci-Unal G, Dokmeci MR, Peppas NA, Khademhosseini A (2014) 25th anniversary article: rational design and applications of hydrogels in regenerative medicine. Adv Mater 26(1):85–124. https://doi.org/10.1002/adma.201303233
112. Hynes WF, Doty NJ, Zarembinski TI, Schwartz MP, Toepke MW, Murphy WL, Atzet SK, Clark R, Melendez JA, Cady NC (2014) Micropatterning of 3D microenvironments for living biosensor applications. Biosensors (Basel) 4(1):28–44. https://doi.org/10.3390/bios4010028
113. Pereira RF, Bartolo PJ (2015) 3D bioprinting of photocrosslinkable hydrogel constructs. J Appl Polym Sci 132(48). https://doi.org/10.1002/app.42458
114. Slaughter BV, Khurshid SS, Fisher OZ, Khademhosseini A, Peppas NA (2009) Hydrogels in regenerative medicine. Adv Mater 21(32–33):3307–3329. https://doi.org/10.1002/adma.200802106
115. Duan B, Kapetanovic E, Hockaday LA, Butcher JT (2014) Three-dimensional printed trileaflet valve conduits using biological hydrogels and human valve interstitial cells. Acta Biomater 10(5):1836–1846. https://doi.org/10.1016/j.actbio.2013.12.005
116. Kim BS, Jang J, Chae S, Gao G, Kong JS, Ahn M, Cho DW (2016) Three-dimensional bioprinting of cell-laden constructs with polycaprolactone protective layers for using various thermoplastic polymers. Biofabrication 8(3):035013. https://doi.org/10.1088/1758-5090/8/3/035013
117. Axpe E, Oyen ML (2016) Applications of alginate-based bioinks in 3D bioprinting. Int J Mol Sci 17(12). https://doi.org/10.3390/ijms17121976
118. Nguyen D, Hagg DA, Forsman A, Ekholm J, Nimkingratana P, Brantsing C, Kalogeropoulos T, Zaunz S, Concaro S, Brittberg M, Lindahl A, Gatenholm P, Enejder A, Simonsson S (2017) Cartilage tissue engineering by the 3D bioprinting of iPS cells in a Nanocellulose/alginate bioink. Sci Rep 7(1):658. https://doi.org/10.1038/s41598-017-00690-y
119. Markstedt K, Mantas A, Tournier I, Martínez Ávila H, Hägg D, Gatenholm P (2015) 3D bioprinting human chondrocytes with Nanocellulose–alginate bioink for cartilage tissue engineering applications. Biomacromolecules 16(5):1489–1496. https://doi.org/10.1021/acs.biomac.5b00188
120. Bakarich SE, Gorkin R, in het Panhuis M, Spinks GM (2014) Three-dimensional printing fiber reinforced hydrogel composites. ACS Appl Mater Interfaces 6 (18):15998–16006. doi:https://doi.org/10.1021/am503878d
121. Wang Y, Wu S, Kuss MA, Streubel PN, Duan B (2017) Effects of hydroxyapatite and hypoxia on Chondrogenesis and hypertrophy in 3D bioprinted ADMSC laden constructs. ACS Biomater Sci Eng 3(5):826–835. https://doi.org/10.1021/acsbiomaterials.7b00101
122. Yang J, Zhang YS, Yue K, Khademhosseini A (2017) Cell-laden hydrogels for osteochondral and cartilage tissue engineering. Acta biomaterialia 57 (supplement C):1-25. https://doi.org/10.1016/j.actbio.2017.01.036
123. Cui X, Breitenkamp K, Lotz M, D'Lima D (2012) Synergistic action of fibroblast growth factor-2 and transforming growth factor-beta1 enhances bioprinted human neocartilage formation. Biotechnol Bioeng 109(9):2357–2368. https://doi.org/10.1002/bit.24488
124. Kesti M, Eberhardt C, Pagliccia G, Kenkel D, Grande D, Boss A, Zenobi-Wong M (2015) Bioprinting complex cartilaginous structures with clinically compliant biomaterials. Adv Funct Mater 25(48):7406–7417. https://doi.org/10.1002/adfm.201503423
125. Lee CH, Cook JL, Mendelson A, Moioli EK, Yao H, Mao JJ (2010) Regeneration of the articular surface of the rabbit synovial joint by cell homing: a proof of concept study. Lancet 376(9739):440–448. https://doi.org/10.1016/S0140-6736(10)60668-X
126. Hung K-C, Tseng C-S, Dai L-G, Hsu S-H (2016) Water-based polyurethane 3D printed scaffolds with controlled release function for customized cartilage tissue engineering. Biomaterials 83(Supplement C):156–168. doi:https://doi.org/10.1016/j.biomaterials.2016.01.019
127. Gao G, Zhang XF, Hubbell K, Cui X (2017) NR2F2 regulates chondrogenesis of human mesenchymal stem cells in bioprinted cartilage. Biotechnol Bioeng 114(1):208–216. https://doi.org/10.1002/bit.26042

# Chapter 11
# The Use of Electrospinning Technique on Osteochondral Tissue Engineering

**Marta R. Casanova, Rui L. Reis, Albino Martins, and Nuno M. Neves**

**Abstract** Electrospinning, an electrostatic fiber fabrication technique, has attracted significant interest in recent years due to its versatility and ability to produce highly tunable nanofibrous meshes. These nanofibrous meshes have been investigated as promising tissue engineering scaffolds since they mimic the scale and morphology of the native extracellular matrix. The sub-micron diameter of fibers produced by this process presents various advantages like the high surface area to volume ratio, tunable porosity, and the ability to manipulate the nanofiber composition in order to get desired properties and functionality. Electrospun fibers can be oriented or arranged randomly, giving control over both mechanical properties and the biological response to the fibrous scaffold. Moreover, bioactive molecules can be integrated with the electrospun nanofibrous scaffolds in order to improve the cellular response. This chapter presents an overview of the developments on electrospun polymer nanofibers including processing, structure, and their applications in the field of osteochondral tissue engineering.

**Keywords** Nanofibrous meshes · Processing parameters · Topographies · Surface functionalization

## 11.1 Introduction

Over the years, electrospinning has received attention from the biomedical community as a cutting-edge technology enabling the fabrication of meshes of fibers with diameters in the micro- to nano-scale that can be used as filtration membranes, catalytic nanofibers, fiber-sensors, and tissue engineering scaffolds [1–3].

---

M. R. Casanova · R. L. Reis · A. Martins · N. M. Neves (✉)
3B's Research Group—Biomaterials, Biodegradable and Biomimetics, Avepark—Parque de Ciência e Tecnologia, Zona Industrial da Gandra, Barco/Guimarães, Portugal

ICVS/3B's—PT Government Associate Laboratory, Braga/Guimarães, Portugal
e-mail: nuno@dep.uminho.pt

Basically, a scaffold needs to be mechanically resistant and support cell growth and proliferation, but gradually degrade along with the formation of new tissue and, finally, be replaced totally by the newly formed tissue [4].

Electrospinning is a versatile and cost-effective polymer processing technique for the production of nonwoven micro- to nano-fibrous meshes. The fiber dimension defines several interesting properties such as area to volume ratio, porosity, and mechanical properties. From a biological point of view, almost every extracellular matrix (ECM) of connective tissue is based on nanofibrous structures (e.g., skin, cartilage, and bone). These similarities make electrospun nanofibers great candidates to provide the cells an environment similar to the native structure of the ECM [4, 5]. Most of the desirable properties of tissue engineered scaffolds can be addressed by electrospun nanofibrous meshes, such as porosity, interconnected pores with adjustable pores size, capabilities for effective surface functionalization, and adjustable surface morphology [6–9].

This chapter provides an in-depth analysis of the principles behind the electrospinning technique, the influence of the processing parameters over the fiber characteristics, and discusses the applications and impact of electrospun nanofibrous meshes in the field of osteochondral tissue engineering (OTE).

## 11.2 Basics of the Electrospinning Technique

The process of using electrostatic forces to produce fibers has been known for over 100 years. Unlike conventional fiber spinning methods (as dry-spinning or melt-spinning), electrospinning has its roots in electrostatic spraying (electrospraying), making use of electrostatic forces to stretch the solution as it solidifies [10, 11]. Already in the '30s, the electrospinning technique was patented by A. Formhals for the commercial fabrication of artificial filaments [12]. However, only in the mid-'90s, researchers began to explore the high potential of this method for producing nanofibers. Nowadays, electrospun fibrous meshes are widely used successfully, especially for biomedical applications.

Electrospinning has many controllable processing parameters that affect the fibers formation and resulting structure [6]. Generally, the production of nanofibers using the electrospinning technique is based on uniaxial stretching of a viscoelastic solution, containing a dissolved polar or nonpolar polymer solution, caused by electrostatic forces [10, 13].

There are basically three main components in the traditional electrospinning setup: a high voltage power source, a capillary tube with a pipette or a needle, and a metallic collector base (Fig. 11.1) [10, 14]. The fiber formation is achieved by forcing the viscous polymeric solution through a spinneret, in most cases the metal tip of a needle, exposed to an electric field, forming initially a droplet. As the intensity of the electric field is increased, the droplet is turned into a conical shape fluid structure called the Taylor cone, which was first described mathematically by G. I. Taylor in 1964 [15]. If the viscosity and surface tension of the polymeric solution are

**Fig. 11.1** The common setup and working principle of electrospinning. Reprinted with permission from [6]. Copyright 2015, Elsevier

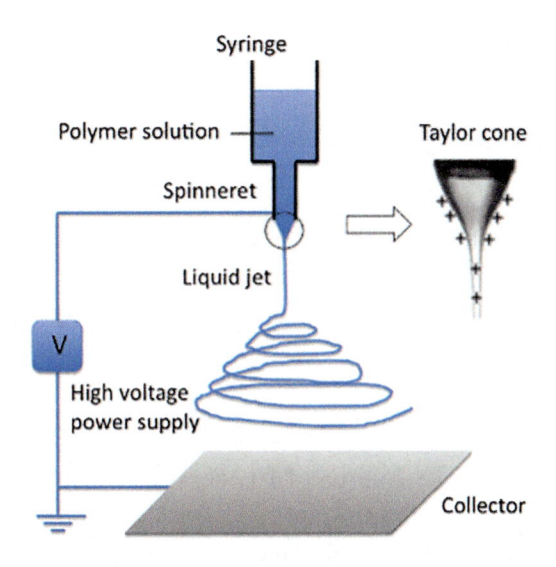

suitable, the breaking of the cone is prevented and a stable jet is formed. The increase in the applied high voltage leads to the elongation of the Taylor cone. When the repulsive force within the charged solution is higher than its surface tension, the cone is elongated, becoming a jet. This fluid filament is directed to a grounded collector. At this time, the solvent is evaporated and the nanofibers are formed and deposited at the surface of the collector as a nonwoven web of small fibers—the nanofibrous mesh [10, 13, 16]. The obtained nonwoven mesh structure consists typically of randomly aligned fibers deposited in successive layers over a static collector. The fibers are stacked one over another while being deposited, building several interconnections due to the presence of residual solvent.

The nanofibers obtained are in fact a single continuous fiber that continues to be deposited on the collector as long as there's enough solution in the syringe to feed the electrospinning jet [16]. Electrospun fibers can also be forced into aligned deposition when using suitable collectors [17, 18]. According to the application, the collector configuration may differ: stationary plates and rotating mandrel. The different collectors provide different electric fields, enabling the production of special morphologies.

The main problems related with the conventional arrangement of the electrospinning apparatus are: (1) the gravitational force, which allows drops to fall down from the needle creating defects on the mesh; (2) the low efficiency, intrinsically associated with the production of only one nanofiber at a time [19]. This is particularly relevant in industrial applications where a continuous and rapid process is essential. Various solutions were studied to optimize this technique combining solvents and polymers with different characteristics. Moreover, new configurations and accessories can be developed to adapt the system for a given application, obtaining meshes with different morphologies, porosity and porosity distribution, diameter, and fiber structure [6, 7, 20–22].

**Table 11.1** Electrospinning parameters and their effects on fiber morphology

| Parameter | Effect on fiber morphology |
|---|---|
| Applied voltage | Decrease in fiber diameter with increase in voltage |
| Flow rate | Decrease in fiber diameter with decrease in flow rate; Generation of beads with too high flow rate |
| Distance between tip and collector | Decrease in fiber diameter with increase in spinning distance; Generation of beads with too small and too large distance |
| Polymer concentration (viscosity) | Increase in fiber diameter with increase of concentration of the polymeric solution |
| Solution conductivity | Decrease in fiber diameter with increase in conductivity |
| Solvent volatility | High humidity results in circular pores on the fibers; Increase in temperature results in decrease in fiber diameter |

## 11.3 Influence of Processing Parameters

Electrospinning has shown to be a quite simple and versatile technique to obtain nonwoven meshes, achieving a consistent number of samples with similar properties, namely a controllable fiber diameter. The properties of the fibers can be modified depending on the choice of materials and solvents used which will affect the solution properties (such as solution viscosity, volatility, concentration, surface tension, and conductivity), solution temperature, applied voltage at the tip of the needle, angle of the spinneret, spinneret tip-to-collector distance, the temperature, air velocity in the electrospinning chamber, and humidity [6, 20, 23–27]. The last three variables (distance, temperature, and humidity) are the ones that control the solvent evaporation during the electrospinning process. These processing parameters listed in Table 11.1 have the following effects over the fiber morphology.

A major disadvantage of electrospinning is the lack of independent control over all processing parameters, since changing one parameter, other parameters are simultaneously influenced. For this reason, it is still a challenge to achieve a sufficient stability in the production of microstructural features on the fibrous meshes. The parameters that affect the electrospinning process can be divided in three major categories:

### 11.3.1 Processing Parameters

To obtain fibers with controllable diameter, the electrical field intensity needs to be adjusted by playing with three processing parameters: the applied voltage, the defined flow rate, and the distance between the needle tip and the grounded collector [26, 28]. The type of collector used, concerning its material, shape, and size, also causes changes in the shape and intensity of the electrical field.

The fibers formation from few microns to tens of nanometers in diameter is also controlled by the intensity of the applied electric field. Only after attaining a

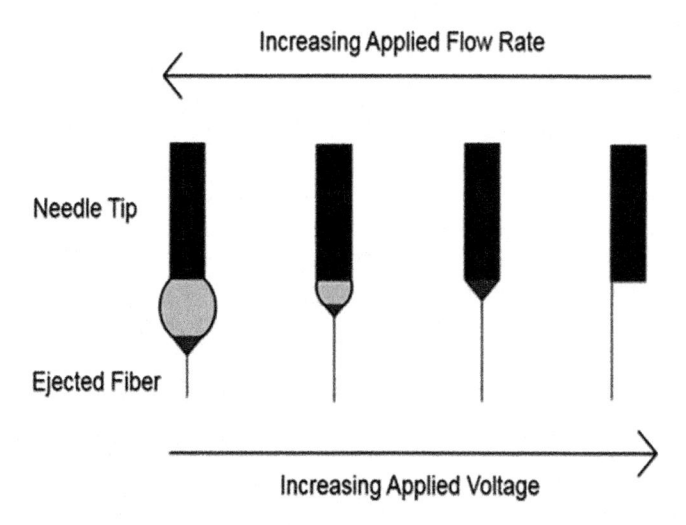

**Fig. 11.2** Formations of the "Taylor cone" (dark gray) and pendant drop (light gray). Adapted with permission from [1]. Copyright 2008, Elsevier

threshold voltage, the fiber formation occurs. This induces the necessary charges on the polymeric solution, along with the generation of an electric field, and initiates the electrospinning process [10, 15].

The shape of the pendant drop at the tip—Taylor cone—is dependent on the processing parameters, being mainly influenced by the electric field intensity and the flow rate, and also by the solution parameters [10, 29]. At relatively low processing voltages, in combination with relatively high flow rates, a pendant drop is formed at the tip of the needle (Fig. 11.2). By increasing the voltage or decreasing the flow rate, the drop retreats until the Taylor cone begins at the tip of the needle. This leads to the most stable processing conditions. By having a relatively high voltage, together with a relatively low flow rate, the Taylor cone is formed inside the capillary of the needle. The ejection of the polymeric solution at these processing conditions is associated with increasing observation of bead-like defects at the surface of the meshes [10, 23, 24, 26]. If the electric potential is further increased, the former continuous jet turns into a spraying of single fiber segments or even single drops ("electrospraying"), which is not desirable for scaffold materials, since it does not produce a self-sustained fiber mesh.

The flow rate of the polymeric solution has a significant impact on the fiber size and on the fiber shape. In general, a lower flow rate is desirable because the solvent has enough time to evaporate, forming uniform and smooth fibers [30]. By its side, high flow rates result in beaded fibers due to the insufficient time for the fibers to stretch before reaching the collector [31, 32]. Another factor that controls the diameter and morphology of the fiber is the distance between the needle tip and the collector. It has been found that a minimum distance is required to give the fibers sufficient time to undergo stretching and dry before reaching the collector. Otherwise, with distances that are either small or too large, beads are obtained [21,

33, 34]. Initially, the straight jet causes the formation of a single fiber, which becomes unstable and suffers stretching in long oscillatory movements. Along the way to the collector, the solvent evaporates and the charge density increases, which causes a further local stretching besides strong charge repulsion against equally charged areas on the fiber. The resulting perpendicular moving or "whipping instability" of the fiber stream is bounded by an enveloping cone towards the collector [35]. All these effects cause further reduction of the fiber diameter, when traveling in the direction of the collector. Furthermore, the collector material should be conductive to create stable potential difference between needle and collector and to control the electric field that controls the fibers' deposition pattern.

## *11.3.2 Solution Parameters*

The solution parameters are mainly determined by the characteristics of the polymer(s) and the solvent(s), being the most frequently investigated ones. The most significant parameters include concentration, molecular weight, viscosity, surface tension, and electric conductivity of the polymeric solution [10, 20, 23, 36].

In the electrospinning process, the concentration of the polymer determines the spinnability of a solution, namely if the fibers form or not [28, 32]. Above a critical concentration, a continuous fibrous structure is obtained. The fiber diameter can be significantly decreased by decreasing the polymer concentration, although there is a minimum limit to obtain uniform nanofibers without beads [9, 37]. High solution concentrations can be difficult to process owing to the high viscosity of the solution. Thus, an optimal range of polymer concentrations exists in which the fibers can be electrospun, when all the other parameters are held constant.

The molecular weight of a polymer also has a significant effect on the electrical and rheological characteristics of the solution [21, 37]. It reflects the number of entanglements of polymer chains in a solution. Therefore, low molecular weight solutions tend to form beads instead of fibers, whereas high molecular weight solutions give fibers with relatively large diameters [6, 21].

The concentration and molecular weight of a polymer determine the viscosity of a solution, playing an important role over the fiber size and morphology during electrospinning [6, 26, 38]. In fact, a very low viscosity solution cannot form uniform and smooth fibers, while very high viscosities can be difficult to obtain continuous jet forming fibers.

The formation of droplets, beads, and fibers depends on the surface tension of the polymeric solution, being therefore a required parameter for fiber formation. Generally, the high surface tension of a polymeric solution limits the fiber formation, because of the unstable jet and dispersion of droplets, while the low surface tension can form uniform and smooth fibers [9, 31].

The solution conductivity is mainly determined by the polymer type, solvent used, and/or the availability of ionizable salts. The electrical conductivity and the solvents volatility are set by the choice of the solution composition, containing specific amounts of solvents and/or additional non-solvents. By increasing the solution conductivity, there is a considerable decrease in the fiber diameter. However, low solution conductivity may cause insufficient elongation of the jet which induce the formation of the undesired beads [10, 23, 39].

### 11.3.3 Environmental Parameters

Apart from the solution and processing parameters, environmental parameters (i.e., temperature and humidity) also affect the electrospinning process. To obtain a better mobility of the polymer molecules, a higher temperature is needed for the electrospinning process [26, 40]. Few studies reported that, by increasing the temperature, the viscosity of the polymeric solution decreased, resulting in smaller fiber diameters.

The humidity has major effects in the electrical conductivity of the air and causes changes in the electrical field, while it affects the evaporating rate of solvents as well. At low humidity, the solvents will evaporate completely, rising the rate of solvent evaporation. The high humidity leads to the appearance of small circular pores at the fibers' surface [6, 41].

## 11.4 Generation of Specific Topographies in Electrospun Fibers

In the standard electrospinning process, the fibers are randomly deposited onto a planar collector, defined as the counter electrode, obtaining a nonwoven fibrous mesh. However, the electrospinning technique is not limited to the production of nonwovens with a random planar fiber orientation. Other fiber morphologies and structures (i.e., parallel and crossed fiber arrays, helical or wavy fibers, and patterned fiber web) can be also obtained via modified electrospinning process or collectors (Fig. 11.3) [17, 20, 42–44].

Uniaxial alignment of nanofibers can be achieved when a cylinder at high rotating speed is used as collector [45, 46]. However, the degree of orientation for nanofibers collected by rotating drums is far from having a perfect alignment due to the inconsistent and incontrollable polymeric jets [47, 48]. Instead, by using special electrode arrangements, a high degree orientation can be achieved. A parallel nanofibers deposition can be obtained by using two parallel flat plates or with frame-shaped electrodes [18]. As using a quadratic arrangement of four electrodes, a cross-shaped deposition of nanofibers can be obtained [17, 49].

Taking advantage of the electrospinning setup versatility, strategies to produce complex mesh structures have been developed [16, 17, 50, 51]. Kidoaki et al. proposed the layer-by-layer electrospinning and mixing electrospinning strategies [50].

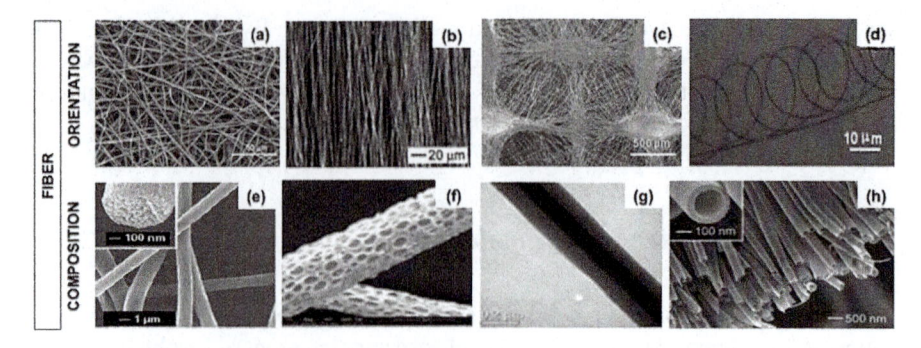

**Fig. 11.3** Morphology of different electrospun nanofibrous structures. (**a**) random orientation; (**b**) parallel and (**c**) patterned nanofiber meshes; (**d**) helical fibers; (**e**) solid nanofiber; (**f**) porous nanofiber; (**g**) core-shell nanofiber; (**h**) hollow nanofiber. Adapted with permission from [9, 17, 36, 43]. Copyright 2004, John Wiley & Sons. Copyright 2011, John Wiley & Sons. Copyright 2014, Elsevier. Copyright 2016, Elsevier

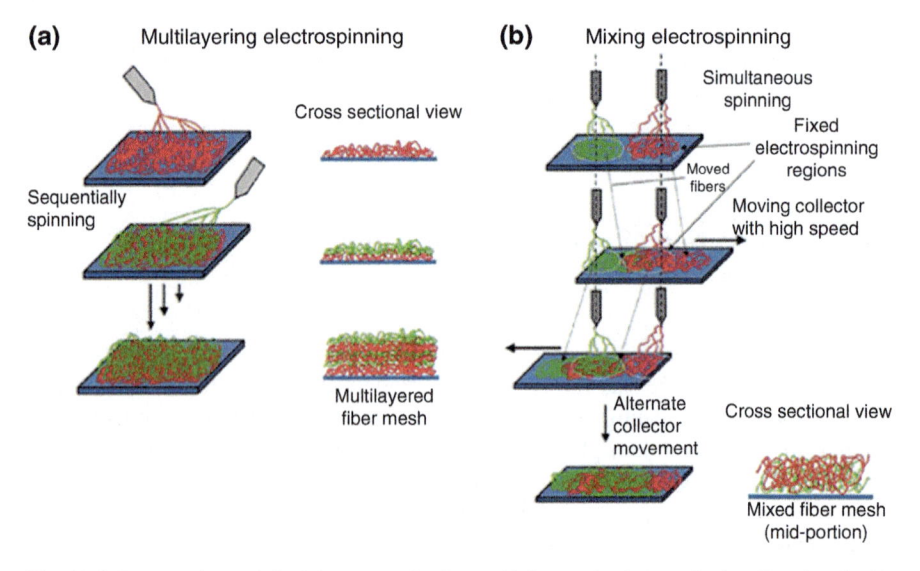

**Fig. 11.4** Proposed material mixing strategies for multiple nozzle electrospinning. Reprinted with permission from [4]. Copyright 2005, Elsevier

In the layer-by-layer strategy, the obtained mesh is either structured hierarchically by sequentially spinning two or more polymeric solutions through separate nozzles over the same static collector (Fig. 11.4a). In the mixing electrospinning strategy, the solutions are spun at the same time over the same moving collector, obtaining a homogenous mixture of multiple fibers (Fig. 11.4b).

Electrospinning generally enables obtaining solid fibers with a smooth surface. However, it has the ability to produce fibers with different structures, such as porous and core-shell or hollow fibers (Fig. 11.3g, h) [20, 52].

The production of a porous structure into the bulk of an electrospun nanofiber can be achieved by two slightly different approaches. One is based on the selective removal of one component used to produce nanofibers made from two phase material, such as a composite or blend material. The other approach involves the use of the phase separation of different polymers during the electrospinning, under the application of proper electrospinning parameters [20, 53].

The use of the conventional electrospinning setup allows the production of the core-shell nanofibers from a polymeric solution containing two polymers that will phase separate as the solvent is evaporating [52]. Coaxial electrospinning setup also allows the production of core-shell nanofibers by coelectrospinning of two different polymeric solutions through a spinneret comprising two coaxial capillaries [54, 55]. This coaxial electrospinning method has been used to encapsulate a wide variety of biological molecules into the core of the nanofibers [56, 57].

## 11.5 Surface Modification of Nanofibrous Meshes

All the above-described methods were focused on the development of scaffolds architecture able to support cell adhesion or migration along the electrospun nanofibrous mesh. However, the surfaces of electrospun nanofibrous meshes may be physically or chemically modified to provide biomimetic microenvironments for the surrounding tissues and cells [9, 44].

Surface modification methods have been applied to electrospun nanofibrous meshes to enhance their biocompatibility or induce specific biological response from attached cells. Several surface modification methods have been implemented on electrospun nanofibers, which can be divided into chemical- and bio-functionalization.

### 11.5.1 Chemical Modification

A wide range of surface modification methods can be used to change the surface chemistry of electrospun nanofibers, aiming to improve the scaffold hydrophilicity and provide a more desirable environment for cellular adhesion and growth.

Plasma treatment has been widely used to modify the surface of many polymeric materials including nanofibrous scaffolds, with the purpose to create polar groups, such as hydroxyl, carboxyl, or amine at their surfaces [58, 59]. The presence of these functional groups results in changes of polymeric surface properties such as wettability, polarity, and bio-adhesion [5, 59, 60]. For example, poly-L-lactic acid (PLLA) nanofibers modified by plasma treatment demonstrated an increase in the surface hydrophilicity and the amount of oxygen-containing groups, which enhance the cell adhesion of porcine mesenchymal stem cells (pMSCs) at earlier culture time intervals [60].

The combination of inorganic compounds (e.g., hydroxyapatite) within nanofibrous scaffolds represents an exclusive advantage in the tissue engineering field, namely in the case of bone and teeth. Hydroxyapatite is known as an osteoconductive and osteoinductive material. These properties could facilitate the integration with native bone and, eventually, promote new bone formation [61, 62]. Likewise, electrospun nanofibrous meshes coated with a biomimetic calcium phosphate layer, mimicking the extracellular microenvironment found in the human bone structure, support and enhance the proliferation of osteoblasts for long culture periods [7]. Collagen, the major organic component of bone, is among the most preferred matrix of electrospun nanofibers to mimic bone. Therefore, several studies on electrospinning of collagen with hydroxyapatite nanoparticles can be found in the literature [63–68].

Another chemical modification method relies on performing graft polymerization to introduce multifunctional properties at the nanofibers' surface by aminolysis and alkali treatment. Generally, the surface graft polymerization method begins with plasma and UV radiation treatment in order to produce free radicals for the polymerization [56, 69]. Aminolysis is a straightforward method for the introduction of amino groups at the surface of polyesters by the cleavage of ester bonds and the simultaneous generation of amide bonds. The most commonly used aminolysis reagent is a diamine, namely hexamethylenediamine. One of the two $NH_2$ groups reacts with the carbonyl carbon while the other amino group remains free at the nanofiber surface, potentially stable for subsequent chemical reactions [69–72].

## 11.5.2 Bio-Functionalization

Considering the physicochemical flexibility of electrospun nanofibrous meshes, the incorporation of biological active factors represents an important research domain in the area of tissue engineering. In principle, the nanofibrous scaffolds will easily absorb such bioactive molecules, due to the high surface area, and then possibly result in optimal local release, providing a smart approach to maintain higher levels of bioactive molecules at the neighboring cells.

A great number of electrospinning-based research works have been reported, incorporating different bioactive molecules: the short peptide sequence RGD [73] and perlecan (a natural heparin sulfate proteoglycan) [51, 74].

From the biological point of view, electrospun nanofibers containing GFs can be considered good candidates for tissue engineering applications. Growth factors (GFs) are biological substances with particular chemical sequences that regulate a variety of cellular processes, namely stimulation of cellular growth, proliferation, as well as trigger their differentiation. Examples of GFs include bone morphogenetic proteins (BMPs) [75–77], vascular endothelial growth factor (VEGF) [71, 78, 79], basic fibroblast growth factor (bFGF) [71, 80], insulin-like growth factor (IGF) [76], and transforming growth factor beta (TGF-β) [71, 81].

## 11.6 Osteochondral Application of Nanofibrous Meshes

Osteochondral defects (i.e., defects which affect both the articular cartilage and underlying subchondral bone) typically derived from traumatic injuries or osteochondritis [82]. Grafting of osteochondral composite tissues, consisting of a superficial cartilaginous layer (corresponding to articular cartilage) and an underlying calcified tissue (corresponding to subchondral bone), represents a promising approach to restore the biological and mechanical functionality of the joint [83]. To repair osteochondral tissue with its multiple cell types organized in a specific pattern, a functional scaffold is required to control the spatial organization of the different cells [84]. In this way, a scaffold with homogeneous properties is not able to support the activity and expression of multiple cells in osteochondral tissue. Therefore, functional gradient osteochondral scaffolds with physical and chemical properties are needed.

Electrospinning has been demonstrated to be a powerful tool for fabricating tissue engineering scaffolds, as it is simple, inexpensive, versatile, and capable of forming ECM-mimicking structures [1]. However, for ultimate use in the field of osteochondral tissue engineering (OTE), cell infiltration hindrance and insufficient mechanical strength of conventional electrospun scaffolds should be investigated. These inherent limitations of electrospinning have gained considerable attention and a variety of approaches for overcoming them have been proposed [84, 85]. To address the issue of insufficient mechanical properties, combining electrospun constructs with either more robust matters/structures or applying different postprocessing modifications have been proposed [86–88]. Moreover, after fabrication, the nanofibers' surface can be modified with a diversity of bioactive molecules due to the surface chemistry flexibility and the high surface area of the nanofibrous scaffold.

A wide range of works have been showing the potential of electrospinning to produce a wide range of nanofibrous scaffolds from which some can be selected in order to create, with the help of a multilayered arrangement, a three-dimensional scaffold with gradients of properties trying to recapitulate the zonal matrix of the bone-cartilage interface [87–92]. Erisken et al. produce a poly (ε-caprolactone) (PCL) mesh with controlled gradation of insulin and β-glycerophosphate (β-GP) concentrations in between the two sides of a nanofibrous scaffold. This was achieved via the combined use of twin-screw extrusion and electrospinning techniques. The use of both insulin and β-GP at graded concentrations led to the differentiation of human adipose-derived stromal cells in a location-dependent manner: chondrogenic differentiation of the stem cells increased at insulin-rich locations and mineralization increased at β-GP-rich locations [90].

Liu et al. fabricated a novel scaffold for OTE by the combination of 3D bioprinting with multi-nozzle electrospinning. For temporally controlled release of gentamicin sulfate (GS) and deferoxamine (DFO), blend electrospun GS/polyvinyl alcohol (PVA) and coaxial electrospun core PVA-DFO/shell PCL fibers were deposited in the scaffold. The composite scaffold has the potential to delivery multiple biomolecules with various release profiles over space and time, achieving special characteristics of osteochondral scaffolds [92].

Mouthuy et al. investigated the potential of the electrospinning technique to build a three-dimensional construct recapitulating the zonal matrix of the bone-cartilage interface. In that context, a wide range of poly (lactic-co-glycolic acid) (PLGA) meshes with different proportions of collagen and hydroxyapatite nanoparticles were obtained by electrospinning. They show that the membranes have a better potential to serve as a scaffold to engineer either bone or cartilage thanks to the presence of collagen and hydroxyapatite nanoparticles [89].

## 11.7 Concluding Remarks

Osteochondral defects are a major clinical problem and the development of more effective scaffolds with the aim of osteochondral regeneration is a challenging topic. Electrospinning techniques are in continuous development to produce nanofibers with similar architecture of natural fibrillar ECM. Indeed, electrospinning constitutes the main technique generally available for the production of continuous polymeric nanofibers. Its inherent high surface to volume ratio, ease of operation, and cost-effectiveness are all appealing features for OTE application. Considering the features of electrospun fibrous scaffolds, they can be considered a promising solution to this problem, facilitating bone-cartilage interface regeneration.

## References

1. Martins A, Reis RL, Neves NM (2008) Electrospinning: processing technique for tissue engineering scaffolding. Int Mater Rev 53(5):257–274. https://doi.org/10.1179/1743280 08x353547
2. Rajesh KP, Natarajan TS (2009) Electrospun polymer nanofibrous membrane for filtration. J Nanosci Nanotechnol 9(9):5402–5405
3. Beringer LT, Xu X, Shih W, Shih WH, Habas R, Schauer CL (2015) An electrospun PVDF-TrFe fiber sensor platform for biological applications. Sensor Actuat A Phys 222:293–300. https://doi.org/10.1016/j.sna.2014.11.012
4. Smith LA, Ma PX (2004) Nano-fibrous scaffolds for tissue engineering. Colloid Surf B 39(3):125–131. https://doi.org/10.1016/j.colsurfb.2003.12.004
5. Martins A, Pinho ED, Faria S, Pashkuleva I, Marques AP, Reis RL, Neves NM (2009) Surface modification of electrospun polycaprolactone nanofiber meshes by plasma treatment to enhance biological performance. Small 5(10):1195–1206. https://doi.org/10.1002/smll.200801648
6. Casper CL, Stephens JS, Tassi NG, Chase DB, Rabolt JF (2004) Controlling surface morphology of electrospun polystyrene fibers: effect of humidity and molecular weight in the electrospinning process. Macromolecules 37(2):573–578. https://doi.org/10.1021/ma0351975
7. Araujo JV, Martins A, Leonor IB, Pinho ED, Reis RL, Neves NM (2008) Surface controlled biomimetic coating of polycaprolactone nanofiber meshes to be used as bone extracellular matrix analogues. J Biomater Sci Polym E 19(10):1261–1278. https://doi.org/10.1163/156856208786052335

8. da Silva MA, Crawford A, Mundy J, Martins A, Araujo JV, Hatton PV, Reis RL, Neves NM (2009) Evaluation of extracellular matrix formation in polycaprolactone and starch-compounded polycaprolactone nanofiber meshes when seeded with bovine articular chondrocytes. Tissue Eng Part A 15(2):377–385. https://doi.org/10.1089/ten.tea.2007.0327
9. Rezvani Z, Venugopal JR, Urbanska AM, Mills DK, Ramakrishna S, Mozafari M (2016) A bird's eye view on the use of electrospun nanofibrous scaffolds for bone tissue engineering: current state-of-the-art, emerging directions and future trends. Nanomedicine 12(7):2181–2200. https://doi.org/10.1016/j.nano.2016.05.014
10. Doshi J, Reneker DH (1995) Electrospinning process and applications of electrospun fibers. J Electrostat 35(2-3):151–160. https://doi.org/10.1016/0304-3886(95)00041-8
11. Ganan-Calvo AM, Davila J, Barrero A (1997) Current and droplet size in the electrospraying of liquids. Scaling laws. J Aerosol Sci 28(2):249–275. https://doi.org/10.1016/S0021-8502(96)00433-8
12. Formhals A (1934) Method and apparatus for spinning. US patent 1975504
13. Gibson P, Schreuder-Gibson H, Rivin D (2001) Transport properties of porous membranes based on electrospun nanofibers. Colloid Surf A 187:469–481. https://doi.org/10.1016/S0927-7757(01)00616-1
14. Ghorani B, Tucker N (2015) Fundamentals of electrospinning as a novel delivery vehicle for bioactive compounds in food nanotechnology. Food Hydrocolloid 51:227–240. https://doi.org/10.1016/j.foodhyd.2015.05.024
15. Taylor G (1964) Disintegration of water drops in an electric field. Proc R Soc Lond Ser A Math Phys Sci 280(1382):383–397. https://doi.org/10.1098/rspa.1964.0151
16. Dabirian F, Ravandi SAH, Pishevar AR, Abuzade RA (2011) A comparative study of jet formation and nanofiber alignment in electrospinning and electrocentrifugal spinning systems. J Electrost 69(6):540–546. https://doi.org/10.1016/j.elstat.2011.07.006
17. Li D, Wang YL, Xia YN (2004) Electrospinning nanofibers as uniaxially aligned arrays and layer-by-layer stacked films. Adv Mater 16(4):361–366. https://doi.org/10.1002/adma.200306226
18. Sharma N, Jaffari GH, Shah SI, Pochan DJ (2010) Orientation-dependent magnetic behavior in aligned nanoparticle arrays constructed by coaxial electrospinning. Nanotechnology 21(8):85707. https://doi.org/10.1088/0957-4484/21/8/085707
19. Teo WE, Ramakrishna S (2006) A review on electrospinning design and nanofibre assemblies. Nanotechnology 17(14):R89–R106. https://doi.org/10.1088/0957-4484/17/14/R01
20. Bognitzki M, Frese T, Steinhart M, Greiner A, Wendorff JH, Schaper A, Hellwig M (2001) Preparation of fibers with nanoscaled morphologies: electrospinning of polymer blends. Polym Eng Sci 41(6):982–989. https://doi.org/10.1002/pen.10799
21. Lee JS, Choi KH, Do Ghim H, Kim SS, Chun DH, Kim HY, Lyoo WS (2004) Role of molecular weight of atactic poly(vinyl alcohol) (PVA) in the structure and properties of PVA nanofabric prepared by electrospinning. J Appl Polym Sci 93(4):1638–1646. https://doi.org/10.1002/app.20602
22. Alves da Silva M, Martins A, Costa-Pinto AR, Monteiro N, Faria S, Reis RL, Neves NM (2017) Electrospun nanofibrous meshes cultured with Wharton's jelly stem cell: an alternative for cartilage regeneration, without the need of growth factors. Biotechnol J 12. https://doi.org/10.1002/biot.201700073
23. Deitzel JM, Kleinmeyer J, Harris D, Tan NCB (2001) The effect of processing variables on the morphology of electrospun nanofibers and textiles. Polymer 42(1):261–272. https://doi.org/10.1016/S0032-3861(00)00250-0
24. Deitzel JM, Kleinmeyer JD, Hirvonen JK, Tan NCB (2001) Controlled deposition of electrospun poly(ethylene oxide) fibers. Polymer 42(19):8163–8170. https://doi.org/10.1016/S0032-3861(01)00336-6
25. Yarin AL, Koombhongse S, Reneker DH (2001) Taylor cone and jetting from liquid droplets in electrospinning of nanofibers. J Appl Phys 90(9):4836–4846. https://doi.org/10.1063/1.1408260

26. Megelski S, Stephens JS, Chase DB, Rabolt JF (2002) Micro- and nanostructured surface morphology on electrospun polymer fibers. Macromolecules 35(22):8456–8466. https://doi.org/10.1021/ma020444a
27. Tan SH, Inai R, Kotaki M, Ramakrishna S (2005) Systematic parameter study for ultrafine fiber fabrication via electrospinning process. Polymer 46(16):6128–6134. https://doi.org/10.1016/j.polymer.2005.05.068
28. Frenot A, Chronakis IS (2003) Polymer nanofibers assembled by electrospinning. Curr Opin Colloid Interface Sci 8(1):64–75. https://doi.org/10.1016/S1359-0294(03)00004-9
29. Reneker DH, Chun I (1996) Nanometre diameter fibres of polymer, produced by electrospinning. Nanotechnology 7(3):216–223. https://doi.org/10.1088/0957-4484/7/3/009
30. Hsu CM, Shivkumar S (2004) Nano-sized beads and porous fiber constructs of poly(epsilon-caprolactone) produced by electrospinning. J Mater Sci 39(9):3003–3013. https://doi.org/10.1023/B:JMSC.0000025826.36080.cf
31. Thompson CJ, Chase GG, Yarin AL, Reneker DH (2007) Effects of parameters on nanofiber diameter determined from electrospinning model. Polymer 48(23):6913–6922. https://doi.org/10.1016/j.polymer.2007.09.017
32. Reneker DH, Yarin AL (2008) Electrospinning jets and polymer nanofibers. Polymer 49(10):2387–2425. https://doi.org/10.1016/j.polymer.2008.02.002
33. Geng XY, Kwon OH, Jang JH (2005) Electrospinning of chitosan dissolved in concentrated acetic acid solution. Biomaterials 26(27):5427–5432. https://doi.org/10.1016/j.biomaterials.2005.01.066
34. Ki CS, Baek DH, Gang KD, Lee KH, Um IC, Park YH (2005) Characterization of gelatin nanofiber prepared from gelatin-formic acid solution. Polymer 46(14):5094–5102. https://doi.org/10.1016/j.polymer.2005.04.040
35. Hohman MM, Shin M, Rutledge G, Brenner MP (2001) Electrospinning and electrically forced jets. I. Stability theory. Phys Fluids 13(8):2201–2220. https://doi.org/10.1063/1.1383791
36. Sun B, Long YZ, Zhang HD, Li MM, Duvail JL, Jiang XY, Yin HL (2014) Advances in three-dimensional nanofibrous macrostructures via electrospinning. Prog Polym Sci 39(5):862–890. https://doi.org/10.1016/j.progpolymsci.2013.06.002
37. Koski A, Yim K, Shivkumar S (2004) Effect of molecular weight on fibrous PVA produced by electrospinning. Mater Lett 58(3–4):493–497. https://doi.org/10.1016/S0167-577x(03)00532-9
38. Bhardwaj N, Kundu SC (2010) Electrospinning: a fascinating fiber fabrication technique. Biotechnol Adv 28(3):325–347. https://doi.org/10.1016/j.biotechadv.2010.01.004
39. Brown TD, Daltona PD, Hutmacher DW (2016) Melt electrospinning today: an opportune time for an emerging polymer process. Prog Polym Sci 56:116–166. https://doi.org/10.1016/j.progpolymsci.2016.01.001
40. Rodoplu D, Mutlu M (2012) Effects of electrospinning setup and process parameters on nanofiber morphology intended for the modification of quartz crystal microbalance surfaces. J Eng Fiber Fabr 7(2):118–123
41. Pelipenko J, Kristl J, Jankovic B, Baumgartner S, Kocbek P (2013) The impact of relative humidity during electrospinning on the morphology and mechanical properties of nanofibers. Int J Pharm 456(1):125–134. https://doi.org/10.1016/j.ijpharm.2013.07.078
42. Dersch R, Liu TQ, Schaper AK, Greiner A, Wendorff JH (2003) Electrospun nanofibers: internal structure and intrinsic orientation. J Polym Sci Pol Chem 41(4):545–553. https://doi.org/10.1002/pola.10609
43. Martins A, da Silva MLA, Faria S, Marques AP, Reis RL, Neves NM (2011) The influence of patterned nanofiber meshes on human mesenchymal stem cell osteogenesis. Macromol Biosci 11(7):978–987. https://doi.org/10.1002/mabi.201100012
44. Martins A, Reis R, Neves N (2012) Critical aspects of electrospun meshes for biomedical applications. In: Neves N (ed) Electrospinning for advanced biomedical applications and therapies, pp 69–87
45. Matthews JA, Wnek GE, Simpson DG, Bowlin GL (2002) Electrospinning of collagen nanofibers. Biomacromolecules 3(2):232–238

46. Kameoka J, Orth R, Yang YN, Czaplewski D, Mathers R, Coates GW, Craighead HG (2003) A scanning tip electrospinning source for deposition of oriented nanofibres. Nanotechnology 14(10):1124–1129. https://doi.org/10.1088/0957-4484/14/10/310. pii: S0957-4484(03)61381-4
47. Zussman E, Rittel D, Yarin AL (2003) Failure modes of electrospun nanofibers. Appl Phys Lett 82(22):3958–3960. https://doi.org/10.1063/1.1579125
48. Buttafoco L, Kolkman NG, Engbers-Buijtenhuijs P, Poot AA, Dijkstra PJ, Vermes I, Feijen J (2006) Electrospinning of collagen and elastin for tissue engineering applications. Biomaterials 27(5):724–734. https://doi.org/10.1016/j.biomaterials.2005.06.024
49. Greiner A, Wendorff JH (2007) Electrospinning: a fascinating method for the preparation of ultrathin fibers. Angew Chem Int Ed Engl 46(30):5670–5703. https://doi.org/10.1002/anie.200604646
50. Kidoaki S, Kwon IK, Matsuda T (2005) Mesoscopic spatial designs of nano- and microfiber meshes for tissue-engineering matrix and scaffold based on newly devised multilayering and mixing electrospinning techniques. Biomaterials 26(1):37–46. https://doi.org/10.1016/j.biomaterials.2004.01.063
51. Casper CL, Yang WD, Farach-Carson MC, Rabolt JF (2007) Coating electrospun collagen and gelatin fibers with perlecan domain I for increased growth factor binding. Biomacromolecules 8(4):1116–1123. https://doi.org/10.1021/bm061003s
52. Li D, Xia YN (2004) Electrospinning of nanofibers: reinventing the wheel? Adv Mater 16(14):1151–1170. https://doi.org/10.1002/adma.200400719
53. Malda J, Rouwkema J, Martens DE, le Comte EP, Kooy FK, Tramper J, van Blitterswijk CA, Riesle J (2004) Oxygen gradients in tissue-engineered PEGT/PBT cartilaginous constructs: measurement and modeling. Biotechnol Bioeng 86(1):9–18. https://doi.org/10.1002/bit.20038
54. Sun ZC, Zussman E, Yarin AL, Wendorff JH, Greiner A (2003) Compound core-shell polymer nanofibers by co-electrospinning. Adv Mater 15(22):1929. https://doi.org/10.1002/adma.200305136
55. Konno M, Kishi Y, Tanaka M, Kawakami H (2014) Core/shell-like structured ultrafine branched nanofibers created by electrospinning. Polym J 46(11):792–799. https://doi.org/10.1038/pj.2014.74
56. Shin YM, Kim KS, Lim YM, Nho YC, Shin H (2008) Modulation of spreading, proliferation, and differentiation of human mesenchymal stem cells on gelatin-immobilized poly(L-lactide-co-epsilon-caprolactone) substrates. Biomacromolecules 9(7):1772–1781. https://doi.org/10.1021/bm701410g
57. Sahoo S, Ang LT, Goh JCH, Toh SL (2010) Growth factor delivery through electrospun nanofibers in scaffolds for tissue engineering applications. J Biomed Mater Res A 93a(4):1539–1550. https://doi.org/10.1002/jbm.a.32645
58. Martins A, Gang W, Pinho ED, Rebollar E, Chiussi S, Reis RL, Leon B, Neves NM (2010) Surface modification of a biodegradable composite by UV laser ablation: in vitro biological performance. J Tissue Eng Regen Med 4(6):444–453. https://doi.org/10.1002/term.255
59. Oberbossel G, Probst C, Giampietro VR, von Rohr PR (2017) Plasma afterglow treatment of polymer powders: process parameters, wettability improvement, and aging effects. Plasma Process Polym 14(3):e1600144. https://doi.org/10.1002/ppap.201600144
60. Liu W, Zhan JC, Su Y, Wu T, Wu CC, Ramakrishna S, Mo XM, Al-Deyab SS, El-Newehy M (2014) Effects of plasma treatment to nanofibers on initial cell adhesion and cell morphology. Colloid Surf B 113:101–106. https://doi.org/10.1016/j.colsurfb.2013.08.031
61. Chen F, Tang QL, Zhu YJ, Wang KW, Zhang ML, Zhai WY, Chang JA (2010) Hydroxyapatite nanorods/poly(vinyl pyrolidone) composite nanofibers, arrays and three-dimensional fabrics: electrospun preparation and transformation to hydroxyapatite nanostructures. Acta Biomater 6(8):3013–3020. https://doi.org/10.1016/j.actbio.2010.02.015
62. Puppi D, Piras AM, Chiellini F, Chiellini E, Martins A, Leonor IB, Neves N, Reis R (2011) Optimized electro- and wet-spinning techniques for the production of polymeric fibrous

scaffolds loaded with bisphosphonate and hydroxyapatite. J Tissue Eng Regen Med 5(4):253–263. https://doi.org/10.1002/term.310
63. Zhang YZ, Venugopal JR, El-Turki A, Ramakrishna S, Su B, Lim CT (2008) Electrospun biomimetic nanocomposite nanofibers of hydroxyapatite/chitosan for bone tissue engineering. Biomaterials 29(32):4314–4322. https://doi.org/10.1016/j.biomaterials.2008.07.038
64. Song JH, Kim HE, Kim HW (2008) Electrospun fibrous web of collagen-apatite precipitated nanocomposite for bone regeneration. J Mater Sci Mater Med 19(8):2925–2932. https://doi.org/10.1007/s10856-008-3420-7
65. Phipps MC, Clem WC, Catledge SA, Xu Y, Hennessy KM, Thomas V, Jablonsky MJ, Chowdhury S, Stanishevsky AV, Vohra YK, Bellis SL (2011) Mesenchymal stem cell responses to bone-mimetic electrospun matrices composed of polycaprolactone, collagen I and nanoparticulate hydroxyapatite. PLoS One 6(2):e16813. https://doi.org/10.1371/journal.pone.0016813
66. Xie J, Lou X, Wang X, Yang L, Zhang Y (2015) Electrospun nanofibers of hydroxyapatite/collagen/chitosan promote osteogenic differentiation of the induced pluripotent stem cell-derived mesenchymal stem cells. J Control Release 213:e53. https://doi.org/10.1016/j.jconrel.2015.05.087
67. Zhou Y, Yao H, Wang J, Wang D, Liu Q, Li Z (2015) Greener synthesis of electrospun collagen/hydroxyapatite composite fibers with an excellent microstructure for bone tissue engineering. Int J Nanomed 10:3203–3215. https://doi.org/10.2147/IJN.S79241
68. Kwon GW, Gupta KC, Jung KH, Kang IK (2017) Lamination of microfibrous PLGA fabric by electrospinning a layer of collagen-hydroxyapatite composite nanofibers for bone tissue engineering. Biomater Res 21:11. https://doi.org/10.1186/s40824-017-0097-3
69. Monteiro N, Martins A, Pires R, Faria S, Fonseca NA, Moreira JN, Reisa RL, Neves NM (2014) Immobilization of bioactive factor-loaded liposomes on the surface of electrospun nanofibers targeting tissue engineering. Biomater Sci 2(9):1195–1209. https://doi.org/10.1039/c4bm00069b
70. Zhu Y, Mao ZW, Gao CY (2013) Aminolysis-based surface modification of polyesters for biomedical applications. RSC Adv 3(8):2509–2519. https://doi.org/10.1039/c2ra22358a
71. Oliveira C, Costa-Pinto AR, Reis RL, Martins A, Neves NM (2014) Biofunctional nanofibrous substrate comprising immobilized antibodies and selective binding of autologous growth factors. Biomacromolecules 15(6):2196–2205. https://doi.org/10.1021/bm500346s
72. Piai JF, da Silva MA, Martins A, Torres AB, Faria S, Reis RL, Muniz EC, Neves NM (2017) Chondroitin sulfate immobilization at the surface of electrospun nanofiber meshes for cartilage tissue regeneration approaches. Appl Surf Sci 403:112–125. https://doi.org/10.1016/j.apsusc.2016.12.135
73. Zhu TH, Yu K, Bhutto MA, Guo XR, Shen W, Wang J, Chen WM, El-Hamshary H, Al-Deyab SS, Mo XM (2017) Synthesis of RGD-peptide modified poly(ester-urethane) urea electrospun nanofibers as a potential application for vascular tissue engineering. Chem Eng J 315:177–190. https://doi.org/10.1016/j.cej.2016.12.134
74. Hartman O, Zhang C, Adams EL, Farach-Carson MC, Petrelli NJ, Chase BD, Rabolt JE (2010) Biofunctionalization of electrospun PCL-based scaffolds with perlecan domain IV peptide to create a 3-D pharmacokinetic cancer model. Biomaterials 31(21):5700–5718. https://doi.org/10.1016/j.biomaterials.2010.03.017
75. Kim BR, Nguyen TBL, Min YK, Lee BT (2014) In vitro and in vivo studies of BMP-2-loaded PCL-gelatin-BCP electrospun scaffolds. Tissue Eng Part A 20(23–24):3279–3289. https://doi.org/10.1089/ten.tea.2014.0081
76. Yin LH, Yang SH, He MM, Chang YC, Wang KJ, Zhu YD, Liu YH, Chang YR, Yu ZH (2017) Physicochemical and biological characteristics of BMP-2/IGF-1-loaded three-dimensional coaxial electrospun fibrous membranes for bone defect repair. J Mater Sci Mater Med 28(6). https://doi.org/10.1007/s10856-017-5898-3
77. Niu BJ, Li B, Gu Y, Shen XF, Liu Y, Chen L (2017) In vitro evaluation of electrospun silk fibroin/nano-hydroxyapatite/BMP-2 scaffolds for bone regeneration. J Biomater Sci Polym E 28(3):257–270. https://doi.org/10.1080/09205063.2016.1262163

78. Guex AG, Hegemann D, Giraud MN, Tevaearai HT, Popa AM, Rossi RM, Fortunato G (2014) Covalent immobilisation of VEGF on plasma-coated electrospun scaffolds for tissue engineering applications. Colloid Surf B 123:724–733. https://doi.org/10.1016/j.colsurfb.2014.10.016
79. Wang K, Zhang QY, Zhao LQ, Pan YW, Wang T, Zhi DK, Ma SY, Zhang PX, Zhao TC, Zhang SM, Li W, Zhu MF, Zhu Y, Zhang J, Qiao MQ, Kong DL (2017) Functional modification of electrospun poly(epsilon-caprolactone) vascular grafts with the fusion protein VEGF-HGFI enhanced vascular regeneration. ACS Appl Mater Inter 9(13):11415–11427. https://doi.org/10.1021/acsami.6b16713
80. Lee H, Lim S, Birajdar MS, Lee SH, Park H (2016) Fabrication of FGF-2 immobilized electrospun gelatin nanofibers for tissue engineering. Int J Biol Macromol 93:1559–1566. https://doi.org/10.1016/j.ijbiomac.2016.07.041
81. Cui X, Liu MH, Wang JX, Zhou Y, Xiang Q (2015) Electrospun scaffold containing TGF-beta 1 promotes human mesenchymal stem cell differentiation towards a nucleus pulposus-like phenotype under hypoxia. IET Nanobiotechnol 9(2):76–84. https://doi.org/10.1049/iet-nbt.2014.0006
82. Gomoll AH, Madry H, Knutsen G, van Dijk N, Seil R, Brittberg M, Kon E (2010) The subchondral bone in articular cartilage repair: current problems in the surgical management. Knee Surg Sports Traumatol Arthrosc 18(4):434–447. https://doi.org/10.1007/s00167-010-1072-x
83. Gao JZ, Dennis JE, Solchaga LA, Awadallah AS, Goldberg VM, Caplan AI (2001) Tissue-engineered fabrication of an osteochondral composite graft using rat bone marrow-derived mesenchymal stem cells. Tissue Eng 7(4):363–371. https://doi.org/10.1089/10763270152436427
84. Wei J, Herrler T, Liu K, Han D, Yang M, Dai CC, Li QF (2016) The role of cell seeding, bioscaffolds, and the in vivo microenvironment in the guided generation of osteochondral composite tissue. Tissue Eng Pt A 22(23–24):1337–1347. https://doi.org/10.1089/ten.tea.2016.0186
85. Khorshidi S, Solouk A, Mirzadeh H, Mazinani S, Lagaron JM, Sharifi S, Ramakrishna S (2016) A review of key challenges of electrospun scaffolds for tissue-engineering applications. J Tissue Eng Regen Med 10(9):715–738. https://doi.org/10.1002/term.1978
86. Erisken C, Kalyon DM, Wang HJ (2010) Viscoelastic and biomechanical properties of osteochondral tissue constructs generated from graded Polycaprolactone and Beta-Tricalcium phosphate composites. J Biomech Eng 132(9):091013. https://doi.org/10.1115/1.4001884
87. Liverani L, Roether JA, Nooeaid P, Trombetta M, Schubert DW, Boccaccini AR (2012) Simple fabrication technique for multilayered stratified composite scaffolds suitable for interface tissue engineering. Mater Sci Eng A 557:54–58. https://doi.org/10.1016/j.msea.2012.05.104
88. Yunos DM, Ahmad Z, Salih V, Boccaccini AR (2013) Stratified scaffolds for osteochondral tissue engineering applications: electrospun PDLLA nanofibre coated bioglass (R)-derived foams. J Biomater Appl 27(5):537–551. https://doi.org/10.1177/0885328211414941
89. Mouthuy PA, Ye H, Triffitt J, Oommen G, Cui Z (2010) Physico-chemical characterization of functional electrospun scaffolds for bone and cartilage tissue engineering. Proc Inst Mech Eng H 224(H12):1401–1414. https://doi.org/10.1243/09544119jeim824
90. Erisken C, Kalyon DM, Wang HJ, Ornek-Ballanco C, Xu JH (2011) Osteochondral tissue formation through adipose-derived stromal cell differentiation on biomimetic polycaprolactone nanofibrous scaffolds with graded insulin and beta-glycerophosphate concentrations. Tissue Eng Part A 17(9–10):1239–1252. https://doi.org/10.1089/ten.tea.2009.0693
91. Yang WX, Yang F, Wang YN, Both SK, Jansen JA (2013) In vivo bone generation via the endochondral pathway on three-dimensional electrospun fibers. Acta Biomater 9(1):4505–4512. https://doi.org/10.1016/j.actbio.2012.10.003
92. Liu YY, Yu HC, Liu Y, Liang G, Zhang T, Hu QX (2016) Dual drug spatiotemporal release from functional gradient scaffolds prepared using 3D bioprinting and electrospinning. Polym Eng Sci 56(2):170–177. https://doi.org/10.1002/pen.24239

# Chapter 12
# Supercritical Fluid Technology as a Tool to Prepare Gradient Multifunctional Architectures Towards Regeneration of Osteochondral Injuries

Ana Rita C. Duarte, Vitor E. Santo, Manuela E. Gomes, and Rui L. Reis

**Abstract** Platelet lysates (PLs) are a natural source of growth factors (GFs) known for its stimulatory role on stem cells which can be obtained after activation of platelets from blood plasma. The possibility to use PLs as growth factor source for tissue healing and regeneration has been pursued following different strategies. Platelet lysates are an enriched pool of growth factors which can be used as either a GFs source or as a three-dimensional (3D) hydrogel. However, most of current PLs-based hydrogels lack stability, exhibiting significant shrinking behavior. This chapter focuses on the application of supercritical fluid technology to develop three-dimensional architectures of PL constructs, crosslinked with genipin. The proposed technology allows in a single step operation the development of mechanically stable porous structures, through chemical crosslinking of the growth factors present in the PL pool, followed by supercritical drying of the samples. Furthermore gradient structures of PL-based structures with bioactive glass are also presented and are described as an interesting approach to the treatment of osteochondral defects.

**Keywords** Supercritical fluid technology · Platelet lysate · Genipin · Polymerization · Gradient structures

---

A. R. C. Duarte (✉) · V. E. Santo · M. E. Gomes · R. L. Reis
3B's Research Group—Biomaterials, Biodegradables and Biomimetics, European Institute of Excellence on Tissue Engineering and Regenerative Medicine, University of Minho, Barco/Guimarães, Portugal

ICVS/3B's—PT Government Associate Laboratory, Braga/Guimarães, Portugal
e-mail: aduarte@dep.uminho.pt

© Springer International Publishing AG, part of Springer Nature 2018
J. M. Oliveira et al. (eds.), *Osteochondral Tissue Engineering*,
Advances in Experimental Medicine and Biology 1058,
https://doi.org/10.1007/978-3-319-76711-6_12

## 12.1 Introduction

Full thickness chondral and osteochondral defects and early osteoarthritis represent one of the most significant challenges facing the global health care community due to the limited healing potential of articular cartilage, resulting in chronic degeneration. Osteochondral tissue is a gradual transition from cartilage to bone in which the key constituents of each tissue undergo an exchange in predominance. Structurally, the osteochondral interface is the connection between a layer of hyaline cartilage and underlying bone and it is crucial for load transfer between bone and cartilage [1, 2]. The most commonly used standard therapies for treating articular cartilage defects are generally successful for pain relief and improved function but do not restore the articular cartilage and subchondral bone completely, leading to degeneration over time [3]. Tissue Engineering (TE) offers an alternative approach to the current treatments, aiming at the regeneration of tissues through the use of cells within a supporting matrix that may also incorporate biomolecules (e.g., growth factors—GFs) that enhance cell function and/or tissue regeneration [4].

The concept of osteochondral TE, a hybrid of bone and cartilage regeneration, has attracted considerable attention, particularly as a technique for promoting superior cartilage integration and as a treatment for osteochondral defects [5]. The engineering of complex tissues, which involve multiple cell types organized in specific patterns, such as the orthopedic interface [6], is still a rather challenging task and it is recognized that the effective regeneration of these tissues has not been fully attained. For osteochondral scaffolds, additional design criteria should be considered to achieve the best possible simultaneous growth of the two independent tissues involved. A single scaffold with a homogeneous structure may not be the ideal support for such applications and therefore a bilayered scaffold combining parts with differing physical and chemical properties might be the most suitable approach to promote the simultaneous individual growth of cartilage and bone on a single integrated implant. This may require the use of biphasic constructs with areas with distinct mechanical, structural, and molecular properties [7, 8]. Some studies have focused on the development of two layers with different architecture and mechanical properties [9, 10], whereas others have focused more on the delivery of bioactive agents distributed in the structure with a concentration gradient [8], to promote the formation of a calcified matrix on the bone side of the interface and a cartilage-type extracellular matrix (ECM) on the opposite side of the interface. Natural and synthetic polymers have been used to produce these scaffolds but another interesting material that can be used to create new cell-laden structures comprises the use of Platelet Lysate (PL). In the past three decades, the increasing knowledge on the physiological roles of platelets in wound healing and tissue injury suggests the potential of using platelets as therapeutic tools [11]. Platelets are anucleate cytoplasmic fragments which form an intracellular storage pool of proteins vital to wound healing. Upon activation of platelets, lysis and subsequent release of several GFs occurs, naturally constituting a pool of bioactive factors at physiological concentrations. This protein concentrate is known to play key roles in tissue healing and stimulation of cell expansion and recruitment. Simultaneously, it can also form a tridimensional network amenable to act as support for cell culture. Previous research studies have reported the development of PL-based scaffolds/hydrogels; however, they typically show high levels of shrinkage and lack long-term stability, requiring combination with other materials to acquire mechanical integrity [12].

In this chapter, we explore a new TE approach targeting the treatment of osteochondral defects. The strategy proposed consists in the combination of a novel autologous PL-based gradient scaffold to promote the regeneration of the orthopedic interface using supercritical fluid technology. Our aim was the development of a PL-based gradient scaffold crosslinked with genipin. The bulk component is based on the genipin-crosslinked PL network produced by supercritical fluid-assisted process. One of the sides of the scaffold (the one that faces the bone layer) is enriched with calcium phosphates (CaP) or Bioglass® to enhance osteoinductivity and mechanical properties. The other side (facing cartilage) is void of ceramic component. Although the use of PL in tissue regeneration strategies is not new, this project presents the novelty of simultaneously using this blood derivative as gradient-inducing scaffolding system for the osteochondral defect, and simultaneously as a source of bioactive proteins to promote the regeneration of the tissues.

Supercritical fluid technology has already proven to be feasible for many pharmaceutical applications and it has also been demonstrated to be a valid alternative to conventional processes for the preparation of tridimensional scaffolds due to its mild processing parameters [13–16].

## 12.2 Supercritical Fluid Technology

The development of 3D architectures for TE and regenerative medicine using supercritical fluid technology can take advantage of different properties of a supercritical fluid and several techniques have already been reported. The development of matrices for the regeneration of osteochondral defects explored in this chapter relies on the principles of the supercritical assisted phase inversion. In this method, two mechanisms occur at the same time: (1) the diffusion of the supercritical fluid into the protein solution, which results on the precipitation of the proteins from the aqueous solution; and (2) the removal of the solvent by the supercritical fluid. The phase separation and precipitation of the proteins to form a porous scaffold is favored by a high solubility between the solvent and the anti-solvent, i.e., a higher affinity of the solvent to the carbon dioxide [17]. The mutual affinity of water and carbon dioxide is very low due to their opposite polarity. To improve polarity of carbon dioxide, a small amount of an entrainer or cosolvent can be mixed with the gas. This can, in some cases, produce dramatic effects on the solvent power, greatly enhancing the solubility and/or affinity between two components [18]. Accordingly, it has been described in the literature, the successful preparation of chitosan membranes from dilute acetic acid aqueous solutions by supercritical assisted phase inversion [19]. On another hand, organic solvents have been reported to promote protein precipitation, at moderate concentration, without affecting the functional properties of the proteins. Ethanol is a particularly interesting solvent and previous studies have demonstrated that processing proteins in the presence of small amounts of ethanol does not compromise protein activity. In particular, we have reported the development of 3D architectures of PDLLA scaffolds with PL-loaded nanoparticles, in the presence of small amounts of ethanol. The results obtained suggest that the activity of the GFs was not compromised and the ability to guide stem cell differentiation was maintained [13, 16].

## 12.3 Development of PL-Based 3D Multifunctional Architectures

Osteochondral TE has shown an increasing development and investment towards the design of suitable strategies to stimulate the regeneration of damaged cartilage and underlying subchondral bone tissue [20–23]. The use of two scaffolds with specific properties for bone and cartilage architectures, combined at the time of implantation as a multilayered structure was one of the first approaches for regeneration of large osteochondral defects. New design approaches have been proposed, including the use of bilayered scaffolds with distinct properties in each side of the construct, and the use of continuous gradient scaffolds. Different gradient structures have been proposed, focusing either on ranges of morphological features, mechanical properties, presentation of bioactive molecules, or combinations of these [8–10, 24, 25].

Most of these structures have been designed using natural and/or synthetic polymers. Due to its bioactivity and structural potential, PL also arises as potential bulk material for the generation of novel scaffolds for osteochondral regeneration. The pool of GFs present in PL comprises several of the signalling molecules known to induce cell expansion and migration. Moreover, some of them are also known to stimulate osteogenic and chondrogenic differentiation of human stem cells [13, 26]. Moreover, PL can be obtained from the patient's own blood, allowing us to pursue an autologous approach and eliminating immunogenic and disease transmission concerns [11]. Most of the strategies involving PL for the generation of 3D constructs either use polymeric networks absorbed and functionalized with PL [27, 28] or calcium-activated PL hydrogels for cell encapsulation. Hydrogels using PL as bulk material typically show fast degradation and relative instability, thus limiting their application potential [12]. Moreover, their mechanical properties are also poor and tend to be suitable for regeneration of soft tissues [11]. The osteochondral region of joints is a particularly mechanically demanding microenvironment and requires the application of biomaterials capable of withstanding an array of physical stimuli [1, 29–31]. To our knowledge, there is not published data regarding the use of PL as bulk material for the development of advanced scaffolds for osteochondral regeneration.

12 Supercritical Fluid Technology as a Tool to Prepare Gradient Multifunctional... 269

**Stage 1**
- $CO_2$ promoted the crosslinking reaction
- Ethanol promoted the precipitation of the proteins, which were crosslinked with genipin

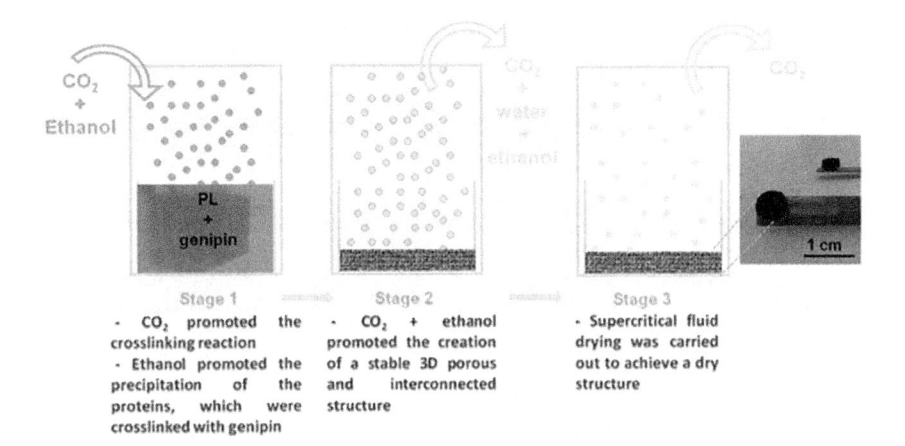

**Stage 2**
- $CO_2$ + ethanol promoted the creation of a stable 3D porous and interconnected structure

**Stage 3**
- Supercritical fluid drying was carried out to achieve a dry structure

**Fig. 12.1** Schematic representation of the supercritical fluid technique designed for the preparation of PL-based 3D architectures

The technique herein reported leads to the generation of a gradient scaffold, produced in a single step procedure, in which the simultaneous precipitation of PL proteins and the establishment of crosslinking bridges between the amino groups of the proteins and genipin takes place. Figure 12.1 presents a summary of the process.

One of the advantages of the use of supercritical fluid technology is the fact that single unit operations can be designed, avoiding the need for several subsequent steps of production. As a result, a 3D porous scaffold is obtained. Briefly, PL mixed with the crosslinker agent (genipin) is placed inside a high-pressure vessel, heated to the desired temperature (35–40 °C), and pressurized with carbon dioxide until the set pressure of the experiment (80–140 bar). The crosslinking reaction is allowed to take place for a pre-determined period of time and afterwards a stream of ethanol + $CO_2$ passes through the vessel in order to promote phase inversion and extract the aqueous solution. Finally, the high-pressure chamber is flushed by adding fresh $CO_2$ under the same conditions in order to extract the residual ethanol and the system is slowly depressurized.

Genipin is a naturally occurring crosslinking agent which was chosen, particularly due to its low cytotoxicity and to the ability to bind proteins or amino-acids between adjacent amino groups, forming blue pigments [32]. PL, on the other hand, is a protein concentrate, thus offering a high number of crosslinking sites to interact with the ester groups of genipin, leading to the formation of secondary amide linkages. The identification of specific proteins involved in the crosslinking reaction is not straightforward due to the enriched composition of PL. It has been reported that the mechanisms of crosslinking reactions with genipin are different at different pH values [33–35]. Among the studies documented, it has been suggested that the acid catalysis is necessary for the crosslinking reaction to occur. The advantages of the crosslinking reaction under dense carbon dioxide atmosphere are related with the acidification of aqueous solutions after the solubilization of $CO_2$ molecules, which,

| Sample | PL LXL | PL MXL | PL HXL |
|---|---|---|---|
| SEM |  |  | 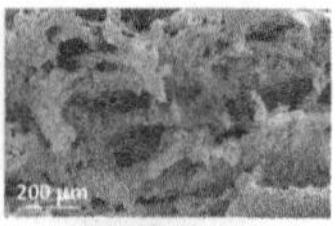 |
| Micro-CT | | | |
| 2D sections | | | |
| Porosity (%) | 17.3 | 14.2 | 12.2 |
| Interconnectivity (%) | 32.3 | 36.4 | 51.6 |
| Mean pore size (µm) | 90.7 | 93.6 | 85.9 |

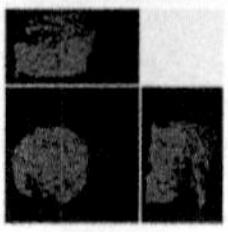

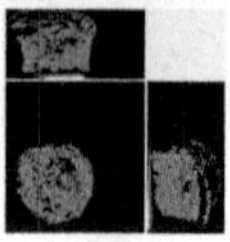

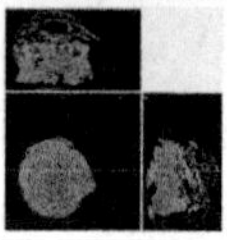

**Fig. 12.2** Morphological characterization of the scaffolds prepared with different genipin concentrations: SEM and micro-CT 2D micrographs of cross sections of the PL scaffolds

depending on the operating conditions, can reach pH values up to 3. Previous work reported in the literature has shown though that the activity of the proteins is not compromised by the low pH at which the process takes place [36]. In fact, we have previously developed genipin-crosslinked PL membranes for cell culture applications, with reduced toxicity [37].

The production of PL scaffolds was pursued by using three different genipin concentrations. Genipin was dissolved in PL suspensions at final concentration range of 0.18–0.25% (w/v) before initiation of the supercritical fluid process. The optimal reaction time was evaluated and it was found to be 1 h. Three conditions were defined according to their classification of crosslinking degree: low ($PL_{LXL}$), medium ($PL_{MXL}$), and high crosslinking degree ($PL_{HXL}$). Morphological analysis of the three categories of architectures was performed by Scanning Electron Microscopy (SEM) and micro-Computed Tomography (micro-CT). Qualitative and quantitative outputs extracted from these characterization methods are shown in Fig. 12.2.

The morphological analysis of the scaffolds provided the first hints on their feasibility to act as structural support for growth and proliferation of seeded cells. The crosslinking degree showed influence on the morphological properties of the 3D structures, as it can be seen from the quantitative measurement of % of porosity, interconnectivity, and mean pore size. Mean pore size ranged between 85.9 and 93.6 µm, typically considered an appropriate dimension for cell migration through the pore. However, the obtained structures were highly compact, with porosity levels ranging from 12.2 to 17.3%, from the highest crosslinked scaffold to the lowest, respectively. This indicates the formation of highly packed architectures, which could be confirmed by SEM micrographs of cross sections of these PL scaffolds. Moreover, interconnectivity levels were also low, curiously being the lowest for the $PL_{LXL}$ condition (32.3%) and the highest for $PL_{HXL}$ (51.6%). These levels of porosity and interconnectivity could discourage further studies; however, we had evidence from previous studies from our laboratory that even PL-enriched structures with

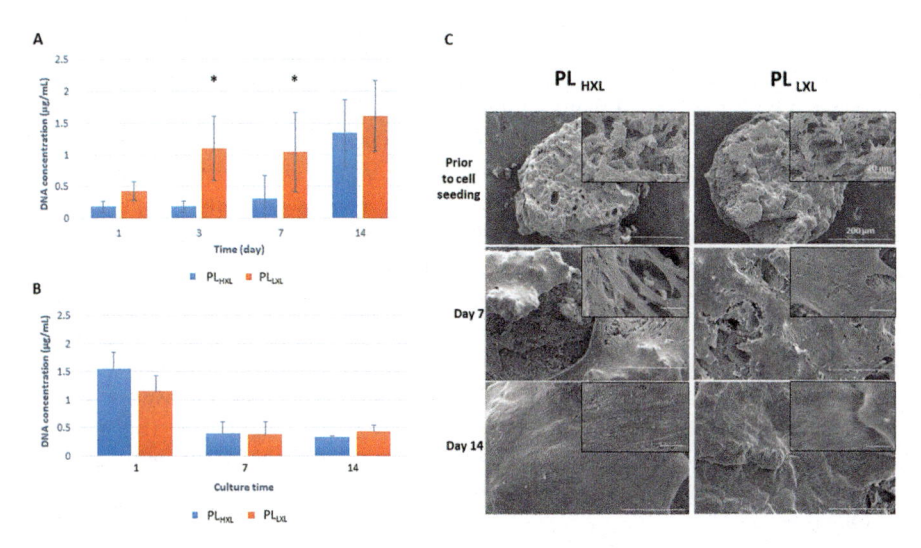

**Fig. 12.3** Cell adhesion and proliferation in PL$_{LXL}$ and PL$_{HXL}$ scaffolds. (**a**) Measurement of DNA concentration of constructs seeded with ATDC5 cells; (**b**) Measurement of DNA concentration of constructs seeded with hASCs; (**c**) SEM micrographs of PL$_{LXL}$ and PL$_{HXL}$ scaffolds prior and after seeding with hASCs. * Represents a statistically significant difference ($p<0.05$) between DNA concentration levels of PL$_{LXL}$ and PL$_{HXL}$ scaffolds

poor initial porosity could act as templates for cell culture. The progressive dissolution of the PL bulk matrix leads to the cumulative formation of pores within the structure, enabling cell migration and colonization of the inner regions of the construct [38]. At the same time, we also expected a delayed dissolution profile due to the covalent reinforcement induced by genipin crosslinking.

PL$_{LXL}$ and PL$_{HXL}$ scaffold formulations were selected to perform the cellular studies. Figure 12.3 depicts the results from the cell seeding of chondrocyte cell line—ATDC5 (Fig. 12.3a) and human adipose-derived stromal cells—hASCs (Fig. 12.3b, c).

The response of cells to the developed PL-based scaffolds was evaluated by overall DNA quantification along culture period and by characterization of cellular morphology and organization during time. Two different cell types were used: one chondrocytic cell line (ATDC5) and human mesenchymal stem cells isolated from adult adipose-derived tissue stromal cells. Figure 12.3a shows a progressive increase of DNA concentration in ATDC5 cultures for both conditions, indicating that PL scaffolds act as templates for cell proliferation. At days 3 and 7 of culture, PL$_{LXL}$ showed significantly higher DNA levels, thus indicating an enhanced cellular proliferation for the lowest crosslinked structures. At day 14, there were no significant differences in cumulative cell number between the scaffolds with distinct degrees of crosslinking.

Figure 12.3b reports DNA concentration of hASCs up to 14 days of culture. For this condition, there were no differences between the highest and the lowest crosslinked structures. At day 1, DNA concentration was the highest for these cultures, decreasing during the first week and remaining constant up to 14 days of *ex vivo* culture. These cells were cultured with chondrogenic differentiation medium supplemented with the typical factors for stimulation of in vitro chondrogenesis, with the exception of Transforming Growth Factor-β (TGF-β), which was available as a component of the PL concentrate. Therefore, the lack of cell expansion throughout the 14 days of culture period was not surprising, as cells were more committed to differentiation processes rather than cell division. Figure 12.3c provides a morphological and qualitative characterization of cell adhesion, proliferation, and colonization of the PL-based constructs. It was possible to observe that hASCs managed to adhere to the surface of the scaffolds, presenting the typical elongated morphology of these cells when adhered onto a substrate. At day 14, a confluent layer of cells was evident for both scaffold formulations, thus confirming that cells were proliferating and that the material was not inducing cytotoxicity to the adhered cells. From these results, both high- and low-crosslinked PL scaffolds were viable templates for cell culture and could be further explored for osteochondral tissue engineering applications.

## 12.4 Development of Gradient PL-Based Architectures for Osteochondral Tissue Regeneration Applications

The main challenge in developing a functional osteochondral implant is related with the different features required for each region of the defect. While the scaffold region exposed to the cartilage side should possess lower mechanical properties, the component that aims to regenerate the subchondral bone requires strong mechanical properties and mineralization capacity, which may not be achieved solely by the crosslinking of PL with genipin. Bioactivity of the structure can be enhanced by the presence of ceramics presenting inherent osteoinductive properties. In this sense, the preparation of biodegradable composites containing hydroxyapatite-based calcium phosphates (CaP) or Bioglass® (BG) is a viable complement to the construct design.

Taking this in consideration, we have selected the condition $PL_{HXL}$ for the generation of gradient scaffolds, due to its superior interconnectivity levels. The incorporation of the inorganic material in the 3D structures only required one additional step to the protocol already described for the preparation of PL scaffolds, and consisted in the dispersion of BG particles within the PL suspension prior to the supercritical fluid processing steps. The incorporation of inorganic material in the 3D structures did not compromise the generation of the structures and led to the natural formation of a bilayered architecture: one BG-enriched region and other BG-poor phase (as presented in Fig. 12.4).

Significant structural changes were observed regarding porosity (53%), interconnectivity (78%), and average pore size (177.9 μm) in the BG-enriched region. The

**Fig. 12.4** SEM micrograph of a cross section of the PL$_{HXL}$-BG scaffold, clearly depicting two distinct regions: one BG-enriched layer on the lower section of the scaffold and one BG-poor region on the upper part of the architecture

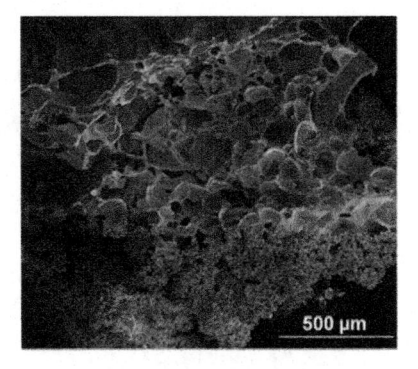

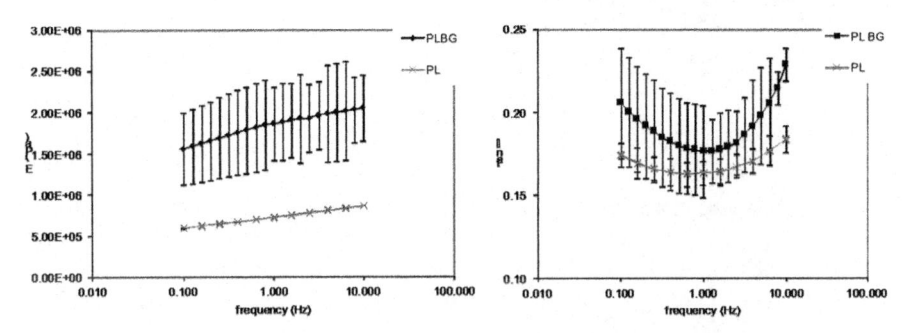

**Fig. 12.5** Mechanical properties of the PL$_{HXL}$-BG and PL$_{HXL}$ scaffolds: Storage modulus ($E'$) curves as function of frequency and loss factor (tan $\delta$) curves as function of frequency

BG-poor layer also presented significant changes in comparison with the PL$_{HXL}$ scaffolds produced in the absence of BG particles.

The impact of incorporation of BG particles and architectural changes of the scaffold on the overall mechanical performance of the constructs was evaluated by Dynamic Mechanical Analysis (DMA). Mechanical analysis of the scaffolds was performed in dynamic compression mode on hydrated samples using DMA measurements. Figure 12.5 shows the isothermal response of the various samples as a function of frequency in terms of storage modulus ($E'$) and the loss factor (tan $\delta$).

The presence of BG particles in the 3D scaffold led to improved mechanical properties, particularly on the storage modulus, throughout the range of frequencies of compression forces imposed in the scaffolds. Whereas for PL$_{HXL}$ scaffolds, the elastic modulus was found between 0.55 and 0.8 MPa, the elastic modulus of PL$_{HXL}$-BG scaffolds was comprised between 1.5 and 2 MPa. This was an expected observation and it has been reported in previous studies as BG particles may also act as reinforcement fillers enhancing the mechanical properties of the 3D structures in which they are dispersed [39, 40].

The following step was focused on the evaluation of the bioactivity of the PL$_{HXL}$ and PL$_{HXL}$ BG samples, thus validating the influence of BG addition on the deposition of mineralized matrix. To attain this, the scaffolds were immersed in a simulated

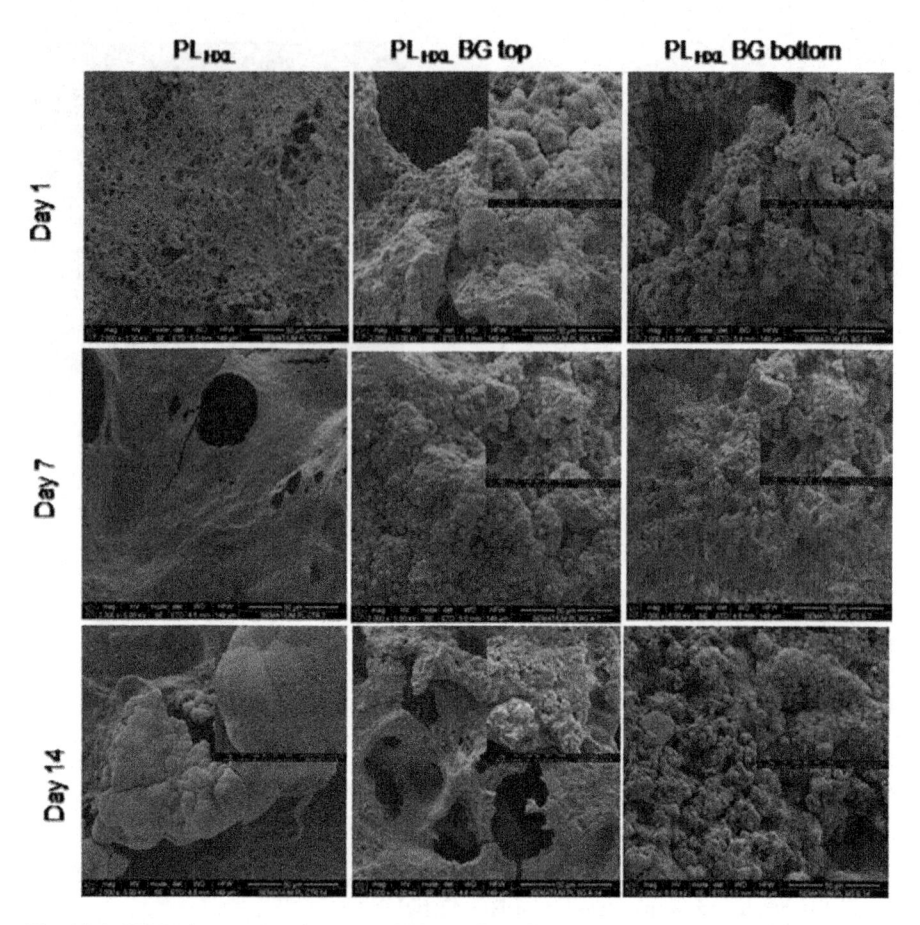

**Fig. 12.6** SEM micrographs of PL$_{HXL}$ and PL$_{HX}$-BG scaffolds after immersion in SBF solution for 1, 7, and 14 days

body fluid (SBF) solution and the formation of CaP crystals was followed after 1, 7, and 14 days. Figure 12.6 shows the SEM micrographs of the scaffolds after immersion in SBF. The PL$_{HXL}$-BG were analyzed in the BG-rich zone (bottom) and BG-poor zone (top).

We could observe that the PL$_{HXL}$ scaffolds were bioactive per se, although the presence of BG accelerated the nucleation and growth of an apatite layer. The typical cauliflower-like crystals, characteristic of hydroxyapatite (HA), could be detected in the PL$_{HXL}$-BG scaffolds after 7 days of immersion in SBF. These findings were checked by infrared spectroscopy, which confirmed the chemical nature of the formed crystals (Fig. 12.7).

The strong FTIR bands characteristic of phosphate and carbonate in hydroxyapatite crystals, namely $v_3$-PO$_4$ at 1040 cm$^{-1}$ and $v_3$-CO$_3$ in the region 1400–1550 cm$^{-1}$, can be observed in the spectra of PL$_{HXL}$ and PL$_{HXL}$ BG, with greater intensity on the PL$_{HXL}$ BG. In the PL scaffolds loaded with BG, it is also possible to identify other

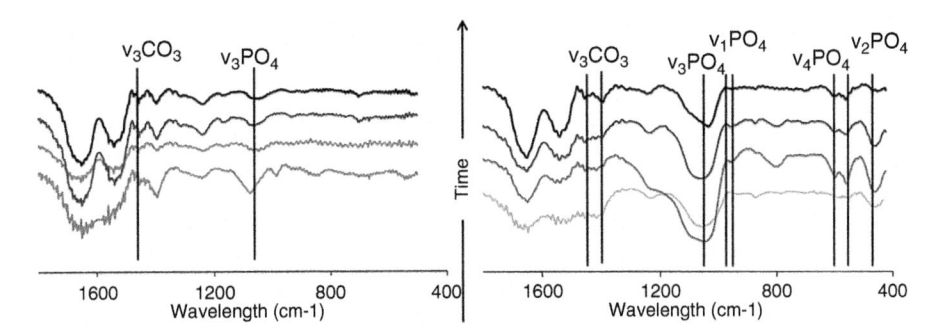

**Fig. 12.7** FTIR analysis of the scaffolds immersed in SBF solution. FTIR spectra on the left is representative of the PL$_{HXL}$ scaffold. The spectra on the right is representative of the PL$_{HXL}$-BG scaffolds

spectral characteristic bands of hydroxyapatite [41], such as $\nu_1$-PO$_4$ at 962 cm$^{-1}$, $\nu_2$-PO$_4$ at 472 cm$^{-1}$, and $\nu_4$-PO$_4$ at 575 and 561 cm$^{-1}$.

The presence of BG particles was shown to be important to tune the rate of formation of the HA layer on the surface of the scaffolds and to be able to synchronize it with the sequence of cellular changes that take place upon new tissue formation.

## 12.5 Conclusions

The development of mechanically stable 3D architectures based on PL as bulk material with potential application in TE and regenerative medicine was discussed in this chapter. Herein, we report a new methodology based on supercritical fluid technology for the development of a stable PL-based scaffold crosslinked with genipin. PL is interesting source of GFs which can be obtained after activation of platelets from the patient's blood plasma, therefore reinforcing the potential on the development of autologous scaffolding architectures based on this protein concentrate. The elimination of immunogenic and disease transmission concerns due to its autologous origin are important advantages on the use of these materials.

There is an interest in developing GF-loaded scaffolds for TE strategies, aiming to boost the regeneration capacity of these materials. Nevertheless, several conventional methodologies for scaffold production and GF immobilization require the use of organic solvents and/or high processing temperatures, which hampers the preparation of scaffolds loaded with active GFs. In order to overcome these limitations, carbon dioxide (CO$_2$) has been used as an agent to form 3D scaffolds.

Our methodology enabled to produce a stable PL scaffold without the addition of an extra polymeric matrix, overcoming the traditional lack of stability and quick shrinkage of 3D hydrogels and scaffolds based on PL. The PL scaffolds crosslinked with genipin were characterized in terms of their morphological, mechanical, chemical, and biological performance by different techniques. Furthermore, the dispersion of bioactive particles of BG within the 3D architecture enhanced the mechanical properties of the scaffold and promoted the growth of a calcium phosphate layer on the surface, similar to HA present in the bone. Moreover, it enabled the automatic formation of a bilayered structure, composed by one BG-enriched region and one BG-poor side. This functionally graded scaffold is suitable for application in osteochondral tissue engineering, as it could be placed in contact with cartilage and subchondral bone tissue. Cellular studies with our proposed structures were also promising, demonstrating that PL scaffolds were suitable templates for cell adhesion, proliferation, and colonization of the 3D structure.

The development of PL-based bilayered scaffolds for osteochondral tissue engineering, with the dual role of acting as a structural template and GFs source for the interface regeneration, represents a major advance in the state of the art in this field.

**Acknowledgements** The research leading to these results has received funding from the project *"Accelerating tissue engineering and personalized medicine discoveries by the integration of key enabling nanotechnologies, marine-derived biomaterials and stem cells,"* supported by Norte Portugal Regional Operational Programme (NORTE 2020), under the PORTUGAL 2020 Partnership Agreement, through the European Regional Development Fund (ERDF).

# References

1. Yang PJ, Temenoff JS (2009) Engineering orthopedic tissue interfaces. Tissue Eng. Part B Rev. 15:127–141
2. Ahmed TAE, Hincke MT (2010) Strategies for articular cartilage lesion repair and functional restoration. Tissue Eng. Part B. Rev. 16:305–329
3. Lefebvre V, Smits P (2005) Transcriptional control of chondrocyte fate and differentiation. Birth Defects Res Part C Embryo Today Rev 75:200–212
4. Malafaya PB, Silva GA, Reis RL (2007) Natural-origin polymers as carriers and scaffolds for biomolecules and cell delivery in tissue engineering applications. Adv Drug Deliv Rev 59:207–233
5. O'Shea TM, Miao X (2008) Bilayered scaffolds for osteochondral tissue engineering. Tissue Eng. Part B. Rev. 14:447–464
6. Chen J et al (2011) Simultaneous regeneration of articular cartilage and subchondral bone in vivo using MSCs induced by a spatially controlled gene delivery system in bilayered integrated scaffolds. Biomaterials 32:4793–4805
7. Mano JF, Reis RL (2007) Osteochondral defects: present situation and tissue engineering approaches. J. Tissue Eng. Regen. Med. 1:261–273
8. Chen FM, Zhang M, Wu ZF (2010) Toward delivery of multiple growth factors in tissue engineering. Biomaterials 31:6279–6308
9. Kon E, Mutini A, Arcangeli E, Delcogliano M, Filardo G, Nicoli Aldini N, Pressato D, Quarto R, Zaffagnini S, Marcacci M (2008) Novel nanostructured scaffold for osteochondral regeneration: pilot study in horses. J. Tissue Eng. Regen. Med. 2:408–417
10. Lu HH, Subramony SD, Boushell MK, Zhang X (2010) Tissue engineering strategies for the regeneration of orthopedic interfaces. Ann. Biomed. Eng. 38:2142–2154
11. Anitua E et al (2006) New insights into and novel applications for platelet-rich fibrin therapies. Trends Biotechnol 24:227–234
12. Haberhauer M et al (2008) Cartilage tissue engineering in plasma and whole blood scaffolds. Adv. Mater. 20:2061–2067

13. Santo VE et al (2012) Enhancement of osteogenic differentiation of human adipose derived stem cells by the controlled release of platelet lysates from hybrid scaffolds produced by supercritical fluid foaming. J. Control. Release 162:19–27
14. Duarte a RC, Mano JF, Reis RL (2009) Perspectives on: supercritical fluid technology for 3D tissue engineering scaffold applications. J. Bioact. Compat. Polym. 24:385–400
15. Duarte ARC et al (2013) Unleashing the potential of supercritical fluids for polymer processing in tissue engineering and regenerative medicine. J. Supercrit. Fluids 79:177–185
16. Santo VE, Duarte ARC, Gomes ME, Mano JF, Reis RL (2010) Hybrid 3D structure of poly(d,l-lactic acid) loaded with chitosan/chondroitin sulfate nanoparticles to be used as carriers for biomacromolecules in tissue engineering. J Supercrit Fluids 54:320–327
17. van de Witte P, Dijkstra PJJ, van den Berg JWA, Feijen J (1996) Phase separation processes in polymer solutions in relation to membrane formation. J. Memb. Sci. 117:1–31
18. Eckert CA, Knutson BL, Debenedetti PG (1996) Supercritical fluids as solvents for chemical and materials processing. Nature 383:313–318
19. Temtem M et al (2009) Supercritical CO2 generating chitosan devices with controlled morphology. Potential application for drug delivery and mesenchymal stem cell culture. J. Supercrit. Fluids 48:269–277
20. Keeney M, Pandit A (2009) The osteochondral junction and its repair via bi-phasic tissue engineering scaffolds. Tissue Eng. Part B. Rev. 15:55–73
21. Gadjanski I, Vunjak-Novakovic G (2015) Challenges in engineering osteochondral tissue grafts with hierarchical structures. Expert Opin. Biol. Ther. 2598:1–17
22. Nukavarapu SP, Dorcemus DL (2012) Osteochondral tissue engineering: current strategies and challenges. Biotechnol Adv. https://doi.org/10.1016/j.biotechadv.2012.11.004
23. Canadas RF, Marques AP, Reis RL, Oliveira JM (2017) In: Oliveira JM, Reis RL (eds) Regenerative strategies for the treatment of knee joint disabilities. Springer International, Berlin, pp 213–233. https://doi.org/10.1007/978-3-319-44785-8_11
24. Yan LP et al (2015) Bilayered silk/silk-nanoCaP scaffolds for osteochondral tissue engineering: In vitro and in vivo assessment of biological performance. Acta Biomater. 12:227–241
25. Yan LP, Oliveira JM, Oliveira AL, Reis RL (2013) Silk fibroin/nano-CaP bilayered scaffolds for osteochondral tissue engineering. Key Eng. Mater. 587:245–248
26. Zaky SH, Ottonello A, Strada P, Cancedda R, Mastrogiacomo M (2008) Platelet lysate favours in vitro expansion of human bone marrow stromal cells for bone and cartilage engineering. J. Tissue Eng. Regen. Med. 2:472–481
27. Santo VE et al (2016) Engineering enriched microenvironments with gradients of platelet lysate in hydrogel fibers. Biomacromolecules 17:1985–1997
28. Babo PS et al (2016) Assessment of bone healing ability of calcium phosphate cements loaded with platelet lysate in rat calvarial defects. J. Biomater. Appl. 31:637–649
29. Yan L, Oliveira JM, Oliveira AL, Reis RL (2015) Current concepts and challenges in osteochondral tissue engineering and regenerative medicine. ACS Biomater. Sci. Eng. 1(4):150220124046001. https://doi.org/10.1021/ab500038y
30. Ribeiro V, Pina S, Oliveira JM, Reis RL (2017) In: Oliveira JM, Reis RL (eds) Regenerative strategies for the treatment of knee joint disabilities. Springer International, Berlin, pp 129–146. https://doi.org/10.1007/978-3-319-44785-8_7
31. Cengiz IF, Oliveira JM, Reis RL (2014) In: Magnenat-Thalmann N, Ratib O, Choi HF (eds) 3D multiscale physiological human. Springer, London, pp 25–47. https://doi.org/10.1007/978-1-4471-6275-9_2
32. Wang C, Lau TT, Loh WL, Su K, Wang D (2011) Cytocompatibility study of a natural biomaterial crosslinker—Genipin with therapeutic model cells. J Biomed Mater Res B Appl Biomater 97:58–65. https://doi.org/10.1002/jbm.b.31786
33. Muzzarelli RAA (2009) Genipin-crosslinked chitosan hydrogels as biomedical and pharmaceutical aids. Carbohydr. Polym. 77:1–9

34. Butler MF, Ng Y, Pudney PDA (2003) Mechanism and kinetics of the crosslinking reaction between biopolymers containing primary amine groups and Genipin. J Polym Sci Part A Polym Chem 41:3941–3953
35. Mu C, Zhang K, Lin W, Li D (2012) Ring-opening polymerization of genipin and its long-range crosslinking effect on collagen hydrogel. J Biomed Mater Res A 101:385–393. https://doi.org/10.1002/jbm.a.34338
36. Chuang M, Johannsen M (2009) Characterization of pH in aqueous CO2—systems. Polym Degrad Stab 97(6):839–848
37. Babo P et al (2014) Platelet lysate membranes as new autologous templates for tissue engineering applications. Inflamm. Regen. 34:033–044
38. Santo VE, Popa EG, Mano JF, Gomes ME, Reis RL (2015) Natural assembly of platelet lysate-loaded nanocarriers into enriched 3D hydrogels for cartilage regeneration. Acta Biomater. 19:56–65
39. Duarte ARC, Caridade SG, Mano JF, Reis RL (2009) Processing of novel bioactive polymeric matrixes for tissue engineering using supercritical fluid technology. Mater. Sci. Eng. C 29:2110–2115
40. Rezwan K, Chen QZ, Blaker JJ, Boccaccini AR (2006) Biodegradable and bioactive porous polymer/inorganic composite scaffolds for bone tissue engineering. Biomaterials 27:3413–3431
41. Rey C, Combes C (2016) Biomineralization and biomaterials. pp 95–127. https://doi.org/10.1016/B978-1-78242-338-6.00004-1

# Part V
# Hydrogels Systems for Osteochondral Tissue Applications

# Chapter 13
# Gellan Gum-Based Hydrogels for Osteochondral Repair

Lígia Costa, Joana Silva-Correia, J. Miguel Oliveira, and Rui L. Reis

**Abstract** Gellan gum (GG) is a widely explored natural polysaccharide that has been gaining attention in tissue engineering (TE) and regenerative medicine field, and more recently in osteochondral TE approaches. Taking advantage of its inherent features such as biocompatibility, biodegradability, similarity with the extracellular matrix and easy functionalization, GG-based hydrogels have been studied for their potential for cartilage and bone tissue regeneration. Several preclinical studies describe the successful outcome of GG in cartilage tissue engineering. By its turn, GG composites have also been proposed in several strategies to guide bone formation. The big challenge in osteochondral TE approaches is still to achieve cartilage and bone regeneration simultaneously through a unique integrated bifunctional construct. The potential of GG to be used as polymeric support to reach both bone and cartilage regeneration has been demonstrated. This chapter provides an overview of GG properties and the functionalization strategies employed to tailor its behaviour to a particular application. The use of GG in soft and hard tissues regeneration approaches, as well in osteochondral integrated TE strategies is also revised.

**Keywords** Gellan gum · Osteochondral · Cartilage · Bone · Bilayered scaffolds

---

L. Costa · J. Silva-Correia
3Bs Research Group—Biomaterials, Biodegradables and Biomimetics, European Institute of Excellence on Tissue Engineering and Regenerative Medicine, University of Minho, Barco GMR, Portugal

ICVS/3B's—PT Government Associate Laboratory, Braga/Guimarães, Portugal

J. M. Oliveira (✉) · R. L. Reis
3Bs Research Group—Biomaterials, Biodegradables and Biomimetics, European Institute of Excellence on Tissue Engineering and Regenerative Medicine, University of Minho, Barco GMR, Portugal

ICVS/3B's—PT Government Associate Laboratory, Braga/Guimarães, Portugal

The Discoveries Centre for Regenerative and Precision Medicine, Headquarters at University of Minho, Avepark, 4805-017 Barco, Guimarães, Portugal
e-mail: miguel.oliveira@dep.uminho.pt

## 13.1 Introduction

Gellan gum (GG) is a linear polysaccharide resulting from the fermentation process of *Sphingomonas elodea* bacteria, being the major component of their extracellular polymeric substance (EPS) [1]. Its great availability and low-cost production has turned GG an industrially relevant polymer widely applied in different areas such as food and pharmaceutical industries. In these fields, it has been employed as gelling agent due to its ability to form heat and acid resistant transparent gels, at low polymer concentrations, in the presence of multivalent cations [2]. Over the past decade, GG has been successfully applied in several tissue engineering and regenerative medicine (TERM) approaches. Being a natural origin hydrogel, GG is biocompatible, and its highly hydrated polymeric network conjugated to its chemical nature confers the capacity to mimic the extracellular matrix (ECM), making this biomaterial a promising candidate for tissue engineering (TE) approaches. Despite presenting attractive properties for TE approaches, GG possesses some limitations that hurdle its use in different applications. Some disadvantages include weak mechanical properties, high processing and gelling window temperatures (90 °C–100 °C; ~50 °C, respectively) and lack of specific motifs for cell adhesion [1]. However, some of these limitations can be overcome since the polymer backbone possesses functional reactive groups amenable of easy modification/functionalization. In fact, the major advantage of this biopolymer is its incredible versatility to being processed through different techniques and modified into new derivatives with different properties, enabling to tailor its physiochemical and biological properties to suit specific requirements of a particular tissue [3]. GG has been largely exploited in soft tissue regeneration approaches, such as cartilaginous tissues. Different studies have been demonstrating the chondrogenic potential of biofunctional GG matrices to support chondrocytes viability and differentiation and to promote the deposition of hyaline-like ECM towards new cartilage formation. Additionally, the injectability, cell encapsulation and *in situ* gelling abilities of GG hydrogels turned this biomaterial an attractive candidate to be used as advanced cell carrier in non-invasive approaches [4, 5]. On the other side, the high affinity of GG for calcium ions has promoted its application in bone regeneration strategies [6–8]. It has been demonstrated that calcium-crosslinkable GG hydrogels can be easily mineralized with calcium phosphate (CaP) by soaking in physiological-like solutions. In this sense, several mineralization methods conjugated with the reinforcement of GG polymeric matrices with inorganic/ceramic resorbable fillers have been the most successfully attempted strategy to create a mineral phase in GG hydrogels, improving their mechanical properties and functionalization to guide new bone formation [6, 9, 10]. Based on the demonstrated potential of GG to engineer soft and hard tissues, this biomaterial has been recently proposed in osteochondral regeneration approaches [11]. GG has been employed as polymeric substrate for developing bilayer scaffolds containing a bone compartment functionalized with mineral components such as hydroxyapatite (HAp), or by adding cell adhesion microcarriers to improve osteoblasts proliferation, viability and functionality [11, 12].

In this chapter, the present state of preclinical research using GG for osteochondral TE is presented. The GG properties and main functionalization strategies that have promoted its application in TE field are also overviewed.

## 13.2 Gellan Gum

GG is an anionic bacterial-derived exopolysaccharide that consists of a tetrasaccharide repeating sequence of two residues of β-D-glucose (Glc), one of β-D-glucuronic acid (GlcA) and one of α-L-rhamnose, organized on a linear structure [D-Glc(β1→4)D-GlcA(β1→4)D-Glc(β1→4)L-Rha(α1→3)]$_n$ (Fig. 13.1) [13]. Several parameters have been described to affect polymer's biosynthesis yields, composition, structure and properties. Thus, several studies describe the optimization of GG large-scale production. Among the factors that affect GG production, the media components, as well as the carbon and nitrogen sources, are critical for gellan broth characteristics. The pH, agitation rate, dissolved oxygen and temperature have also been reported as critical points for GG production [1, 14, 15]. This biopolymer was discovered in 1978 by Kelco (San Diego, USA), during a large-scale screening programme to

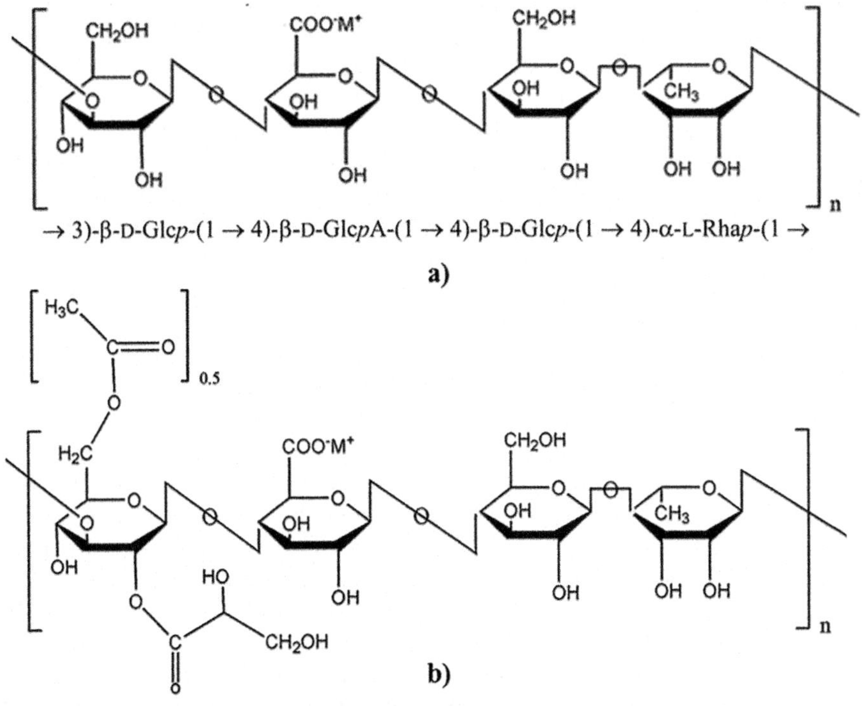
CH2OH
COO-M+
OH
CH3
n
→ 3)-β-D-Glcp-(1 → 4)-β-D-GlcpA-(1 → 4)-β-D-Glcp-(1 → 4)-α-L-Rhap-(1 →
a)
H3C
C=O
0.5
H2C
HO
b)

**Fig. 13.1** Chemical structure of deacylated (**a**) and native acetylated form (**b**) of gellan gum. Adapted with permission from Prajapati et al. [2]. Copyright 2013, Elsevier

identify polysaccharides with useful mechanical properties [16]. GG initial application was restricted to food and cosmetic industries as a multifunctional gelling, stabilizing and suspending agent [16]. GG was approved by FDA and EU as E418 for these purposes and since then several commercial products based on this polymer has been introduced in the market. Gelzan®, KELCOGEL®, Gel-Gro®, GELRITE® and Phytagel® are the trade names of some commercial products of GG with applications as gelling agent for clinical and non-clinical applications [2]. This polysaccharide can present different degrees of acetylation: (1) the acetylated form (high acyl gellan gum, HAGG), which is the native state of the biopolymer, and (2) the deacetylated form (low acyl gellan gum, LAGG), which is the type of GG commonly used in the commercial products. Generally, GG has a molecular weight around ~500 kDa [17]. The acyl and glyceryl substituents of the HAGG bind to the same Glc residue on GG backbone and, in general, this form presents two acyl groups: one molecule of L-glycerate and 0,5 of $O$-acetate per repeating unit [16].

The low acyl form is obtained by alkali treatment at high temperatures and results in acyl groups' hydrolysis. The acyl substituents removal results in a change from the soft, elastic and thermoreversible gels to more firm, brittle and higher thermostable gels [2]. The range of different properties obtained with the two isoforms of GG offers the possibility to modify the final rheological properties by blending different ratios of LAGG and HAGG. The gel-forming ability of GG at lower concentrations and in the presence of certain cations conjugated to its biodegradability, non-toxicity and high water holding capacity make GG one of the most extensively studied natural polysaccharides [18]. These distinct properties, associated with a great availability and low production costs, widespread its application to different areas. In the biomedical field, GG has been explored as a nasal, ocular, and colon-targeted drug delivery system due also to its versatility as encapsulating agent [18]. In the biotechnology industry, this biomaterial has also been largely employed as an alternative to agar for microbiological media [2].

More recently, due to the previously mentioned attractive features, GG has been exploited for injectable/scaffolding TE and regenerative medicine approaches. Besides that, its processing ability under mild conditions, allowing for cell encapsulation, is a great advantage that is being explored in many investigations [1]. Several authors have been proposing GG for cartilage TE approaches due its biocompatibility/non-cytotoxicity when loaded with chondrocytes, bone marrow cells, articular chondrocytes or adipose-derived stem cells [4, 5, 19–21]. The structural similarity of GG hydrogels with ECM, as well as their elastic moduli that resemble the common tissue, also supports their potential application in cartilaginous tissues repair. The injectability and in situ gelling ability of GG derivatives have also been exploited for the development of cell-laden carriers, thus avoiding invasive procedures. GG derivatives have also been proposed for bone regeneration [8, 10, 22]. In fact, the interest on GG in TERM field has incredibly increased in the last years, mainly due to its easily tunable properties via chemical modification/functionalization or blending with other polymers to mimic/reproduce the native features of a specific tissue [23].

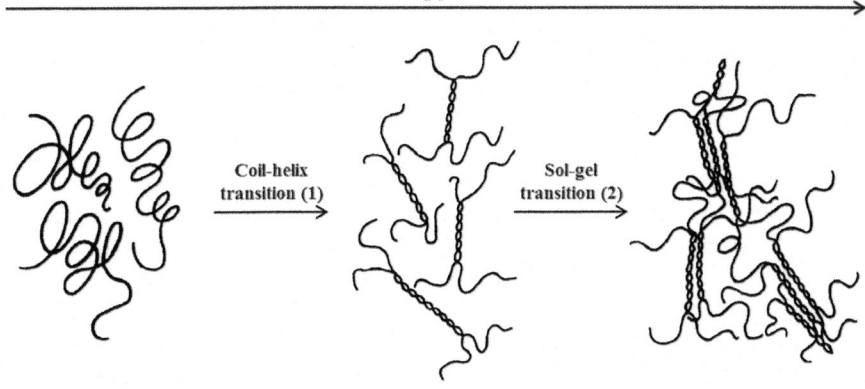

**Fig. 13.2** Schematic representation of the distinct structures/conformations adopted by GG chains during the gelation process. Adapted with permission from Miyoshi [24]. Copyright 1996, Elsevier

## 13.2.1 Structure and Properties

The unique properties of GG and its capacity to form thermoreversible gels are related with the gelation mechanism. The conformational transition from coil to double helix is a prerequisite to gel formation and occurs upon temperature decrease [16]. The subsequent aggregation of the helical GG chains on the junction zones leads to the formation of a tridimensional network. Stabilization of the helices at temperatures higher than *Tm* (coil-helix transition temperature) occurs via aggregation through intra- and inter-chain hydrogen bonds along the helix between hydroxymethyl groups of the 4-linked glucosyl units in one chain and carboxylate groups in the other, giving a thermal hysteresis between the gelation and melting [16] (Fig. 13.2).

This helical structure model adopted by the GG in a solid state was firstly proposed by Upstill et al. [25] using X-ray diffraction data from polycrystalline and well-oriented lithium salt GG fibres. Later, Chandrasekaran et al. [26] showed that the polymer exists in the solid state as an antiparallel packing of a co-axial three-fold, left-handed and half-staggered double helix with a pitch of 5.64 nm. X-ray diffraction data of the native form of GG suggests that helices are stabilized by interchain associations involving the glycerate groups, with the acetyl substituents positioned on the periphery of the helix. The electrostatic repulsion between the acyl groups inhibits end-to-end type intermolecular associations, resulting in a decrease of the degree of continuity and homogeneity of the gelled system and this also reflects on the subsequent softer and elastic properties of HAGG gels. The conformational changes experimented during the gelation process have been elucidated by several techniques including light scattering, nuclear magnetic resonance (NMR), small-angle, X-ray scattering, circular dichroism (CD), differential scanning calorimetry (DSC) and viscosity studies [27]. These works have demonstrated that the GG gelation is dependent on several factors: the concentration and molecular weight of the polymer, cations type and concentration, and pH [16]. The gelation of GG is an ionotropic process; thus, the hydrogel network formation is mediated by cations wherein the divalent cations are more effective on the gelation process and result in stronger and thermally stable gels. This is explained by the direct binding between pairs of carboxylate groups from the glucuronic acid molecules mediated by divalent cations, while monovalent cations just induce aggregation by suppressing the electrostatic repulsion between the ionized carboxylate groups on GG chains [23]. The cations effect on the aggregation of chains is greater as larger is the size of the cations species [28]. The high acyl form adopts the helical geometry at much higher temperatures as compared to the low acyl form and its resultant properties are less dependent on the concentration of ions in solution. In the presence of cations, although the same incremental effect on the gelation temperature was reported for both forms of GG by Huang et al., no significant differences were noted between different types of divalent and monovalent cations [29]. Taking advantage of the ionic gelation of GG, Smith and co-workers [19] created self-sustainable GG crosslinked by the addition of cell culture media, suggesting their potential use as cell-laden hydrogels. Horinaka et al. [28] studied the pH effect on the conformation of GG helices through optical rotation and fluorescence anisotropy measurements by analysing different pH conditions. The results demonstrated that an acidic pH enhances the intermolecular aggregation of gellan chains [28]. This effect is attributed to alterations on the conformation of the gellan chains in the aqueous system, promoted by the shielding effect on the electrostatic repulsion between the carboxyl groups and change in the anionic nature of the GG chains [28]. Thus, in the presence of salt ions, the gelling and melting temperatures of GG gels suffer a shift to higher temperatures, and an increase of the number of junction zones is observed. Consequently, the resultant hydrogels are more heat resistant and show enhanced mechanical properties, such as higher elastic modulus and rigidity [30]. Ogawa et al. [31] studied the effect of the molar mass on the coil-to-double-helix transition in aqueous solutions by light scattering and CD measurements, and viscosity and DSC studies. The results showed that the increase of molecular weight accelerates the coil-helix transition. The purification methods can also affect the gelation and the mechanical properties of the resultant hydrogels since the commercial GG contains divalent cation contaminants (mainly $Ca^{2+}$ and $Mg^{2+}$), known to form strong associations with the cation binding sites [32]. A study performed by Kirchmajer et al. [13] compared commercial and purified sodium salt gellan gum (Na-GG), showing that the commercial GG possesses inorganic cations which affects its dissolution, the gelation behaviour, and rheological properties. The dissolution of purified salt form of GG was completely achieved at lower temperatures, as compared to the commercial GG. In terms of gelation, the purified Na-GG form exhibited lower gel transition temperatures for all ionic-crosslinking methods tested. Regarding their rheological properties, both commercial GG and their purified form showed enhanced mechanical properties when crosslinked with calcium.

## 13.2.2 GG Functionalization Strategies

The success of TE approaches is directly dependent on employing high performance biomaterials with superior mechanical stability and integrity capable of supporting cell proliferation/differentiation, while their biodegradation enables the restoration of functional native tissue. For engineering complex tissues, hybrid biomaterials are being developed by means of blending different polymers to fulfil all demanding requirements [1]. Although GG has been suggested as a promising biomaterial in TE field, there are some limitations of the native GG form that need to be overcome. The mechanical weakness, poor stability at long-term physiological conditions, high gelling temperatures which makes cell encapsulation infeasible and the scarcity of specific attachment sites for cells to adhere, are the major constraints attributed to GG hydrogels. Besides all the limitations that reflect on the GG properties and consequently on its physicochemical and biological performance, researchers have been attempting several strategies to tailor GG properties to a particular application, being the most employed, chemical (by covalent functionalization) or physical modification (by interpenetrating network formation) and blending with other polymers or inorganic materials [1, 3].

Gellan gum is an incredible versatile biopolymer due to the free carboxyl and hydroxyl groups in its repeating unit that can be easily modified into new derivatives with different properties with respect to the native polymer. In fact, the chemical modification through covalent shifts can represent an advantage in terms of mechanical stability comparing to ionic-crosslinking, since the diffusion of the cations observed at long term under physiological conditions affects GG hydrogels' structural stability. Several studies involving the chemical modification of the GG backbone have been performed in order to enhance GG properties for TE purposes. Gong et al. [33] optimized the gelling point of GG via chemical scissoring by $NaIO_4$ oxidation followed by Smith degradation. This approach relies on the hypothesis that a decrease on the molecular weight could difficult the assembling and aggregation of GG chains, thus hindering the gelation process and consequently decreasing the $T_{gelation}$. In this study, a novel injectable hydrogel with a lower gelling point of 37.5 °C was developed for cartilaginous regeneration with superior long-term performance when cultured with human epidermis fibroblasts (hEFBs). Another study performed by Tang et al. [34] confirmed that oxidization has an evident effect on the gelation temperature by destroying partially the crosslinking points. In this study, a complex GG hydrogel with improved mechanical properties (compression modulus of 278 kPa) and a gelation temperature close to physiological conditions was synthesized by GG oxidation via $NaIO_4$-based cleavage reaction and further complexation with carboxymethyl (CM) chitosan by $Ca^{2+}$ crosslinking and Schiff reaction. In the same way, the swelling ability and degradation behaviour of produced hydrogels was affected by the oxidation extent and CM-chitosan concentration. The CM-chitosan gels also demonstrated an improved biological performance when encapsulated with chondrocytes, as compared with the oxidized and non-oxidized form, which was attributed to a reduction of free aldehyde groups on CM-chitosan.

The functionalization of GG by methacrylation introduces a new gelation mechanism for GG through photo-polymerization, which represents a great advance to address injectability in TE purposes by enabling cell encapsulation and in situ gelation [22, 35, 36]. The methacrylation has been adopted by several authors to modulate the mechanical properties of GG [35, 37]. Silva-Correia and co-workers have explored the potential of GG-based hydrogels for IVD tissue engineering applications. In this work, the methacrylation of low acyl GG using glycidyl methacrylate enhanced the mechanical performance and allowed to develop an injectable vehicle for nucleus pulposus regeneration (NP) [22]. Both ionic- and photo-crosslinkable methacrylated gellan gum (GG-MA) were obtained, although photo-crosslinked hydrogels have showed better results in terms of mechanical properties and degradation behaviour. The biocompatibility studies also showed that GG-MA hydrogels are non-cytotoxic to L929 cells, suggesting their potential for cell-based regeneration purposes [22, 38]. Coutinho et al. [37] also tuned the physical and mechanical properties of high and low acyl GG through methacrylation. In this study, methacrylic anhydride groups were introduced in the GG chain and different degrees of methacrylation were experimented. The results showed that the mechanical and degradation behaviour of the GG-MA could be highly tuned using different crosslinking mechanisms. The biocompatibility tests proved that the produced hydrogels are able to encapsulate NIH-3T3 fibroblast cells, thus confirming GG-MA potential use in a wide range of TE applications [37]. By its turn, Pacelli et al. [39] proposed a new synthetic route for methacrylation with the aim to produce an in situ photo-crosslinked injectable hydrogel. In this work, a GG derivative was developed by introducing methacrylic groups and further combining the GG-MA with polyethylene glycol dimethacrylate (PEG-DMA). The new proposed synthetic strategy avoids the requirement of high amounts of methacrylic anhydride, by changing the water solvent reaction per anhydrous DMSO and using 4-DMAP as nucleophilic catalyst. This new method revealed a better control on the reproducibility of the methacrylation, which was successfully confirmed by FTIR and $^1$H NMR spectroscopy studies. The mechanical properties of GG-MA grafted with PEG-DMA were improved as confirmed by rheological studies. The results also evidenced a relation between the concentration/molecular weight of PEG-DMA used and the final strength of the hydrogel. The new developed system proved to be injectable and possess in situ photo-crosslinking ability. Thus, the use of GG-MA hydrogels grafted with PEG-DMA as a delivery system for therapeutic agents was hypothesized since the material also revealed the capacity of modulating the release of small molecules like sulindac, nonsteroidal anti-inflammatory drug (NSAID) of the arylalkanoic acid class, is non-cytotoxic to human fibroblast cell line (WI-38). The same authors designed a novel nanocomposite (NC) hydrogel by combining laponite with GG-MA photo-chemical hydrogels. The laponite® XLG was used as a nanofiller to modulate the final properties of methacrylated hydrogels. This nanoclay was also used as a linker for the rigid polymeric chains of GG-MA. The strategy was to reinforce the mechanical properties and establish a good balance between the swelling behaviour and biocompatibility for developing of a novel device for biomedical applications. The mechanical properties of the developed hybrid materials were investigated, as well as their biocompatibility and swelling and diffusion ability. The results showed that the nanoclay incorporation promoted the formation of stronger and stable gels that maintained their structural integrity after thermal treatment, which represents an advantage when sterile systems like wound dressing coats are envisioning. The swelling and diffusion tests showed that the nanoclay can also modulate the swelling ability of the polymeric network and interact with the ofloxacin, slowing down its release over time in the first 8 h. Finally, the cytotoxic tests based on a red neutral assay showed that GG-MA with 1% w/v of laponite is biocompatible for human fibroblast cell line (WI-38) [40]. With a different goal, Ming-Wei Lee et al. [41] proposed a photo-crosslinked GG-based film with anti-barrier properties for clinical applications. In this work, GG was grafted with the photo-functional group cinnamyl bromide in a molar ratio of 5:1 of cinnamate to GG carboxyl residues. The NMR and FTIR characterization confirmed that the covalent grafting was successfully achieved and the *in vitro* and *in vivo* assays showed that the produced film effectively prevented tissue adhesion demonstrating a great potential for use in medical field. Du et al. [42] proposed a new method for GG functionalization through double bonds via thiolation. In this work, a novel injectable GG hydrogel able to be chemically and physically crosslinked in situ was developed. The procedure was based on grafting of gellan with cysteine at room temperature following carbodiimide chemistry and crosslinking via disulphide under mild conditions. The results from structural characterization, CD and rheology demonstrated that the produced thiolated-gellan maintains its unique 3D conformation, but with a lower phase transition temperature under physiological conditions. It was also demonstrated that the GG derivative was non-toxic to ARPE-19 cells, showing to be a promising vehicle for use as injectable hydrogel in the biomedical and bioengineering fields.

Based on the same strategy, O. Novac [43] reported a novel GG derivative: *N*-(2-aminoethyl)-2-acetamidyl gellan gum (GCM-EDA). The derivative was synthetized by carboxymethylation via nucleophilic substitution of primary hydroxyl groups of the β-D-glucose unit, followed by reaction with tert-butyl *N*-(2-aminoethyl) carbamate (*N*-Boc-EDA) using 1-ethyl-3-(3-dimethylaminopropyl) carbodiimide (EDAC) as an activator. The last step of the procedure consisted on the deprotection with trifluoroacetic acid. The structural characterization of the GCM-EDA through spectroscopic techniques revealed the success of the chemical synthesis, while rheological analysis evidenced a direct relation between the alkalinity conditions of the carboxymethylation reaction and the dynamic viscosity of GCM derivatives. The MTT cytotoxicity assay demonstrated the biocompatibility of this material in the presence of bEnd3 cells, showing the potential of this new GG derivative for drug delivery applications in the central nervous system and treatment of cerebral vascular tumours.

It was recognized that one of the major and common disadvantages of most polysaccharide-based hydrogels, including GG, is the lack of affinity for anchorage dependent cells (ADCs) which limits their applications in TE as ADC's loading carriers. This critical limitation has been attributed to the hydrophilic and anionic nature of GG that naturally repulse cells. Numerous strategies envisioning the improvement of adhesion, proliferation and differentiation of ADCs have been made in the last years and are mainly based on the incorporation of proteins or peptide sequences in the GG backbone. Wang et al. [44] proposed a covalent coating of gellan microspheres with gelatin layers to create cell binding ligands for human ADCs. Gellan microspheres were prepared from Phytagel®, based on a water-in-oil (W/O) emulsion process followed by a series of redox (oxidation–reduction) and crosslinking treatments to tailor their dimensions and injectability. The gellan microspherical surfaces were then grafted with gelatin, and cell loading tests were conducted with human dermal fibroblasts (HDFs) and human foetal osteoblasts (hFOBs). Morphological characterization proved that both HDFs and hFOBs attached well and grew rapidly on the microspheres surfaces. Early osteogenic differentiation was also noted by positive ALP production. Based on the same principle, Ferris et al. [45] combined purified-GG with arginine-glycine-aspartic acid (RGD) sequence that is known to enhance integrin-mediated cell attachment. The efficiency of conjugation was evaluated by radiolabelling using the well-established chloramine-T method and the results showed that a 40% conjugation efficiency was achieved. Rheological studies revealed the formation of weaker gels in terms of mechanical strength that could be explained by two reasons: (1) the reduction of available carboxylate groups and (2) a steric hindrance caused by peptide bulk that difficulted the formation of interchain associations. The biological performance of the RGD-GG hydrogels was also investigated by seeding and encapsulation of two cell lines: the anchorage dependent murine skeletal muscle C2C12 cell line and the rat pheochromocytoma PC12 cell line, which could be maintained in adherent and suspended cultures. Both cell lines showed an increase in cellular metabolic activity, quantified through MTS assay and both proliferated when seeded and encapsulated in the hydrogels, although only the C2C12 cell line was able to differentiate in the both experiments (on surfaces and in the bulk of the RGD-GG hydrogels).

Physical modification performed at the molecular level or involving the macrostructure of the hydrogel has been used as another strategy to modulate the final properties of GG for TE applications. Silva et al. [46] introduced a novel concept based on cell-compatible spongy-like GG hydrogels that can allow avoiding highly cost and time-consuming approaches aiming at enhancing cells adhesion to hydrogels. The authors assumed that by varying specific parameters of the sequential processing stages of hydrogels it is possible to tune the physical/mechanical properties of GG hydrogels. In this sense, several variations were experimented, such as the time of stabilization of the GG hydrogels before freezing, the freezing time and methodology (performed at 196 °C in liquid nitrogen, at −20 or −80 °C in the freezer). The spongy-like hydrogels were obtained after rehydration of the freeze-dried polymeric networks. Morphological characterization and rheological assays performed with spongy-like hydrogels and their precursors demonstrated the influence of changing the parameters along the processing methodology, which was especially pronounced in pore size and compressive modulus. The microstructural rearrangements noted by pore wall thickening and pore size augmentation were attributed to ice crystal formation during the freezing stage. This microarchitecture conjugated with a significantly lower water content had a positive effect on cell adhesive properties of GG spongy-like hydrogels, that has been observed to be dependent on the cell type. The improvement of mechanical and cell adhesive properties of the spongy-like hydrogels, associated to their off-the-shelf availability in an intermediate dried state, turns this concept a potential strategy to address diverse TERM applications particularly as soft tissue analogues.

## 13.3 Gellan Gum for Osteochondral Tissue Repair

Osteochondral (OC) lesions, commonly caused by joint diseases such as osteoarthritis, consist in areas of articular injury or degeneration involving cartilage and subchondral bone damage [47]. Despite the current clinical strategies to treat cartilage OC defects, that include autologous chondrocyte implantation, microfracture, and mosaicplasty, the long-term biofunctionality of the repaired tissue cannot be ensured [48]. Given the widespread prevalence of joint diseases and the huge negative impact that represents for patients' life quality, there is a pressing need for the development of alternative clinical treatments that effectively address the pathology. Nevertheless advances have been made in OC repair approaches, mimicking the high complex organization and composition of articular cartilage and osteochondral interface tissues remains a challenging issue. The different healing abilities and biomechanical properties of the distinct tissues involved in OC tissue enhance the complexity to OC regeneration strategies [49]. Moreover, the unique properties of the interface structure, that play a critical role in the maintenance of cartilage integration, have also been a concerning issue in the establishment of regenerative OC approaches. TERM approaches has emerged as one of the most promising tools to reproduce the complex biological organization of OC complex [50]. Based on the development of biomimetic and bioactive scaffolds conjugated with growth factors and/or stem cells, the hierarchical structure of the OC complex has been attempted to be reproduced by using different systems, from the simple monophasic scaffolds to more complex structures like the bilayered, multiphasic or gradient scaffolds with continuous interfaces [48]. Ideally, a functional OC scaffold should present two or more regions differing in composition, microarchitecture (including pore size and porosity) and mechanical gradients to support the different cell types that composed the OC unit. The intermediate region between the cartilage- and bone-like layers should create a smooth transition to avoid scaffold delamination, while facilitating stress transfer [51]. The scaffolds should maintain their structural integrity when implanted on the defect site and degrade in a fashion that matches neo-tissue growth. During regeneration, the scaffolds should also prevent the invasion of synovial fluid to the subchondral bone and the vascularization in the chondral layer [52]. The selection of a biocompatible, non-immunogenic, biodegradable and mechanically stable material is the first consideration in the fabrication of a scaffold system for OC regeneration. The materials currently used as OC implantable grafts range from natural and synthetic polymers to inorganic materials (ceramics and bioactive glasses) or even metallic materials. As there isn't yet a single material that can fulfil all requirements to mimic this complex structure, the combination of different materials by blending, copolymerization or doping with inorganic materials (e.g. bioactive ceramics) has been a recurrent methodology to achieve it. By using these methods, researchers intend to improve the mechanical properties of OC scaffolds, enhance their integration in the native tissue and to guide the cells to a specific phenotype. To address these requirements, and due to its similarity to ECM, GG has been proposed for OC repair strategies. The ability to form a hydrophilic 3D-polymeric matrix that mimics the native ECM, its biodegradability and biocompatibility make GG an appealing versatile biomaterial for OC tissue engineering strategies. In fact, several studies attempting the chemical modification of GG hydrogels have showed lower gelation temperatures, improved mechanical properties and biological performances crucial to their use as vehicles for OC and cartilage regeneration [49]. Additionally, the capacity to enable the incorporation of growth factors and their release in a controlled manner represents a great advantage for the control of growth, morphology and differentiation of stem cells. Moreover, this biomaterial is amenable of processing by advanced technologies as 3D printing allowing personalized approaches to the fabrication of custommade constructs respecting their shape, composition, geometry and internal architecture [49].

### 13.3.1 Scaffolds for Regeneration of Cartilage

Considering the critical role of load bearing tissues as articular cartilage in the biomechanics of joint structures and the frequency of pathological situations involving cartilage damage/degeneration with huge economic impacts in the healthcare systems, it's demanding the development of new alternative treatments to the conventional autogenic or allogenic cartilage transplants. TE arises as one of the most promising approaches for the treatment of cartilaginous tissues, by proposing scaffold-based strategies to guide cartilage regeneration. The strategy relies on the combination of a 3D template to fill the tissue lesion and support cell growth, interactions and differentiation by using signalling molecules, while its biodegradation opens space for new tissue ingrowth. However, cartilaginous tissues present limited intrinsic self-repair ability due to its avascular nature [53]. The structural similarity of GG to native glycosaminoglycan, due to the presence of glucuronic acid residues in this backbone, and its elastic moduli close to the native tissue, encouraged its application for cartilage engineering purposes. In fact, several investigations have demonstrated the ability of GG-derived hydrogels to support cellular viability of different cell types. Oliveira et al. performed a pioneer work, proposing GG as a new biomaterial for cartilage tissue engineering applications. In a first study, Oliveira et al. [21] showed the flexibility of this viscoelastic biomaterial to be processed into different structures (discs, membranes, fibres, particles and scaffolds) and its capacity to support chondrocytes encapsulation, viability and high proliferation during 2 weeks in culture. In a further study, the same authors proposed an optimized form of GG hydrogel as an injectable and in situ gelling system. The structural stability, injectability and in situ gelation capacity of the developed system were demonstrated through rheological studies. The biological performance of the GG as injectable carrier was evaluated through in vitro and in vivo studies. The in vitro studies were performed by culturing the human articular chondrocytes encapsulated in GG discs in expansion and differentiation medium for a total of 56 days, whereas the in vivo studies were conducted by ectopic implantation in Balb/c mice during 21 days. Results showed no immunogenicity of the developed system and a good integration within the host tissues. Moreover, the biocompatibility was also evidenced through the observation that the injectable system was able to hold cell viability and hyaline-like ECM deposition [5]. To evaluate the in vivo potential of the developed system, Oliveira et al. [4] tested the injectable cell-laden GG hydrogel to regenerate a full-thickness articular cartilage defect in rabbit. The assay was conducted for 8 weeks and different populations of autologous cells were tested: chondrogenic (ASC+GF) and non-chondrogenic pre-differentiated rabbit adipose stem cells (ASC) and rabbit articular chondrocytes (AC) compared to the controls gellan gum and empty defect. Cell-laden approaches showed better results in terms of integration and regeneration of cartilage similar to the surrounding native tissue. Particularly, ASC+GF in the presence of transforming growth factor beta1 (TGF-$\beta$1) and bone morphogenetic protein 2 (BMP-2) revealed to be the best cell-laden approach to achieve a quality tissue as confirmed by PCR analysis. Based on the assumption that synovium-derived mesenchymal stem cells (SMSC) have superior chondrogenic capacity, Fan et al. [54] reported a cell-based approach to achieve cartilage regeneration using an injectable modified GG hydrogel. The chondrogenic potential of the developed system was assessed by *in vitro* chondrogenic culture of the cell-laden constructs for 21 and 42 days followed by RT-PCR and histological/immunohistochemical analysis for detection of cartilage-related gene markers and cartilage-specific ECM components, respectively. The viability of the SMSC was also investigated through WST assay. The results demonstrated that SMSC have high potential for chondrogenic differentiation, which was enhanced in a 3D environment and in the presence of TGF-$\beta$1. These results were corroborated by the high expression levels of cartilage-related genes (collagen type II, aggrecan, biglycan and SOX 9) and deposition of cartilaginous matrix proteins. Therefore, this study confirmed the potential of this cell source for future application in cell-laden approaches for clinical cartilage repair. Considering the high mechanical strength required to engineer load bearing tissues, Shin et al. [55] developed a double-network (DN) hydrogel produced in a two-step crosslinking mechanism. In the first, a rigid and brittle GG-MA network was formed and in the second, a soft and ductile network made of gelatin methacrylamide (GelMA) was produced. The mechanical strength of the DN GG-MA/GelMA hydrogels was assessed through compression tests which evidenced the improved mechanical properties of the DN system as comparing to the single networks. The best formulation in terms of compressive modulus and failure strain and stress was achieved by combining a large mass ratio of the second to the first network. Moreover, a balance of methacrylation range of the second network is crucial to allow an intermediate crosslinking density that enables the formation of an effective network and by other hand that simultaneously avoids the formation of a high crosslinked network unable to dissipate energy. The DN hydrogels also allowed for cell encapsulation with NIH-3T3 fibroblasts during processing. The biocompatibility of this hybrid hydrogel conjugated with unique mechanical properties resembling cartilaginous tissues turns this material a promising candidate to be used as cell-laden scaffold for load bearing tissues regeneration. Envisioning a patient-specific approach for regeneration of complex cartilaginous tissues, Kesti et al. [56] proposed a novel bioink based on a blend of GG and alginate combined with BioCartilage (Arthrex), a clinical commercial product containing particles from cartilage ECM. The bioink was produced from a 6% purified GG solution and 4% alginate blend in a ratio of 50:50 and printed with the shape and architecture of full-sized grafts followed by ionic-crosslinking with $SrCl_2$ in DMEM medium. The printability of the developed bioink and the compatibility of the process for viable cell encapsulation were investigated. Rheological analysis demonstrated the printability of the bioink that exhibited a shear thinning behaviour critical for extrusion bioprinting. Moreover, the bioink was able to be printed at low pressures favouring cell viability during the process. The mechanical tests revealed that, although the bioink has a lower order of magnitude of tensile modulus as compared to the native articular cartilage, the grafts exhibited an elastic behaviour similar to the native tissue. The biological performance was also investigated through in vitro culturing of the bioink alone and bioink+Biocartilage with chondrocytes isolated from a full-thickness articular cartilage defect of bovine for 8 weeks. The role of TGF-β3 in chondrogenic differentiation was also investigated. The histological data and viability assays (i.e. Calcein AM and DNA quantification) showed that bioink+BioCartilage cultured in supplemented conditions with TGF-β3 supported chondrocyte proliferation and cartilage ECM deposition. This study presented a promising strategy to improve the bioactivity of bioinks, offering a better proximity with the native tissue features, which could be a future option to recreate functional specific tissues using scaffold-guiding regeneration approaches. Based on a previous study that showed the ability of GG as viscosity enhancer and jellifying promoter of gelatin-methacryloyl (gelMA) hydrogels and considering their ability to support chondrogenic viability and differentiation, Mouser et al. [57, 58] proposed a bioink made from UV crosslinked gelatin-methacryloyl (geMA)/GG blend for cartilage grafts bioprinting. Rheological and mechanical tests revealed that the addition of GG to gelMA hydrogels improved their printability by inducing a yielding behaviour, and promoted an increase on the overall construct stiffness. Additionally, the biological performance of the bioink laden with primary chondrocytes was also investigated through biochemical and histological assays. The results showed that all tested formulations supported chondrogenesis and deposition of cartilaginous matrix, being 10/0.5% gelMA/gellan blend the most suitable formulation to generate chondrocyte-laden 3D printed constructs. Concerning that the mechanical and structural stability of bioprinted GG-based constructs is a critical factor for cartilage regeneration approaches, Yu et al. [59] performed a study to evaluate the influence of the surface area per mass, i.e. porosity, on the mechanical performance and degradation behaviour of GG-printed scaffolds. The degradation test performed in simulated body fluid showed that GG-printed scaffolds tend to become stiffer and harder during the degradation test, being this effect potentiated by the increase of their surface area and porosity. This can be attributed to the crosslinking with cations present in SBF solution that increases hydrogels' compressive modulus and strength. However, this implies a faster degradation rate that may not be able to support the slow native cartilage regeneration. Therefore, this study demonstrated the importance of creating a balance between surface area and porosity of GG-based scaffolds that enables the maintenance of good mechanical performance without simultaneously compromising the structural stability of constructs developed for cartilage tissue regeneration.

## 13.3.2 Scaffolds for Regeneration of Bone

Contrary of the limited self-repair of the cartilaginous tissues, bone has been the most widely investigated tissue in TERM field due to their highest potential for regeneration. Originally, the application of hydrogels has been directed for soft tissues regeneration, but recently, successful attempts have been made to tailor hydrogels for hard tissue engineering. In fact, hydrogels offer some advantageous features that the non-swelling polymers and conventional ceramics cannot provide. Their unique highly hydrated nature, biocompatibility, biodegradation and injectability are some of them that have prompted their exploitation for bone TE applications. However, the lack of mineralization ability of inert hydrogels has been a major hurdle in their application in bone-substituting approaches [60]. In this sense, inspired by the hierarchical nature of bone, nanocomposites made from biopolymeric matrices (e.g. natural biodegradable polymers) doped with bioresorbable inorganic and ceramic fillers have been attracting great deal of interest to create a suitable and bioactive mechanical support with osteogenic potential to guide bone formation in TE strategies [9]. GG has been recently proposed in several bone TE approaches as an inexpensive calcium-crosslinkable polysaccharide. Douglas et al. [7, 8, 61, 62] performed an extensive work on the development of enzymatic mineralized GG matrices to guide bone formation. In fact, hydrogels mineralization has been an advantageous attempted strategy for bone regeneration, since it improved the stiffness and enhanced in this manner the bioactivity of the matrix regarding their osteogenic potential. In one of their first studies, Douglas et al. [8] proposed the enzymatic mineralization of GG hydrogels by incorporating ALP enzyme, that has a key role on bone mineralization, with calcium phosphate (CaP). The effect of polydopamine (PDA), a surface functionalization agent that has showed to enhance mineralization and favour cell attachment and osteogenic differentiation, was also investigated [63]. In this study, the potential of ALP incorporation to direct apatite-like mineral formation and improve mechanical properties through an increase in stiffness was demonstrated in the presence of calcium glycerophosphate (CaGP) solution. The positive effect of PDA on mineralization and osteoblastic cell attachment and proliferation was also evidenced. In other study, Douglas et al. [7]

proposed a functionalized GG-inorganic composite with antibacterial activity and enhanced cell adhesive properties. The mineralization was achieved by using the same method, through incorporating ALP during gelation, followed by incubation in a solution containing calcium (Ca). In this study, the authors incorporated zinc in the mineralization medium, since it was previously shown that this element can enhance cell adhesion and proliferation and promote osteogenic differentiation. Moreover, zinc has been shown to exhibit antibacterial properties when used as additive of some hydrogel matrices and stimulate angiogenesis, which is a critical factor in bone healing processes [64–67]. Antibacterial activity was observed in the zinc-containing mineralized hydrogel matrices, as well as their ability to support MC3T3-E1 osteoblast-like cells adhesion and proliferation, being the most promising formulation made from equal ratios of Ca:Zn 0.025:0.025 (mol/dm$^3$). Recently, Douglas et al. [61] introduced GG-CaCO$_3$/Mg composites obtained through enzymatic mineralization of GG with urease, in a mineralization medium containing urea and different ratios of calcium and magnesium ions. Mineral formation was assessed by scanning electron microscopy (SEM), and it was shown that mineral formation depends on the Ca/Mg ratio of the mineralization solution, thus varying GG mineral composition from calcite to magnesian calcite and hydromagnesite with the increase of Mg content. It was observed that magnesium enrichment of mineralization medium results in a decrease of crystallinity of the mineral formed and in the compressive strength values of Mg-composite systems. Similarly, an increase in magnesium content in the mineral phase had no positive effect on MC3T3-E1 osteoblast-like cell viability, and a pronounced cytotoxic effect on the composites containing a non-mineral hydromagnesite phase was observed. On contrary, GG-calcite composites were successfully reinforced and promoted osteoblast-like cell adhesion and growth. Beyond the different GG enzymatic mineralization strategies developed by Douglas and co-authors, the same group proposed an injectable self-gelling GG composite for bone tissue engineering [6]. The enrichment of GG with bioglass particles was performed in order to enhance mineralization with calcium phosphate (CaP) and improve the biological performance of the gelling system, envisioning a composite with antibacterial properties for bone regeneration applications. Three bioglass formulations were tested: one calcium-rich, one calcium-poor and one possessing similar composition to the widely used 45S5 bioglass. It was observed that the reinforcement of GG hydrogels promoted mineralization and an enhancement on mechanical properties revealed by an increase of compressive modulus of the samples containing higher and lower levels of calcium. It was also observed a stimulation of the differentiation of rat mesenchymal stem cells (rMSCs) in the formulation containing a similar composition to 45S5 bioglass. The same authors also attempted the reinforcement of a hydrogel self-gelling system with α-tricalcium phosphate (α-TCP) [68]. The strategy relies on the premise that α-TCP would serve as a vehicle for internal gelation of the GG through released Ca$^{2+}$, that acts as ionic-crosslinker for GG network and at the same time would promote the formation of crystals of calcium-deficient hydroxyapatite (CDHA) by the hydrolysis of this inorganic compound. The developed composites were characterized in terms of chemical composition and mechanical performance. The results showed that gelation occurred within 30 min and α-TCP was converted to CDHA that appears non-homogeneously distributed along the composite. These results suggest the potential of composites as alternatives to the conventional α-TCP bone cements; however, further in vitro and in vivo studies of the developed composites are needed to evaluate their biological performance. Based on the similar strategy, Gantar et al. [10] proposed the reinforcement of spongy-like GG hydrogels with bioglass (BAG). As expected, an improvement of the microstructure and the mechanical properties of the composite materials was reported, being this effect potentiated by larger amounts of bioactive-glass particles incorporated on GG matrices. Bioglass incorporation also promoted the deposition of an apatite layer when soaked in simulated body fluid. No significant differences on the biological performance in terms of cell viability and adhesion were observed between the GG and the GG-BAG spongy-like hydrogels during human-adipose stem cells (hASCs) culturing for 72 h. Thus, longer-term culture tests are required to confirm the potential of BAG reinforcement of GG spongy-like hydrogels for improve cell adhesion and osteogenic differentiation. Moreover, further improvements on the strategy are required to achieve an appropriate mechanical behaviour capable to support biomechanical loads. More recently, Manda et al. [69] developed GG–hydroxyapatite (HAp) spongy-like hydrogels to reproduce the organic (GG) and inorganic (HAp) phases of the bone tissue. The reinforced GG spongy-like hydrogels showed improved mechanical properties, noted by an increase in the storage modulus to 70–80 KPa. A mineral phase created by homogenous dispersion of HAp particles on the polymeric network was also reported. Micro-CT analysis revealed high porosities of the composites and interconnectivity that fills the requirements for an ideal cell-laden scaffold to guide bone formation. The bioactivity of the spongy-like reinforced hydrogels was significantly potentiated in the spongy-like hydrogels crosslinked with $CaCl_2$. Concerning their biological performance, adipose-derived stem cells showed to adhere and spread in HAp spongy-like hydrogels network under osteogenic culture conditions [69]. Bellini et al. [70] developed an in situ gelling system composed by hyaluronic acid (HA), GG and calcium chloride, for bone osteochondral defects repair. The innovative strategy proposed in this work relies on the application of two biocompatible polymers, one with osteoinductive properties acting like a filler to bone defect and the other acting as a cap to avoid the leakage of the first from the defect region. Based on the osteoinductive properties previously demonstrated by HA, this polymer containing calcium chloride was selected as carrier for autologous bone marrow cells to fill the bone defect and, by its turn, GG hydrogel was applied as a cap to cover the defect. The gelation was promoted at the interface by the salt in the suspension. The mechanical performance and structural stability of the developed system were characterized, as well as the adhesion potential to fill bone defects in an in vitro artificial pig bone defect. The osteogenic potential of the system was also assessed by seeding HA/calcium solution with primary osteoblasts and covering with GG solution, during 14 days. An inhomogeneous but stable gelling system was reached in 5 min and a typical gel behaviour was observed by all the formulations tested. A proportional relation between the elasticity or degradation behaviour and the calcium content was noted and the inverse was observed regarding HA content. Intermediate calcium content formulations revealed the best results in the in vitro adhesion tests, exhibiting the formation of a stronger gelling system able to resist at shaking and water flow. Finally, it was observed that the HA–Ca–GG gelling system supported osteoblast viability and progression noted by accumulation of ECM particles and mineralized nodules characteristic of osteoblastic cultures.

### 13.3.3 Bilayered Scaffolds

The popularization of bioprinting technologies has opening new opportunities in osteochondral tissue engineering field. In fact, bioprinting technologies is a unique potential approach capable of precisely recapitulating the heterogeneous cellular composition and anisotropic ECM organization of OC native tissues. Moreover, this advanced technology allows the processing and fabrication of cell-based hydrogels commonly called bioinks that enable to incorporate labile biological materials, such as genes and growth factors, under physiological conditions, offering an improved environment for cell survival, proliferation, migration and biosynthetic activity towards regeneration [53]. In addition, 3D printing enables the construction of personalized regenerative implants, respecting the precise shape and internal architecture of OC defect by means of medical imaging (MRI) and reverse engineering approaches. Therefore, bioprinting technologies have receiving considerable attention in the last years as a potential approach to reach an integrated system of biodirective matrices to guide heterotypic differentiation for OC complex regeneration [71]. Combining bioprinting and microcarrier (MC) technologies, Levato et al. [12] developed a bilayered osteochondral model composed by an osteogenic layer consisting in a MC-laden bioink and a cartilage region created from MC-free bioink, being the bioink composed of GelMA-GG hydrogel. MCs composed by polylactic acid (PLA) loaded with mesenchymal stem cells (MSCs) were fabricated using a green solvent-based method, followed by functionalization with human recombinant collagen type-I to improve cell response, and then cultured under static and dynamic conditions. The effect of MCs incorporation on the mechanical and printing properties of the GelMA-GG bioink was studied through dynamical mechanical analysis (DMA). The viability, morphology and osteogenic potential of MC-laden bioink were also assessed. The results showed that PLA MCs reinforce the mechanical properties of the soft GelMA-GG matrix without compromising the printability of the composite material. Moreover, MCs facilitated MSCs adhesion promoting osteogenic differentiation and bone matrix deposition. This study demonstrated the potential of cell-laden MCs incorporation on hydrogel-based bioinks for the development of advanced constructs for guiding bone and cartilage tissue regeneration. By its turn, Pereira et al. [11] proposed an integrated biphasic GG scaffold composed of a cartilage-like layer containing 2 wt.% of low acyl GG and a bone-like layer by doping an equal solution with different amounts of HAp (5%, 10%, 15%

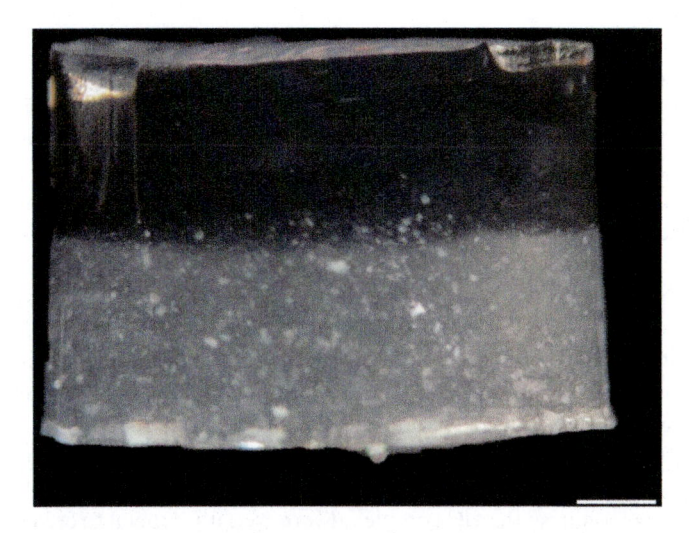

**Fig. 13.3** Macroscopic appearance of the bilayered scaffolds composed of LAGG $_{[2\%]}$/LAGG $_{[2\%]}$(HAp)$_{[15\%]}$. Scale: 1 cm

and 20%) (Fig. 13.3). The bioactivity of the developed bilayered system was investigated by soaking in simulated body fluid solution for 14 days and assessment of apatite formation by SEM, FTIR and X-ray diffraction analysis. An integrated cohesive bilayered structure was successfully achieved and it was possible to observe by SEM that the different ratios of HAp did not affect pore size. The in vitro bioactivity test revealed the formation of a thick apatite-like layer restricted to the bone-mimicking compartment, after 14 days of incubation. Thus, bilayer hydrogels could represent a promising scaffolding strategy to regenerate the different tissues of an osteochondral defect in an integrated system. However, further studies are needed to evaluate the biological performance and functionality of the integrated system in vivo, as well their potential to regenerate cartilaginous tissues.

## 13.4 Concluding Remarks

Natural polymer-based hydrogels have stimulated the interest of scientific community as promising materials to suit TERM applications. Among the natural hydrogels that have been successfully applied in different TE contexts, GG has receiving considerable attention due to its incredible versatility. In fact, the possibility of easily tailoring the GG properties, such as their mechanical and biological performance, opens great perspectives for their application in TE field. As reviewed in this chapter, substantial pre-clinical studies have demonstrated the potential of this biomaterial to support and induce chondrogenic and osteogenic phenotypes for being used in cartilage and bone regeneration approaches. The similarity of this biomaterial with ECM, its mechanical properties resembling cartilaginous tissues, biocompatibility, injectability and encapsulation abilities, conjugated to its flexibility of being processed into several structures through advanced technologies, such as custom-made rapid prototyping technologies, have justified its exploitation as cell-laden support to cartilage regeneration approaches. The same features have encouraged its application in bone repair approaches, being necessary in this case to create an inorganic mineral phase imbedded in the hydrogel polymeric network, to confer osteoinductivity and adequate mechanical properties promoting a more favourable microenvironment to bone healing. As previously presented, GG-composite materials have also demonstrated their potential in several bone regeneration strategies. However, their application as a biofunctional integrated system for osteochondral complex repair/regeneration remains a challenging subject, existing only a few studies reporting the application of GG-based hydrogels as scaffolding approaches to achieve simultaneously cartilage and bone repair. Actually, it is recognized the complexity of mimicking in one single integrated system the heterogeneity and the different biofunctional architectures and properties of the different tissues that compose the OC complex. Moreover, it is crucial to create a smooth transition that mimics the unique properties of the interface region and simultaneously allow the compatibility between the phases, enabling the construction of a cohesive structure able to be integrated within the host tissues. The conjugation of a versatile biomaterial, such as GG, as a polymeric support able to guide a specific cell phenotype in the presence of biological cues, with the current advanced microscale processing techniques (for the design and control of hierarchical architectures) could be a promising strategy to create an integrated engineered scaffold for future clinical application in OC defects repair.

**Acknowledgments** The authors acknowledge the Portuguese Foundation for Science and Technology (FCT) for the funds obtained through the project B-FABULUS (PTDC/BBB-ECT/2690/2014) and the financial support provided to Joaquim M. Oliveira (IF/01285/2015) and Joana Silva-Correia (IF/00115/2015) under the programme "Investigador FCT."

# References

1. Stevens LR, Gilmore KJ, Wallace GG, in het Panhuis M (2016) Tissue engineering with gellan gum. Biomater Sci 4:1276–1290. https://doi.org/10.1039/C6BM00322B
2. Prajapati VD, Jani GK, Zala BS, Khutliwala TA (2013) An insight into the emerging exopolysaccharide gellan gum as a novel polymer. Carbohydr Polym 93:670–678. https://doi.org/10.1016/j.carbpol.2013.01.030
3. Bacelar AH, Silva-Correia J, Oliveira JM, Reis RL (2016) Recent progress in gellan gum hydrogels provided by functionalization strategies. J Mater Chem B 4:6164–6174. https://doi.org/10.1039/C6TB01488G
4. Oliveira JT, Gardel LS, Rada T et al (2010) Injectable gellan gum hydrogels with autologous cells for the treatment of rabbit articular cartilage defects. J Orthop Res 28:1193–1199. https://doi.org/10.1002/jor.21114

5. Oliveira JT, Santos TC, Martins L et al (2010) Gellan gum injectable hydrogels for cartilage tissue engineering applications: in vitro studies and preliminary in vivo evaluation. Tissue Eng Part A 16:343–353. https://doi.org/10.1089/ten.tea.2009.0117
6. Douglas TEL, Piwowarczyk W, Pamula E et al (2014) Injectable self-gelling composites for bone tissue engineering based on gellan gum hydrogel enriched with different bioglasses. Biomed Mater 9:45014. https://doi.org/10.1088/1748-6041/9/4/045014
7. Douglas TEL, Pilarz M, Lopez-Heredia M et al (2015) Composites of gellan gum hydrogel enzymatically mineralized with calcium-zinc phosphate for bone regeneration with antibacterial activity. J Tissue Eng Regen Med 11:1610–1618. https://doi.org/10.1002/term.2062
8. Douglas T, Wlodarczyk M, Pamula E et al (2014) Enzymatic mineralization of gellan gum hydrogel for bone tissue-engineering applications and its enhancement by polydopamine. J Tissue Eng Regen Med 8:906–918. https://doi.org/10.1002/term.1616
9. Pina S, Oliveira JM, Reis RL (2015) Natural-based nanocomposites for bone tissue engineering and regenerative medicine: a review. Adv Mater 27:1143–1169. https://doi.org/10.1002/adma.201403354
10. Gantar A, da Silva LP, Oliveira JM et al (2014) Nanoparticulate bioactive-glass-reinforced gellan-gum hydrogels for bone-tissue engineering. Mater Sci Eng C 43:27–36. https://doi.org/10.1016/j.msec.2014.06.045
11. Pereira DR, Canadas RF, Silva-Correia J et al (2014) Gellan gum-based hydrogel bilayered scaffolds for osteochondral tissue engineering. Key Eng Mater 587:255–260. https://doi.org/10.4028/www.scientific.net/KEM.587.255
12. Levato R, Visser J, Planell JA et al (2014) Biofabrication of tissue constructs by 3D bioprinting of cell-laden microcarriers. Biofabrication 6:35020. https://doi.org/10.1088/1758-5082/6/3/035020
13. Kirchmajer DM, Steinhoff B, Warren H et al (2014) Enhanced gelation properties of purified gellan gum. Carbohydr Res 388:125–129. https://doi.org/10.1016/j.carres.2014.02.018
14. Banik RM, Santhiagu A (2006) Improvement in production and quality of gellan gum by Sphingomonas paucimobilis under high dissolved oxygen tension levels. Biotechnol Lett 28:1347–1350. https://doi.org/10.1007/s10529-006-9098-3
15. Banik R, Santhiagu A, Upadhyay S (2007) Optimization of nutrients for gellan gum production by Sphingomonas paucimobilis ATCC-31461 in molasses based medium using response surface methodology. Bioresour Technol 98:792–797. https://doi.org/10.1016/j.biortech.2006.03.012
16. Morris ER, Nishinari K, Rinaudo M (2012) Gelation of gellan—a review. Food Hydrocoll 28:373–411. https://doi.org/10.1016/j.foodhyd.2012.01.004
17. Fialho AM, Moreira LM, Granja AT et al (2008) Occurrence, production, and applications of gellan: current state and perspectives. Appl Microbiol Biotechnol 79:889–900. https://doi.org/10.1007/s00253-008-1496-0
18. Osmałek T, Froelich A, Tasarek S (2014) Application of gellan gum in pharmacy and medicine. Int J Pharm 466:328–340. https://doi.org/10.1016/j.ijpharm.2014.03.038
19. Smith AM, Shelton RM, Perrie Y, Harris JJ (2007) An initial evaluation of gellan gum as a material for tissue engineering applications. J Biomater Appl 22:241–254. https://doi.org/10.1177/0885328207076522
20. Oliveira JT, Santos TC, Martins L et al (2009) Performance of new gellan gum hydrogels combined with human articular chondrocytes for cartilage regeneration when subcutaneously implanted in nude mice. J Tissue Eng Regen Med 3:493–500. https://doi.org/10.1002/term.184
21. Oliveira JT, Martins L, Picciochi R et al (2010) Gellan gum: a new biomaterial for cartilage tissue engineering applications. J Biomed Mater Res A 93:852–863. https://doi.org/10.1002/jbm.a.32574
22. Silva-Correia J, Oliveira JM, Caridade SG et al (2011) Gellan gum-based hydrogels for intervertebral disc tissue-engineering applications. J Tissue Eng Regen Med 5:e97–e107. https://doi.org/10.1002/term.363

23. Ferris CJ, Gilmore KJ, Wallace GG, in het Panhuis M (2013) Modified gellan gum hydrogels for tissue engineering applications. Soft Matter 9:3705. https://doi.org/10.1039/c3sm27389j
24. Miyoshi E (1996) Rheological and thermal studies of gel-sol transition in gellan gum aqueous solutions. Carbohydr Polym 30:109–119. https://doi.org/10.1016/S0144-8617(96)00093-8
25. Upstill C, Atkins EDT, Attwool PT (1986) Helical conformations of gellan gum. Int J Biol Macromol 8:275–288. https://doi.org/10.1016/0141-8130(86)90041-3
26. Chandrasekaran R (1997) Molecular architecture of polysaccharide helices in oriented fibers. Adv Carbohydr Chem Biochem 52:311–439
27. Ogawa E, Takahashi R, Yajima H, Nishinari K (2005) Thermally induced coil-to-helix transition of sodium gellan gum with different molar masses in aqueous salt solutions. Biopolymers 79:207–217. https://doi.org/10.1002/bip.20349
28. Horinaka J, Kani K, Hori Y, Maeda S (2004) Effect of pH on the conformation of gellan chains in aqueous systems. Biophys Chem 111:223–227. https://doi.org/10.1016/j.bpc.2004.06.003
29. Huang Y, Singh PP, Tang J, Swanson BG (2004) Gelling temperatures of high acyl gellan as affected by monovalent and divalent cations with dynamic rheological analysis. Carbohydr Polym 56:27–33. https://doi.org/10.1016/j.carbpol.2003.11.014
30. Picone CSF, Cunha RL (2011) Influence of pH on formation and properties of gellan gels. Carbohydr Polym 84:662–668. https://doi.org/10.1016/j.carbpol.2010.12.045
31. Ogawa E, Takahashi R, Yajima H, Nishinari K (2006) Effects of molar mass on the coil to helix transition of sodium-type gellan gums in aqueous solutions. Food Hydrocoll 20:378–385. https://doi.org/10.1016/j.foodhyd.2005.03.016
32. Doner LW (1997) Rapid purification of commercial gellan gum to highly soluble and gellable monovalent cation salts. Carbohydr Polym 32:245–247. https://doi.org/10.1016/S0144-8617(96)00168-3
33. Gong Y, Wang C, Lai RC et al (2009) An improved injectable polysaccharide hydrogel: modified gellan gum for long-term cartilage regeneration in vitro. J Mater Chem 19:1968. https://doi.org/10.1039/b818090c
34. Tang Y, Sun J, Fan H, Zhang X (2012) An improved complex gel of modified gellan gum and carboxymethyl chitosan for chondrocytes encapsulation. Carbohydr Polym 88:46–53. https://doi.org/10.1016/j.carbpol.2011.11.058
35. Nguyen KT, West JL (2002) Photopolymerizable hydrogels for tissue engineering applications. Biomaterials 23:4307–4314. https://doi.org/10.1016/S0142-9612(02)00175-8
36. Silva-Correia J, Oliveira JM, Oliveira JT, Sousa RA, Reis RL (2011) Photo-crosslinked Gellan gum-based hydrogels: methods and uses thereof. WO2011/119059, Priority date: 105030 26.03.2010 PT
37. Coutinho DF, Sant SV, Shin H et al (2010) Modified gellan gum hydrogels with tunable physical and mechanical properties. Biomaterials 31:7494–7502. https://doi.org/10.1016/j.biomaterials.2010.06.035
38. Silva-Correia J, Gloria A, Oliveira MB et al (2013) Rheological and mechanical properties of acellular and cell-laden methacrylated gellan gum hydrogels. J Biomed Mater Res Part A 101:3438–3446. https://doi.org/10.1002/jbm.a.34650
39. Pacelli S, Paolicelli P, Dreesen I et al (2015) Injectable and photocross-linkable gels based on gellan gum methacrylate: a new tool for biomedical application. Int J Biol Macromol 72:1335–1342. https://doi.org/10.1016/j.ijbiomac.2014.10.046
40. Pacelli S, Paolicelli P, Moretti G et al (2016) Gellan gum methacrylate and laponite as an innovative nanocomposite hydrogel for biomedical applications. Eur Polym J 77:114–123. https://doi.org/10.1016/j.eurpolymj.2016.02.007
41. Lee M, Tsai H, Wen S, Huang C (2012) Photocrosslinkable gellan gum film as an anti-adhesion barrier. Carbohydr Polym 90:1132–1138. https://doi.org/10.1016/j.carbpol.2012.06.064
42. Du H, Hamilton P, Reilly M, Ravi N (2012) Injectable in situ physically and chemically crosslinkable gellan hydrogel. Macromol Biosci 12:952–961. https://doi.org/10.1002/mabi.201100422

43. Novac O, Lisa G, Barbu E et al (2013) Synthesis and characterization of N-(2-aminoethyl)-2-acetamidyl gellan gum with potential biomedical applications. Carbohydr Polym 98:174–177. https://doi.org/10.1016/j.carbpol.2013.04.085
44. Wang C, Gong Y, Lin Y et al (2008) A novel gellan gel-based microcarrier for anchorage-dependent cell delivery. Acta Biomater 4:1226–1234. https://doi.org/10.1016/j.actbio.2008.03.008
45. Ferris CJ, Stevens LR, Gilmore KJ et al (2015) Peptide modification of purified gellan gum. J Mater Chem B 3:1106–1115. https://doi.org/10.1039/C4TB01727G
46. da Silva LP, Cerqueira MT, Sousa RA et al (2014) Engineering cell-adhesive gellan gum spongy-like hydrogels for regenerative medicine purposes. Acta Biomater 10:4787–4797. https://doi.org/10.1016/j.actbio.2014.07.009
47. Lopa S, Madry H (2014) Bioinspired scaffolds for osteochondral regeneration. Tissue Eng Part A 20:2052–2076. https://doi.org/10.1089/ten.tea.2013.0356
48. Seo S, Mahapatra C, Singh RK et al (2014) Strategies for osteochondral repair: Focus on scaffolds. J Tissue Eng 5:204173141454185. https://doi.org/10.1177/2041731414541850
49. Yang J, Zhang YS, Yue K, Khademhosseini A (2017) Cell-laden hydrogels for osteochondral and cartilage tissue engineering. Acta Biomater 57:1–25. https://doi.org/10.1016/j.actbio.2017.01.036
50. Rodrigues MT, Gomes ME, Reis RL (2011) Current strategies for osteochondral regeneration: from stem cells to pre-clinical approaches. Curr Opin Biotechnol 22:726–733. https://doi.org/10.1016/j.copbio.2011.04.006
51. Yousefi A-M, Hoque ME, Prasad RGSV, Uth N (2015) Current strategies in multiphasic scaffold design for osteochondral tissue engineering: a review. J Biomed Mater Res A 103:2460–2481. https://doi.org/10.1002/jbm.a.35356
52. Correia SI, Pereira H, Silva-Correia J et al (2013) Current concepts: tissue engineering and regenerative medicine applications in the ankle joint. J R Soc Interface 11:20130784–20130784. https://doi.org/10.1098/rsif.2013.0784
53. Ozbolat IT (2017) Bioprinting of osteochondral tissues: a perspective on current gaps and future trends. Int J Bioprinting 3:1–12. https://doi.org/10.18063/IJB.2017.02.007
54. Fan J, Gong Y, Ren L et al (2010) In vitro engineered cartilage using synovium-derived mesenchymal stem cells with injectable gellan hydrogels. Acta Biomater 6:1178–1185. https://doi.org/10.1016/j.actbio.2009.08.042
55. Shin H, Olsen BD, Khademhosseini A (2012) The mechanical properties and cytotoxicity of cell-laden double-network hydrogels based on photocrosslinkable gelatin and gellan gum biomacromolecules. Biomaterials 33:3143–3152. https://doi.org/10.1016/j.biomaterials.2011.12.050
56. Kesti M, Eberhardt C, Pagliccia G et al (2015) Bioprinting complex cartilaginous structures with clinically compliant biomaterials. Adv Funct Mater 25:7406–7417. https://doi.org/10.1002/adfm.201503423
57. Mouser VHM, Melchels FPW, Visser J et al (2016) Yield stress determines bioprintability of hydrogels based on gelatin-methacryloyl and gellan gum for cartilage bioprinting. Biofabrication 8:35003. https://doi.org/10.1088/1758-5090/8/3/035003
58. Melchels FPW, Dhert WJA, Hutmacher DW, Malda J (2014) Development and characterisation of a new bioink for additive tissue manufacturing. J Mater Chem B 2:2282. https://doi.org/10.1039/c3tb21280g
59. Yu I, Kaonis S, Chen R (2017) A study on degradation behavior of 3D printed gellan gum scaffolds. Procedia CIRP 65:78–83. https://doi.org/10.1016/j.procir.2017.04.020
60. Gkioni K, Leeuwenburgh SCG, Douglas TEL et al (2010) Mineralization of hydrogels for bone regeneration. Tissue Eng Part B Rev 16:577–585. https://doi.org/10.1089/ten.teb.2010.0462
61. Douglas TEL, Łapa A, Samal SK et al (2017) Enzymatic, urease-mediated mineralization of gellan gum hydrogel with calcium carbonate, magnesium-enriched calcium carbonate and magnesium carbonate for bone regeneration applications. J Tissue Eng Regen Med 11:3556–3566. https://doi.org/10.1002/term.2273

62. Douglas TEL, Krawczyk G, Pamula E et al (2016) Generation of composites for bone tissue-engineering applications consisting of gellan gum hydrogels mineralized with calcium and magnesium phosphate phases by enzymatic means. J Tissue Eng Regen Med 10:938–954. https://doi.org/10.1002/term.1875
63. Ding YH, Floren M, Tan W (2016) Mussel-inspired polydopamine for bio-surface functionalization. Biosurf Biotribol 2:121–136. https://doi.org/10.1016/j.bsbt.2016.11.001
64. Mourino V, Cattalini JP, Boccaccini AR (2012) Metallic ions as therapeutic agents in tissue engineering scaffolds: an overview of their biological applications and strategies for new developments. J R Soc Interface 9:401–419. https://doi.org/10.1098/rsif.2011.0611
65. Yang F, Dong W, He F et al (2012) Osteoblast response to porous titanium surfaces coated with zinc-substituted hydroxyapatite. Oral Surg Oral Med Oral Pathol Oral Radiol 113:313–318. https://doi.org/10.1016/j.tripleo.2011.02.049
66. Thian ES, Konishi T, Kawanobe Y et al (2013) Zinc-substituted hydroxyapatite: a biomaterial with enhanced bioactivity and antibacterial properties. J Mater Sci Mater Med 24:437–445. https://doi.org/10.1007/s10856-012-4817-x
67. Imai K, Nishikawa T, Tanaka A et al (2010) In vitro new capillary formation with eight metal ions of dental biomaterials. J Oral Tissue Eng 8:74–79
68. Douglas TEL, Schietse J, Zima A et al (2017) Novel self-gelling injectable hydrogel/alpha-tricalcium phosphate composites for bone regeneration: Physiochemical and microcomputer tomographical characterization. J Biomed Mater Res A. https://doi.org/10.1002/jbm.a.36277
69. Manda MG, da Silva LP, Cerqueira MT, et al (2017) Gellan gum-hydroxyapatite composite spongy-like hydrogels for bone tissue engineering. J Biomed Mater Res Part A 1–26. https://doi.org/10.1002/jbm.a.36248
70. Bellini D, Cencetti C, Meraner J et al (2015) An in situ gelling system for bone regeneration of osteochondral defects. Eur Polym J 72:642–650. https://doi.org/10.1016/j.eurpolymj.2015.02.043
71. Mouser VHM, Levato R, Bonassar LJ et al (2017) Three-dimensional bioprinting and its potential in the field of articular cartilage regeneration. Cartilage 8:327–340. https://doi.org/10.1177/1947603516665445

# Chapter 14
# Silk Fibroin-Based Hydrogels and Scaffolds for Osteochondral Repair and Regeneration

Viviana P. Ribeiro, Sandra Pina, J. Miguel Oliveira, and Rui L. Reis

**Abstract** Osteochondral lesions treatment and regeneration demands biomimetic strategies aiming physicochemical and biological properties of both bone and cartilage tissues, with long-term clinical outcomes. Hydrogels and scaffolds appeared as assertive approaches to guide the development and structure of the new osteochondral engineered tissue. Moreover, these structures alone or in combination with cells and bioactive molecules bring the mechanical support after in vitro and in vivo implantation. Moreover, multilayered structures designed with continuous interfaces furnish appropriate features of the cartilage and subchondral regions, namely microstructure, composition, and mechanical properties. Owing the potential as scaffolding materials, natural and synthetic polymers, bioceramics, and composites have been employed. Particularly, significance is attributed to the natural-based biopolymer silk fibroin from the *Bombyx mori* silkworm, considering its unique mechanical and biological properties. The significant studies on silk fibroin-based

---

Viviana P. Ribeiro and Sandra Pina contributed equally to this work

V. P. Ribeiro (✉) · S. Pina
3B's Research Group—Biomaterials, Biodegradables and Biomimetics, University of Minho, Headquarters of the European Institute of Excellence on Tissue Engineering and Regenerative Medicine, Avepark, Zona Industrial da Gandra, Barco, Guimarães 4805-017, Portugal

ICVS/3B's—PT Government Associate Laboratory, Braga/Guimarães, Portugal
e-mail: viviana.ribeiro@dep.uminho.pt

J. M. Oliveira · R. L. Reis
3B's Research Group—Biomaterials, Biodegradables and Biomimetics, University of Minho, Headquarters of the European Institute of Excellence on Tissue Engineering and Regenerative Medicine, Avepark, Zona Industrial da Gandra, Barco, Guimarães 4805-017, Portugal

ICVS/3B's—PT Government Associate Laboratory, Braga/Guimarães, Portugal

The Discoveries Centre for Regenerative and Precision Medicine, Headquarters at University of Minho, Avepark, Zona Industrial da Gandra, Barco, Guimarães 4805-017, Portugal

© Springer International Publishing AG, part of Springer Nature 2018
J. M. Oliveira et al. (eds.), *Osteochondral Tissue Engineering*,
Advances in Experimental Medicine and Biology 1058,
https://doi.org/10.1007/978-3-319-76711-6_14

structures, namely hydrogels and scaffolds, towards bone, cartilage, and osteochondral tissue repair and regeneration are overviewed herein. The developed biomimetic strategies, processing methodologies, and final properties of the structures are summarized and discussed in depth.

**Keywords** Silk fibroin · Hydrogels · Scaffolds · Osteochondral regeneration

## 14.1 Introduction

Tissue engineering (TE) field has been evolved as a way to compensate the limited supply of donor tissue and organ transplants related to a high morbidity and mortality [1]. TE approaches involve different research areas simultaneously, including cell biology, materials science, and clinical evaluation, with the final purpose of creating a suitable microenvironment that mimics the tissue at the host site in a desired and faster regeneration [2]. Such microenvironment is typically composed by three-dimensional (3D) porous scaffolds, on which cells are stimulated to grow and organize to form an extracellular matrix (ECM) used to initiate the regenerative process [3]. These 3D constructs provide the chemical and mechanical support for in vitro ECM formation, being gradually degraded, resorbed, or metabolized after in vivo implantation. Therefore, apart from an essential biocompatibility, the scaffolds must possess an equivalent degradation profile to the host tissue, while keeping the mechanical properties and structural integrity promoted by the forming ECM [4, 5].

Natural- and synthetic-based polymers, ceramic materials, and composites have been proposed for scaffolding strategies in TE approaches [6, 7]. Natural-based polymers have emerged as preferred sources for the development of scaffolds with better biocompatibility and lower risk of metabolized degradation products, while the synthetic polymers are more stable and easier to process and modify [8, 9]. For example, collagen, gelatin, and chitosan include some of the most investigated natural polymers in TE field. However, these scaffold materials may present poor mechanical properties associated to a rapid degradation profile. Structural proteins like elastin, fibrin, silk, and albumin have also been used as sutures, and more recently for scaffolds production, and as drug delivery agents [10, 11]. Among them, special interest has been attributed to silk protein produced by a wide range of arthropods and lepidopteran insects, including spiders, scorpions, mites, flies, and silkworms, which possess a large molecular weight of 200–350 kDa, or more. It has been used for centuries in textiles production and as clinical sutures (good skin affinity). Their availability for large-scale processing was also economically advantageous for use in TE applications [12]. From the different sources of silk proteins, *Bombyx mori* silk produced by silkworms became the most investigated for diverse TE applications, holding impressive mechanical properties, biocompatibility, biodegradability, low immunogenicity, and suitable processability [12]. It is synthesized in a liquid state in the epithelial cells of the insects' glands, secreted in the lumen and converted (spun) into a liquid-to-solid state (fibers) when in contact with the external air, being mechanically drawn in the form of cocoons. The spun fibers are composed by two animal-based proteins: a core protein named fibroin, surrounded by a glue-like protein named sericin [13]. Even it has been found that sericin may contain some biocompatibility problems [12], several studies have been proposing both silk fibroin (SF) [14–17] and silk sericin [18–20] for diverse biomedical and TE applications. SF has been recognized for presenting favorable biocompatibility, tunable molecular structure, and remarkable mechanical properties with controllable degradation rates, and for that reason it remains the most extensively studied silk protein as promising candidate for several structural [21], biomedical [16, 17], and TE applications [14, 15, 22]. Until now, several forms have been used to fabricate scaffolds made of SF, including films [23], membranes [16], fibers [16], textiles [15, 22], sponges [14], and hydrogels [24], used for the regeneration of soft tissues, like skin [23], ligament and tendon [25], and different hard tissues, including bone [15], cartilage [26], and osteochondral (OC) tissue [14].

The natural OC tissue existing in all human body joints is composed by two main tissues, the articular cartilage and subchondral bone, connected by a stable interface that unifies both elements as a single and complex tissue [27]. The great challenge of OC TE is the design of structures able to meet all the physicochemical and biological requirements for the repair/replacement of bone and cartilage tissues, and at the same time ensure a good compatibility between these two phases.

OC defects or damage can happen in any joint in human body, and affect both the articular cartilage, the underlying subchondral bone, as well its interface [27]. Moreover, the OC tissue repair involves a deep understanding of how the OC interface is combined in terms of structure, and mechanical and biological properties. Over the past years, several studies have been reported towards the repair/regeneration of these tissues, by creating single-phase structures able to fit into the defected area [26, 28–32].

Considering the heterogeneity of OC tissue, innovative 3D structures comprising different mechanical properties and biological performances, according to the target OC tissue layer, are strongly required. The recent development of bilayered scaffolds and improved multi-phased, or stratified, scaffolds with distinct subchondral bone and cartilage layers have been applied for this purpose [33, 34]. In general, OC TE strategies can be categorized into monophasic, biphasic, and triphasic depending on the physicochemical and cellular/biological characteristics of the scaffolds (Table 14.1) [35]. Bilayered hydrogels [36] and complex bilayered scaffolds have been reported for OC regeneration applications [37]. Recent studies in the field are leading to promising approaches to use SF-based biomaterials for OC tissue repair and regeneration strategies [38, 39]. The most recent and relevant studies focused on SF-based structures, namely hydrogels and scaffolds, targeting bone, cartilage, and OC tissue repair and regeneration are overviewed. Additionally, it is summarized developed strategies, processing methodologies, and final properties of the structures.

**Table 14.1** Osteochondral TE scaffolding design approaches

|  | Physicochemical properties |  | Biological properties |  |
|---|---|---|---|---|
| Monophasic | One material with same porosity |  | One cell type |  |
| Biphasic | Two layers with different material composition and porosity |  | Two different cell types, with bioactive molecules/growth factors |  |
| Triphasic | Three layers with different material composition and porosity |  | Three different cell types, with bioactive molecules/growth factors |  |

## 14.2 Characteristics of Silk Fibroin

Silk fibroin (SF) from *Bombyx mori* is composed by three protein components. A heavy (H) and a light (L) chain polypeptides of ~ 350 kDa and ~ 26 kDa, respectively, form the H-L complex linked by a single disulfide bond at the C-terminus of the H-chain (Fig. 14.1a). This H-L complex is also non-covalently linked to a glycoprotein (P25) of ~ 25 kDa in a ratio of 6:6:1 to form micellar units [40]. The H-chains are composed by hydrophobic domains containing highly ordered amino acid sequences repeats, capable of organizing themselves together into β-sheet or crystalline structures through intramolecular or intermolecular forces, including hydrogen bonding, van der Waals forces, and hydrophobic interactions, forming the basis for the tensile strength of SF (Fig. 14.1b) [41, 42]. The L-chains are smaller and less ordered than the H-chains, relatively elastic, and its sequence is not involved in the formation of the crystalline region in SF (amorphous region) (Fig. 14.1c) [40]. The hydrophobic repetitive domains that compose a H-chain are also interspaced by hydrophilic regions (Fig. 14.1a) [43]. These repeating units are known as Ala-Gly-Ser-Gly-Ala-Gly, where glycine (Gly, ~ 43–46%), alanine (Ala, ~ 25–30%), and serine (Ser, ~ 12%) are the three simplest and most abundant amino acids. The next most abundant amino acids in H-chain are tyrosine (~ 5%), the larger amino acid with a polar side chain (semi-crystalline), valine (~ 2%), followed by aspartic acid, phenylalanine, glutamic acid, threonine, isoleucine, leucine, proline, arginine, lysine, and histidine, present in much smaller percentages (less than 2%) [40, 44].

The complexity of SF can be demonstrated by their four different structures (silk I, silk II, silk III, and random coil), formed through different physicochemical stimuli, and that can be transformed to each other under proper conditions [45]. Silk I is formed alternatively by α-helix and β-sheet main conformations, while silk II is rich in β-sheet content and corresponds to the main structural configuration of SF providing high mechanical and physicochemical properties. Silk III is formed by a threefold α-helix crystal structure, and the random coil structure usually exists in the SF solution [44, 46]. By controlling the crystalline and amorphous domains of SF structures (size, number, distribution, orientation, and spatial arrangement), it is

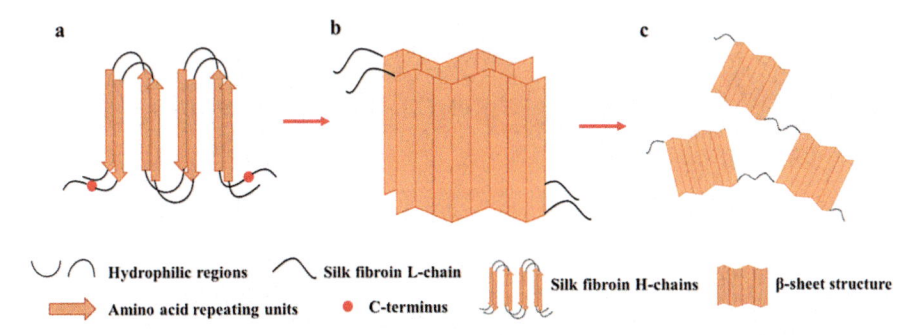

**Fig. 14.1** Schematic illustration of the natural SF protein composition: (**a**) H-L complex formation. (**b**) H-chains organizing themselves together into β-sheet structures. (**c**) β-sheet structures linked by amorphous domains

possible to produce SF-based matrices with distinct mechanical properties, degradation profile, and aqueous processability, which makes this protein attractive for distinct biomedical and TE applications [12].

## 14.3 Silk Fibroin-Based Hydrogels

Hydrogels formation follows several distinctive requirements that mimic the natural ECM microenvironment of tissues [47]. The hydrophilic nature of the ECM is represented by the hydrophilic crosslinking of polymer-based hydrogels, formed by the reaction of one, or more monomers connected by hydrogen bonds or/and van der Waals interactions between the chains [48]. One of the most important advantages of hydrogel systems is their aqueous environment that not only protects cells, but can also be sensitive for drugs and biological agents incorporation, transport, and delivery at the injury site [47]. Moreover, they can present tailored mechanical properties, degradation profiles, and swelling abilities according to their final applications [24, 49]. Facing the traditional scaffolding strategies, hydrogel networks have also been proposed as injectable systems not only for TE strategies, but also for other clinical applications [50].

The naturally derived hydrogels have desirable biological properties as compared to synthetic hydrogels; however, they can present rapid degradation profiles for hard tissues regeneration, not to mention the chemical and molecular instability, which usually limits the reproducibility of natural-based materials [51]. Such limitations can be overcome through hydrogels formed from regenerated aqueous SF solutions, that when submitted to different physical and chemical treatments, including mechanical agitation, ultrasonication, thermal treatment, pH adjustments, organic solvents, crosslinking using ionic species (Ca$^{2+}$ ions) or biological agents (enzymes), acquire a sol-gel transition (Fig. 14.2) [24, 52–54]. During the structural transition process, an interconnected network is formed in the aqueous solution

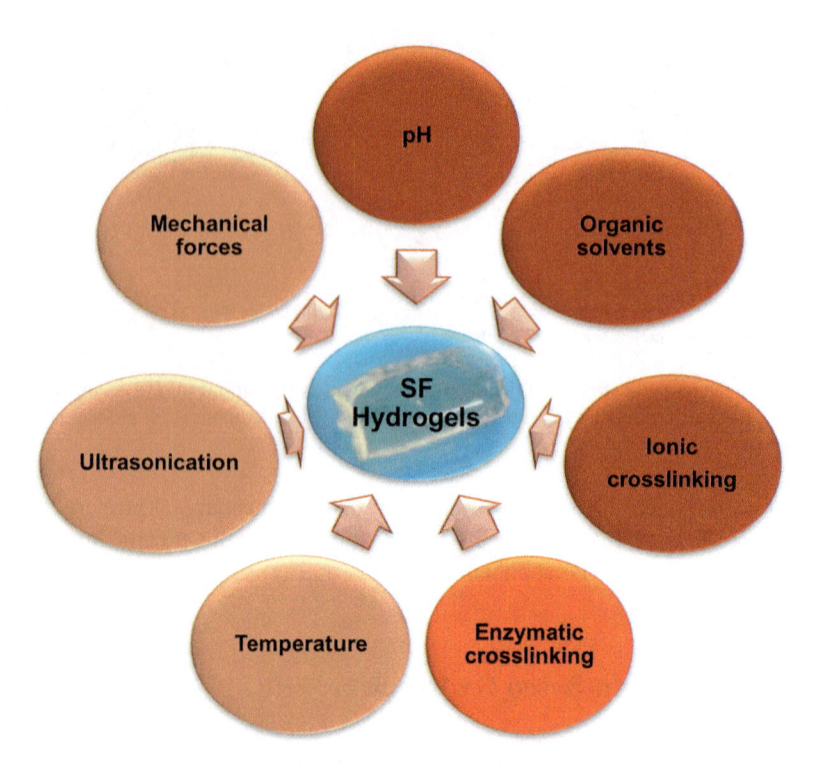

Fig. 14.2 Schematic illustration of the physicochemical processes used to prepare SF hydrogels

either by the β-sheet aggregates formation (transition from random coil to β-sheet) [55] or by the crosslinking of fibroin molecules [56]. Yan et al. [50] proposed a horseradish peroxidase (HRP)-mediated crosslinking approach to produce novel SF hydrogels in a random coil conformation, that can undergo intrinsic conformational changes from amorphous to β-sheet (Fig. 14.3). These hydrogels can be adjusted in terms of gelation time, mechanical properties, and degradation profile, only by changing the SF concentration and the crosslinker (HRP/$H_2O_2$) ratio. This will allow producing different hydrogel networks according to its final application. Moreover, the enzymatic crosslinking of SF was conducted at physiological conditions, envisioning their application as injectable systems for drug delivery purposes. In a recent study [57], the potential applications of SF-based hydrogels were magnified by the production of agarose-SF sponges, processed by freeze-drying agarose-SF blended hydrogels.

The traditional approach for TE and regenerative medicine involve the culture of cells, withdrawn from the host tissue, into a pre-established structural matrix and subsequent implantation into the defect site [1]. Structures designed to mimic the natural ECM microenvironment of the replacing tissues usually involve the engineering of matrices at several levels, depending on the physical, chemical, and biological properties of the host tissue, and SF has been proven to be a suitable material to engineer different tissues [4, 12, 58]. Combining polymeric scaffolds for osteogenesis induction with a cartilaginous-like hydrogel matrix has been one of the most

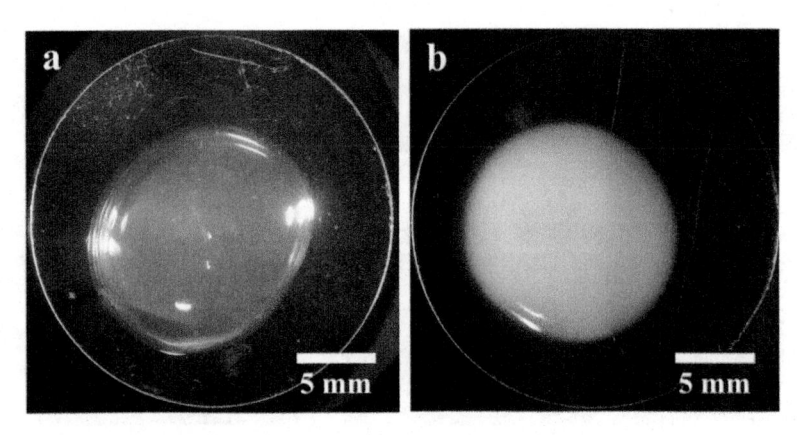

**Fig. 14.3** Macroscopic images of the horseradish peroxidase-mediated crosslinked SF hydrogels, analyzed (**a**) in a random coil conformation at day 1 and (**b**) after β-sheet conformational transition at day 14

studied strategies for OC tissue engineering [37, 59]. Furthermore, the mechanical strength and high biocompatibility of SF make this material a rational choice as injectable fillers or as scaffolds for bone TE purposes [60, 61]. Fini et al. [60] have tested the in vitro behavior of injectable SF hydrogels, through osteoblast culture, and after in vivo implantation into critical-sized defects of rabbit distal femurs. The proposed SF hydrogels were obtained by adding citric acid into aqueous SF solution, and a commercial synthetic poly(D,L lactide-glycolide) (PLGA) copolymer was used as control material [62]. *In vitro* tests showed a significant increase of cell proliferation in the SF hydrogels and a higher bone remodeling capability was observed after in vivo implantation into the femoral defects, as compared to the synthetic PLGA control. Sonication-induced SF hydrogels were also proposed by Zhang et al. [63] as injectable bone replacement biomaterials (Fig. 14.4). These hydrogels were combined with the osteogenic-related growth factors, vascular endothelial growth factor (VEGF), and bone morphogenic protein-2 (BMP-2), and evaluated as vehicles for the encapsulation and release of biological agents. The dual factors were slowly released from the injectable SF hydrogels promoting angiogenesis and new bone formation after in vivo implantation in rabbit maxillary sinus. The authors concluded that the proposed SF hydrogels can be used as injectable matrices in a minimally invasive approach to fill and regenerate bone tissue. Moreover, the possibility of being used as vehicles to deliver multiple growth factors was also a great achievement.

Considering the high incidence of articular cartilage-related injuries [64], hydrogels are particularly desired matrices for cartilage TE, since these are water-swollen materials capable of retaining and absorbing large volumes of water and maintaining sufficient mechanical properties to support loading forces. Moreover, hydrogels have shown to be capable of encapsulating cells, biomolecules, and growth factors, for controlled drug delivery approaches after implantation in cartilage defects [65, 66]. SF-based hydrogels have also shown great potential for cartilage regeneration

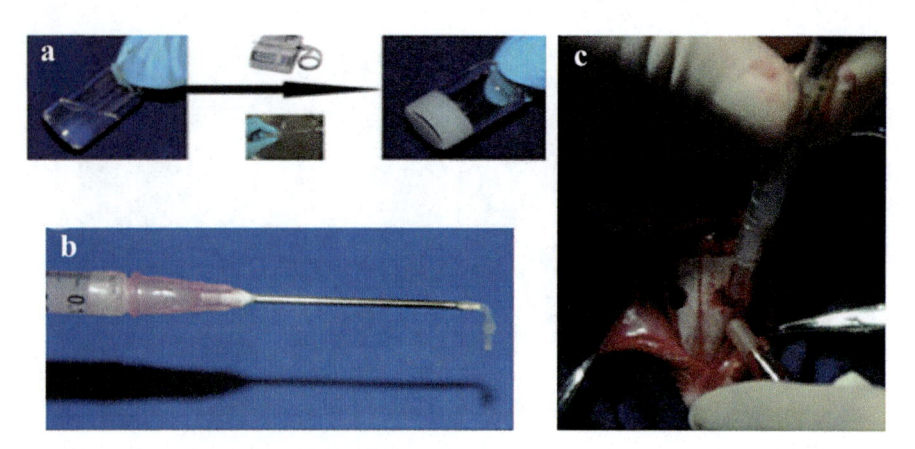

**Fig. 14.4** Photographs of the sonication-induced SF hydrogels and rabbit maxillary sinus surgery: (**a**) Aqueous SF solution transforming into SF hydrogels by ultrasonication. (**b**) Injectable property of the SF hydrogels. (**c**) SF hydrogel injection into the rabbit maxillary sinus cavity. Adapted with permission [63]. Copyright 2011, Elsevier

applications [67]. Sonication-induced SF hydrogels were proposed by Chao et al. [68] as an alternative approach to the commonly accepted agarose hydrogels [69] that yield the ability to sustain immature chondrocytes with biomechanical properties comparable to the native cartilage tissue. However, the non-degradability and lack of possibilities to modify the agarose's structure, composition, and mechanical properties increased the author's interest of using biocompatible, biodegradable, and highly tuned SF hydrogels to prepare cartilaginous constructs. These hydrogels presented a variety of structural and mechanical properties according to the SF extraction method, concentration, and gelation conditions. Moreover, the rapid encapsulation of chondrocytes and full maintenance of cell viability for 42 days with ECM formation (collagen and glycosaminoglycans) suggests that these hydrogels can be used as 3D models of cartilage tissue formation and maturation. In a different study, Park et al. [70] proposed novel sonication-induced SF composite hydrogels with fibrin/hyaluronic acid for nucleus pulposus cartilage formation. The authors demonstrated that the composite hydrogels allowed the chondrogenic differentiation in five different groups made of fibrin/hyaluronic acid and different SF concentrations. Importantly, the mechanical strength measurements also showed that SF induced a stronger mechanical support for cartilage tissue on composite hydrogels than fibrin/hyaluronic acid hydrogels alone. Yodmuang et al. [71] have proposed SF-based composites made by combining silk microfibers with SF hydrogels as a potential support material for cartilage TE. SF fiber-agarose hydrogel composite materials were used as control condition, showing that the 100% SF-based composites presented better and similar mechanical properties to those of native cartilage tissue (Fig. 14.5). The SF fiber reinforcement significantly influenced the mechanical and biological properties of composite materials supporting chondrocytes maturation and cartilage matrix deposition. Once again, the versatility of SF as a composite material came to overcome the limitations presented by the "gold

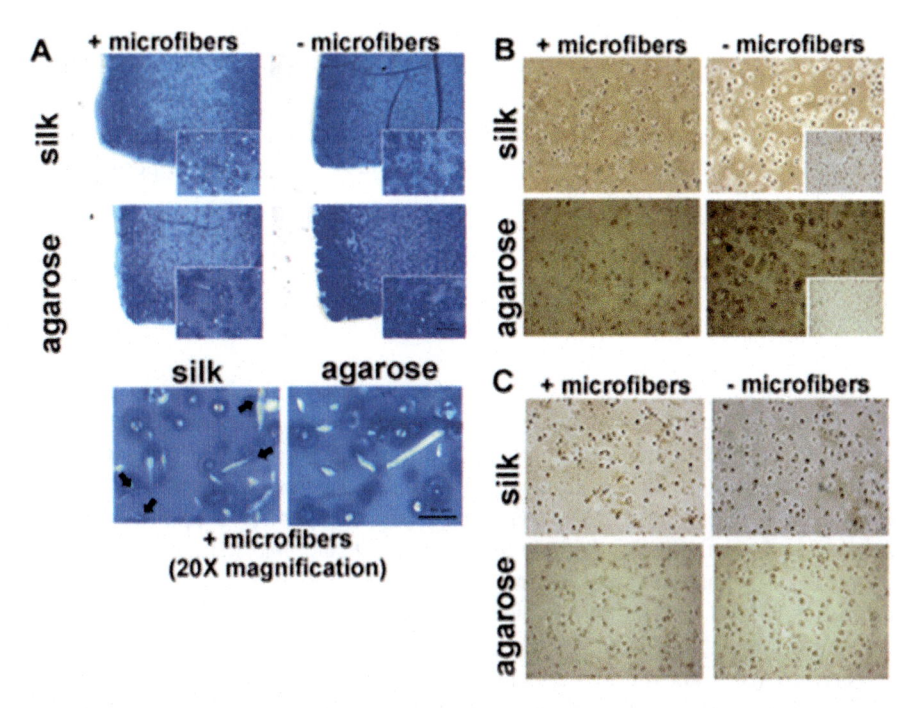

**Fig. 14.5** Histological analysis of the SF- and agarose-based hydrogels combined with silk microfibers. (**A**) Alcian blue staining used to visualize the glycosaminoglycans (GAGs) content within the hydrogels. The black arrows indicate the localization of GAGs surrounding the silk microfibers within the SF-based hydrogels. Immunohistochemical analysis of (**B**) type II collagen and (**C**) type I collagen content. The inset images represent the negative staining. Adapted with permission [71]. Copyright 2014, Elsevier

standard" agarose-based biomaterials. The same recognition was done by Singh et al. [57] that demonstrated higher levels of cartilaginous tissue formation (glycosaminoglycans and collagen matrix deposition) within microporous hydrogels of SF blended with agarose, as compared with the microporous hydrogels of agarose used as control.

Lately, bilayered hydrogels combining different polymeric materials [36, 72], or encapsulating growth factors and/or cell populations, have also shown promising results in OC tissue repair/regeneration [73–75]. SF biomaterials can be particularly attractive for these strategies due to the self-assembly properties and controlled processing of the β-sheet crystalline content, which enable to modulate the degradation rate and mechanical properties of SF structures according to the target OC tissue [76]. In a recent study, a 3D bio-printing method was used to create SF-gelatin (SF-G) bioinks incorporated with human mesenchymal stem cells [77]. Enzymatic and physical crosslinking methods were applied after cell incorporation for post-printing stabilization, showing that both developed constructs supported multilineage differentiation and specific tissue formation according to the applied crosslinking method. These results provide a proof-of-concept for the fabrication of 3D heterogeneous tissue constructs using different crosslinking methods of SF-G hydrogels with different mechanical properties and biological effects, especially required for OC tissue regeneration. Moreover, the possibility of creating a printed construct for a target tissue in a patient-specific approach would be the answer for personalized OC therapy and regeneration.

## 14.4 Silk Fibroin-Based Scaffolds

Scaffolds are 3D porous matrices developed to provide a defined microenvironment that promotes tissue repair and regeneration. Ideally, scaffolds should be able to: (1) stimulate cell-biomaterial interactions, cell attachment, growth, and migration, (2) facilitate transport of mass, nutrients, and regulatory factors for cell survival, proliferation, and differentiation, (3) afford structural mechanical support, as tensile strength and elasticity, (4) degrade at a controlled rate, and (5) present minimal degree of inflammation or toxicity in vivo [78]. Further, scaffolds have desired characteristics for cell transfer into a defect site and to limit cell loss, instead of simple cells injection to the defects.

Layered scaffolds aiming bone and OC TE applications, where both underlying bone and cartilage tissues are damaged, are able to promote regeneration with specific properties and biological requirements [49, 55–57]. The strategy is the fabrication of stratified scaffolds consisting of separate osteogenic and chondrogenic regions, which can be manufactured in a single integrated implant, or fabricated independently and joined together with sutures or sealants. It is pretended to ensure a good compatibility between the regions by keeping the porous structure and the mechanical strength. Porosity and pore sizes, respectively, of 50–90% and 300 μm are required for an improved osteogenesis, whereas a pore size between 90 and 120 μm is recommended for chondrogenesis [79]. Several technologies have been applied to produce scaffolds with organized porosity and pore size, namely foam replica, salt-leaching/solvent casting, freeze-drying, phase separation, gas foaming, rapid prototyping, supercritical fluid technology, additive manufacturing, photolithography, microfluidics, and electrospinning [14, 26, 31, 33, 80–86]. These techniques also allow envisioning the encapsulation of pharmaceutical agents/drugs and cells.

Considering the unique properties of SF for biomedical applications as abovementioned, the fabrication of useful scaffold SF-based systems, as well as constructs, has been extensively investigated with very positive results, to repair and regenerate the bone, cartilage, and OC tissues [26, 31, 39, 87, 88]. Saha et al. [39] evaluated the osteo-/chondro-inductive ability of acellular mulberry and non-mulberry SF scaffolds as an implantable platform in OC therapeutics. It was shown that SF scaffolds of *Antheraea mylitta* (non-mulberry) were more chondro-inductive, while those of *Bombyx mori* (mulberry) were more osteo-inductive in similar conditions. The in vitro culture in chondro- and osteo-inductive media, showed that non-mulberry constructs seeded with human bone marrow stromal cells exhibited chondrocyte-like cells behavior up to 8 weeks of culture, whereas mulberry constructs seeded with human bone marrow stromal cells formed bone-like nodules. In vivo neomatrix formation on the scaffolds, absorbed with transforming growth factor β-3 or recombinant human BMP-2, was demonstrated after implantation for 8 weeks, in OC defects of the knee joints of Wistar rats. The neomatrix formed comprised collagen and glycosaminoglycans except in mulberry silk without growth factors, where a predominantly collagenous matrix was observed.

A different strategy was reported by Chen et al. [89] where they used SF sponge scaffolds seeded with rabbit bone marrow stromal cells (BMSC) aiming to engineer a multilayered OC construct. BMSC-seeded scaffolds were first cultured separately in osteogenic and chondrogenic stimulation media. Then, the differentiated pieces were combined with RADA self-assembling peptides and subsequently co-cultured. It was shown that after co-culture, the GAG production in the chondrogenic region was downregulated compared with the chondrogenic control group, while the GAG production in the osteogenic region was greater than from the osteogenic control group. Furthermore, in the intermediate region of co-cultured samples, hypertrophic chondrogenic gene markers collagen type X and MMP-13 were found on both chondrogenic and osteogenic sections. However, significant differences of gene expression profile were found in distinct zones of the constructs co-cultured, and the intermediate region had significantly higher hypertrophic chondrocyte gene expression. Moreover, results showed that specific stimulation from osteogenic and chondrogenic BMSCs affected both layers inducing the formation of an OC interface. Another interesting work investigated the capability of regenerated SF/natural degummed silk fiber composite scaffolds, combined with fibrin glue, with and without autologous chondrocytes for the repair of OC lesions [90, 91]. The scaffolds showed very good mechanical properties and porosity due to the incorporation of silk fibers into the SF. In vivo biocompatibility tests of the scaffolds after implantation in OC defects of rabbit knees demonstrated good healing, regular chondrocyte arrangement, great connection to native tissues, and complete degradation 36 weeks post-implantation.

In order to improve the rapid degradation and low mechanical strength of pure chondroitin sulfate (CS), Zou et al. [92] developed a scaffold combining SF with CS, using salt-leaching, freeze-drying, and crosslinking methodologies, for cartilage tissue repair. The scaffolds exhibited a porous and interconnected structure with pores sizes of approximately 100–300 μm (Fig. 14.6I). In vitro biocompatibility tests using human articular chondrocytes (hACs) cultured on the scaffolds showed the formation of clusters inside the pores of SF scaffold, but better adhesion in SF/CS scaffold (Fig. 14.6II). After 12 weeks post-implantation in a rabbit OC defect model, the defects in SF scaffolds were repaired with fibrocartilage tissue and cartilage-like tissue, generated a thinner layer compared to the surrounding normal cartilage tissue. The defects in SF/CS scaffolds were repaired by thicker cartilage-like tissue, and well-organized subchondral bone (Fig. 14.6III). It was also observed that SF/CS scaffolds maintained better chondrocyte phenotype than SF scaffold, and silk-CS scaffolds reduced chondrocyte inflammatory response that

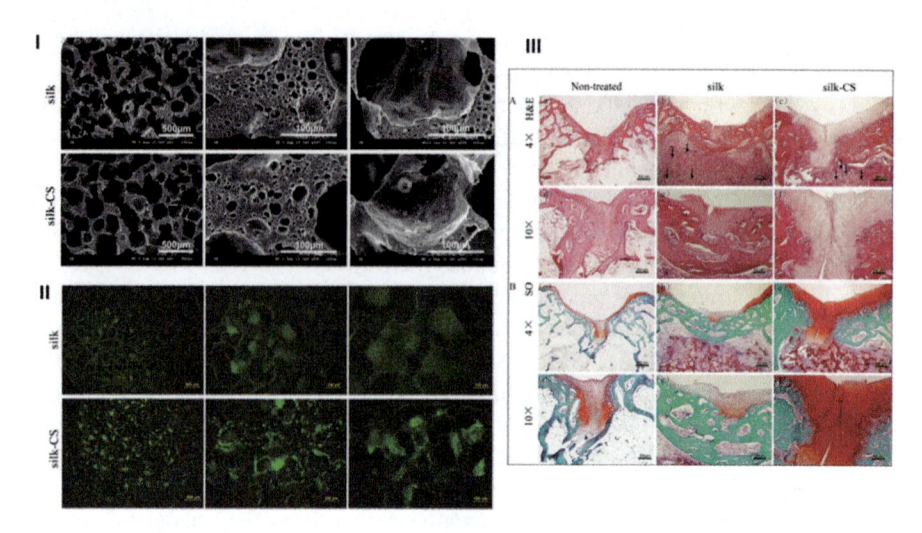

Fig. 14.6 (I) Microstructure of SF and SF/CS scaffolds observed by scanning electron microscopy. (II) Cytocompatibility of SF and SF/CS scaffolds with hACs (scale bars: 500, 200, and 100 μm). (III) Histological evaluation after 12 weeks post-implantation: (A) Hematoxylin and eosin (H&E) histological images. (a, d) Non-treated group; (b, e) SF group; (c, f) SF/CS group. Arrows indicate rudimental scaffold materials. (B) Safranin-O staining of histological sections. (a, d) Non-treated group; (b, e) SF group; (c, f) SF/CS group. (Aa–c) and (Ba–c), scale bar: 500 mm. (Ad–f) and (Bd–f), scale bar: 200 mm. (A, B) Images of the bottom row represent the top row images at higher magnification. Adapted with permission [92]. Copyright 2017, Elsevier

was induced by interleukin (IL)-1β, consistent with the well-reported anti-inflammatory activities of CS.

Alternative approaches for OC TE involve SF-based structures combined with other functional materials, as the case of calcium phosphates (CaPs), can significantly enhance its biofunctionalities, and hence improved advantages of the final composites [14, 31, 87]. Yan et al. [14] prepared bilayered scaffolds for OC defects regeneration, consisting of SF and SF/CaP, respectively, for cartilage and bone regions using salt-leaching/freeze-drying techniques (Fig. 14.7A). The scaffolds showed improved micro/macrostructure able to promote cell attachment and proliferation, as well as enhanced osteoconductivity and mechanical strength by the incorporation of calcium phosphate in SF. Also, good adhesion and proliferation of rabbit bone marrow mesenchymal stromal cells (RBMSCs) cultured on the scaffolds during 7 days were observed. The scaffolds were implanted subcutaneously and in critical sizes of OC defects in the rabbit knee for 4 weeks, showing a fully integration into the host tissue with no inflammation (Fig. 14.7B). Moreover, the ingrowths of the subchondral bone in the bottom domain and the regeneration of cartilage in the surface area of the implant were observed. The quantitative results of CaP content and porosity of different regions showed much higher void space in the defect controls than in the defects with implant, with CaP content of 20% higher in SF/CaP layer than in the SF. From Fig. 14.7C, it can be observed that the connective tissues were tightly integrated in the implants, and filled the inner pores of the scaffolds,

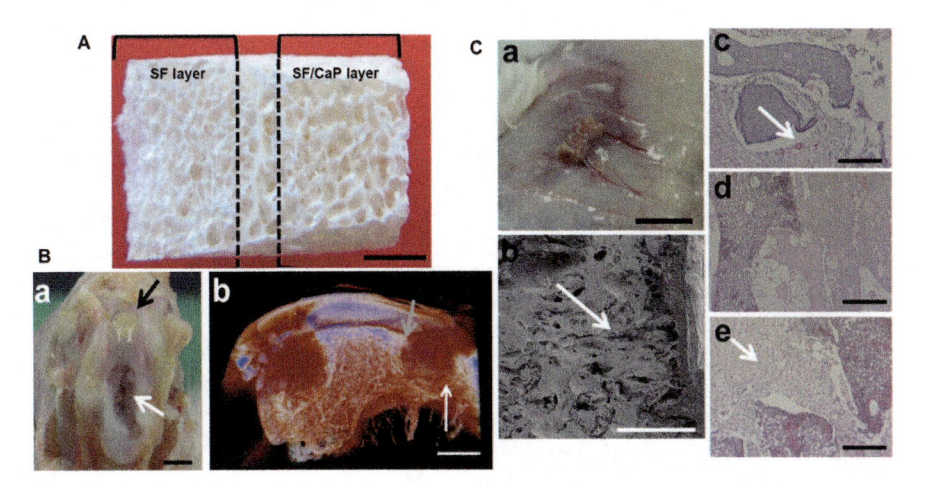

14 Silk Fibroin-Based Hydrogels and Scaffolds for Osteochondral Repair... 317

**Fig. 14.7** (**A**) Macroscopic image of the bilayered scaffolds (scale bar: 3 mm). (**B**) Images of the bilayered explants after implantation in rabbit OC defects: (a) Macroscopic image of the explants, the black arrow indicates the implanted scaffold and the white arrow indicates the defect control (scale bar: 5 mm); (b) micro-CT analysis of the explants, the grey arrow indicates neocartilage, and the white arrow indicates new subchondral bone formation (scale bar: 4 mm). (**C**) Images of the bilayered explants after subcutaneous implantation in rabbit: (a) Macroscopic image of the explants (scale bar: 1 cm); (b) SEM image of the explants (scale bar: 1 mm), the arrow indicates the interface; (c-e) H&E staining of the SF layer, interface and SF/CaP layer in the explants, respectively (scale bar: 200 mm). The arrow in (c) indicates vessels, and the arrow in (e) indicates fibroblasts. Adapted with permission [14]. Copyright 2014, Elsevier

with visible vessels formed inside the scaffolds, and some fibroblasts presented in the SF/CaP layer.

Recently, in a work developed by our group, biofunctional scaffolds composed of SF and β-tricalcium phosphate, incorporating different ions, reported enhanced mechanical properties, improved cell proliferation, and higher osteogenic potential, which can also be used to engineer the bone layer in OC applications [93]. The scaffolds showed macropores highly interconnected with a size around 500 μm, and microporous structure pores with a size range of 1–10 μm (Fig. 14.8A). The biomineralized SF scaffolds, after immersion in SBF for 15 days, showed globule-like structures of apatite crystals, while the incorporation of ceramic powders into SF leads to the formation of porous spherulites-like structures (Fig. 14.8B). Interestingly, in vitro assays using hASCs presented different responses on cell proliferation/differentiation when varying the ionic agents in the biofunctional scaffolds (Fig. 14.8C). The incorporation of Zn into the scaffolds led to improved proliferation, while the Sr- and Mn-doped scaffolds presented higher osteogenic potential as demonstrated by DNA quantification and ALP activity. The combination of Sr with Zn led to an influence on cell proliferation and osteogenesis when compared with single ions.

As mentioned earlier, biomimetic OC multilayered systems, with specific microstructures and properties, have the potential clinical benefit in promoting bone and cartilage tissue repair and replacement. Taking an OC approach, Çakmak et al. [94]

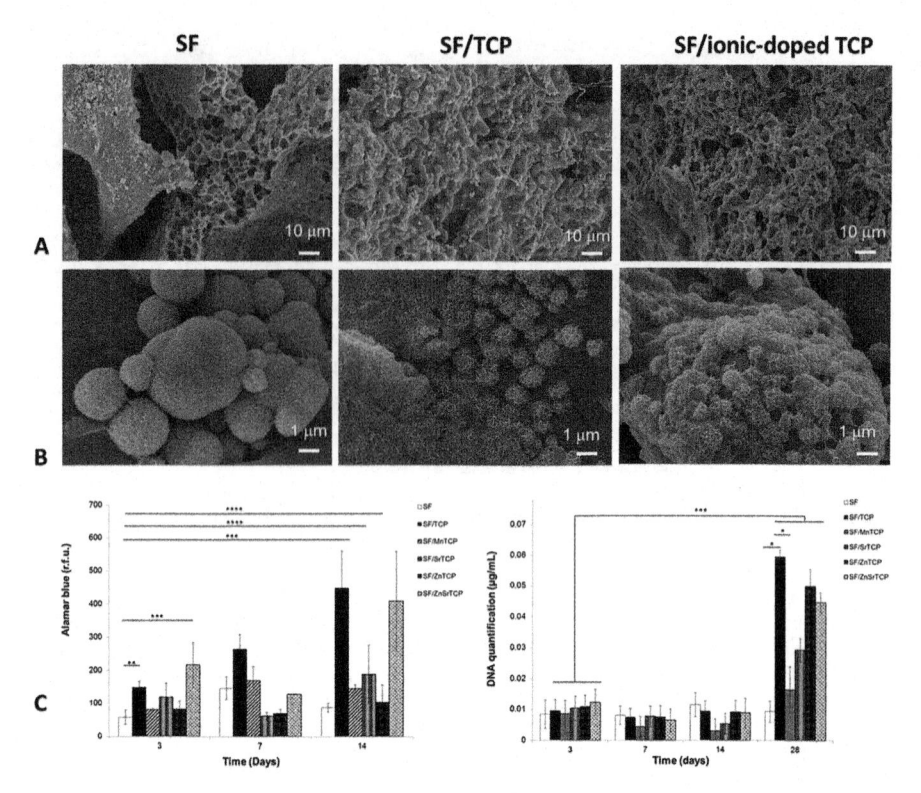

**Fig. 14.8** Scanning electron micrographs of SF, SF/TCP and SF/ionic-doped TCP scaffolds. (**A**) Before and (**B**) after 15 days of mineralization. (**C**) Viability and proliferation of hASCs seeded in the scaffolds: Alamar blue assay of hASCs cultured for 14 days (left), and DNA quantification at different time points (right). *significant differences compared with SF and SF/MnTCP and with SF/MnTCP and SF/SrTCP ($p$ <0.05), **significant differences compared between SF and SF/TCP ($p$ <0.005), ***significant differences compared between SF at day 3 and the different compositions at day 28 ($p$ <0.0005), ****significant differences compared between with SF and SF/SrTCP and SF/ZnTCP ($p$ <0.0001). Adapted with permission [93]. Copyright 2017, S. Karger AG, Basel

designed a SF-based trilayered scaffold suitable for both bone and cartilage, fabricated by salt-leaching process. For this purpose, the bone side was prepared with 4% (w/v) SF plus 5% (w/w) hydroxyapatite, the interface was obtained from 4% (w/v) SF, and for the cartilage layer were used arginine-glycine-aspartic acid-serine (RGDS)-containing peptide amphiphile hydrogels. The final mean pore size obtained for bone and OC interface layers was, respectively, 416 ± 87 μm and 194 ± 67 μm. Osteogenic and chondrogenic activity were evaluated by hBMSCs cultured in the SF scaffold in osteogenic media, while hACs were encapsulated and cultured inside the PA-RGDS in chondrogenic media, without using selective growth factors. After 2 weeks of growing separately, the bone and cartilage layers were combined with the interface layer by the soft silk scaffolds, followed by co-cultured in an OC cocktail medium. Results showed that the presence of hACs in the co-cultures significantly increases the osteogenic differentiation of hBMSCs, whereas hACs produces a significant amount of glycosaminoglycans (GAGs) for

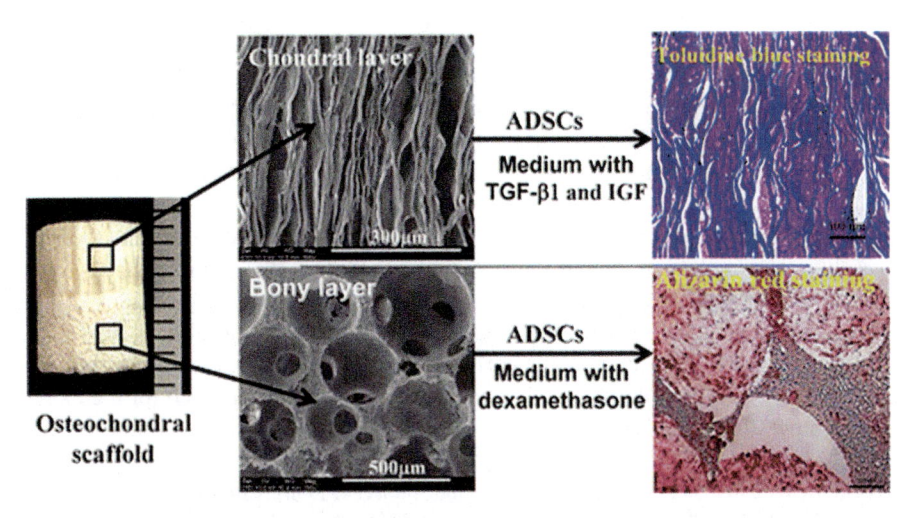

**Fig. 14.9** Macroscopic image of the OC scaffold showing the microstructure of bone and chondral layers. Adapted with permission [38]. Copyright 2014, American Chemical Society

the cartilage region. Moreover, the effect of hBMSCs on chondrogenic differentiation of hACs was less effective than that of hACs on hBMSCs, and the hACs in the co-culture preserved the amount of synthesized chondrogenic ECM. Ding et al. [38] developed a trilayered scaffold combining SF/hydroxyapatite and paraffin-sphere leaching with a modified temperature gradient-guided thermal-induced phase separation (TIPS) technique. The bone layer and interface are constituted respectively by a porous and dense structure (Fig. 14.9). Live/dead tests indicated good biocompatibility for supporting the growth, proliferation, and infiltration of adipose-derived stem cells (ADSCs) in the scaffolds. Histological and immunohistochemical stainings confirmed that the ADSCs could be induced to differentiate towards chondrocytes or osteoblasts in vitro at chondral and bony layers in the presence of chondrogenic or osteogenic culture medium, respectively. Moreover, the intermediate layer could play an isolating role for preventing the cells within the chondral and bone layers from mixing with each other.

## 14.5 Concluding Remarks and Research Efforts

Many progresses have been made over the past few decades in order to fully treat and replace the damaged or non-functional OC tissues. TE is an essential approach for that, which can combine different biomaterials, bioactive molecules, and cells. Repairing OC lesions remains a formidable challenge due to the high complexity of native OC tissue, and the limited self-repair capability of cartilage. Innovative strategies, such as the ones aforementioned, present solutions for specific OC challenges, and take important roles in cell proliferation and differentiation, envisioning the formation of new tissues. Such approaches will provide the production of hybrid constructs that act as bioresorbable temporary implants and resemble the physical characteristics of the ECM. Hydrogels are defined as possessing high water content and viscoelastic nature. On the other hand, scaffolds have mechanical strength necessary to temporarily offer structural support until new tissue ingrowth.

Among all the natural biopolymers presently available, SF has shown remarkable potential for biomedical applications due to its favorable structural and mechanical properties, as well good biocompatibility, biodegradability, and thermal stability. The design of 3D structures involving SF results in biomaterials with structural integrity for self-healing and load-bearing for future applications in OC TE. Furthermore, multiphasic structures, with distinct subchondral bone and cartilage regions and well-integrated interface, can overcome the common problems experienced with monolayers scaffolds, and be an effective approach for the effective regeneration of OC tissue.

**Acknowledgments** The authors thank to the project FROnTHERA (NORTE-01-0145-FEDER-000023), supported by Norte Portugal Regional Operational Programme (NORTE 2020), under the PORTUGAL 2020 Partnership Agreement, through the European Regional Development Fund (ERDF). The financial support from the Portuguese Foundation for Science and Technology to Hierarchitech project (M-ERA-NET/0001/2014), for the fellowship grant (SFRH/BPD/113806/2015) and for the fund provided under the program Investigador for J. M. Oliveira (IF/00423/2012 and IF/01285/2015) are also greatly acknowledged.

# References

1. Nerem RM, Sambanis A (1995) Tissue engineering: from biology to biological substitutes. Tissue Eng 1(1):3–13
2. Langer R, Vacanti JP, Vacanti CA, Atala A, Freed LE, Vunjak-Novakovic G (1995) Tissue engineering: biomedical applications. Tissue Eng 1(2):151–161
3. Hubbell JA (1995) Biomaterials in tissue engineering. Nat Biotechnol 13(6):565–576
4. Ma PX (2008) Biomimetic materials for tissue engineering. Adv Drug Deliv Rev 60(2):184–198
5. Furth ME, Atala A, Van Dyke ME (2007) Smart biomaterials design for tissue engineering and regenerative medicine. Biomaterials 28(34):5068–5073
6. Shin H, Jo S, Mikos AG (2003) Biomimetic materials for tissue engineering. Biomaterials 24(24):4353–4364
7. Seal B, Otero T, Panitch A (2001) Polymeric biomaterials for tissue and organ regeneration. Mater Sci Eng R Rep 34(4):147–230
8. Lutolf M, Hubbell J (2005) Synthetic biomaterials as instructive extracellular microenvironments for morphogenesis in tissue engineering. Nat Biotechnol 23(1):47–55
9. Nair LS, Laurencin CT (2005) Polymers as biomaterials for tissue engineering and controlled drug delivery. In: Tissue engineering I. Springer, Berlin, pp 47–90
10. Nair LS, Laurencin CT (2007) Biodegradable polymers as biomaterials. Prog Polym Sci 32(8):762–798
11. Malafaya PB, Silva GA, Reis RL (2007) Natural–origin polymers as carriers and scaffolds for biomolecules and cell delivery in tissue engineering applications. Adv Drug Deliv Rev 59(4):207–233
12. Kundu B, Rajkhowa R, Kundu SC, Wang X (2013) Silk fibroin biomaterials for tissue regenerations. Adv Drug Deliv Rev 65(4):457–470

13. Mondal M (2007) The silk proteins, sericin and fibroin in silkworm, Bombyx mori Linn., a review. Caspian J Environ Sci 5(2):63–76
14. Yan L-P, Silva-Correia J, Oliveira MB, Vilela C, Pereira H, Sousa RA, Mano JF, Oliveira AL, Oliveira JM, Reis RL (2015) Bilayered silk/silk-nanoCaP scaffolds for osteochondral tissue engineering: in vitro and in vivo assessment of biological performance. Acta Biomater 12:227–241
15. Ribeiro VP, Silva-Correia J, Nascimento AI, da Silva MA, Marques AP, Ribeiro AS, Silva CJ, Bonifácio G, Sousa RA, Oliveira JM (2017) Silk-based anisotropical 3D biotextiles for bone regeneration. Biomaterials 123:92–106
16. Chirila TV, Barnard Z, Harkin DG, Schwab IR, Hirst LW (2008) Bombyx mori silk fibroin membranes as potential substrata for epithelial constructs used in the management of ocular surface disorders. Tissue Eng Part A 14(7):1203–1211
17. Z-x C, X-m M, K-h Z, L-p F, A-l Y, He C-l, Wang H-s (2010) Fabrication of chitosan/silk fibroin composite nanofibers for wound-dressing applications. Int J Mol Sci 11(9):3529–3539
18. Kundu B, Kundu SC (2012) Silk sericin/polyacrylamide in situ forming hydrogels for dermal reconstruction. Biomaterials 33(30):7456–7467
19. Mandal BB, Ghosh B, Kundu S (2011) Non-mulberry silk sericin/poly (vinyl alcohol) hydrogel matrices for potential biotechnological applications. Int J Biol Macromol 49(2):125–133
20. Zhang Y-Q (2002) Applications of natural silk protein sericin in biomaterials. Biotechnol Adv 20(2):91–100
21. Sirichaisit J, Young R, Vollrath F (2000) Molecular deformation in spider dragline silk subjected to stress. Polymer 41(3):1223–1227
22. Ribeiro VP, Almeida LR, Martins AR, Pashkuleva I, Marques AP, Ribeiro AS, Silva CJ, Bonifácio G, Sousa RA, Reis RL (2016) Influence of different surface modification treatments on silk biotextiles for tissue engineering applications. J Biomed Mater Res B Appl Biomater 104(3):496–507
23. Luangbudnark W, Viyoch J, Laupattarakasem W, Surakunprapha P, Laupattarakasem P (2012) Properties and biocompatibility of chitosan and silk fibroin blend films for application in skin tissue engineering. ScientificWorldJournal 2012:697201
24. Yan LP, Oliveira JM, Oliveira AL, Reis RL (2017) Core-shell silk hydrogels with spatially tuned conformations as drug-delivery system. J Tissue Eng Regen Med 11(11):3168–3177
25. Sahoo S, Toh SL, Goh JC (2010) A bFGF-releasing silk/PLGA-based biohybrid scaffold for ligament/tendon tissue engineering using mesenchymal progenitor cells. Biomaterials 31(11):2990–2998
26. Yan L-P, Oliveira JM, Oliveira AL, Caridade SG, Mano JF, Reis RL (2012) Macro/microporous silk fibroin scaffolds with potential for articular cartilage and meniscus tissue engineering applications. Acta Biomater 8(1):289–301
27. Nukavarapu SP, Dorcemus DL (2013) Osteochondral tissue engineering: current strategies and challenges. Biotechnol Adv 31(5):706–721
28. Jeong CG, Zhang H, Hollister SJ (2012) Three-dimensional polycaprolactone scaffold-conjugated bone morphogenetic protein-2 promotes cartilage regeneration from primary chondrocytes in vitro and in vivo without accelerated endochondral ossification. J Biomed Mater Res A 100(8):2088–2096
29. Malda J, Woodfield T, Van Der Vloodt F, Wilson C, Martens D, Tramper J, Van Blitterswijk C, Riesle J (2005) The effect of PEGT/PBT scaffold architecture on the composition of tissue engineered cartilage. Biomaterials 26(1):63–72
30. Zhou J, Xu C, Wu G, Cao X, Zhang L, Zhai Z, Zheng Z, Chen X, Wang Y (2011) In vitro generation of osteochondral differentiation of human marrow mesenchymal stem cells in novel collagen–hydroxyapatite layered scaffolds. Acta Biomater 7(11):3999–4006
31. Yan L-P, Salgado AJ, Oliveira JM, Oliveira AL, Reis RL (2013) De novo bone formation on macro/microporous silk and silk/nano-sized calcium phosphate scaffolds. J Bioact Compat Polym 28(5):439–452
32. Tan H, Chu CR, Payne KA, Marra KG (2009) Injectable in situ forming biodegradable chitosan–hyaluronic acid based hydrogels for cartilage tissue engineering. Biomaterials 30(13):2499–2506

33. Oliveira JM, Rodrigues MT, Silva SS, Malafaya PB, Gomes ME, Viegas CA, Dias IR, Azevedo JT, Mano JF, Reis RL (2006) Novel hydroxyapatite/chitosan bilayered scaffold for osteochondral tissue-engineering applications: scaffold design and its performance when seeded with goat bone marrow stromal cells. Biomaterials 27(36):6123–6137
34. Chen J, Chen H, Li P, Diao H, Zhu S, Dong L, Wang R, Guo T, Zhao J, Zhang J (2011) Simultaneous regeneration of articular cartilage and subchondral bone in vivo using MSCs induced by a spatially controlled gene delivery system in bilayered integrated scaffolds. Biomaterials 32(21):4793–4805. https://doi.org/10.1016/j.biomaterials.2011.03.041
35. Jeon JE, Vaquette C, Klein TJ, Hutmacher DW (2014) Perspectives in Multiphasic Osteochondral Tissue Engineering. Anat Rec 297(1):26–35. https://doi.org/10.1002/ar.22795
36. Pereira DR, Canadas RF, Silva-Correia J, Marques AP, Reis RL, Oliveira JM (2014) Gellan gum-based hydrogel bilayered scaffolds for osteochondral tissue engineering. In: Key engineering materials. Trans Tech, Stafa-Zurich, pp 255–260
37. Rodrigues MT, Lee SJ, Gomes ME, Reis RL, Atala A, Yoo JJ (2012) Bilayered constructs aimed at osteochondral strategies: the influence of medium supplements in the osteogenic and chondrogenic differentiation of amniotic fluid-derived stem cells. Acta Biomater 8(7):2795–2806
38. Ding X, Zhu M, Xu B, Zhang J, Zhao Y, Ji S, Wang L, Wang L, Li X, Kong D, Ma X, Yang Q (2014) Integrated trilayered silk fibroin scaffold for osteochondral differentiation of adipose-derived stem cells. ACS Appl Mater Interfaces 6(19):16696–16705. https://doi.org/10.1021/am5036708
39. Saha S, Kundu B, Kirkham J, Wood D, Kundu SC, Yang XB (2013) Osteochondral tissue engineering in vivo: a comparative study using layered silk fibroin scaffolds from mulberry and nonmulberry silkworms. PLoS One 8(11):e80004
40. Koh L-D, Cheng Y, Teng C-P, Khin Y-W, Loh X-J, Tee S-Y, Low M, Ye E, Yu H-D, Zhang Y-W (2015) Structures, mechanical properties and applications of silk fibroin materials. Prog Polym Sci 46:86–110
41. Simmons AH, Michal CA, Jelinski LW (1996) Molecular orientation and two-component nature of the crystalline fraction of spider dragline silk. Science 271(5245):84
42. Zhang Q, Yan S, Li M (2009) Silk fibroin based porous materials. Materials 2(4):2276–2295
43. McGill M, Coburn JM, Partlow BP, Mu X, Kaplan DL (2017) Molecular and macro-scale analysis of enzyme-crosslinked silk hydrogels for rational biomaterial design. Acta Biomater 63:76–84
44. Hu Y, Zhang Q, You R, Wang L, Li M (2012) The relationship between secondary structure and biodegradation behavior of silk fibroin scaffolds. Adv Mater Sci Eng 2012
45. Wilson D, Valluzzi R, Kaplan D (2000) Conformational transitions in model silk peptides. Biophys J 78(5):2690–2701
46. Valluzzi R, Gido SP, Zhang W, Muller WS, Kaplan DL (1996) Trigonal crystal structure of Bombyx mori silk incorporating a threefold helical chain conformation found at the air–water interface. Macromolecules 29(27):8606–8614
47. Geckil H, Xu F, Zhang X, Moon S, Demirci U (2010) Engineering hydrogels as extracellular matrix mimics. Nanomedicine 5(3):469–484
48. Guarino V, Gloria A, Raucci MG, Ambrosio L (2012) Hydrogel-based platforms for the regeneration of osteochondral tissue and intervertebral disc. Polymer 4(3):1590–1612
49. Jeon O, Bouhadir KH, Mansour JM, Alsberg E (2009) Photocrosslinked alginate hydrogels with tunable biodegradation rates and mechanical properties. Biomaterials 30(14):2724–2734
50. Yan L-P, Silva-Correia J, Ribeiro VP, Miranda-Gonçalves V, Correia C, da Silva MA, Sousa RA, Reis RM, Oliveira AL, Oliveira JM (2016) Tumor growth suppression induced by biomimetic silk fibroin hydrogels. Sci Rep 6:31037
51. Huang S, Fu X (2010) Naturally derived materials-based cell and drug delivery systems in skin regeneration. J Control Release 142(2):149–159
52. Chen X, Li W, Zhong W, Lu Y, Yu T (1997) pH sensitivity and ion sensitivity of hydrogels based on complex-forming chitosan/silk fibroin interpenetrating polymer network. J Appl Polym Sci 65(11):2257–2262

53. Guziewicz N, Best A, Perez-Ramirez B, Kaplan DL (2011) Lyophilized silk fibroin hydrogels for the sustained local delivery of therapeutic monoclonal antibodies. Biomaterials 32(10):2642–2650
54. Yucel T, Cebe P, Kaplan DL (2009) Vortex-induced injectable silk fibroin hydrogels. Biophys J 97(7):2044–2050
55. Kim U-J, Park J, Li C, Jin H-J, Valluzzi R, Kaplan DL (2004) Structure and properties of silk hydrogels. Biomacromolecules 5(3):786–792
56. Teixeira LSM, Feijen J, van Blitterswijk CA, Dijkstra PJ, Karperien M (2012) Enzyme-catalyzed crosslinkable hydrogels: emerging strategies for tissue engineering. Biomaterials 33(5):1281–1290
57. Singh YP, Bhardwaj N, Mandal BB (2016) Potential of agarose/silk fibroin blended hydrogel for in vitro cartilage tissue engineering. ACS Appl Mater Interfaces 8(33):21236–21249
58. Kasoju N, Bora U (2012) Silk fibroin in tissue engineering. Adv Healthc Mater 1(4):393–412
59. Zhang W, Lian Q, Li D, Wang K, Hao D, Bian W, He J, Jin Z (2014) Cartilage repair and subchondral bone migration using 3D printing osteochondral composites: a one-year-period study in rabbit trochlea. Biomed Res Int 2014:746138
60. Fini M, Motta A, Torricelli P, Giavaresi G, Aldini NN, Tschon M, Giardino R, Migliaresi C (2005) The healing of confined critical size cancellous defects in the presence of silk fibroin hydrogel. Biomaterials 26(17):3527–3536
61. Kim HJ, Kim U-J, Kim HS, Li C, Wada M, Leisk GG, Kaplan DL (2008) Bone tissue engineering with premineralized silk scaffolds. Bone 42(6):1226–1234
62. Gentile P, Chiono V, Carmagnola I, Hatton PV (2014) An overview of poly (lactic-co-glycolic) acid (PLGA)-based biomaterials for bone tissue engineering. Int J Mol Sci 15(3):3640–3659
63. Zhang W, Wang X, Wang S, Zhao J, Xu L, Zhu C, Zeng D, Chen J, Zhang Z, Kaplan DL (2011) The use of injectable sonication-induced silk hydrogel for VEGF 165 and BMP-2 delivery for elevation of the maxillary sinus floor. Biomaterials 32(35):9415–9424
64. Tetteh ES, Bajaj S, Ghodadra NS, Cole BJ (2012) The basic science and surgical treatment options for articular cartilage injuries of the knee. J Orthopaed Sports Phys Ther 42(3):243–253
65. Ribeiro V, Pina S, Oliveira JM, Reis RL (2017) Fundamentals on Osteochondral Tissue engineering. In: Regenerative strategies for the treatment of knee joint disabilities. Springer, Berlin, pp 129–146
66. Weiss P, Fatimi A, Guicheux J, Vinatier C (2010) Hydrogels for cartilage tissue engineering. In: Biomedical applications of hydrogels handbook. Springer, Berlin, pp 247–268
67. Floren M, Migliaresi C, Motta A (2016) Processing techniques and applications of silk hydrogels in bioengineering. J Funct Biomater 7(3):26
68. Chao PHG, Yodmuang S, Wang X, Sun L, Kaplan DL, Vunjak-Novakovic G (2010) Silk hydrogel for cartilage tissue engineering. J Biomed Mater Res B Appl Biomater 95(1):84–90
69. Mauck RL, Soltz MA, Wang CC, Wong DD, Chao P-HG, Valhmu WB, Hung CT, Ateshian GA (2000) Functional tissue engineering of articular cartilage through dynamic loading of chondrocyte-seeded agarose gels. J Biomech Eng 122(3):252–260
70. Park S-H, Cho H, Gil ES, Mandal BB, Min B-H, Kaplan DL (2011) Silk-fibrin/hyaluronic acid composite gels for nucleus pulposus tissue regeneration. Tissue Eng Part A 17(23–24):2999–3009
71. Yodmuang S, McNamara SL, Nover AB, Mandal BB, Agarwal M, Kelly T-AN, Chao P-hG, Hung C, Kaplan DL, Vunjak-Novakovic G (2015) Silk microfiber-reinforced silk hydrogel composites for functional cartilage tissue repair. Acta Biomater 11:27–36
72. Holland TA, Bodde EW, Baggett LS, Tabata Y, Mikos AG, Jansen JA (2005) Osteochondral repair in the rabbit model utilizing bilayered, degradable oligo (poly (ethylene glycol) fumarate) hydrogel scaffolds. J Biomed Mater Res A 75(1):156–167
73. Guo X, Park H, Liu G, Liu W, Cao Y, Tabata Y, Kasper FK, Mikos AG (2009) In vitro generation of an osteochondral construct using injectable hydrogel composites encapsulating rabbit marrow mesenchymal stem cells. Biomaterials 30(14):2741–2752

74. Kim K, Lam J, Lu S, Spicer PP, Lueckgen A, Tabata Y, Wong ME, Jansen JA, Mikos AG, Kasper FK (2013) Osteochondral tissue regeneration using a bilayered composite hydrogel with modulating dual growth factor release kinetics in a rabbit model. J Control Release 168(2):166–178
75. Lu S, Lam J, Trachtenberg JE, Lee EJ, Seyednejad H, van den Beucken JJ, Tabata Y, Wong ME, Jansen JA, Mikos AG (2014) Dual growth factor delivery from bilayered, biodegradable hydrogel composites for spatially-guided osteochondral tissue repair. Biomaterials 35(31):8829–8839
76. Grayson WL, Chao P-HG, Marolt D, Kaplan DL, Vunjak-Novakovic G (2008) Engineering custom-designed osteochondral tissue grafts. Trends Biotechnol 26(4):181–189
77. Das S, Pati F, Choi Y-J, Rijal G, Shim J-H, Kim SW, Ray AR, Cho D-W, Ghosh S (2015) Bioprintable, cell-laden silk fibroin–gelatin hydrogel supporting multilineage differentiation of stem cells for fabrication of three-dimensional tissue constructs. Acta Biomater 11:233–246
78. Thorrez L, Shansky J, Wang L, Fast L, VandenDriessche T, Chuah M, Mooney D, Vandenburgh H (2008) Growth, differentiation, transplantation and survival of human skeletal myofibers on biodegradable scaffolds. Biomaterials 29(1):75–84. https://doi.org/10.1016/j.biomaterials.2007.09.014
79. Karageorgiou V, Kaplan D (2005) Porosity of 3D biomaterial scaffolds and osteogenesis. Biomaterials 26:5474–5491
80. van de Witte P, Dijkstra P, vanden Berg J, Feijen J (1996) Phase separation processes in polymer solutions in relation to membrane formation. J Membr Sci 117:1–31
81. Marrella A, Cavo M, Scaglione S (2017) Rapid prototyping for the engineering of osteochondral tissues. In: Oliveira JM, Reis RL (eds) Regenerative strategies for the treatment of knee joint disabilities. Springer International, Cham, pp 163–185. https://doi.org/10.1007/978-3-319-44785-8_9
82. Oliveira AL, Sampaio SC, Sousa RA, Reis RL (2006) Controlled mineralization of nature-inspired silk fibroin/hydroxyapatite hybrid bioactive scafolds for bone tissue engineering applications. Paper presented at the 20th European Conference on Biomaterials, Nantes, France, 27 September–1 October
83. Abdelaal O, Darwish S (2011) Fabrication of tissue engineering scaffolds using rapid prototyping techniques. World Acad Sci Eng Technol 59:577–585
84. Chae T, Yang H, Leung V, Ko F, Troczynski T (2013) Novel biomimetic hydroxyapatite/alginate nanocomposite fibrous scaffolds for bone tissue regeneration. J Mater Sci Mater Med 24:1885–1894
85. Cui L, Zhang N, Cui W, Zhang P, Chen X (2015) A novel nano/micro-fibrous scaffold by melt-spinning method for bone tissue engineering. J Bionic Eng 12(1):117–128
86. Cardea S, Scognamiglio M, Reverchon E (2016) Supercritical fluid assisted process for the generation of cellulose acetate loaded structures, potentially useful for tissue engineering applications. Mater Sci Eng C 59:480–487
87. Yan L, Oliveira J, Oliveira A, Reis R (2014) Silk fibroin/nano-CaP bilayered scaffolds for osteochondral tissue engineering. Key Eng Mater 587:245. https://doi.org/10.4028/www.scientific.net/KEM.587.245
88. Yan LP, Oliveira JM, Oliveira AL, Reis RL (2015) In vitro evaluation of the biological performance of macro/micro-porous silk fibroin and silk-nano calcium phosphate scaffolds. J Biomed Mater Res B Appl Biomater 103(4):888–898. https://doi.org/10.1002/jbm.b.33267
89. Chen K, Shi P, Teh TKH, Toh SL, Goh JCH (2016) In vitro generation of a multilayered osteochondral construct with an osteochondral interface using rabbit bone marrow stromal cells and a silk peptide-based scaffold. J Tissue Eng Regen Med 10(4):284–293. https://doi.org/10.1002/term.1708
90. Kazemnejad S, Khanmohammadi M, Mobini S, Taghizadeh-Jahed M, Khanjani S, Arasteh S, Golshahi H, Torkaman G, Ravanbod R, Heidari-Vala H, Moshiri A, Tahmasebi M-N, Akhondi M-M (2016) Comparative repair capacity of knee osteochondral defects using regenerated silk fiber scaffolds and fibrin glue with/without autologous chondrocytes during 36 weeks in rabbit model. Cell Tissue Res 364(3):559–572. https://doi.org/10.1007/s00441-015-2355-9

91. Mobini S, Hoyer B, Solati-Hashjin M, Lode A, Nosoudi N, Samadikuchaksaraei A, Gelinsky M (2013) Fabrication and characterization of regenerated silk scaffolds reinforced with natural silk fibers for bone tissue engineering. J Biomed Mater Res A 101A(8):2392–2404. https://doi.org/10.1002/jbm.a.34537
92. Zhou F, Zhang X, Cai D, Li J, Mu Q, Zhang W, Zhu S, Jiang Y, Shen W, Zhang S, Ouyang HW (2017) Silk fibroin-chondroitin sulfate scaffold with immuno-inhibition property for articular cartilage repair. Acta Biomater 63(Supplement C):64–75. https://doi.org/10.1016/j.actbio.2017.09.005
93. Pina S, Canadas RF, Jiménez G, Perán M, Marchal JA, Reis RL, Oliveira JM (2017) Biofunctional ionic-doped calcium phosphates: silk fibroin composites for bone tissue engineering scaffolding. Cells Tissues Organs 204(3–4):150–163
94. Çakmak S, Çakmak AS, Kaplan DL, Gümüşderelioğlu M (2016) A silk fibroin and peptide amphiphile-based co-culture model for osteochondral tissue engineering. Macromol Biosci 16(8):1212–1226. https://doi.org/10.1002/mabi.201600013

# Chapter 15
# In Situ Cross-Linkable Polymer Systems and Composites for Osteochondral Regeneration

María Puertas-Bartolomé, Lorena Benito-Garzón, and Marta Olmeda-Lozano

**Abstract** Injectable hydrogels have demonstrated being a promising strategy for cartilage and bone tissue engineering applications, owing to their minimal invasive injection procedure, easy incorporation of cells and bioactive molecules, improved contact with the surrounding tissues and ability to match defects with complex irregular shapes, characteristics of osteoarthritic pathology. These unique properties make them highly suitable bioscaffolds for treating defects which are otherwise not easily accessible without and invasive surgical procedure. In this book chapter it has been summarized the novel appropriate injectable hydrogels for cartilage and bone tissue engineering applications of the last few years, including the most commonly used materials for the preparation, both natural and synthetic, and their fabrication techniques. The design of a suitable injectable hydrogel with an adequate gelation time that gathers perfect bioactive, biocompatible, biodegradable and good mechanical properties for clinical repair of damaged cartilage and bone tissue is a challenge of significant medical interest that remain to be achieved.

**Keywords** Hydrogels · In situ cross-linking · Biopolymers · Composites · Osteochondral

---

M. Puertas-Bartolomé (✉)
Institute of Polymer Science and Technology-ICTP-CSIC, C/Juan de la Cierva 3, Madrid, Spain

CIBER-BNN, Health Institute Carlos III, C/Monforte de Lemos 3-5, Madrid, Spain
e-mail: mpuertas@ictp.csic.es

L. Benito-Garzón
Faculty of Medicine, University of Salamanca, C/Alfonso X el Sabio, Salamanca, Spain

M. Olmeda-Lozano
University Hospital Infanta Elena, Madrid, Spain

© Springer International Publishing AG, part of Springer Nature 2018
J. M. Oliveira et al. (eds.), *Osteochondral Tissue Engineering*,
Advances in Experimental Medicine and Biology 1058,
https://doi.org/10.1007/978-3-319-76711-6_15

## 15.1 Introduction

Cartilage and bone damage is an important pathology that affects to a large part of the population and can be caused by a trauma, arthritis or even a sport-related injury [1, 2]. However, clinical repair of the damaged cartilage and bone tissue is an important challenge that remains to be achieved, and thus, developing a novel biomaterial to repair and regenerate the damaged cartilage and bone tissue permanently is of significant medical interest.

The biology of cartilage and bony extracellular matrix (ECM) is crucial in achieving an effective cartilage and bone tissue regeneration. Both cartilage and bone ECM provide very specialized and complex 3D microenvironments crucial for chondrocytes and osteoblasts growth and maintenance [3, 4]. On the one hand, cartilage is a fiber-reinforced composite material with chondrocytes within a specialized and unique ECM composed of glycosaminoglycans (GAG), structural and functional proteins and glycoproteins [5–7]. Due to lack of vascularization, innervation, lymphatic networks, and progenitor, it is limited in terms of healing and recovery of damage cartilage tissue [8–13]. On the other hand, bone is a connective tissue highly vascularized and biomineralized with high mechanical properties [3, 14] whose ECM is composed of collagen I fibers and carbonated hydroxyapatite, complemented with proteoglycans, glycoproteins, and sialoproteins [15, 16]. However, common techniques for bone repair such as autografting and allografting are limited [17–19]. Therefore, the complexity of the cartilage and bony ECM, essential for their specific functions, becomes the development of novel ECM-mimicking biomaterials a great challenge in cartilage and bone engineering.

Hydrogels have attracted strong attention for applications in tissue engineering and regenerative medicine owing to their three-dimensional ECM-mimicking polymeric network, swellability and their porous framework, allowing for cell transplantation and proliferation [20–22]. Particularly, they have attracted widespread interest among all the biomaterials for their use as scaffolds for cartilage and bone tissue engineering owing to their similarity to the cartilage and bony ECM that provides an appropriate microenvironment facilitating the cell adhesion, migration, proliferation, and differentiation as well as an effective delivery of nutrients and growth factors (Fig. 15.1) [23–30]. Injectable hydrogels in situ formed upon injection have also gained increased attention recently for osteochondral regeneration. These systems allow an aqueous mixture of gel precursors and bioactive agents (therapeutic drugs, cells, or bioactive factors) to be mixed and injected using a syringe into the defect and, once administered, the gel precursors are cross-linked through physical interactions or chemical bonds in situ [3, 31, 32]. Principally, injectable hydrogels are highly desirable for cartilage and bone clinical trials owing to advantages such as minimally invasive injection procedure replacing implantation surgery, easy incorporation of cells and bioactive molecules, good alignment with the surrounding tissues, and ability to match defects with complex irregular geometries, shape, and size that are hallmarks of osteoarthritic pathology [32–36]. Therefore, these unique properties make injectable hydrogels highly suitable bioscaffolds for treating

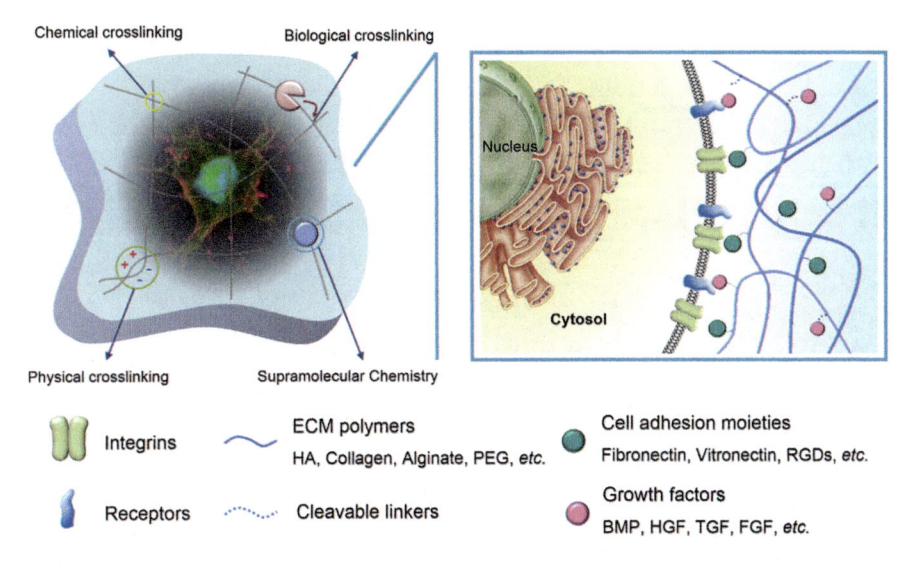

**Fig. 15.1** Interaction between cell receptors and biomolecules conjugated to extracellular matrix in situ hydrogels. Reprinted with permission from Prog Polym Sci 39(12):1973–1986 Copyright (2014) Elsevier

defects which are otherwise not easily accessible without an invasive surgical procedure, supporting cartilage, and bone regeneration.

Both biomaterials and appropriate fabrication techniques play crucial roles in designing suitable injectable hydrogels as scaffolds for cartilage and bone tissue-engineering. Nowadays, a large number of hydrogels derived from different natural or synthetic polymers have been exploited for reconstruction of deficient osteochondral interface or articular cartilage tissues [37–42]. Natural polymers such as ECM proteins and polysaccharides have been widely used owing to some advantages such as a variety of chemical functionalities that can confer mucoadhesive or antibacterial properties, affinity towards other biomolecules, in particular proteins, inherent biocompatibility, biodegradability, and nonimmunogenicity [43–52]. Also numerous hydrogels based on synthetic polymers such as poly(ethylene glycol) (PEG), poly(L-glutamic acid), and polyacrylamide (HPAM) have been reported due to the highly tunable mechanical and physicochemical properties [53–55]. Besides, hybrid hydrogels consisting of natural and synthetic biopolymers combine the biocompatibility, biodegradability, and tunable mechanical strength that made them also an interesting option for osteochondral and cartilage tissue regeneration [36, 56, 57]. Furthermore, selective and fast cross-linking reactions under physiological conditions are prerequisites for in situ formation of injectable hydrogels, conferring mechanical stability and structural integrity to the hydrogel matrix [58]. Physically injectable hydrogels are spontaneously formed by weak secondary forces such as hydrophobic or ionic interactions, normally induced by change of environmental factors including temperature, pH, force, concentration, and ions. Otherwise, chemical hydrogels are formed by covalently cross-linking, using nontoxic chemical

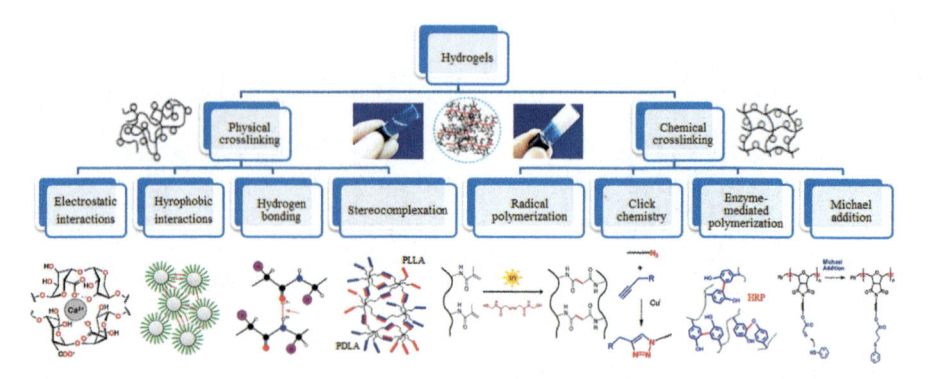

**Fig. 15.2** Schematic representation of different methods for production of physically and chemically cross-linked hydrogels. Reprinted with permission from Biomacromolecules 17(11):3441–3463 Copyright (2016) American Chemical Society

cross-linkers or enzymes [59, 60]. By controlling the density of cross-linking, the mechanical properties and biodegradation rates of the hydrogels can be highly tunable [61]. The scheme describing different methods for production of physically and chemically cross-linked hydrogels is illustrated in Fig. 15.2.

Therefore, the development of a suitable injectable hydrogel for cartilage and bone engineering is urgently needed. In this book chapter have been discussed different strategies of the last few years for developing novel injectable hydrogels for cartilage and bone tissue repair, focusing on the biomaterials of preparation.

## 15.2 In Situ Hydrogels Prepared with Different Biomaterials

### 15.2.1 Chitosan Based In Situ Hydrogels

Chitosan is a linear polysaccharide composed of D-glucosamine and N-acetylglucosamine units obtained by deacetylation of natural chitin, found in shells of shrimps and crabs. Chitosan is currently an attractive material for tissue-engineering applications due to its great biocompatibility, biodegradability, and hemostatic and antibacterial activity. Particularly, its structural similarity to cartilage glycosaminoglycan makes chitosan-based hydrogels great candidates as injectable materials for cartilage repair [47, 62–65].

However, native chitosan is insoluble in water because of its strong intermolecular hydrogen bonding, it is soluble in acetic acid solution but it requires exhaustive washing steps and restricts its biomedical applications. To overcome this limitation, different chemical modifications can be carried out in order to obtain water-soluble chitosan derivatives. For example, Sharma et al. carried out the chemical coupling of an amino acid to the native chitosan, obtaining a water-soluble histidine-modified chitosan [66]. Another common method is the conjugation of glycolic acid to the chitosan yielding chitosan-based derivatives with a higher water solubility until a maximum pH of 7.8 [65]. In the study of Rong Jin et al. this coupling method is used to obtain a chitosan–glycolic acid (GA) soluble at water that allows the preparation of an injectable hydrogel under physiological conditions for cartilage regeneration [67]. The hydrogel system consists on chitosan–glycolic acid/tyrosine (Ch-GA/Tyr) conjugates enzymatically cross-linked using either tyrosinase or horseradish peroxidase (HRP). This kind of cross-linking is being of high interest in the last years because of its low cytotoxicity and mild cross-linking conditions [65, 68]. Both enzymes provide a stable cross-linking, but the proliferation of chondrocytes is better for the hydrogels cross-linked with tyrosinase so they are more suited for cartilage regeneration. This enzyme, in the presence of oxygen, catalyzes the reaction from phenol to quinone which can react with other quinones or with the nucleophilic amino groups obtaining the network [69]. Moreover, the gelation time can be modulated by changing the concentration of enzyme and the degree of tyrosine substitution. Thus, the tyrosinase-cross-linked Ch-GA/Tyr conjugates have a high potential of application for cartilage tissue engineering. In another study, Kamoun has also prepared a water-soluble chitosan derivative conjugating succinic anhydride to the chitosan [70]. In this case, a new in situ self-forming hydrogel is obtained with this N-succinic chitosan and dialdehyde starch cross-linking the aldehyde groups with the functional amine groups through a Schiff-base reaction without using any additional cross-linker. The properties of this new class of hybrid hydrogels make them preferable as a covalent in situ forming scaffold for cartilage tissue engineering.

Glycol-chitosan derivatives have been currently used in order to create injectable materials for cartilage tissue engineering. In the study of Sung Cho et al. a new photocrosslinkable thermogelling polymer based on methacrylated hexanoyl glycol chitosan (M-HGC) has been developed for cartilage regeneration [71, 72]. The novelty of this work lies in the use of thermosensitive polymer based on a polysaccharide that shows thermogelling property around body temperature, while usually they are based on synthetic block copolymers. This new system can be chemically cross-linked through the methacryl groups by UV radiation obtaining a stable and elastic hydrogel with improved mechanical properties, without using any additional cross-linker. Based on the biodegradability, good chondrocyte viability after 7 days, and modulated thermogellation, the M-HGC hydrogels have proven to be promising candidates as injectable carriers for cartilage regeneration. Cao et al. have obtained an in situ injectable hydrogel system using also glycol chitosan chemically cross-linked under physiological conditions with a functionalized poly(ethylene glycol) (poly(EO-co-Gly)-CHO containing pendant aldehyde groups [73]. In this case the cross-linking consists of a Schiff's base formation between aldehyde pendant groups of the poly(EO-co-Gly)-CHO and the amino groups of glycol chitosan obtaining stable covalent benzoic-imine bonds. The concentration of cross-linker, the gelation time, network morphology, and mechanical properties can be modulated. High viability of chondrocytes encapsulated in the hydrogels was demonstrated after 14 days, and the phenotype was well preserved, proving that these injectable hydrogels are promising materials for cartilage tissue engineering (Fig. 15.3).

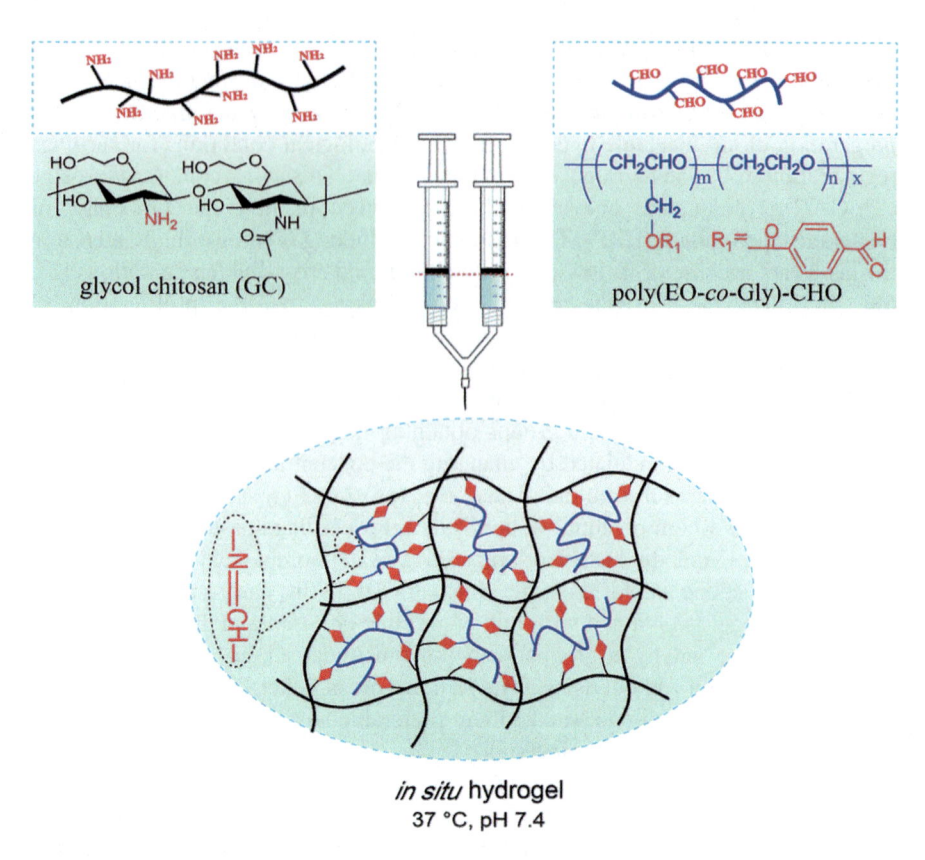

**Fig. 15.3** Schematic illustration of the injectable hydrogels formation by Schiff base cross-linking between aqueous solutions of glycol chitosan and poly (EO-co-Gly)-CHO. Adapted with permission from J Mater Chem B 3(7):1268–1280 Copyright (2014) The Royal Society of Chemistry

All these research works presented based on chitosan focus on cartilage regeneration repair due to its structure similar to glycosaminoglycans present in the cartilage extracellular matrix and the limited healing ability of the cartilage. However, articular osteochondral defects involve not only to cartilage but to underlying subchondral bone. Therefore, it is a great challenge to design new materials to regenerate simultaneously cartilage and bone, obtaining complex tissues. Zhang et al. have designed a single scaffold consisting of two different regions, poly(L-glutamic acid) (PLGA)/chitosan (CS) hydrogel and PLGA/CS/n-hydroxyapatite-graft-poly(L-glutamic acid) (nHA-g-PLGA) supporting both hyaline cartilage and subchondral bone regeneration respectively [74]. Adipose derived stem cells were seeded into the osteochondral scaffolds, and the cell/scaffolds were tested in vivo in rabbit articular osteochondral defects confirming a good regeneration process both in cartilage and subchondral bone.

Thermosensitive hydrogels have also attracted great attention for biomedical applications, due to their similarity to ECM and their support for cell proliferation migration and differentiation [75, 76]. Different natural and synthetic polymers have been used to synthesize this kind of polymers [77, 78] but those with natural polymers are particularly attractive due to their similarity with ECM, such as polysaccharides or proteins [79, 80]. Chitosan has been commonly used in the preparation of thermosensitive polymers in combination with phosphates derivatives such a as glycerophosphate [77], ammonium hydrogen phosphate [81], and dipotassium hydrogen orthophosphate [82]. Specifically, in 2000, Chenite et al. developed a commercially available system called BST-CarGel® with highly deacetylated Chitosan and glycerophosphate (GP) which has been used clinically for cartilage repair and whose ownership belongs nowadays to Piramal Life Sciences. This thermosensitive system turns into gel at body temperature, so it gels in situ when injected in vivo with a complicated gelation mechanism [83, 84]. Extensive studies have been carried out with the system BST-CarGel® in the area of cartilage tissue engineering. For example, it has been demonstrated that BST-CarGel® with bone marrow stimulation significantly promotes hyaline cartilage regeneration [85–87]. Also a large number of studies have demonstrated that both CS/α-GP- and CS/β-GP-based hydrogels are biodegradable and biocompatible owning tunable physicochemical properties [88–90]. However, the gelation time for CS/ β-GP-based hydrogels is too long and could result in an ectopic cartilage tissue formation causing pain. To decrease this gelation time, the β-GP concentration can be increased, but this leads to significant cytotoxicity limiting the biomedical applications. In order to overcome this limitation, Assaad et al. have reported a method to decrease the cytotoxicity combining β-GP with $NaHCO_3$ and phosphate buffer [91]. Also Deng et al. have used $NaHCO_3$ in order to obtain shorter gelation times and better mechanical properties, improving the biocompatibility [92]. Additionally, other works have been reported in which the CH/GP is combined with other biopolymers in order to obtain a new system with improved mechanical properties, degradation speed, and stability [47, 93, 94]. Furthermore, a large number of clinical studies have been carried out, as will be explained in the Sect. 15.3 [95–98] proving that BST-CarGel® is an effective treatment for cartilage repair significantly better than microfracture (MFX) alone.

## *15.2.2 Hyaluronic Acid Based In Situ Hydrogels*

Hyaluronic acid is a high molecular weight linear polysaccharide composed of disaccharide units of glucuronic acid (HA) and *N*-acetylglucosamine found in soft tissue and synovial fluid such as cartilage ECM. HA interacts with chondrocytes through the receptor CD44 playing an important role in cartilage formation, mesenchymal cell migration, condensation, and chondrogenic differentiation, and maintaining cartilage homeostasis. Furthermore, HA provides lubrication and resistance to articulating surfaces, promotes wound healing and tissue regeneration, and increases osteoblastic bone formation, and therefore, it is a very attractive material for cartilage tissue repair [99–101].

However, HA is not good as filler for bone defects because it forms a solution that can leak out from the defect. This is an important restriction in the use of HA in biomedicine, so it is necessary to modify the HA rheological properties by combining it with other polymers in order to overcome this limitation [102]. In the study of Bellini et al., a three-component hydrogel consisting of HA, gellan gum, and calcium chloride was reported [103]. This stable system has been designed in a smart combination suitable for filling and capping bone defects avoiding any leakage. The gellan gum is ionically cross-linked with calcium chloride without any additional cross-linker which could be cytotoxic, and changing the hydrogel composition they can modulate the gelation time, mechanical properties, and degradation times. Human primary osteoblasts are cultured into the hydrogels showing a good proliferation and osteoblastic progression, so these hydrogels are suitable materials for bone regeneration of osteochondral defects. In the study of Yu et al., HA is mixed this time with a synthetic polymer, PEG, using a new strategy that includes two cross-linking methods [104]. First, a fast gelation is obtained by enzymatic cross-linking, and subsequently, a Diels–Alder cross-linking provides the hydrogel antifatigue and shape memory properties. ATDC-5 cells were encapsulated in the hydrogel demonstrating high proliferation and metabolic viability, and thus, these results support a promising strategy for different cartilage tissue engineering applications. Afterwards, this group has also developed a similar HA/PEG system obtaining a faster gelation time by Diels–Alder cross-linking [105]. It is demonstrated that changing the gelation time the cell encapsulation viability can be improved. HA-based hydrogels have been also prepared in combination with chitosan, taking advantage of its similar structure to GAG. Park et al. have obtained an injectable HA/methacrylated glycol chitosan hydrogel by photocrosslinking with a riboflavin photoinitiator and visible light [48]. Chondrocytes are successfully encapsulated in the hydrogel and irradiated, maintaining high viability and round shape, and with increased deposition of cartilaginous ECM. Thus, this Ch/HA-based hydrogel has the requirements and the potential to be used in cartilage tissue engineering.

Another common strategy to improve the mechanical properties of the HA-based hydrogels is to modify the HA rheological properties by using HA derivatives. Palumbo et al. have used two hyaluronic acid derivatives bearing ethylenediamine (EDA) or EDA and octadecyl groups $C_{18}$ to obtain a stable hydrogel cross-linked with divinylsulfone functionalized inulin through an azo-Michael reaction [46]. It is proved that the presence of pendant $C_{18}$ chains improve the mechanical properties and also decrease its susceptibility to hyaluronidase hydrolysis. Furthermore, the chemistry of this cross-linking is a safe strategy for chondrocytes encapsulation which results in high proliferation and viability until 28 days. In the study of Calogerto et al. a similar system has been prepared with the amino derivative HA-EDA and α-elastin-grafted HA-EDA (HA-EDA-g-α-elastin) chemically cross-linked via Michael-type addition forming stable hydrogels [106]. Chondrocytes encapsulated in the hydrogel recognize α-elastin through specific interactions that favor a better metabolic activity and allow the improvement of the elastic modulus of the hydrogel by acting as biological cross-linkers. In addition, degradation profiles indicate that this injectable hydrogel scaffold possesses desired properties for the treatment of articular cartilage damage under physiological conditions. Also in the study of Fenn et al., methacrylated-HA hydrogels are obtained controlling the degree of modification and the molecular weight, in order to achieve the optimal mechanical properties for osteochondral regeneration [107]. Visible light cross-linked is presented as a viable methodology for natural-based in situ hydrogels and a safe alternative to UV photocrosslinked hydrogels, which can compromise the viscosity or the stiffness of the material. Furthermore, in vitro test with human mesenchymal stem cells (MSCs) shows that these hydrogels does not present significant cytotoxic effects.

Another option to overcome the poor mechanical properties of the HA is to combine it with other polymers. A new strategy is reported in the work of Domingue et al. by reinforcing the hydrogel with cellulose nanocrystals as nanofillers [108]. The nanocomposite hydrogel include adipic acid dihydrazide-modified HA and aldehyde-modified HA reinforced by aldehyde-modified cellulose nanocrystals. The system exhibits good cell supportive properties due to the improved structural integrity. Also hASC encapsulated in the hydrogel are able to spread well exhibiting noticeable proliferative activity. Therefore, this work supports a valuable strategy for modulate the structural and biological properties of the hydrogel, expanding their range of applications (Fig. 15.4).

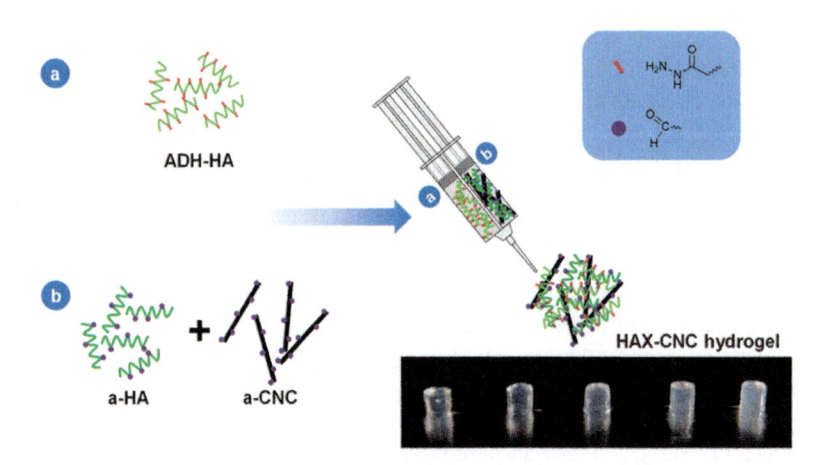

**Fig. 15.4** Scheme of injectable hyaluronic acid-based hydrogels (HAX) reinforced with CNCs, prepared using a double-barrel syringe. (**a**) Hydrazide-functionalized hyaluronic acid (ADH-HA) and (**b**) aldehyde functionalized hyaluronic acid (a-HA), with or without aldehyde-modified CNCs (a-CNCs), are mixed and cross-linked obtaining stable hydrogels. Reprinted with permission from Bioconjug Chem 26(8):1571–1581 Copyright (2015) American Chemical Society

## *15.2.3 Collagen/Gelatin-Based In Situ Hydrogels*

Collagen is the most abundant protein of the body and it can be found in cartilage, bone, skin, connective tissue and ligaments. Recently, natural collagen-based hydrogels have been widely used to design scaffolds for different biomedical applications. ECM of cartilage is rich in collagen, so it has been specifically used lately in cartilage tissue engineering, due to their biocompatibility, biodegradability, the low antigenic property and the ability of inducing chondrogenic differentiation of mesenchymal stem cells (MSCs) promoting the secretion of cartilage specific ECM [10, 50, 109, 110]. In the study of Cooper et al. a collagen interpenetrating polymer network has been designed to mimic cartilage glycosaminoglycans and improve the lubrication, as a therapy for osteoarthritis diseases [111]. The system consists on the in situ formation of a photopolymerized network with native collagen fibrils which reduce the solid-solid interfacial friction, demonstrating its potential as a treatment to enhance the tribological function of cartilage in the articulations. Also Yuan et al. have prepared an injectable physically cross-linked composite combining type I and type II collagen [50]. It is demonstrated that chondrocytes embedded in the hydrogel can maintain their phenotype and secrete cartilage-specific ECM. Furthermore, the ECM secretion of these chondrocytes can be regulated changing the compressive modulo of the hydrogel.

Collagen-based in situ hydrogels are usually combined with other biomaterials. In the study of Kontturi et al. type II collagen and HA are integrated creating an injectable hydrogel similar to cartilage tissue [112]. The biomaterials are chemically cross-linked with poly(ethylene glycol) ether tetrasuccinimidyl glutarate (4SPEG), noncytotoxic for chondrocytes [113], and complemented with the transformed growth factor β1, obtaining a strong and stable network. Chondrocytes are successfully encapsulated in the hydrogel system maintaining viability and characteristics, so this hydrogel is a potential cell delivery system for cartilage tissue engineering. In the work of Choi et al. type II collagen, chondroitin sulfate (CS) and chitosan are integrated in an injectable hydrogel enhancing the physical stability of Col or Ch by themselves [29]. The hydrogel system consists on methacrylated glycol chitosan bearing the cartilaginous ECM, and riboflavin, which allows in situ gelation by UV–Vis photopolymerization. The effectively encapsulation and proliferation of chondrocytes and MSCs and their production of cartilage matrix support that these hydrogels are potential systems to promote tissue regeneration. Another study carried out by Moreira et al. has combined collagen with chitosan, in this case to prepare bioactive thermoresponsive injectable hydrogels [114]. The chitosan-based hydrogel is chemically conjugated with collagen and reinforced with bioactive glass nanoparticles, improving the mechanical properties. Cell studies demonstrate that this is an innovative and promising thermogelling material for bone and cartilage bioapplications.

From the degradation of collagen, it is obtained the Gelatin, a biocompatible and biodegradable natural protein. It has been recently used to prepare injectable hydrogels because of its cartilage reparation properties. However, gelatin-based

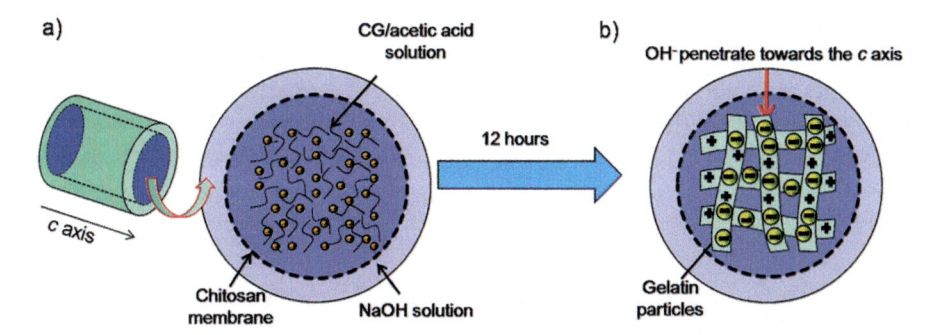

**Fig. 15.5** Schematic representation of formation process of chitosan–gelatin (CG) gel with in situ precipitation method: (**a**) chitosan membrane filled with CG/acetic acid solution and put into the NaOH solution; (**b**) the gel is formed with the OH from NaOH penetrating towards the c axis after 12 h, then positively charged chitosan protonated and established strong electrostatic interactions with the negatively charged gelatin. Reprinted with permission from RSC Adv 5:55640–55647 Copyright (2015) The Royal Society of Chemistry

hydrogels, similar to collagen-based hydrogels, possess low mechanical properties and shrinkage of gels [115, 116]. Currently, enzymatic cross-linking method has attracted high interest in the preparation of injectable hydrogels for biomedical applications and it has been demonstrated to enhance chondrocyte proliferation in vitro and collagen and GAG production [65, 68]. Wang et al. have reported an enzymatic cross-linked gelatin-based injectable hydrogel for ectopic cartilage formation and early-stage osteochondral defect repair [117]. The system is composed of gelatin–hydroxyphenylpropionic acid (Gtn-HPA) and cross-linked by oxidative coupling of HPA moieties catalyzed by $H_2O_2$ and HRP. Varying the concentration ratio Gtn-HPA/$H_2O_2$ the stiffness of the hydrogels can be controlled, maintaining fast gelation time. It is demonstrated that changes is this stiffness allow the modulation of chondrocyte cellular functions, and cartilage tissue histogenesis and repair. In a different study Shen et al. have created a new though porous hydrogel combining gelatin with chitosan via in situ precipitation (described in Fig. 15.5) [49]. The porosity and degradation rate of the hydrogel can be tuned by changing the composition, obtaining a hydrogel with improved mechanical properties for cartilage engineering.

A thermoresponsive injectable hydrogel has been designed for Oh et al. with gelatin-graft-poly(*N*-isopropylacrylamide) [118]. In this case, the system is prepared by gelling oil-in-water high internal phase emulsions using the thermoresponsive property of the copolymer units obtaining a high porosity system. This new strategy of fabrication provides scaffolds suitable for cell encapsulation and supporting its spreading and proliferation. An innovative study has been developed by Puertas et al. to obtain another thermosensitive hydrogel with bioadhesive properties composed of gelatin and a catechol containing copolymer [119]. The system exhibits good biocompatibility, antioxidant behavior, and anti-inflammatory properties. Likewise, excellent bone bioadhesive properties have been demonstrated attributed to the catechol moiety, showing that this system possesses a great potential of application for bone tissue regeneration processes. Also in the field of bone tissue engineering, Visser et al. have created a cell carrier with a gelatin methacrylamide hydrogel and multipotent stromal cells and cartilage-derived matrix particles embedded on it [120]. It is found that this system allows for the formation of endochondral bone in the degradable hydrogel, offering a new strategy for relevant-size bone grafts.

## 15.2.4 Alginate-Based In Situ Hydrogels

Alginate is natural polysaccharide consisting of guluronic and maluronic acids. Biopolymers that mimic the ECM properties have been widely used for tissue engineering because their potential promoting cellular adhesion, migration and proliferation, likewise, alginate has been commonly used to create injectable hydrogels for cartilage tissue engineering applications [121–123]. Alginate-based hydrogels can form rapid and stable biocompatible scaffolds; however, the mechanical properties of the hydrogel are not strong enough to support the regenerated tissue. Consequently, alginate is usually modified or combined with other biomaterials to overcome this drawback. Furthermore, it can be also blended with other polymers to improve its cell adhesion ability [45].

Balakrishnan et al. have developed a novel hydrogel consisting of oxidized alginate, gelatin, and borax by self-cross-linking or auto-polymerization [124], without using any additional cytotoxic component [43]. The adhesive hydrogel is well integrated with the cartilage and chondrocytes encapsulated demonstrates favorable viability, migration, and proliferation within the matrix while maintaining their phenotype. Therefore, this is a promising cell-attractive adhesive material for cartilage tissue engineering for the treatment of OA. Zeng et al. also use an alginate/gelatin based hydrogel [125]. In this work a novel microcavitary hydrogel is obtained with gelatin microspheres within an alginate matrix and it is demonstrated that the size of microcavity affects the proliferation of encapsulated chondrocytes, preferring the smallest size. Park et al. have created a biodegradable hydrogel with alginate and hyaluronic acid. In this case alginate has been partially oxidized with sodium periodate in order to improve the biodegradability of alginate in physiological conditions [45]. Chondrocytes were encapsulated in the system and it was injected subcutaneously into mice, showing GAG secretion, chondrogenic marker genes expression and effective cartilage regeneration. Oxidized alginate has been also used in the study of Yan et al. with oxidized poly(L-glutamic acid), obtaining a novel biodegradable hydrogel by self-cross-linking through Schiff base formation in physiological conditions [59]. This novel material has demonstrated successful injectability, rapid in vivo gelling, chondrocytes encapsulation and viability, satisfactory mechanical stability, and favorable ectopic cartilage formation. Similarly, in the study of Zheng et al. alginate has been blended creating a triple-phased hydrogel with nano-hydroxyapatite (nano-HA), collagen and alginate ionically cross-linked with $Ca^{2+}$, which is an easy method extensively use for

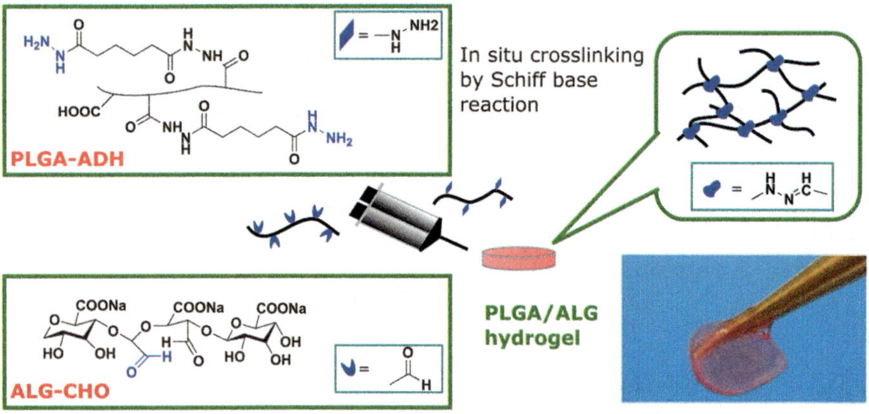

**Fig. 15.6** Schematic representation of in situ cross-linking mechanism of PLGA/ALG injectable hydrogel using dual syringe and obtaining a stable system. Reprinted with permission from Biomacromolecules 15(12):4495–4508 Copyright (2014) American Chemical Society

alginate-based hydrogels formation [126–129]. This triphased composite possess improved mechanical properties, high viability and proliferation and upregulated hyaline cartilage markers, thus it is a valuable strategy for osteochondral tissue engineering. However, alginate hydrogels ionically cross-linked have some disadvantages. Specifically, due to the difficulty in the control of the cross-linking, the mechanical and biophysical properties are uncontrolled. Therefore, Lee et al. describe a bioinspired approach for the preparation of alginate hydrogels based on dopamine conjugated to the alginate backbone and a posterior oxidative cross-linking [130]. The system exhibits moldable physical and mechanical properties and a higher biocompatibility due to the lack of cations in the hydrogel. This approach can provide a new method to develop tunable biocompatible hydrogels for osteochondral therapies (Fig. 15.6).

## 15.2.5 Fibrin-Based In Situ Hydrogels

Fibrin is a natural fibrous protein that has a key role in blood clotting widely used for cartilage tissue engineering as a cell matrix, enhancing cell attachment, mobility, proliferation and differentiation [131, 132]. However, fibrin presents low mechanical stiffness and fast degradation so it is commonly used in combination with other materials.

Benavides et al. have developed a fibrin-based injectable hydrogel containing poly(ethylene glycol) (PEG), which increase the longevity of the hydrogel, and human amniotic fluid-derived stem cells[133]. This novel hydrogel is a fibrin-driven platform for vascular formation and promotes in situ neovascularization of human amniotic fluid stem derived cells from. Almeida et al. have developed an injectable fibrin-based hydrogel functionalized with cartilage ECM microparticles and the transforming growth factor-$\beta_3$ as a new therapy for joint regeneration [134]. It is demonstrated that fibrin-based hydrogels can induce in vivo production of cartilage tissue by freshly isolated stromal cells, so these results support a new potential method for cell-based therapies for articular cartilage regeneration. Hwang et al. have developed a different fibrin-based hydrogel consisting of alginate particles and cells in a fibrin matrix [135]. This system enhances cellular viability, mobility, and proliferation, and contributes to volume retention and vascularization in vivo, so it is an attractive approach for soft tissue reparation.

### 15.2.6 Elastin-Based In Situ Hydrogels

Currently, in situ hydrogels based on ECM proteins such as collagen, fibrin, or elastin are promising materials for cartilage tissue engineering. Elastin improves elasticity and cellular interactions during the tissue regeneration [136], but it has low mechanical properties, and the use of cytotoxic cross-linking agents is necessary.

In the study of Fathi et al. elastin has been covalently cross-linked with the copolymer poly(NIPAAmco-(PLA/HEMA)-co-OEGMA-co-NAS) obtaining a thermoresponsive and highly cytocompatible system with improved mechanical properties and structural stability [137]. Furthermore, the degradation rate and gelation properties can be modulated by changing the composition of copolymer, to meet the specific requirements for cartilage regeneration.

### 15.2.7 Chondroitin Sulfate-Based In Situ Hydrogels

Chondroitin sulfate is a sulfated glycosaminoglycan composed of repeating units of D-glucuronic acid and *N*-acetylgalactosamine. It plays an important structural role in cartilage providing resistance to compression so it has been widely used for cartilage tissue engineering [138, 139].

Wiltsey et al. have reported a family of injectable hydrogels consisting of poly(N-isopropylacrylamide)-graft-chondroitin sulfate which forms an adhesive interface with surrounding disc tissue [140]. It has been demonstrated that the material possesses good mechanical properties, improved adhesive tensile strength and high cytocompatibility with human embryonic kidney 293 cells, thus, these injectable hydrogels can be used for nucleus pulposus regeneration. In the study of Chen et al. an injectable pullulan/chondroitin sulfate composite has been developed [141]. The hydrogel is enzymatically cross-linked with horseradish peroxidase and hydrogen peroxide, which makes the system easily adaptable and minimally invasive (Fig. 15.7). The presence of chondroitin sulfate facilitates chondrogenesis and enhances cell proliferation, thus showing promise for regenerating cartilage tissue.

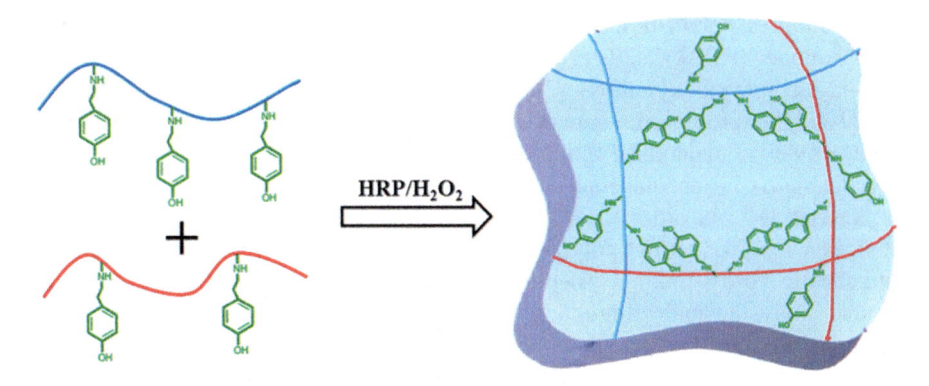

**Fig. 15.7** Hydrogel formation from carboxymethyl pullulan-tyramine (CMP-Tas) and chondroitin sulfate-tyramine (CS-TA) via HRP-mediated cross-linking in the presence of $H_2O_2$. Adapted with permission from Sci Rep 6:20014 Copyright (2016) Springer Nature

## *15.2.8 Synthetic Biomaterials-Based In Situ Hydrogels*

Due to the wide range of possibilities of modification of their chemical structure and molecular composition, hydrogels based on synthetic polymers exhibit highly tunable mechanical properties, biodegradability, biocompatibility, and biochemical characteristics. However, cell–material interaction and biocompatibility has to be taken into account throughout the design of these materials. Several degradable synthetic polymers have been studied for the development of injectable hydrogels for cartilage tissue engineering; these polymers include Poly(ethylene glycol) PEG [142], poly(L-glutamic acid) [54], poly(vinyl alcohol) [143, 144], $\alpha,\beta$-poly-(N-hydroxyethyl)-DL-aspartamide [145], poly(N-isopropyl acrylamide) (PNIPAAm) [57], PEG-poly(N-isopropyl acrylamide) (PNIPAAm) [146], poly(lactic acid) PLLA [147] and methoxy polyethylene glycol–poly($\epsilon$-caprolactone) [148].

Koushki et al. reported a biphasic hydrogel based on partially hydrolyzed polyacrylamide (HPAM) and nanocrystalline hydroxyapatite (nHAp) enzymatically cross-linked with chromium acetate [55]. Mechanical and physical properties of the hydrogel system can be tuned by varying the molecular weight and the cross-link agent concentration. In the above-said work, the incorporation of nHAp provides a dual function to the hydrogel for differentiation towards both soft and hard tissue, resulting in an increased gelation rate, improved mechanical properties, and increased surface wettability. However, a cytotoxic behavior for mesenchymal stem cells is evidenced for high chromium acetate concentrations. One of the concerns associated with the enzyme-mediated cross-linking reaction is the retention of the enzyme inside the hydrogel matrix, which inevitably occurs during gelation. HRP-catalyzed cross-linking is a potential safety strategy [149] to solve this issue. For example, Ren et al. have developed in situ bioinspired polymers based on glycopolypeptide hydrogels through enzymatic cross-linking reaction in the presence of HRP and $H_2O_2$ [2]. These systems are designed to act as an analogue of proteoglycans present in the ECM of native cartilage and they are synthesized by conjugation of poly(γ-propargyl-L-glutamate) (PPLG) with 3-(4-hydroxyphenyl) propanamide (HPPA) and azido-modified mannose (PPLG-g-Man/HPPA) or 1-(2-(2-methoxyethoxy)ethoxy)-2-azidoethane (PPLG-g-MEO$_3$/HPPA). Results reveal a higher proliferation rate of chondrocytes cultured in vitro in PPLG-g-Man/HPPA as well as maintained chondrocyte phenotype and production of cartilaginous specific matrix in subcutaneous model.

Nevertheless, in order to avoid additional chemical cross-linking agents Yan et al. have reported a novel poly(L-glutamic acid)-based injectable hydrogel [54]. In situ chemical cross-linking system using Schiff base reaction has been carried out. In this study the hydrazide-modified poly(L-glutamic acid) (PLGA–ADH) and aldehyde-modified poly(L-glutamic acid) (PLGA–CHO) are prepared using carbodiimide hydrochloride (EDC) activation and sodium periodate (NaIO$_4$) oxidation. Both the hydrazide modification degree of PLGA–ADH and oxidation degree of PLGA–CHO can be adjusted by the amount of activators and NaIO$_4$, respectively. Furthermore, the gelation time, equilibrium swelling, degradation rate, microscopic morphology, and rheological properties can be easily controlled by the solid content of the hydrogels, –NH$_2$/–CHO molar ratio and oxidation degree of PLGA–CHO. These materials have demonstrated successful injectability, rapid in vivo gelling, chondrocytes encapsulation and viability, satisfactory mechanical stability, and favorable ectopic cartilage formation.

Synthetic hydrogels derived from PEG offer a broad field for tuning a wide range of scaffold properties while simultaneously maintaining the chondrocyte phenotype and supporting cartilage-specific matrix production [150]. Hydrogel degradation is a critical design parameter for matrix elaboration, where nondegrading hydrogels restrict matrix deposition and evolution to the nearby cells. Skaalure et al. [53] have developed a new cartilage-specific, degradable hydrogel based on poly(ethylene glycol) (PEG). Novel photoclickable thiol-ene PEG hydrogels are presented such as enzymatically degradable peptide with the incorporation of a cross-linker based on aggrecan. Degradability of the aggrecanase sensitive hydrogel is investigated by encapsulating bovine chondrocytes, which exhibit differences in metabolic activity. The efficacy of the degradable hydrogel is dependent on the cell source. Nonetheless, the aggrecanase-sensitive hydrogel is degraded and preserve the chondrocyte phenotype. Kwon et al. [148] have formulated an injectable in situ forming hydrogel from methoxy polyethylene glycol (MPEG) and poly (*-caprolactone) (PCL). MPEG-PLC (MP) diblock polymers undergo a solution-to-gel phase transition at body temperature, so can be premixed with chondrocytes before being injected. This system is able to form cartilage in a subcutaneous site in vivo, which accumulated with increasing implantation time.

## 15.3 Clinical Applications of In Situ Hydrogels

Articular cartilage is a very specialized connective tissue that covers the joints whose function is acting as a shock absorber of large loads providing a slip surface with low friction. Pathologies that affect hyaline cartilage are the main cause of pain and dysfunction, producing a medium or long term degeneration of the joint. Hyaline cartilage has the characteristics of being avascular, alymphatic, and aneural tissue, so its capacity for regeneration and self-repair is very limited.

In recent years, multiple surgical procedures have been developed with the aim of repairing osteochondral lesions. The most common technique is microfractures, involving making small holes in the area of injury until the subchondral bone in order to create a blood clot that can fill the defect allowing the access of progenitor cells from the bone marrow [151]. However, it usually results on the creation of fibrocartilage, which does not meet the mechanical characteristics of articular hyaline cartilage, and is kept in the injury area for a short period of time. Good clinical results are obtained only for young patients with small chondral lesions, low body mass index and an acute onset of symptoms, thus, finding a good procedure for clinical repair of damaged cartilage and bone tissue is an important challenge that remains to be achieved.

The use of injectable hydrogels is increasingly widespread in clinical trial for cartilage and bone tissue engineering, as explained before, due to its numerous advantages such as minimally invasive injection procedure or ability to fill the chondral defect area regardless of its shape and geometry [152–154]. Because of these systems, the need for traditional open surgeries has been reduced, reducing also the medical care expenses and the time of surgical procedures as well as decreasing postoperative pain and improving recovery time. In addition, these hydrogels can serve as vehicles for therapeutic agents such as growth factors, drugs, or certain cells that stimulate tissue regeneration [3, 31, 32, 35]. Injectable hydrogels can be combined with microfracture technique in order to increase the stability of the blood clot formed and avoiding the creation of fibrocartilage. To reach that goal several injectable hydrogels are being used in clinical practice for osteochondral lesions repair [98].

In June 2013, the use of the GelrinC hydrogel for the treatment of chondral and osteochondral lesions was approved (Regents Biomaterials). GelrinC is an injectable and bioabsorbable hydrogel composed of polyethylene glycol diacrylate (PEG-DA) covalently bound to denatured human fibrinogen (DHF), whose mechanical properties and degradation rate can be adjusted by modifying PEGylation. After standard microfracture, the hydrogel is injected as a liquid, conforming to the size, shape, and depth of the lesion. Then, after short exposure to UV-A light in situ, the hydrogel becomes a semisolid implant tightly integrated with the surrounding tissue and bone, preventing the uncontrolled infiltration of mesenchymal stem cells into the defect and the formation of fibrous tissue. The GelrinC scaffold formed will be reabsorbed and replaced by regenerated cartilage in a period of 6–12 months [42, 155]. After testing its efficacy and safety, the GelrinC hydrogel has been used in clinical practice. In the work of Jeuken et al. [156] patients with knee chondral lesions are treated with GelrinC hydrogel. Microfractures are performed and the GelrinC solution is implanted on them. After 90 s of exposure to UVA light, the hydrogel becomes an elastomeric implant allowing the filling of the defect, regardless of the geometry or depth of the defect. Then, the repair tissue composition is evaluated by Magnetic Resonance Imaging MRI: proportion of GAG, water and orientation and architecture of collagen fibers. In this image study, the complete filling of the lesion is visualized at 6 months postoperatively, and it is maintained until 18 months. After 24 months, the MRI values are similar to those of the native articular cartilage. Thus, an integration of the product with the surrounding cartilage and bone tissue has been observed, demonstrating the safety and efficacy of formation of new hyaline cartilage from the periphery of the inward lesion.

Another product used in clinical practice with the aim of restoring chondral defects associated with microfractures is a chitosan and glycerol phosphate based hydrogel called BST-CarGel (Piramal Life Sciences, Laval, Quebec, Canada), structurally similar to GAG found in articular cartilage (explained above in Sect. 15.2.1). Several in vivo studies support the use of this BST-CarGel for chondral pathology. In the work of Spiller et al., an experimental in vivo study is carried out by applying BST-CarGel in chondral lesions on sheep knees [39]. They injected hydrogel loaded with chondrocytes and that not loaded with the same, observing a complete repair for lesions with BST-CarGel loaded with chondrocytes, and partial repair for lesions with nonloaded CarGel. In other study performed by Chevrier et al., similar chitosan and glycerol phosphate hydrogels are loaded with autologous blood from New Zealand white rabbits [95]. The in vivo experiment is carried out performing a microfracture until the subchondral bone and injecting the BST-CarGel. In the knees treated with the hydrogel three changes can be observed in the healing process to be taken into account: there is an increase in the local inflammatory response in an initial phase as well as the proportion of bone marrow cells; an increase in vascularization and repair tissue is observed; and finally, an increase of the intramembranous bone formation and subchondral bone remodeling. These results are very similar to those obtained by Mathieu et al.'s study [96], where chondral lesions are also performed on rabbit knees and treated with BST-CarGel and autologous blood. In this work another group is studied consisting of the BST-CarGel hydrogel containing blood factors such as thrombin, tissue factor or recombinant human factor VIIa. In both groups an increase in angiogenesis and tissue repair is observed.

Once the safety and efficacy of the product have been demonstrated, several studies have been published in medical clinical practice. In the study of Steinwachs et al. [86], microfractures are performed in patients with a single chondral knee lesion and then the BST-CarGel is injected by arthroscopic approach, reducing the aggressiveness of the surgical procedure and decreasing the recovery time. Six months after arthroscopic treatment the MRI results indicate complete filling of the defect and remission of the subchondral bone edema. The performing of the microfracture and the hydrogel implantation is explained in Fig. 15.8. Also BST-CarGel has been used in other joint lesions. For example, the study of Vilá y Rico et al. describe the arthroscopic technique for BST-CarGel in combination with bone marrow stimulation (BMS) techniques for the treatment of osteochondral talus [157], and in the study of Mas et al. BST-CarGel in combination with a hip arthroscopic microfracture procedure is performed for isolated full-thickness acetabular cartilage defects treatment [87].

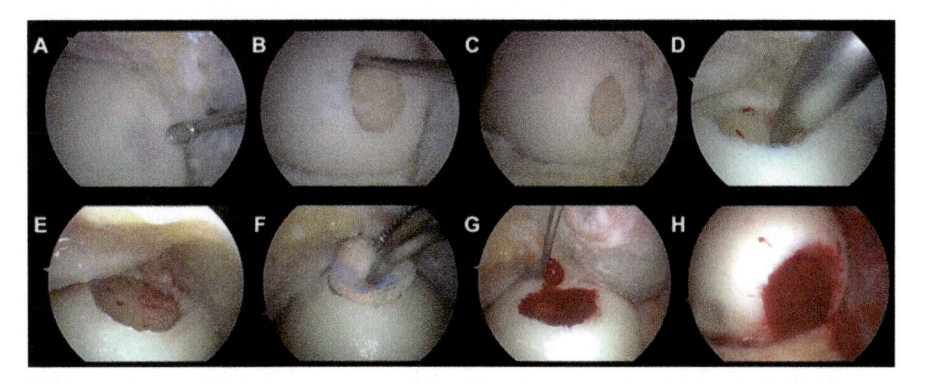

**Fig. 15.8** Key steps of arthroscopic treatment of cartilage lesions with microfracture and BST-CarGel. (**a–c**) The cartilage defect is debrided down to the subchondral bone, with removal of the calcified layer, and to a stable rim of healthy cartilage. (**d**) Microfractures are performed with an awl. (**e**) The defect is arranged in a horizontal position, and the liquid is drained. (**f**) The defect can be further dried with a swab. (**g**) The BST-CarGel is injected into the defect and (**h**) forms a stable clot in the defect after 15 min. Reprinted with permission from Arthrosc Tech 3(3):e399–e402 Copyright (2014) Elsevier

There are currently studies supporting the use of CarGel associated with microfractures for the treatment of chondral lesions [98], such as the study by Shive et al. where they compare the use of BST-CarGel in microfractures with the microfractures treatment alone, assessing structural, clinical and functional results at 5 years post-surgery. It was observed that lesions treated with BST-CarGel presented greater reparative capacity but no changes in the pain or functional scales, so authors conclude that a better cartilage repair, bulkier and of higher quality is produced in the lesions treated with BST-CarGel. These results are very similar to those obtained in the study by Stanish et al. [97] where they also performed a randomized controlled clinical trial of 80 patients with chondral lesions on the femoral condyle divided into two treatment groups: microfractures or microfractures with BST-CarGel. As in the previous study, at MRI level, greater filling of the chondral lesion and higher quality are observed at 12 months postoperatively, while at functional level and life quality, do not present statistically significant changes in both treatment groups. Therefore, these commercial injectable hydrogels favor the repair of chondral lesions by a tissue similar to the primary hyaline cartilage; however, finding a perfect biomaterial to repair osteochondral lesions also aimed at functional level and better quality of life is still a challenge.

## 15.4 Conclusions and Future Trends

In this chapter novel injectable hydrogels for cartilage and bone tissue engineering applications over the recent years are summarized, including the most commonly used materials for the preparation, both natural and synthetic, and their fabrication techniques. Injectable hydrogels are definitely a promising strategy for cartilage and bone tissue engineering applications, owing to their minimal invasive properties and ability to fill irregular defects; however, there are still many challenges to be addressed in order to effectively achieve cartilage and bone repair and regeneration in vivo. The design of a suitable scaffold that gathers perfect bioactive, biocompatible, biodegradable, and mechanical properties is the major challenge of developing injectable hydrogels, and that is why the material used for preparation and the fabrication method chosen are quite important. Natural materials are commonly used for the preparation of injectable hydrogels due to their similarity to the natural cartilage and bone ECMs, thus obtaining perfect biocompatibility and biodegradability; however, their mechanical properties are usually not favorable, limiting their applications. In contrast, injectable hydrogels obtained with synthetic materials possess suitable mechanical properties and stability but poor biocompatible biodegradable and bioactive properties. Likewise, the gelation time plays a key role in the design of injectable scaffolds. In order to fulfill clinical requirements, a suitable injectable hydrogel should be easily injected and rapidly cross-linked in a reasonable time, short enough to prevent the unwanted diffusion of gel precursors; nevertheless, excessively rapid and premature gelation may lead to needle clogging or an increase in the viscosity of the injectable solution. Selective and fast physical or chemical cross-linking reactions under physiological conditions are prerequisites for in situ formation of injectable hydrogels, so the fabrication technique also must be prudently chosen and requires further research. Chemically cross-linked injectable hydrogels show excellent mechanical properties and stability but can have hostile effects in vivo because of chemical reactions, while physically cross-linked hydrogels show a slow response time and poor stability, despite their easy preparation. [38]. In conclusion, the development of novel biomaterials and methodologies to obtain injectable hydrogels integrating perfect gelation time, good mechanical properties, and physiological stability as well as good biocompatibility and biodegradability is an important challenge that remains to be achieved for the clinical application of the hydrogels in cartilage and bone tissue engineering.

## References

1. Walker KJ, Madihally SV (2015) Anisotropic temperature sensitive chitosan-based injectable hydrogels mimicking cartilage matrix. J Biomed Mater Res B Appl Biomater 103(6):1149–1160
2. Ren K, He C, Xiao C, Li G, Chen X (2015) Injectable glycopolypeptide hydrogels as biomimetic scaffolds for cartilage tissue engineering. Biomaterials 51(Supplement C):238–249. https://doi.org/10.1016/j.biomaterials.2015.02.026
3. Amini AA, Nair LS (2012) Injectable hydrogels for bone and cartilage repair. Biomed Mater 7(2):024105. https://doi.org/10.1088/1748-6041/7/2/024105
4. Ahadian S, Sadeghian RB, Salehi S, Ostrovidov S, Bae H, Ramalingam M, Khademhosseini A (2015) Bioconjugated hydrogels for tissue engineering and regenerative medicine. Bioconjug Chem 26(10):1984–2001

5. Benders KE, van Weeren PR, Badylak SF, Saris DB, Dhert WJ, Malda J (2013) Extracellular matrix scaffolds for cartilage and bone regeneration. Trends Biotechnol 31(3):169–176
6. Brown BN, Badylak SF (2014) Extracellular matrix as an inductive scaffold for functional tissue reconstruction. Transl Res 163(4):268–285
7. Zhang X, Zhu J, Liu F, Li Y, Chandra A, Levin LS, Beier F, Enomoto-Iwamoto M, Qin L (2014) Reduced EGFR signaling enhances cartilage destruction in a mouse osteoarthritis model. Bone Res 2:14015
8. Vilela C, Correia C, Oliveira JM, Sousa RA, Espregueira-Mendes J, Reis RL (2015) Cartilage repair using hydrogels: a critical review of in vivo experimental designs. ACS Biomater Sci Eng 1(9):726–739
9. Liao J, Shi K, Ding Q, Qu Y, Luo F, Qian Z (2014) Recent developments in scaffold-guided cartilage tissue regeneration. J Biomed Nanotechnol 10(10):3085–3104
10. Yuan T, Zhang L, Li K, Fan H, Fan Y, Liang J, Zhang X (2014) Collagen hydrogel as an immunomodulatory scaffold in cartilage tissue engineering. J Biomed Mater Res B Appl Biomater 102(2):337–344
11. Buckwalter JA (1998) Articular cartilage: injuries and potential for healing. J Orthopaed Sports Phys Ther 28(4):192–202
12. Zhang W, Ouyang H, Dass CR, Xu J (2016) Current research on pharmacologic and regenerative therapies for osteoarthritis. Bone Res 4:15040
13. Frisch J, K Venkatesan J, Rey-Rico A, Madry H, Cucchiarini M (2015) Current progress in stem cell-based gene therapy for articular cartilage repair. Curr Stem Cell Res Ther 10(2):121–131
14. Tan R, Feng Q, She Z, Wang M, Jin H, Li J, Yu X (2010) In vitro and in vivo degradation of an injectable bone repair composite. Polym Degrad Stab 95(9):1736–1742
15. Alford AI, Kozloff KM, Hankenson KD (2015) Extracellular matrix networks in bone remodeling. Int J Biochem Cell Biol 65:20–31
16. Cordonnier T, Sohier J, Rosset P, Layrolle P (2011) Biomimetic materials for bone tissue engineering–state of the art and future trends. Adv Eng Mater 13(5)
17. Tomlinson RE, Silva MJ (2013) Skeletal blood flow in bone repair and maintenance. Bone Res 1(4):311
18. Flierl MA, Smith WR, Mauffrey C, Irgit K, Williams AE, Ross E, Peacher G, Hak DJ, Stahel PF (2013) Outcomes and complication rates of different bone grafting modalities in long bone fracture nonunions: a retrospective cohort study in 182 patients. J Orthopaed Surg Res 8(1):33
19. Marenzana M, Arnett TR (2013) The key role of the blood supply to bone. Bone Res 1(3):203
20. Ahmed EM (2015) Hydrogel: Preparation, characterization, and applications: A review. J Adv Res 6(2):105–121. https://doi.org/10.1016/j.jare.2013.07.006
21. Hunt JA, Chen R, van Veen T, Bryan N (2014) Hydrogels for tissue engineering and regenerative medicine. J Mater Chem B 2(33):5319–5338. https://doi.org/10.1039/C4TB00775A
22. Mora-Boza A, Puertas-Bartolomé M, Vázquez-Lasa B, San Román J, Pérez-Caballer A, Olmeda-Lozano M (2017) Contribution of bioactive hyaluronic acid and gelatin to regenerative medicine. Methodologies of gels preparation and advanced applications. Eur Polym J 95:11–26
23. Zhang L, Xia K, Lu Z, Li G, Chen J, Deng Y, Li S, Zhou F, He N (2014) Efficient and facile synthesis of gold nanorods with finely tunable plasmonic peaks from visible to near-IR range. Chem Mater 26(5):1794–1798
24. Deng Y, Wang M, Jiang L, Ma C, Xi Z, Li X, He N (2013) A comparison of extracellular excitatory amino acids release inhibition of acute lamotrigine and topiramate treatment in the hippocampus of PTZ-kindled epileptic rats. J Biomed Nanotechnol 9(6):1123–1128
25. Shin SR, Li Y-C, Jang HL, Khoshakhlagh P, Akbari M, Nasajpour A, Zhang YS, Tamayol A, Khademhosseini A (2016) Graphene-based materials for tissue engineering. Adv Drug Deliv Rev 105:255–274

26. Fan C, Wang D-A (2015) A biodegradable PEG-based micro-cavitary hydrogel as scaffold for cartilage tissue engineering. Eur Polym J 72:651–660
27. Fan J, He N, He Q, Liu Y, Ma Y, Fu X, Liu Y, Huang P, Chen X (2015) A novel self-assembled sandwich nanomedicine for NIR-responsive release of NO. Nanoscale 7(47):20055–20062
28. Lu Z, Huang Y, Zhang L, Xia K, Deng Y, He N (2015) Preparation of gold nanorods using 1, 2, 4-Trihydroxybenzene as a reducing agent. J Nanosci Nanotechnol 15(8):6230–6235
29. Choi B, Kim S, Lin B, Wu BM, Lee M (2014) Cartilaginous extracellular matrix-modified chitosan hydrogels for cartilage tissue engineering. ACS Appl Mater Interfaces 6(22):20110–20121. https://doi.org/10.1021/am505723k
30. Yazdimamaghani M, Vashaee D, Assefa S, Walker K, Madihally S, Köhler G, Tayebi L (2014) Hybrid macroporous gelatin/bioactive-glass/nanosilver scaffolds with controlled degradation behavior and antimicrobial activity for bone tissue engineering. J Biomed Nanotechnol 10(6):911–931
31. Sivashanmugam A, Arun Kumar R, Vishnu Priya M, Nair SV, Jayakumar R (2015) An overview of injectable polymeric hydrogels for tissue engineering. Eur Polym J 72(Supplement C):543–565. https://doi.org/10.1016/j.eurpolymj.2015.05.014
32. Li Y, Rodrigues J, Tomas H (2012) Injectable and biodegradable hydrogels: gelation, biodegradation and biomedical applications. Chem Soc Rev 41(6):2193–2221. https://doi.org/10.1039/C1CS15203C
33. Johnson TD, Christman KL (2013) Injectable hydrogel therapies and their delivery strategies for treating myocardial infarction. Expert Opin Drug Deliv 10(1):59–72. https://doi.org/10.1517/17425247.2013.739156
34. Nguyen MK, Lee DS (2010) Injectable biodegradable hydrogels. Macromol Biosci 10(6):563–579. https://doi.org/10.1002/mabi.200900402
35. Bae KH, Wang L-S, Kurisawa M (2013) Injectable biodegradable hydrogels: progress and challenges. J Mater Chem B 1(40):5371–5388. https://doi.org/10.1039/C3TB20940G
36. Radhakrishnan J, Subramanian A, Sethuraman S (2017) Injectable glycosaminoglycan–protein nano-complex in semi-interpenetrating networks: a biphasic hydrogel for hyaline cartilage regeneration. Carbohydr Polym 175(Supplement C):63–74. https://doi.org/10.1016/j.carbpol.2017.07.063
37. Ko DY, Shinde UP, Yeon B, Jeong B (2013) Recent progress of in situ formed gels for biomedical applications. Prog Polym Sci 38(3):672–701. https://doi.org/10.1016/j.progpolymsci.2012.08.002
38. Liu M, Zeng X, Ma C, Yi H, Ali Z, Mou X, Li S, Deng Y, He N (2017) Injectable hydrogels for cartilage and bone tissue engineering. Bone Res 5:17014. https://doi.org/10.1038/boneres.2017.14
39. Spiller KL, Maher SA, Lowman AM (2011) Hydrogels for the repair of articular cartilage defects. Tissue Eng Part B Rev 17(4):281–299. https://doi.org/10.1089/ten.teb.2011.0077
40. Yang J-A, Yeom J, Hwang BW, Hoffman AS, Hahn SK (2014) In situ-forming injectable hydrogels for regenerative medicine. Prog Polym Sci 39(12):1973–1986. https://doi.org/10.1016/j.progpolymsci.2014.07.006
41. Eslahi N, Abdorahim M, Simchi A (2016) Smart polymeric hydrogels for cartilage tissue engineering: a review on the chemistry and biological functions. Biomacromolecules 17(11):3441–3463. https://doi.org/10.1021/acs.biomac.6b01235
42. Chuah YJ, Peck Y, Lau JEJ, Hee HT, Wang D-A (2017) Hydrogel based cartilaginous tissue regeneration: recent insights and technologies. Biomater Sci 5(4):613–631. https://doi.org/10.1039/C6BM00863A
43. Balakrishnan B, Joshi N, Jayakrishnan A, Banerjee R (2014) Self-crosslinked oxidized alginate/gelatin hydrogel as injectable, adhesive biomimetic scaffolds for cartilage regeneration. Acta Biomater 10(8):3650–3663. https://doi.org/10.1016/j.actbio.2014.04.031
44. Seol YJ, Park JY, Jeong W, Kim TH, Kim SY, Cho DW (2015) Development of hybrid scaffolds using ceramic and hydrogel for articular cartilage tissue regeneration. J Biomed Mater Res A 103(4):1404–1413. https://doi.org/10.1002/jbm.a.35276

45. Park H, Lee KY (2014) Cartilage regeneration using biodegradable oxidized alginate/hyaluronate hydrogels. J Biomed Mater Res A 102(12):4519–4525. https://doi.org/10.1002/jbm.a.35126
46. Palumbo FS, Fiorica C, Di Stefano M, Pitarresi G, Gulino A, Agnello S, Giammona G (2015) In situ forming hydrogels of hyaluronic acid and inulin derivatives for cartilage regeneration. Carbohydr Polym 122:408–416. https://doi.org/10.1016/j.carbpol.2014.11.002
47. Naderi-Meshkin H, Andreas K, Matin MM, Sittinger M, Bidkhori HR, Ahmadiankia N, Bahrami AR, Ringe J (2014) Chitosan-based injectable hydrogel as a promising in situ forming scaffold for cartilage tissue engineering. Cell Biol Int 38(1):72–84. https://doi.org/10.1002/cbin.10181
48. Park H, Choi B, Hu J, Lee M (2013) Injectable chitosan hyaluronic acid hydrogels for cartilage tissue engineering. Acta Biomater 9(1):4779–4786. https://doi.org/10.1016/j.actbio.2012.08.033
49. Shen Z-S, Cui X, Hou R, Li Q, Deng H-X, Fu J (2015) Tough biodegradable chitosan–gelatin hydrogels via in situ precipitation for potential cartilage tissue engineering. RSC Adv 5:55640–55647. https://doi.org/10.1039/C5RA06835E
50. Yuan L, Li B, Yang J, Ni Y, Teng Y, Guo L, Fan H, Fan Y, Zhang X (2016) Effects of composition and mechanical property of injectable collagen I/II composite hydrogels on chondrocyte behaviors. Tissue Eng Part A 22(11-12):899–906. https://doi.org/10.1089/ten.TEA.2015.0513
51. Dispenza C, Todaro S, Bulone D, Sabatino MA, Ghersi G, San Biagio PL, Lo Presti C (2017) Physico-chemical and mechanical characterization of in-situ forming xyloglucan gels incorporating a growth factor to promote cartilage reconstruction. Mater Sci Eng C 70(Part 1):745–752. https://doi.org/10.1016/j.msec.2016.09.045
52. Wang R, Leber N, Buhl C, Verdonschot N, Dijkstra PJ, Karperien M (2014) Cartilage adhesive and mechanical properties of enzymatically crosslinked polysaccharide tyramine conjugate hydrogels. Polym Adv Technol 25(5):568–574. https://doi.org/10.1002/pat.3286
53. Skaalure SC, Chu S, Bryant SJ (2015) An enzyme-sensitive PEG hydrogel based on aggrecan catabolism for cartilage tissue engineering. Adv Healthc Mater 4(3):420–431. https://doi.org/10.1002/adhm.201400277
54. Yan S, Zhang X, Zhang K, Di H, Feng L, Li G, Fang J, Cui L, Chen X, Yin J (2016) Injectable in situ forming poly(l-glutamic acid) hydrogels for cartilage tissue engineering. J Mater Chem B 4(5):947–961. https://doi.org/10.1039/C5TB01488C
55. Koushki N, Tavassoli H, Katbab AA, Katbab P, Bonakdar S (2015) A new injectable biphasic hydrogel based on partially hydrolyzed polyacrylamide and nano hydroxyapatite, crosslinked with chromium acetate, as scaffold for cartilage regeneration. AIP Conf Proc 1664(1):070002. https://doi.org/10.1063/1.4918437
56. Schuurman W, Levett PA, Pot MW, van Weeren PR, Dhert WJA, Hutmacher DW, Melchels FPW, Klein TJ, Malda J (2013) Gelatin-methacrylamide hydrogels as potential biomaterials for fabrication of tissue-engineered cartilage constructs. Macromol Biosci 13(5):551–561. https://doi.org/10.1002/mabi.201200471
57. Mellati A, Kiamahalleh MV, Madani SH, Dai S, Bi J, Jin B, Zhang H (2016) Poly(N-isopropylacrylamide) hydrogel/chitosan scaffold hybrid for three-dimensional stem cell culture and cartilage tissue engineering. J Biomed Mater Res A 104(11):2764–2774. https://doi.org/10.1002/jbm.a.35810
58. Overstreet DJ, Dutta D, Stabenfeldt SE, Vernon BL (2012) Injectable hydrogels. J Polym Sci B 50(13):881–903. https://doi.org/10.1002/polb.23081
59. Yan S, Wang T, Feng L, Zhu J, Zhang K, Chen X, Cui L, Yin J (2014) Injectable in situ self-cross-linking hydrogels based on poly(l-glutamic acid) and alginate for cartilage tissue engineering. Biomacromolecules 15(12):4495–4508. https://doi.org/10.1021/bm501313t
60. Patenaude M, Smeets NMB, Hoare T (2014) Designing injectable, covalently cross-linked hydrogels for biomedical applications. Macromol Rapid Commun 35(6):598–617. https://doi.org/10.1002/marc.201300818

61. Yang J, Zhang YS, Yue K, Khademhosseini A (2017) Cell-laden hydrogels for osteochondral and cartilage tissue engineering. Acta Biomater 57(Supplement C):1–25. https://doi.org/10.1016/j.actbio.2017.01.036
62. Tan H, Chu CR, Payne KA, Marra KG (2009) Injectable in situ forming biodegradable chitosan–hyaluronic acid based hydrogels for cartilage tissue engineering. Biomaterials 30(13):2499–2506
63. Di Martino A, Sittinger M, Risbud MV (2005) Chitosan: a versatile biopolymer for orthopaedic tissue-engineering. Biomaterials 26(30):5983–5990
64. Hu X, Zhang Z, Wang G, Yao L, Xia Y, Sun A, Gu Y, He N, Li Z, Yang W (2015) Preparation of chitosan-sodium sodium tripolyphosphate nanoparticles via reverse microemulsion-ionic gelation method. J Bionanosci 9(4):301–305
65. Jin R, Teixeira LM, Dijkstra PJ, Karperien M, Van Blitterswijk C, Zhong Z, Feijen J (2009) Injectable chitosan-based hydrogels for cartilage tissue engineering. Biomaterials 30(13):2544–2551
66. Morris VB, Sharma CP (2010) Folate mediated histidine derivative of quaternised chitosan as a gene delivery vector. Int J Pharm 389(1):176–185
67. Jin R, Lin C, Cao A (2014) Enzyme-mediated fast injectable hydrogels based on chitosan-glycolic acid/tyrosine: preparation, characterization, and chondrocyte culture. Polym Chem 5(2):391–398. https://doi.org/10.1039/C3PY00864A
68. Nguyen DH, Tran NQ, Nguyen CK (2013) Tetronic-grafted chitosan hydrogel as an injectable and biocompatible scaffold for biomedical applications. J Biomater Sci Polym Ed 24(14):1636–1648
69. Lee SH, Lee Y, Lee S-W, Ji H-Y, Lee J-H, Lee DS, Park TG (2011) Enzyme-mediated crosslinking of Pluronic copolymer micelles for injectable and in situ forming hydrogels. Acta Biomater 7(4):1468–1476
70. Kamoun EA (2016) N-succinyl chitosan–dialdehyde starch hybrid hydrogels for biomedical applications. J Adv Res 7(1):69–77. https://doi.org/10.1016/j.jare.2015.02.002
71. Li Z, Cho S, Kwon IC, Janát-Amsbury MM, Huh KM (2013) Preparation and characterization of glycol chitin as a new thermogelling polymer for biomedical applications. Carbohydr Polym 92(2):2267–2275
72. Cho IS, Cho MO, Li Z, Nurunnabi M, Park SY, Kang SW, Huh KM (2016) Synthesis and characterization of a new photo-crosslinkable glycol chitosan thermogel for biomedical applications. Carbohydr Polym 144:59–67. https://doi.org/10.1016/j.carbpol.2016.02.029
73. Cao L, Cao B, Lu C, Wang G, Yu L, Ding J (2015) An injectable hydrogel formed by in situ cross-linking of glycol chitosan and multi-benzaldehyde functionalized PEG analogues for cartilage tissue engineering. J Mater Chem B 3(7):1268–1280. https://doi.org/10.1039/C4TB01705F
74. Zhang K, He S, Yan S, Li G, Zhang D, Cui L, Yin J (2016) Regeneration of hyaline-like cartilage and subchondral bone simultaneously by poly(l-glutamic acid) based osteochondral scaffolds with induced autologous adipose derived stem cells. J Mater Chem B 4(15):2628–2645. https://doi.org/10.1039/c5tb02113h
75. Ni P, Ding Q, Fan M, Liao J, Qian Z, Luo J, Li X, Luo F, Yang Z, Wei Y (2014) Injectable thermosensitive PEG–PCL–PEG hydrogel/acellular bone matrix composite for bone regeneration in cranial defects. Biomaterials 35(1):236–248
76. Peniche H, Reyes-Ortega F, Aguilar MR, Rodríguez G, Abradelo C, García-Fernández L, Peniche C, Román JS (2013) Thermosensitive macroporous cryogels functionalized with bioactive chitosan/bemiparin nanoparticles. Macromol Biosci 13(11):1556–1567
77. Zhou HY, Jiang LJ, Cao PP, Li JB, Chen XG (2015) Glycerophosphate-based chitosan thermosensitive hydrogels and their biomedical applications. Carbohydr Polym 117(Supplement C):524–536. https://doi.org/10.1016/j.carbpol.2014.09.094
78. Klouda L, Mikos AG (2008) Thermoresponsive hydrogels in biomedical applications. Eur J Pharm Biopharm 68(1):34–45
79. Martínez A, Blanco M, Davidenko N, Cameron R (2015) Tailoring chitosan/collagen scaffolds for tissue engineering: effect of composition and different crosslinking agents on scaffold properties. Carbohydr Polym 132:606–619

80. Mogoşanu GD, Grumezescu AM (2014) Natural and synthetic polymers for wounds and burns dressing. Int J Pharm 463(2):127–136
81. Nair LS, Starnes T, Ko J-WK, Laurencin CT (2007) Development of injectable thermogelling chitosan–inorganic phosphate solutions for biomedical applications. Biomacromolecules 8(12):3779–3785
82. Ta HT, Han H, Larson I, Dass CR, Dunstan DE (2009) Chitosan-dibasic orthophosphate hydrogel: a potential drug delivery system. Int J Pharm 371(1):134–141
83. Chenite A, Chaput C, Wang D, Combes C, Buschmann M, Hoemann C, Leroux J, Atkinson B, Binette F, Selmani A (2000) Novel injectable neutral solutions of chitosan form biodegradable gels in situ. Biomaterials 21(21):2155–2161
84. Chenite A, Buschmann M, Wang D, Chaput C, Kandani N (2001) Rheological characterisation of thermogelling chitosan/glycerol-phosphate solutions. Carbohydr Polym 46(1):39–47
85. y Rico JV, Dalmau A, Chaqués FJ, Asunción J (2015) treatment of osteochondral lesions of the talus with bone marrow stimulation and chitosan–glycerol phosphate/blood implants (BST-CarGel). Arthrosc Tech 4(6):e663–e667
86. Steinwachs MR, Waibl B, Mumme M (2014) Arthroscopic treatment of cartilage lesions with microfracture and BST-CarGel. Arthrosc Tech 3(3):e399–e402
87. Tey M, Mas J, Pelfort X, Monllau JC (2015) Arthroscopic treatment of hip chondral defects with bone marrow stimulation and BST-CarGel. Arthrosc Tech 4(1):e29–e33
88. Dang QF, Yan JQ, Lin H, Liu CS, Chen XG, Ji QX, Li J, Liu Y (2015) Biological evaluation of chitosan-based in situ-forming hydrogel with low phase transition temperature. J Appl Polym Sci 132(10)
89. Douglas TE, Skwarczynska A, Modrzejewska Z, Balcaen L, Schaubroeck D, Lycke S, Vanhaecke F, Vandenabeele P, Dubruel P, Jansen JA (2013) Acceleration of gelation and promotion of mineralization of chitosan hydrogels by alkaline phosphatase. Int J Biol Macromol 56:122–132
90. Wu G, Yuan Y, He J, Li Y, Dai X, Zhao B (2016) Stable thermosensitive in situ gel-forming systems based on the lyophilizate of chitosan/α,β-glycerophosphate salts. Int J Pharm 511(1):560–569. https://doi.org/10.1016/j.ijpharm.2016.07.050
91. Assaad E, Maire M, Lerouge S (2015) Injectable thermosensitive chitosan hydrogels with controlled gelation kinetics and enhanced mechanical resistance. Carbohydr Polym 130:87–96
92. Deng A, Kang X, Zhang J, Yang Y, Yang S (2017) Enhanced gelation of chitosan/beta-sodium glycerophosphate thermosensitive hydrogel with sodium bicarbonate and biocompatibility evaluated. Mater Sci Eng C Mater Biol Appl 78:1147–1154. https://doi.org/10.1016/j.msec.2017.04.109
93. Dang Q, Liu K, Zhang Z, Liu C, Liu X, Xin Y, Cheng X, Xu T, Cha D, Fan B (2017) Fabrication and evaluation of thermosensitive chitosan/collagen/alpha, beta-glycerophosphate hydrogels for tissue regeneration. Carbohydr Polym 167:145–157. https://doi.org/10.1016/j.carbpol.2017.03.053
94. Mirahmadi F, Tafazzoli-Shadpour M, Shokrgozar MA, Bonakdar S (2013) Enhanced mechanical properties of thermosensitive chitosan hydrogel by silk fibers for cartilage tissue engineering. Mater Sci Eng C Mater Biol Appl 33(8):4786–4794. https://doi.org/10.1016/j.msec.2013.07.043
95. Chevrier A, Hoemann CD, Sun J, Buschmann MD (2007) Chitosan-glycerol phosphate/blood implants increase cell recruitment, transient vascularization and subchondral bone remodeling in drilled cartilage defects. Osteoarthritis Cartilage 15(3):316–327. https://doi.org/10.1016/j.joca.2006.08.007
96. Mathieu C, Chevrier A, Lascau-Coman V, Rivard GE, Hoemann CD (2013) Stereological analysis of subchondral angiogenesis induced by chitosan and coagulation factors in microdrilled articular cartilage defects. Osteoarthritis Cartilage 21(6):849–859. https://doi.org/10.1016/j.joca.2013.03.012

97. Stanish WD, McCormack R, Forriol F, Mohtadi N, Pelet S, Desnoyers J, Restrepo A, Shive MS (2013) Novel scaffold-based BST-CarGel treatment results in superior cartilage repair compared with microfracture in a randomized controlled trial. J Bone Joint Surg 95(18):1640–1650
98. Shive MS, Stanish WD, McCormack R, Forriol F, Mohtadi N, Pelet S, Desnoyers J, Methot S, Vehik K, Restrepo A (2015) BST-CarGel (R) treatment maintains cartilage repair superiority over microfracture at 5 years in a multicenter randomized controlled trial. Cartilage 6(2):62–72. https://doi.org/10.1177/1947603514562064
99. Muzzarelli RA, Greco F, Busilacchi A, Sollazzo V, Gigante A (2012) Chitosan, hyaluronan and chondroitin sulfate in tissue engineering for cartilage regeneration: a review. Carbohydr Polym 89(3):723–739
100. Astachov L, Vago R, Aviv M, Nevo Z (2011) Hyaluronan and mesenchymal stem cells: from germ layer to cartilage and bone. Front Biosci 16:261–276
101. Kim IL, Mauck RL, Burdick JA (2011) Hydrogel design for cartilage tissue engineering: a case study with hyaluronic acid. Biomaterials 32(34):8771–8782
102. Yeom J, Hwang BW, Yang DJ, Shin H-I, Hahn SK (2014) Effect of osteoconductive hyaluronate hydrogels on calvarial bone regeneration. Biomater Res 18(1):8
103. Bellini D, Cencetti C, Meraner J, Stoppoloni D, D'Abusco AS, Matricardi P (2015) An in situ gelling system for bone regeneration of osteochondral defects. Eur Polym J 72:642–650. https://doi.org/10.1016/j.eurpolymj.2015.02.043
104. Yu F, Cao X, Li Y, Zeng L, Yuan B, Chen X (2014) An injectable hyaluronic acid/PEG hydrogel for cartilage tissue engineering formed by integrating enzymatic crosslinking and Diels-Alder "click chemistry". Polym Chem 5(3):1082–1090. https://doi.org/10.1039/C3PY00869J
105. Yu F, Cao X, Li Y, Zeng L, Zhu J, Wang G, Chen X (2014) Diels–Alder crosslinked HA/PEG hydrogels with high elasticity and fatigue resistance for cell encapsulation and articular cartilage tissue repair. Polym Chem 5(17):5116–5123
106. Fiorica C, Palumbo FS, Pitarresi G, Gulino A, Agnello S, Giammona G (2015) Injectable in situ forming hydrogels based on natural and synthetic polymers for potential application in cartilage repair. RSC Adv 5(25):19715–19723. https://doi.org/10.1039/C4RA16411C
107. Fenn SL, Oldinski RA (2016) Visible light crosslinking of methacrylated hyaluronan hydrogels for injectable tissue repair. J Biomed Mater Res B Appl Biomater 104(6):1229–1236. https://doi.org/10.1002/jbm.b.33476
108. Domingues RM, Silva M, Gershovich P, Betta S, Babo P, Caridade SG, Mano JF, Motta A, Reis RL, Gomes ME (2015) Development of injectable hyaluronic acid/cellulose nanocrystals bionanocomposite hydrogels for tissue engineering applications. Bioconjug Chem 26(8):1571–1581. https://doi.org/10.1021/acs.bioconjchem.5b00209
109. Parmar PA, Chow LW, St-Pierre J-P, Horejs C-M, Peng YY, Werkmeister JA, Ramshaw JA, Stevens MM (2015) Collagen-mimetic peptide-modifiable hydrogels for articular cartilage regeneration. Biomaterials 54:213–225
110. Zhang L, Yuan T, Guo L, Zhang X (2012) An in vitro study of collagen hydrogel to induce the chondrogenic differentiation of mesenchymal stem cells. J Biomed Mater Res A 100(10):2717–2725
111. Cooper BG, Lawson TB, Snyder BD, Grinstaff MW (2017) Reinforcement of articular cartilage with a tissue-interpenetrating polymer network reduces friction and modulates interstitial fluid load support. Osteoarthritis Cartilage 25(7):1143–1149. https://doi.org/10.1016/j.joca.2017.03.001
112. Kontturi LS, Jarvinen E, Muhonen V, Collin EC, Pandit AS, Kiviranta I, Yliperttula M, Urtti A (2014) An injectable, in situ forming type II collagen/hyaluronic acid hydrogel vehicle for chondrocyte delivery in cartilage tissue engineering. Drug Deliv Transl Res 4(2):149–158. https://doi.org/10.1007/s13346-013-0188-1
113. Barhoumi A, Salvador-Culla B, Kohane DS (2015) NIR-triggered drug delivery by collagen-mediated second harmonic generation. Adv Healthc Mater 4(8):1159–1163

114. Moreira CD, Carvalho SM, Mansur HS, Pereira MM (2016) Thermogelling chitosan–collagen–bioactive glass nanoparticle hybrids as potential injectable systems for tissue engineering. Mater Sci Eng C 58:1207–1216
115. Song K, Li L, Li W, Zhu Y, Jiao Z, Lim M, Fang M, Shi F, Wang L, Liu T (2015) Three-dimensional dynamic fabrication of engineered cartilage based on chitosan/gelatin hybrid hydrogel scaffold in a spinner flask with a special designed steel frame. Mater Sci Eng C 55:384–392
116. Santoro M, Tatara AM, Mikos AG (2014) Gelatin carriers for drug and cell delivery in tissue engineering. J Control Release 190:210–218
117. Wang L-S, Du C, Toh WS, Wan ACA, Gao SJ, Kurisawa M (2014) Modulation of chondrocyte functions and stiffness-dependent cartilage repair using an injectable enzymatically crosslinked hydrogel with tunable mechanical properties. Biomaterials 35(7):2207–2217. https://doi.org/10.1016/j.biomaterials.2013.11.070
118. Oh BH, Bismarck A, Chan-Park MB (2015) Injectable, interconnected, high-porosity macroporous biocompatible gelatin scaffolds made by surfactant-free emulsion templating. Macromol Rapid Commun 36(4):364–372. https://doi.org/10.1002/marc.201400524
119. Puertas-Bartolomé M, Fernández-Gutiérrez M, García-Fernández L, Vázquez-Lasa B, San Román J (2018) Biocompatible and bioadhesive low molecular weight polymers containing long-arm catechol-functionalized methacrylate. Eur Polym J 98:47
120. Visser J, Gawlitta D, Benders KE, Toma SM, Pouran B, van Weeren PR, Dhert WJ, Malda J (2015) Endochondral bone formation in gelatin methacrylamide hydrogel with embedded cartilage-derived matrix particles. Biomaterials 37:174–182. https://doi.org/10.1016/j.biomaterials.2014.10.020
121. Follin B, Juhl M, Cohen S, Pedersen AE, Gad M, Kastrup J, Ekblond A (2015) Human adipose-derived stromal cells in a clinically applicable injectable alginate hydrogel: phenotypic and immunomodulatory evaluation. Cytotherapy 17(8):1104–1118
122. Ruvinov E, Cohen S (2016) Alginate biomaterial for the treatment of myocardial infarction: progress, translational strategies, and clinical outlook: from ocean algae to patient bedside. Adv Drug Deliv Rev 96:54–76
123. Venkatesan J, Bhatnagar I, Manivasagan P, Kang K-H, Kim S-K (2015) Alginate composites for bone tissue engineering: a review. Int J Biol Macromol 72:269–281
124. Rojo L, Vázquez B, Deb S, San Román J (2009) Eugenol derivatives immobilized in auto-polymerizing formulations as an approach to avoid inhibition interferences and improve biofunctionality in dental and orthopedic cements. Acta Biomater 5(5):1616–1625
125. Zeng L, Yao Y, Wang DA, Chen X (2014) Effect of microcavitary alginate hydrogel with different pore sizes on chondrocyte culture for cartilage tissue engineering. Mater Sci Eng C Mater Biol Appl 34:168–175. https://doi.org/10.1016/j.msec.2013.09.003
126. Han Y, Zeng Q, Li H, Chang J (2013) The calcium silicate/alginate composite: preparation and evaluation of its behavior as bioactive injectable hydrogels. Acta Biomater 9(11):9107–9117. https://doi.org/10.1016/j.actbio.2013.06.022
127. Jaikumar D, Sajesh K, Soumya S, Nimal T, Chennazhi K, Nair SV, Jayakumar R (2015) Injectable alginate-O-carboxymethyl chitosan/nano fibrin composite hydrogels for adipose tissue engineering. Int J Biol Macromol 74:318–326
128. Zheng L, Jiang X, Chen X, Fan H, Zhang X (2014) Evaluation of novel in situ synthesized nano-hydroxyapatite/collagen/alginate hydrogels for osteochondral tissue engineering. Biomed Mater 9(6):065004. https://doi.org/10.1088/1748-6041/9/6/065004
129. Gothard D, Smith EL, Kanczler JM, Black CR, Wells JA, Roberts CA, White LJ, Qutachi O, Peto H, Rashidi H (2015) In vivo assessment of bone regeneration in alginate/bone ECM hydrogels with incorporated skeletal stem cells and single growth factors. PLoS One 10(12):e0145080
130. Lee C, Shin J, Lee JS, Byun E, Ryu JH, Um SH, Kim D-I, Lee H, Cho S-W (2013) Bioinspired, calcium-free alginate hydrogels with tunable physical and mechanical properties and improved biocompatibility. Biomacromolecules 14(6):2004–2013. https://doi.org/10.1021/bm400352d

131. Choi JW, Choi BH, Park SH, Pai KS, Li TZ, Min BH, Park SR (2013) Mechanical stimulation by ultrasound enhances chondrogenic differentiation of mesenchymal stem cells in a fibrin-hyaluronic acid hydrogel. Artif Organs 37(7):648–655
132. Snyder TN, Madhavan K, Intrator M, Dregalla RC, Park D (2014) A fibrin/hyaluronic acid hydrogel for the delivery of mesenchymal stem cells and potential for articular cartilage repair. J Biol Eng 8(1):10
133. Benavides OM, Brooks AR, Cho SK, Petsche Connell J, Ruano R, Jacot JG (2015) In situ vascularization of injectable fibrin/poly(ethylene glycol) hydrogels by human amniotic fluid-derived stem cells. J Biomed Mater Res A 103(8):2645–2653. https://doi.org/10.1002/jbm.a.35402
134. Almeida HV, Eswaramoorthy R, Cunniffe GM, Buckley CT, O'Brien FJ, Kelly DJ (2016) Fibrin hydrogels functionalized with cartilage extracellular matrix and incorporating freshly isolated stromal cells as an injectable for cartilage regeneration. Acta Biomater 36:55–62. https://doi.org/10.1016/j.actbio.2016.03.008
135. Hwang C, Ay B, Kaplan D, Rubin J, Marra K, Atala A, Yoo J, Lee S (2013) Assessments of injectable alginate particle-embedded fibrin hydrogels for soft tissue reconstruction. Biomed Mater 8(1):014105
136. Ozsvar J, Mithieux SM, Wang R, Weiss AS (2015) Elastin-based biomaterials and mesenchymal stem cells. Biomater Sci 3(6):800–809
137. Fathi A, Mithieux SM, Wei H, Chrzanowski W, Valtchev P, Weiss AS, Dehghani F (2014) Elastin based cell-laden injectable hydrogels with tunable gelation, mechanical and biodegradation properties. Biomaterials 35(21):5425–5435. https://doi.org/10.1016/j.biomaterials.2014.03.026
138. Liao J, Qu Y, Chu B, Zhang X, Qian Z (2015) Biodegradable CSMA/PECA/graphene porous hybrid scaffold for cartilage tissue engineering. Sci Rep 5
139. Dwivedi P, Bhat S, Nayak V, Kumar A (2014) Study of different delivery modes of chondroitin sulfate using microspheres and cryogel scaffold for application in cartilage tissue engineering. Int J Polym Mater Polym Biomater 63(16):859–872
140. Wiltsey C, Kubinski P, Christiani T, Toomer K, Sheehan J, Branda A, Kadlowec J, Iftode C, Vernengo J (2013) Characterization of injectable hydrogels based on poly (N-isopropylacrylamide)-g-chondroitin sulfate with adhesive properties for nucleus pulposus tissue engineering. J Mater Sci Mater Med 24(4):837–847
141. Chen F, Yu S, Liu B, Ni Y, Yu C, Su Y, Zhu X, Yu X, Zhou Y, Yan D (2016) An injectable enzymatically crosslinked carboxymethylated pullulan/chondroitin sulfate hydrogel for cartilage tissue engineering. Sci Rep 6:20014. https://doi.org/10.1038/srep20014
142. Jin R, Moreira Teixeira LS, Krouwels A, Dijkstra PJ, van Blitterswijk CA, Karperien M, Feijen J (2010) Synthesis and characterization of hyaluronic acid–poly(ethylene glycol) hydrogels via Michael addition: an injectable biomaterial for cartilage repair. Acta Biomater 6(6):1968–1977. https://doi.org/10.1016/j.actbio.2009.12.024
143. Bonakdar S, Emami SH, Shokrgozar MA, Farhadi A, Ahmadi SAH, Amanzadeh A (2010) Preparation and characterization of polyvinyl alcohol hydrogels crosslinked by biodegradable polyurethane for tissue engineering of cartilage. Mater Sci Eng C 30(4):636–643. https://doi.org/10.1016/j.msec.2010.02.017
144. Bichara DA, Zhao X, Bodugoz-Senturk H, Ballyns FP, Oral E, Randolph MA, Bonassar LJ, Gill TJ, Muratoglu OK (2011) Porous poly(vinyl alcohol)-hydrogel matrix-engineered biosynthetic cartilage. Tissue Eng Part A 17(3-4):301–309. https://doi.org/10.1089/ten.TEA.2010.0322
145. Sun S, Cao H, Su H, Tan T (2009) Preparation and characterization of a novel injectable in situ cross-linked hydrogel. Polym Bull 62(5):699–711. https://doi.org/10.1007/s00289-009-0048-9
146. Alexander A, Ajazuddin KJ, Saraf S, Saraf S (2014) Polyethylene glycol (PEG)–Poly(N-isopropylacrylamide) (PNIPAAm) based thermosensitive injectable hydrogels for biomedical applications. Eur J Pharm Biopharm 88(3):575–585. https://doi.org/10.1016/j.ejpb.2014.07.005

147. Fabbri M, Soccio M, Costa M, Lotti N, Gazzano M, Siracusa V, Gamberini R, Rimini B, Munari A, García-Fernández L (2016) New fully bio-based PLLA triblock copoly (ester urethane) s as potential candidates for soft tissue engineering. Polym Degrad Stab 132:169–180
148. Kwon JS, Yoon SM, Kwon DY, Kim DY, Tai GZ, Jin LM, Song B, Lee B, Kim JH, Han DK, Min BH, Kim MS (2013) Injectable in situ-forming hydrogel for cartilage tissue engineering. J Mater Chem B 1(26):3314–3321. https://doi.org/10.1039/C3TB20105H
149. Lee BP, Dalsin JL, Messersmith PB (2002) Synthesis and gelation of DOPA-modified poly(ethylene glycol) hydrogels. Biomacromolecules 3(5):1038–1047. https://doi.org/10.1021/bm025546n
150. Nicodemus GD, Skaalure SC, Bryant SJ (2011) Gel structure has an impact on pericellular and extracellular matrix deposition, which subsequently alters metabolic activities in chondrocyte-laden PEG hydrogels. Acta Biomater 7(2):492–504. https://doi.org/10.1016/j.actbio.2010.08.021
151. Sharma B, Fermanian S, Gibson M, Unterman S, Herzka DA, Cascio B, Coburn J, Hui AY, Marcus N, Gold GE, Elisseeff JH (2013) Human cartilage repair with a photoreactive adhesive-hydrogel composite. Sci Transl Med 5(167):167ra166. https://doi.org/10.1126/scitranslmed.3004838
152. Liu H, Liu J, Qi C, Fang Y, Zhang L, Zhuo R, Jiang X (2016) Thermosensitive injectable in-situ forming carboxymethyl chitin hydrogel for three-dimensional cell culture. Acta Biomater 35:228–237. https://doi.org/10.1016/j.actbio.2016.02.028
153. Radhakrishnan J, Subramanian A, Krishnan UM, Sethuraman S (2017) Injectable and 3D bioprinted polysaccharide hydrogels: from cartilage to osteochondral tissue engineering. Biomacromolecules 18(1):1–26. https://doi.org/10.1021/acs.biomac.6b01619
154. Kondiah PJ, Choonara YE, Kondiah PP, Marimuthu T, Kumar P, du Toit LC, Pillay V (2016) A review of injectable polymeric hydrogel systems for application in bone tissue engineering. Molecules 21(11). https://doi.org/10.3390/molecules21111580
155. Jeuken JMR, Roth AK, Peters JR, Donkelaar C, Thies JC, van Rhijn LW, Emans J (2016) Polymers in cartilage defect repair of the knee: current status and future prospects. Polymer 8:219
156. Trattnig S, Ohel K, Mlynarik V, Juras V, Zbyn S, Korner A (2015) Morphological and compositional monitoring of a new cell-free cartilage repair hydrogel technology—GelrinC by MR using semi-quantitative MOCART scoring and quantitative T2 index and new zonal T2 index calculation. Osteoarthritis Cartilage 23(12):2224–2232. https://doi.org/10.1016/j.joca.2015.07.007
157. Vila YRJ, Dalmau A, Chaques FJ, Asuncion J (2015) Treatment of osteochondral lesions of the talus with bone marrow stimulation and chitosan-glycerol phosphate/blood implants (BST-CarGel). Arthrosc Tech 4(6):e663–e667. https://doi.org/10.1016/j.eats.2015.07.008

# Part VI
# Translation of Osteochondral Tissue Products

# Chapter 16
# Stem Cells in Osteochondral Tissue Engineering

Eleonora Pintus, Matteo Baldassarri, Luca Perazzo, Simone Natali, Diego Ghinelli, and Roberto Buda

**Abstract** Mesenchymal stem cells (MSCs) are pluripotent stem cells with the ability to differentiate into a variety of other connective tissue cells, such as chondral, bony, muscular, and tendon tissue. Bone marrow-derived MSCs are pluripotent cells that can differentiate among others into osteoblasts, adipocytes and chondrocytes.

Bone marrow-derived cells may represent the future in osteochondral repair. A one-step arthroscopic technique is developed for cartilage repair, using a device to concentrate bone marrow-derived cells and collagen powder or hyaluronic acid membrane as scaffolds for cell support and platelet gel.

The rationale of the "one-step technique" is to transplant the entire bone-marrow cellular pool instead of isolated and expanded mesenchymal stem cells allowing cells to be processed directly in the operating room, without the need for a laboratory phase. For an entirely arthroscopic implantation are employed a scaffold and the instrumentation previously applied for ACI; in addition to these devices, autologous platelet-rich fibrin (PRF) is added in order to provide a supplement of growth factors. Results of this technique are encouraging at mid-term although long-term follow-up is still needed.

**Keywords** Cartilage repair · Ankle · Stem cells · Bone marrow · Arthroscopy

## 16.1 Introduction

Osteochondral lesions are defects of the cartilaginous surface and underlying subchondral bone; the most involved joints are ankle and knee joint. In the ankle, the majority of such lesions occur within the talar dome, with the tibial plafond more rarely involved. Traditionally, talar dome lesions have been described as occurring

E. Pintus · M. Baldassarri · L. Perazzo · S. Natali · D. Ghinelli · R. Buda (✉)
I Clinic of Orthopaedics and Traumatology, Rizzoli Orthopaedic Institute, Bologna, Italy
e-mail: roberto.buda@ior.it

predominantly in the posteromedial or anterolateral region of the talar dome. In the knee, these defects can be found more frequently at the level of medial femoral condyle, but lateral femoral condyle, patello femoral joints, and tibial plafond can be involved [1, 2]. Lesions of the articular cartilage have a large variety of causes; most frequently these lesions have traumatic origin and, being particularly related to sport activities, they are now in strong increase. These lesions frequently occur in young and active patients (20–30 years), with high functional request [3]. Returning the damaged articular cartilage back to a functionally normal state has been a major challenge for orthopedic surgeons.

Chondral tissue shows poor healing abilities and limited regenerative capacity; therefore, the damage may became irreversible and lead to chronic disability status with pain, recurrent swelling, limited function, and, finally, to an early osteoarthritis [4]. Cartilage lesions can be classified following ICRS criteria as either full or partial thickness depending on whether they extend or not to the subchondral bone. Partial thickness articular cartilage defects are unable to heal by themselves. In this type of lesions, subchondral bone presents a barrier between the defect and the bone marrow cells. Instead, in full thickness cartilage defects, contact with the pluripotent mesenchymal stem cells (MSCs) is available. Spontaneous repair consists of the production of a fibrocartilaginous tissue that fills the gap. This tissue is a weak substitute of hyaline cartilage that gradually degenerates with time. The main efforts for cartilage repair are aimed to filling of the cartilage defect with a tissue that has the same mechanical properties with hyaline cartilage and the integration of this tissue with the original articular cartilage [5].

The greatest success repairing these lesions arthroscopically has occurred using the cell, scaffold, and growth factor trilogy in the repair strategy. It is believed that every good repair requires these three components to ensure adequate repair and regeneration of the lesion. Thanks to technical advancements, regenerative techniques are quickly moving from traditional autologous chondrocyte implantation (ACI), that required of two operations and high costs, to bone marrow-derived cll transplantation (BMDCT), a technique capable to provide a repair of the lesion by hyaline cartilage in a one-step procedure, in conjunction with platelet gel and engineered scaffolds able to support multipotent cells growth and differentiation. In the "one-step technique" the entire bone-marrow cellular pool is transplanted instead of isolated and expanded mesenchymal stem cells, so MSCs are implanted with all the mononuclear cells and high regenerative potential factors present in the bone marrow. This allows cells to be processed directly in the operating room, without the need for a laboratory phase. A scaffold and the instrumentation previously used for ACI are employed for an entirely arthroscopic implantation, and autologous platelet-rich fibrin (PRF) is added in order to provide a supplement of growth factors.

Using a device to concentrate bone marrow-derived cells and collagen powder or hyaluronic acid membrane as scaffolds for cell support and platelet gel, a one-step arthroscopic technique may represent the future in osteochondral repair [6].

## 16.2 Mesenchymal Stem Cells

Mesenchymal stem cells are heterogeneous and multipotent cell populations with the ability to differentiate into a variety of other connective tissue cells, such as chondral, bony, muscular, and tendon tissue, responding to local microenviromental stimuli such as cytokines and growth factors, which are released in response of tissue injury or disease. These cells can be expanded and induced, either in vitro or in vivo, to terminally differentiate into osteoblasts, chondrocytes, adipocytes, tenocytes, myotubes, neural cells, and hematopoietic-supporting stroma.

Mesenchimal stem cells reside in specialized microenvironments known as niches. The niche is essential to support MSCs function and to maintain a correct balance between self-renewal and differentiation. The BM is a complex network of endothelial cells (including sinusoids, arterioles, and transition zone vessels) and mesenchymal stromal cells (including mesenchymal stem and progenitor cells, as well as osteolineage cells, chondrocytes and adipocytes) [7].

MSCs can provide substitute cells for those cells that expire and can account for the support of turnover dynamics in young adults. In case of severe tissue damages, MSCs can be attracted to the injured site where they secrete bioactive factors that function to assist the repair and regeneration process. The MSCs could be mobilized from the marrow or other depots or can be culture-expanded MSCs that are delivered to the damage site either by direct or systemic injection. Once at the site of lesion, these cells produce inhibition factors for scarring and apoptosis, promote angiogenesis, and stimulate host progenitors to divide and differentiate into functional regenerative units [8, 9].

At the end of nineteenth century, Cohnheim conjectured the presence of regenerative cells, hypotizing that bone marrow-derived fibroblasts were involved in wound healing throughout the body. In the 1960s and 1970s, Friedenstein described the ability of stromal cells isolated from bone marrow to regenerate or support ectopic bone, stroma, and hematopoietic tissues [10]. Subsequently, in 1970, Caplan's group [11] provided the first evidence of chondrogenic, osteogenic, and muscular differentiation potential of these cells and introduced the term "mesenchymal stem cells" in the early 1990s [12]. In the late 1980s and early 1990s, the heterogeneous population of mesenchymal stem cells from bone marrow was explored and found to be linked to the development of various mesenchymal tissues, as well as identifying the first surface antigens expressed by MSCs (cluster of differentiation CD73 and CD 105) [13–15].

Several authors [16, 17] found that most of MSCs are of perivascular origin; therefore, there is a direct correlation between MSCs frequency and blood vessel density in stromal vascularized tissue [18]. It is hypotized that pericytes are the source of MSCs, so the extensive distribution of perivascular precursors of MSCs explains their capacity to secrete chemokines locally in response to infection or disease in all vascularized districts of the body [19, 20].

The secretion of growth factors and other chemokines from MSCs induces cells proliferation and angiogenesis. MSCs express mitogenic proteins, such as transforming growth factor alfa and beta (TGF-α and TGF-β), epithelial growth factor (EGF), and insulin-like growth factor-1 (IGF-1) to increase fibroblast, epithelial, and endothelial cell division. MSCs are also responsible to reduction of scar tissue formation thanks to local cells secreting paracrine factors keratinocyte growth factor and macrophage inflammatory proteins [21, 22].

The anti-inflammatory and immunomodulatory capacity of MSC is very important in the restoration of localized or systemic conditions for normal healing and tissue regeneration. In many types of musculoskeletal trauma, inflammatory conditions at the site of lesion block the natural repair processes by local progenitor and mature cells. MSCs prevent proliferation and function of many inflammatory immune cells, including T cells, natural killer cells, B cells, monocytes, macrophages, and dendritic cells. In particular, MSCs restore balance in the types of helper T cells and macrophages: they indirectly promote the transition of $T_H1$ to $T_H2$ cells, improving tissue regeneration in cartilage and muscle, and a shift from macrophages M1, pro-inflammatory and tissue growth inhibition, to M2, anti-inflammatory and tissue healing [23, 24].

BMDCT were the first type of MSCs to be identified, and the ease of collection and relatively high quantity of MSCs still make bone marrow a commonly used source of mesenchymal stem cells. MSCs can be used as a cell suspension expanded by culture or just as a bone marrow concentrate. In addition of bone marrow, MSCs have been discovered and individuated in many adult and fetal tissues, such as dermis, fat, synovial fluid, umbrical cord blood, placenta, and amniotic fluid, with similar phenotypic characteristics but different propensities in proliferation and differentiation potentials [25]. Adipose-derived MSCs (ADSCs) obtained from lipoaspirates offer a great advantage as a cell source for cartilage tissue engineering, due to their abundance, easy availability, and their potential to differentiate into cartilage, besides bone, tendons, skeletal muscle, and fat [26]. Kim and Im [27] reported a lower chondrogenic potential of ADSCs when compared with BMSCs, but this disadvantage might be overcome by using a combination of transforming growth factor beta 2 (TGF-β2) and bone morphogenetic proteins (BMPs) or high doses of TGF-β2 and IGF-I in combination. In a recent review, Im [28], comparing current knowledge on BMSCs and ADSCs, reported that in vitro characterization showed that these two cells have many points of likeness, but ADSCs can be maintained in vitro longer, and exhibit more stable population doubling, higher proliferation and lower senescence than BMSCs. Most in vitro and in vivo studies demonstrated that BMSCs have greater osteogenic potentials, and ADSCs are more angio-inductive. Interesting results shows that BMSCs and ADSCs in combination synergistically improve bone and vessel formation. These findings suggest that this synergism could be utilized to reduce implanted cell numbers, and, consequently, the cost of cell therapy. In the clinical application of stem cells, it is unclear whether cells themselves integrate into a host to improve regeneration or if the main regenerative effects of stem cells come from released paracrine factors, so a full explanation of this phenomenon needed further investigation [28].

Several studies have been done regarding CD markers expressed on ADSCs and BMSCs. A few CD markers have been found to be differently expressed on ADSCs and BMSCs. Some of these have described higher CD106 expression in BMSCs [29, 30], while other studies found no difference [31, 32]. CD146, a cell surface marker of pericytes, is reported to be expressed two time more in early passage BMSCs than in early passage ADSCs. [33]; on the other hand, CD34, which is not detected in BMSCs, is expressed on ADSCs during early culture, but decreases after extensive passage. CD36, another differently expressed surface marker, is positive for ADSCs but negative for BMSCs [29].

In the last years, synovial-derived stem cells (SDSC) have attracted considerable attention as source of stem cells for cartilage tissue engineering because they display greater chondrogenic and less osteogenic potential than MSCs derived from bone marrow or periosteum [34, 35], but it is necessary to have other studies to confirm these data.

The availability and versatility of these remarkable cells make them an excellent treatment option for a wide variety of clinical pathologies [36–38].

The Mesenchymal and Tissue Stem Cell Committee of the International Society for Cellular Therapy has established the following minimal set of standard criteria to provide a uniform characterization of such cells: (1) they must be plastic-adherent when maintained in standard culture conditions; (2) they must express CD105, CD73, and CD90 and lack surface expression of CD45, CD34, CD14 (or CD11b), CD97α (or CD19), and HLA-DR; (3) and they must be capable of differentiating to chondrocytes, osteoblasts, and adipocytes in vitro [39].

Progress in the field of regenerative medicine demonstrated that mesenchymal stem cells (MSCs) could replicate and regenerate both bone as well as cartilaginous tissue, therefore without any need for a laboratory treatment. Owing to the developmental plasticity of mesenchymal stem cells, there is great interest in their application to replace damaged tissues. Combined with modern advances in gene therapy and tissue engineering, they have the potential to improve the quality of life for many. The development of strategies to exploit the potential of stem cells to augment bone formation to replace or restore the function of traumatized, diseased, or degenerated bone is a major clinical and socioeconomic need [9, 40, 41].

## 16.3 Surgical Techniques

### 16.3.1 *Indications*

For the ankle, the arthroscopic treatment is indicated for focal osteochondral lesions of the talar dome or tibial plafond classified as grade 3–4 according to International Cartilage Repair Society classification (ICRS) (area of the lesion $\geq 1.0$ cm$^2$, depth of the lesion <5 mm) [42]. For lesions of depth higher than 5 mm demineralized bone matrix or autologous bone grafting is needed. Patients older than 60 years, patients with osteoarthritis or kissing lesions of the ankle, and patients with rheumatoid arthritis should not be treated with this technique. Malalignment of the lower limb and the presence of joint laxity are considered relative contraindications to be corrected if present.

For the knee, the treatment is indicated in patients who had grade-III or grade-IV osteochondral lesions (according to the classification system of the International Cartilage Repair Society) involving the femoral condyle or patello femoral joint, with clinical symptoms of pain, swelling, locking, or giving-way [43].

## 16.3.2 Surgical Procedure

The surgical technique for the BMDCT consists of several phases, all to be performed during the same surgical session.

### 16.3.2.1 Platelet-Rich Fibrin Gel Production

The autologous platelet gel is used in order to provide directly in situ additional growth factors. The platelet gel is a very effective "accelerator" for healing processes [44]. The secretory granules of platelets, the $\alpha$ granules, contain platelet-derived growth factors AA, BB, and AB, transforming growth factors $\beta 1$ (TGF-$\beta 1$) and $\beta 2$, platelet-derived epidermal growth factor, platelet-derived angiogenesis factor, insulin growth factor 1, and platelet factor 4, which influence bone regeneration [45]. Moreover, the platelet-rich fibrin (PRF) is rich in fibrin and is able to coagulate faster than PRP, providing an additional stability to the implant due to its jelly consistency.

The PRF is produced with an automatic system the day before the operation or the same day; 120 mL of venous blood are harvested with a needle size 16 connected to a bowl previously prepared with the anticoagulant solution. The bowl is then inserted inside the Vivostat® System (Vivolution A/S, 3460 Birkeroed, Denmark) and processed. At the end of the machine cycle, a syringe containing 6 mL of PRF is extracted either to be stored at −35 °C or used immediately. In the case of storage, the PRF needs to be slowly heated to room temperature for 30 min before its use.

### 16.3.2.2 Bone Marrow Aspiration

The bone marrow is aspirated from the posterior iliac crest after preparation of a sterile surgical field with the patient lying prone and already under spinal or general anesthesia (Fig. 16.1). The equipment for the bone marrow harvesting, concentration, and implant is part of a dedicated kit for osteochondral regeneration developed by Istituto Ortopedico Rizzoli and Novagenit (Mezzolombardo, Trento, Italy).

The bone marrow harvesting is performed with a marrow needle (size 11 G × 100 mm), inserted 3 cm deep into the iliac crest marrow. A total of 60 mL of bone marrow is collected as a result of subsequent aspirations and is placed in a bag

**Fig. 16.1** Surgical field showing aspiration of the bone marrow from the posterior iliac crest

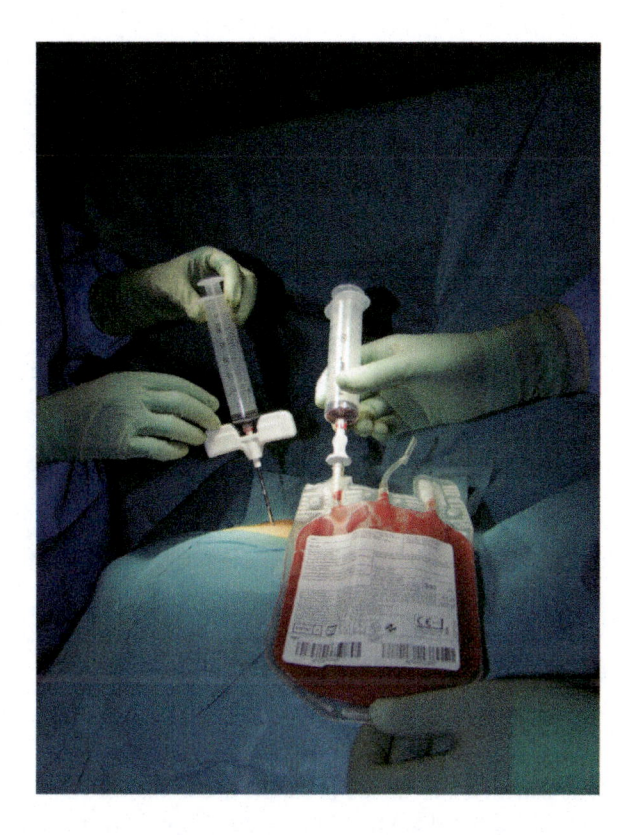

preloaded with 500 UI of epsodilave in 10 mL of saline solution; 5 mL of bone marrow is aspired into a 20 mL plastic syringe, repeating the procedure with several perforations into the iliac crest through the same skin opening. The harvesting is made in little steps on different locations on the crest in order to maximize the collection of stromal cells useful for the regeneration and reduce the diluting effect arising from the aspiration of the peripheral blood.

### 16.3.2.3 Bone Marrow Concentration

The previously extracted bone marrow volume is then reduced by eliminating the plasma and the red cells, therefore increasing its stem cells concentration. This procedure is performed directly at the end of the aspiration phase in the operating room using a cells separator-concentrator (Res-Q, ThermoGenesis, Rancho Cordova, CA) and its related sterile and disposable kit. In 12 min of multiple centrifugation cycles, the cell separator provides 6 mL of concentrated cells, containing the mesenchymal stem cells and other cell populations which constitute the nucleated bone marrow microenvironment.

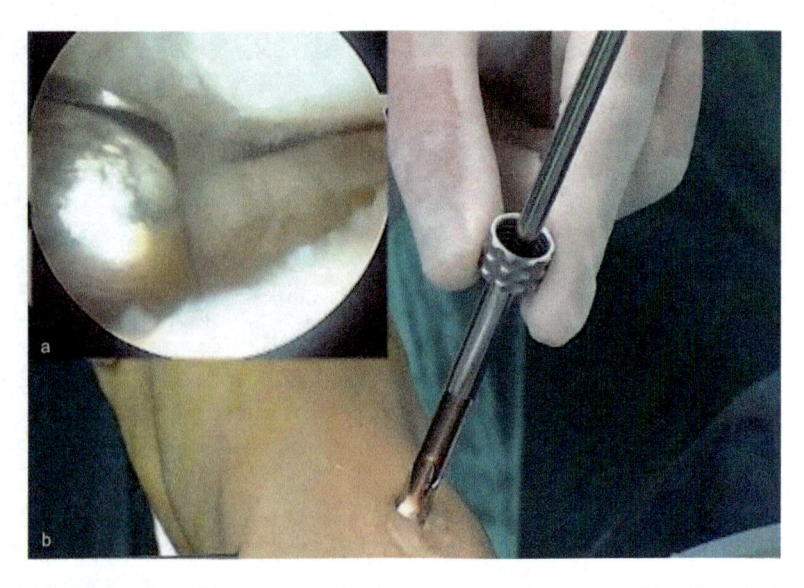

**Fig. 16.2** (**a**) Arthroscopic view: the lesion is inspected and shaved until healthy subchondral bone bed is reached; (**b**) Intraoperative view: a standard ankle arthroscopy is performed through anteromedial and anterolateral accesses

### 16.3.2.4 Surgical Approach

The patient is positioned in supine decubitus with a tourniquet at the leg to be operated. A standard ankle o knee arthroscopy is performed through anteromedial and anterolateral accesses.

The cartilage lesion is identified, and a flipped cannula is inserted into the lesion to enable insertion of the surgical instrumentations and to retract the fat pad from the operative field.

The lesion is inspected, so articular fibrosis, intra-articular loose bodies or osteophytes are to be removed. The osteochondral lesion is debrided, resulting in an area with regular healthy cartilage margins for biomaterial implantation (Fig. 16.2).

The size of lesion is measured with the aid of a millimeter probe and recorded, in order to the biomaterial to be implanted is then prepared in the same size and shape of the lesion.

A hyaluronic acid membrane (Hyalofast®, Fidia Advanced Biopolymers, Italy) or collagen membrane (IOR-G1, Novagenit, Mezzolombardo, TN, Italy) is used for cell support. Approximately 2 mL of marrow concentrate is loaded onto the highly hydrophilic membrane and fastly absorbed, together with 1 mL of PRF using a dedicated spray pen. The biomaterial is accurately clipped, following the size and the dimension of the lesion previously measured using the sizers provided by a special instrument set (Fig. 16.3). The cannula is then inserted through the arthroscopic access closer to the lesion with the help of the trocar, then the trocar is extracted and the joint distension is stopped. The fluid is completely removed from the joint. The

**Fig. 16.3** Surgical instruments originally designed for autologous chondrocyte implantation and then adapted for bone marrow-derived cell transplantation

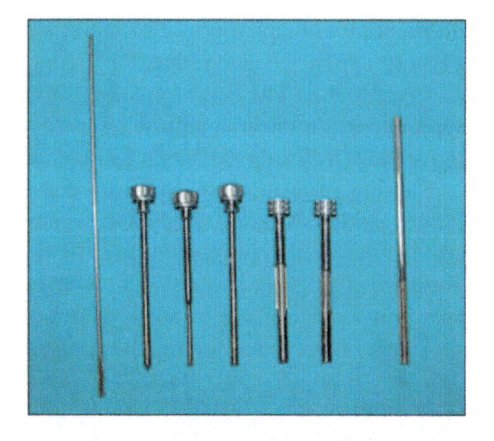

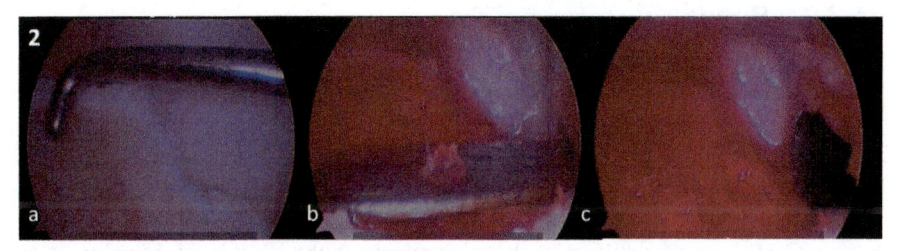

**Fig. 16.4** Artroscopic view of knee (1) and ankle (2) showing (**a**) osteochondral lesion (**b**) biomaterial positioned in the lesion site (**c**) a layer of PRF sprayer over the biomaterial to provide the growth factor

final biomaterial is applied in the window cut out of the cannula and guided to the edge of the lesion. At this point, the cannula is removed and the biomaterial is made to adhere perfectly to the lesion through the use of a flat probe (Fig. 16.4).

At the end, the PRF is applied to cover the lesion, in order to provide a high concentration of growth factors and to further promote the stability of the implant due to coagulation of the PRF. The implant stability is checked performing ankle flexion/extension movements. The skin accesses are closed with a 3-0 absorbable suture wire covered by a bandage, replaced the following day by a flat medication.

### *16.3.3 Postoperative Treatment and Rehabilitation Protocol*

Patients are usually dismissed the day after surgery. Rehabilitation treatment should be personalized for each patient, depending on the patient's clinical status, after an evaluation by a team comprising an orthopedist, a physiatrist, and a physiotherapist. Continuous passive motion (CMP) is immediately performed the day after surgery and gradually increased as tolerated. This movement, through compression and joint decompression, facilitates mesenchymal cells proliferation and their differentiation in the sense of chondrocyte, stimulates the synthesis of molecules of the cartilage matrix and reduces the risk of adhesions inside the joint.

Initially the CPM is performed at slow speed (1 cycle per min) for 6–8 h a day. Joint excursion must be adjusted according to the pain threshold, seeking both the movement of flexion extension that the one of prono-supination.

Starting from the 2nd day post-op the use of a passive mobilization device is recommended. During the first 6 weeks post-op walking is allowed using two crutches without applying any load on the operated leg, following by partial load. Total weightbearing is allowed at 8–10 weeks. Low-impact sport activity such as swimming and cycling are allowed 4 months after the surgery, while high-impact sport activities such as tennis, soccer, and running at 10–12 months.

## 16.3.4 Results

Giannini and his group published several studies of BMC implantation associated with scaffolds, in ankle and knee joint defects. In the 2009 study [6] on talar osteochondral lesions, they showed clinical and radiological improvements at 24 months after BMC implantation in collagen or hyaluronic acid matrix. In other studies [4, 46, 47] they reported a partial worsening between 24 and 48 months of follow-up, but the final result was still satisfactory compared to the basal level. In many comparative evaluations performed, they found positive clinical and MRI outcome in groups of patients treated with one-step BMC HA or collagen matrix implantation, versus open autologous condrocyte implantation (ACI) or arthroscopic matrix-assisted autologous chondrocyte transplantation (MACT) for osteochondral lesions of the talus (OLTs) [1, 48]. Hannon et al. [49] compared retrospective outcomes after arthroscopic bone marrow stimulation (BMS) with and without concentrated bone marrow aspirate (cBMA) as a biological adjunct to the surgical treatment of osteochondral lesions (OCLs) of the talus and concluded that cBMA/BMS group had higher improvements in FAOS, SF-12 PCS and MOCART scores than BMS-alone group; they also demonstrated that groups with cBMA had improved integration of the repair tissue with MRI demonstrating less fissuring and fibrillation.

BMC implantation with scaffold is also explored for the treatment of osteochondral lesions of the knee (OLKs). The first results using BMC on HA matrix is reported by Buda et al., who found positive MRI and histology improvements at short-term follow-up [43, 50]. Gobbi et al. [51] observed superior outcomes using BMC instead of chondrocytes for the treatment of large patellofemoral defects. Similar results were obtained also using BMC on collagen scaffolds, as described by Gigante et al. [52]. After all, Skowronski et al. [53] documented stable mid-term outcomes after the treatment of large chondral lesions. A recent study of Gobbi et al. [54] analyzed use of multipotent stem cells and Hyaluronan-based scaffold for the treatment of full-thickness chondral defects of the knee in patients older than 45 years evaluated for 4 years, compared to a younger control group; they demonstrated that functional outcomes are comparable to younger patients at final follow-up and the results are not affected by age.

However, a long-term comparative study with a larger sample and with a detailed radiological analysis is desirable in order to ultimately assess the potential of this technique for young and old patients. Current evidence suggests that cBMA can improve cartilage repair in OLT, but future clinical research and clinical trials are necessary for better comparison of outcomes with other biological adjuncts.

## 16.4 Conclusions

Our many years of experience in the field of osteochondral repair has shown that the use of mesenchymal cells can be a valuable aid for the orthopedic surgeon. Specifically, the medullary concentrate taken from the iliac crest contains an adequate amount of progenitor cells and a high quantity of growth factors. This element, if used on biocompatible scaffolds and with minimally invasive techniques, allows to obtain good results both from a clinical and functional point of view. We believe that in the light of the results obtained by this method it can represent a valid starting point for developing increasingly sophisticated bioengineering techniques.

## References

1. Buda R, Castagnini F, Cavallo M, Ramponi L, Vannini F, Giannini S (2016) "One-step" bone marrow-derived cells transplantation and joint debridement for osteochondral lesions of the talus in ankle osteoarthritis: clinical and radiological outcomes at 36 months. Arch Orthop Trauma Surg 136:107–116
2. Elias I, Raikin SM, Schweitzer ME, Besser MP, Morrison WB, Zoga AC (2009) Osteochondral lesions of the distal tibial plafond: localization and morphologic characteristics with an anatomical grid. Foot Ankle Int 30(6):524–529
3. Pritsch M, Horoshovski H, Farine I (1986) Arthroscopic treatment of osteochondral lesions of the talus. J Bone Joint Surg Am 68(6):862–865
4. Giannini S, Buda R, Battaglia M, Cavallo M, Ruffilli A, Ramponi L, Pagliazzi G, Vannini F (2013) One-step repair in talar osteochondral lesions: 4-year clinical results and t2-mapping capability in outcome prediction. Am J Sports Med 41(3):511–518
5. Beris AE, Lykissas MG, Papageorgiou CD, Georgoulis AD (2005) Advances in articular cartilage repair. Injury 36(Suppl 4):S14–S23
6. Giannini S, Buda R, Vannini F, Cavallo M, Grigolo B (2009) One step bone marrow-derived cell transplantation in talar osteochondral lesions. Clin Orthop Relat Res 467(12):3307–3320
7. Beerman I, Luis TC, Singbrant S, Lo Celso C, Méndez-Ferrer S (2017) The evolving view of the hematopoietic stem cell niche. Exp Hematol 50:22–26
8. Caplan AI, Dennis JE (2006) Mesenchymal stem cells as trophic mediators. J Cell Biochem 98:1076–1084
9. Owen M, Friedenstein AJ (1988) Stromal stem cells: marrow-derived osteogenic precursors. Ciba Found Symp 136:42–60
10. Friedenstein AJ, Gorskaja JF, Kulagina NN (1976) Fibroblast precursors in normal and irradiated mouse hematopoietic organs. Exp Hematol 4:267–274

11. Caplan AI, Koutroupas S (1973) The control of muscle and cartilage development in the chick limb: the role of differential vascularization. J Embryol Exp Morphol 29(3):571–583
12. Spencer ND, Gimble JM, Lopez MJ (2011) Mesenchymal stromal cells: past, present, and future. Vet Surg 40(2):129–139
13. Haynesworth S, Baber M, Caplan A (1992a) Cell surface antigens on human marrow-derived mesenchymal cells are detected by monoclonal antibodies. Bone 13:69–80
14. Haynesworth S, Goshima J, Goldberg V, Caplan A (1992b) Characterization of cells with osteogenic potential from human bone marrow. Bone 13:81–88
15. Owen M (1985) Lineage of osteogenic cells and their relationship to the stromal system. J Bone Miner Res 3:1–25
16. Caplan AI (2008) All MSCs are pericytes? Cell Stem Cell 3:229–230
17. Crisan M, Yap S, Casteilla L, Chen C-W, Corselli M, Park TS et al (2008) A perivascular origin for mesenchymal stem cells in multiple human organs. Cell Stem Cell 3:301–313
18. Da Silva Meirelles L, Sand TT, Harman RJ, Lennon DP, Caplan AI (2009) MSC frequency correlates with blood vessel density in equine adipose tissue. Tissue Eng Part A 15:221–229
19. Bai L, Lennon DP, Caplan AI, DeChant A, Hecker J, Kranso J, Zaremba A, Miller RH (2012) Hepatocyte growth factor mediates MSCs stimulated functional recovery in animal models of MS. Nat Neurosci 15:862–870
20. Honczarenko M, Le Y, Swierkowski M, Ghiran I, Glodek AM, Silberstein LE (2006) Human bone marrow stromal cells express a distinct set of biologically functional chemokine receptors. Stem Cells 24:1030–1041
21. Doorn J, Van de Peppel J, Van Leeuwen JPTM, Groen N, Van Blitterswijk CA, De Boer J (2011) Pro-osteogenic trophic effects by PKA activation in human mesenchymal stromal cells. Biomaterials 32:6089–6098
22. Haynesworth SE, Baber M, Caplan A (1996) Cytokine expression by human marrow-derived mesenchymal progenitor cells in vitro: effects of dexamethasone and IL-1 alpha. J Cell Physiol 166:585–592
23. Ezquer F, Ezquer M, Contador D, Ricca M, Simon V, Conget P (2012) The antidiabetic effect of mesenchymal stem cells is unrelated to their transdifferentiation potential but to their capability to restore Th1/Th2 balance and to modify the pancreatic microenvironment. Stem Cells 30:1664–1674
24. Mokarram N, Merchant A, Mukhatyar V, Patel G, Bellamkonda RV (2012) Effect of modulating macrophage phenotype on peripheral nerve repair. Biomaterials 33:8793–8801
25. Lodi D, Iannitti T, Palmieri B (2011) Stem cells in clinical practice: applications and warnings. J Exp Clin Cancer Res 30:9
26. Schäffler A, Büchler C (2007) Concise review: adipose tissue-derived stromal cells-basic and clinical implications for novel cell-based therapies. Stem Cells 25(4):818–827
27. Kim HJ, Im GI (2009) Chondrogenic differentiation of adipose tissue-derived mesenchymal stem cells: greater doses of growth factor are necessary. J Orthop Res 27(5):612–619
28. Im GI (2017) Bone marrow-derived stem/stromal cells and adipose tissue-derived stem/stromal cells: their comparative efficacies and synergistic effects. J Biomed Mater Res A 105(9):2640–2648
29. Bourin P, Bunnell BA, Casteilla L, Dominici M, Katz AJ, March KL, Redl H, Rubin JP, Yoshimura K, Gimble JM (2013) Stromal cells from the adipose tissue-derived stromal vascular fraction and culture expanded adipose tissue-derived stromal/stem cells: a joint statement of the International Federation for Adipose Therapeutics and Science (IFATS) and the International Society for Cellular Therapy (ISCT). Cytotherapy 15:641–648
30. Shafiee A, Seyedjafari E, Soleimani M, Ahmadbeigi N, Dinarvand P, Ghaemi N (2011) A comparison between osteogenic differentiation of human unrestricted somatic stem cells and mesenchymal stem cells from bone marrow and adipose tissue. Biotechnol Lett 33:1257–1264
31. Ikegame Y, Yamashita K, Hayashi S-I, Mizuno H, Tawada M, You F, Yamada K, Tanaka Y, Egashira Y, Nakashima S (2011) Comparison of mesenchymal stem cells from adipose tissue and bone marrow for ischemic stroke therapy. Cytotherapy 13:675–685

32. Niemeyer P, Kornacker M, Mehlhorn A, Seckinger A, Vohrer J, Schmal H, Kasten P, Eckstein V, Südkamp NP, Krause U (2007) Comparison of immunological properties of bone marrow stromal cells and adipose tissue-derived stem cells before and after osteogenic differentiation in vitro. Tissue Eng 13:111–121
33. Rider DA, Dombrowski C, Sawyer AA, Ng GH, Leong D, Hutmacher DW, Nurcombe V, Cool SM (2008) Autocrine fibroblast growth factor 2 increases the multipotentiality of human adipose-derived mesenchymal stem cells. Stem Cells 26:1598–1608
34. Jones BA, Pei M (2012) Synovium-derived stem cells: a tissue-specific stem cell for cartilage engineering and regeneration. Tissue Eng Part B Rev 18(4):301–311
35. Kubosch EJ, Heidt E, Bernstein A, Bottiger K, Schmal H (2016) The trans-well coculture of human synovial mesenchymal stem cells with chondrocytes leads to self-organization, chondrogenic differentiation, and secretion of TGFbeta. Stem Cell Res Ther 7(1):64
36. Iyer S, Rojas M (2008) Anti-inflammatory effects of mesenchymal stem cells: novel concept for future therapies. Expert Opin Biol Ther 8:569–582
37. Minguell JJ, Erices A, Conget P (2001) Mesenchymal stem cells. Exp Biol Med 226:507–520
38. Murphy MB, Moncivais K, Caplan A (2013) Mesenchymal stem cells: enviromentally responsive therapeutics for regenerative medicine. Exp Mol Med 45:e54
39. Dominici M, Le Blanc K, Mueller I, Slaper-Cortenbach I, Marini F, Krause D, Deans R, Keating A, Prockop DJ, Horwitz E (2006) Minimal criteria for defining multipotent mesenchymal stromal cells. The International Society for Cellular Therapy position statement. Cytotherapy 8(4):315–317
40. Galois J, Freyria AM, Herbage D, Mainard D (2005) Ingénierie tissulaire du cartilage: état des lieux et perspectives. Cartilage tissue engineering: state-of-the-art and future approaches. Pathol Biol 53:590–598
41. Oreffo RO, Cooper C, Mason C et al (2005) Mesenchymal stem cells: lineage, plasticity, and skeletal therapeutic potential. Stem Cell Rev 1:169–178
42. Giannini S, Buda R, Faldini C, Vannini F, Bevoni R, Grandi G, Grigolo B, Berti L (2005) Surgical treatment of osteochondral lesions of the talus in young active patients. J Bone Joint Surg Am 87(Suppl 2):28–41
43. Buda R, Vannini F, Cavallo M, Grigolo B, Cenacchi A, Giannini S (2010) Osteochondral lesions of the knee: a new one-step repair technique with bone-marrow-derived cells. J Bone Joint Surg Am 92(Suppl 2):2–11
44. Sánchez AR, Sheridan PJ, Kupp LI (2012) Is platelet-rich plasma the perfect enhancement factor? A current review. Int J Oral Maxillofac Implants 18(1):93–103
45. Nair MB, Varma HK, John A (2009) Platelet-rich plasma and fibrin glue-coated bioactive ceramics enhance growth and differentiation of goat bone marrow-derived stem cells. Tissue Eng Part A 15(7):1619–1631
46. Buda R, Cavallo M, Castagnini F, Cenacchi A, Natali S, Vannini F, Giannini S (2015a) Treatment of hemophilic ankle arthropathy with one-step arthroscopic bone marrow-derived cells transplantation. Cartilage 6:150–155
47. Buda R, Vannini F, Castagnini F, Cavallo M, Ruffilli A, Ramponi L, Pagliazzi G, Giannini S (2015b) Regenerative treatment in osteochondral lesions of the talus: autologous chondrocyte implantation versus one-step bone marrow derived cells transplantation. Int Orthop 39:893–900
48. Giannini S, Buda R, Cavallo M, Ruffilli A, Cenacchi A, Cavallo C, Vannini F (2010) Cartilage repair evolution in post-traumatic osteochondral lesions of the talus: from open field autologous chondrocyte to bone-marrow-derived cells transplantation. Injury 41:1196–1203
49. Hannon CP, Ross KA, Murawski CD, Deyer TW, Smyth NA, Hogan MV, Do HT, O'Malley MJ, Kennedy JG (2016) Arthroscopic bone marrow stimulation and concentrated bone marrow aspirate for osteochondral lesions of the talus: a case-control study of functional and magnetic resonance observation of cartilage repair tissue outcomes. Arthroscopy 32(2):339–347
50. Buda R, Vannini F, Cavallo M, Baldassarri M, Luciani D, Mazzotti A, Pungetti C, Olivieri A, Giannini S (2013) One-step arthroscopic technique for the treatment of osteochondral

lesions of the knee with bone-marrow-derived cells: three years results. Musculoskelet Surg 97(2):145–151
51. Gobbi A, Chaurasia S, Karnatzikos G, Nakamura N (2015) Matrix-induced autologous chondrocyte implantation versus multipotent stem cells for the treatment of large patellofemoral chondral lesions: a nonrandomized prospective trial. Cartilage 6:82–97
52. Gigante A, Cecconi S, Calcagno S, Busilacchi A, Enea D (2012) Arthroscopic knee cartilage repair with covered microfracture and bone marrow concentrate. Arthrosc Tech 1:e175–e180
53. Skowronski J, Skowronski R, Rutka M (2013) Large cartilage lesions of the knee treated with bone marrow concentrate and collagen membrane-results. J Orthop Traumatol Rehabil 15:69–76
54. Gobbi A, Scotti C, Karnatzikos G, Mudhigere A, Castro M, Peretti G (2017) One-step surgery with multipotent stem cells and Hyaluronan-based scaffold for the treatment of full-thickness chondral defects of the knee in patients older than 45 years. Knee Surg Sports Traumatol Arthrosc 25:2494–2501

# Chapter 17
# Osteochondral Tissue Engineering: Translational Research and Turning Research into Products

Victoria Spencer, Erica Illescas, Lorenzo Maltes, Hyun Kim, Vinayak Sathe, and Syam Nukavarapu

**Abstract** Osteochondral (OC) defect repair is a significant clinical challenge. Osteoarthritis results in articular cartilage/subchondral bone tissue degeneration and tissue loss, which in the long run results in cartilage/ostecochondral defect formation. OC defects are commonly approached with autografts and allografts, and both these options have found limitations. Alternatively, tissue engineered strategies with biodegradable scaffolds with and without cells and growth factors have been developed. In order to approach regeneration of complex tissues such as osteochondral, advanced tissue engineered grafts including biphasic, triphasic, and gradient configurations are considered. The graft design is motivated to promote cartilage and bone layer formation with an interdigitating transitional zone (i.e., bone–cartilage interface). Some of the engineered OC grafts with autologous cells have shown promise for OC defect repair and a few of them have advanced into clinical trials. This chapter presents synthetic osteochondral designs and the progress that has been made in terms of the clinical translation.

**Keywords** Bone–cartilage interface · Engineered structures · Local microstructure control · Osteochondral tissue development · Bench to bedside translation

---

V. Spencer · E. Illescas · L. Maltes · H. Kim
Department of Biomedical Engineering, University of Connecticut, Storrs, CT, USA

V. Sathe
Department of Orthopaedic Surgery, University of Connecticut Health, Storrs, CT, USA

S. Nukavarapu (✉)
Department of Biomedical Engineering, University of Connecticut, Storrs, CT, USA

Department of Orthopaedic Surgery, University of Connecticut Health, Storrs, CT, USA

Department of Materials Science and Engineering, University of Connecticut, Storrs, CT, USA
e-mail: syam@uchc.edu

## 17.1 Introduction

Osteoarthritis (OA) is a joint disease that gradually wears down the protective articular cartilage until it worsens to affect the underlying subchondral bone. Affecting approximately 31 million in the USA, the progressive degradation of this cartilage and bone tissue results in chronic pain, swelling of the joint, and consequently limited mobility and stiffness. It is estimated that by 2030, about 67 million adults in the USA will suffer from some type of symptomatic OA [1]. OA has a single end stage, characterized by the loss of cartilage and bone tissue products, but the pathophysiological pathways leading to this point can vary. Due to the avascular nature of cartilage, it lacks sufficient self-repair mechanisms, thereby hindering attempts at regeneration.

Although, osteoarthritis induced OC defects mainly originate from mechanical trauma, genetic and hormonal factors can also induce inflammation leading to OA induced OC defects. However, it is important to note that OC defects can be formed without the progression of OA. Malalignment of bones, injuries due to repetitive trauma, or hyperloading beyond physiological limits can also instigate such defects. Hence secondary to the elderly, athletes are very likely to require treatment due to OC tissue damage. Ischemia and avascular necrosis are leading factors to osteochondritis, which is another major cause for OC defects. Unlike OA, which commences primarily as an articular surface problem, osteochondritis dissecans (OCD) begins with the degradation of subchondral bone, which leaves the overlying articular cartilage susceptible to damage, seen as cracks on the surface of the bone [2]. Consequently, when left untreated, chronic OA and OCD can damage enough tissue, leading to various grades of osteochondral defects, requiring the need to employ tissue engineering strategies.

When discussing osteochondral (OC) defects it is important to understand its heterogeneous, multilayered tissue structure. OC tissue is sorted into non-calcified cartilage, calcified cartilage, and subchondral bone. The non-calcified cartilage region, which we refer to as articular cartilage is then further broken down into three zones: superficial, middle, and deep. Although all three of these zones contain similar extracellular components, the composition and organization between them vary in each zone. The extracellular content of articular cartilage mainly consists of chondroitin and keratin sulfate GAGs, aggrecan proteoglycans, and collagen fibers [3]. Collagen fibers in the superficial zone are oriented parallel to the surface and the corresponding chondrocytes are flattened in shape, following the tangential direction of the collagen fibers, providing cartilage with its tensile properties. Meanwhile, in the deep zone, collagen fibers form bundles oriented perpendicular to the surface as the cells in this region arrange themselves in a similar manner (Fig. 17.1). This orientation provides the tissue with the greatest resistance against compressive forces [3, 4]. It is this spatial and structural organization that provides articular cartilage with its unique mechanical properties, but for OC defects, replicating this intricate network remains a challenge.

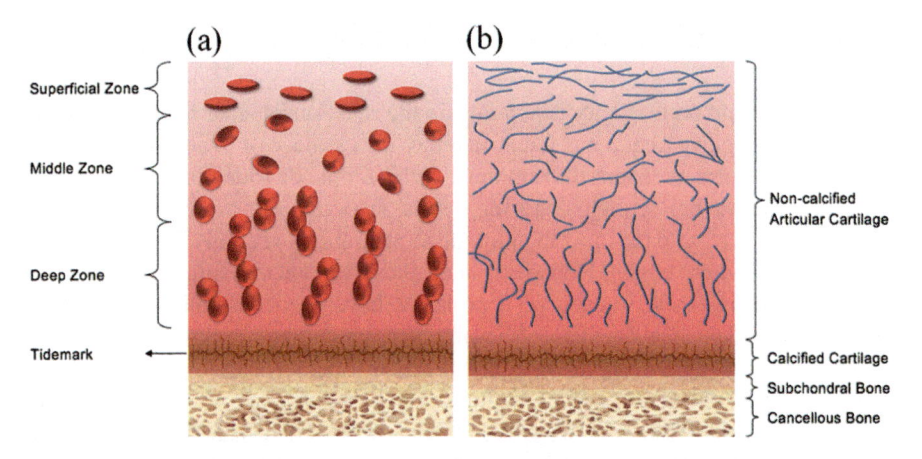

**Fig. 17.1** Osteochondral tissue structure. Structural/spatial organization of (**a**) chondrocytes and (**b**) collagen fibers in non-calcified articular cartilage. Tidemark and subchondral bone layers are shown in both the images. In order to repair OC defects, one has to develop synthetic grafts with structure to support articular cartilage, subchondral bone, and cartilage–bone interface layer regeneration similar to the native tissue

However, OC defects are more than articular cartilage damage and one has to consider the damage to the underlying subchondral bone tissue as well [5]. Subchondral bone is a highly vascularized tissue with its own marrow, constantly supplying the tissue with nutrients, oxygen, and stem cells as part of the active remodeling process [6]. On the contrary, articular cartilage is hyaline cartilage that is avascular and therefore has a very limited regeneration capacity. This overlying tissue gathers nutrients from being in proximal contact with the nutrient supply of subchondral bone. In addition to being avascular, articular cartilage is also aneural, meaning that superficial damage is not always associated with immediate symptoms. In OC tissue, the interfacial zone refers to the small region surrounding the tidemark where the transition from non-calcified hyaline cartilage to calcified cartilage and further subchondral bone occurs. This tidemark produces a reduced pore space, preventing the invasion of vascularizing cells and ensuring cartilage tissue is not in directly exposed to blood, thereby preventing chondrocyte hypertrophy and apoptosis. Often, in an attempt to repair itself, the native stem cell population in bone will generate fibrocartilage at the cartilage defect site. However, seeing as fibrocartilage is not morphologically or biomechanically similar to the hyaline cartilage it is replacing, the body cannot rely on this type of cartilage as a means to restore normal joint function, hence the demand for alternative treatment options [7].

In this chapter, we present engineered graft types developed for osteochondral defect repair and regeneration. The chapter will also present some of the translational efforts in terms of preclinical and clinical studies along with the products available for use in the clinic.

## 17.2 Current Treatment Options and Challenges

Current treatment options vary depending on the cause and degree of severity of the OC defect. Palliative treatments focus on removing cartilage or bone tissue that may be the direct cause of reported symptoms. Arthroscopic debridement, for example, focuses on removing synovitis-causing debris, smoothing down any rough tissue that may interfere with the movement of the joint, and synovectomy to get rid of any inflamed synovium tissue. However, since these treatments simply remove the problem-causing tissue, reparative approaches have been designed, focusing on replacing the damaged cartilage. Microfracture procedures remove the calcified tissue layer, exposing the subchondral bone and drilling through to access the bone marrow, providing access to biological healing molecules and mesenchymal stem cells from bone marrow. This allows the bone marrow stem cells to differentiate into chondrocyte-like cells, developing a cartilage layer of tissue at the defect site. However, this approach often results in fibrocartilage formation which is mechanically inferior to articular cartilage. For larger OC defects, tissue grafts are used to replace the lost tissue. However, since allografts come from a donor, they pose the risk of disease transmission and immune reaction, which can lead to rejection of the graft. Introducing a chronic immune reaction at an OC defect site simply increases the risk of OA development. Slower bone remodeling post-implantation has also been reported with OC allografts. When the patient has enough healthy tissue, an autograft may be used since it comes from the patient's own tissue. However, due to this, autografts not only require multiple surgeries, but also result in donor-site morbidity [3].

Alternatively, tissue regeneration strategies centered on chondrocyte implantation have been considered. Autologous chondrocyte implantation (ACI) uses healthy cartilage harvested from the patient for autologous chondrocyte isolation and their ex vivo expansion. The chondrocytes are then seeded onto the defect site and localized with a periosteal flap. A product developed using this method is referred to as Carticel, which is currently in clinical use. However, such regenerative treatments also have limitations as the procedure involves a multistep process including chondrocyte isolation, expansion, and implantation, and the outcomes of the method are also questionable. Long term postoperative analysis of the defect site has shown varying results, with much of the site being filled with fibrocartilage and Collagen I. Hypertrophic and calcified chondrocytes were also seen with the ACI treatments, caused by osteoinduction of the seeded cells by the periosteal flap used. This warrants the development of other tissue engineering strategies involving specialized scaffolds that can support bone and cartilage layer regeneration along with the development of native-like bone–cartilage interface.

## 17.3 Tissue Engineered Osteochondral Scaffolds

The three principal components of tissue engineering are scaffolds, cells, and growth factors. Scaffolds for tissue engineering are three-dimensional and porous structures that can support cell in-growth and the required tissue formation. For OC tissue engineering, the optimal scaffold, in addition to being able to support cell attachment, proliferation and in-growth, the scaffold must also withstand functional site loading. The knee is subject to a myriad of cyclic forces generated upon walking, bending or running. It is integral that the tissues of the osteochondral regions adequately support functional loading in conjunction with the absorption of applied loading to establish the regeneration of a structurally and functionally effective joint. In addition to loading requirements, an employed scaffold must also be able to regenerate tissue that is similar to that of the native tissue [3, 8–10]. The arrangement of these native tissues is also quite paramount for the establishment of an effective scaffold. The osteochondral region can be divided into three zones consisting of cartilaginous and osseous regions separated by a transitional zone, also referred to as an interface. All three zones have their own unique physiological properties including cellular makeup and ECM composition and structure, which set one tissue apart from that of the adjacent one. Thus, a scaffold designed to mimic the natural stratification of the OC tissue layers would render maximum success at repairing osteochondral defects. Some of the common osteochondral scaffold designs are shown in Fig. 17.2 and detailed in the following sections.

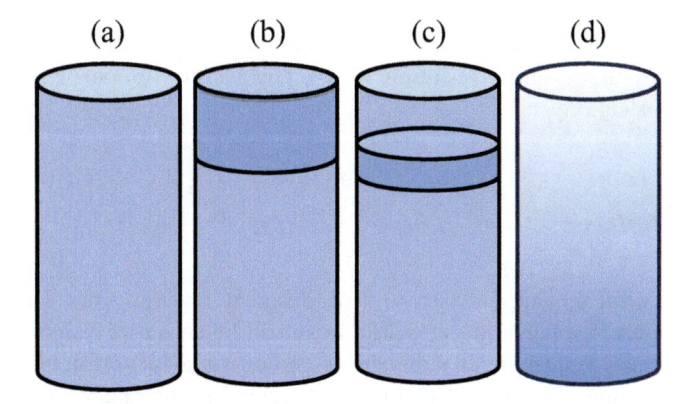

**Fig. 17.2** Osteochondral scaffold configurations. (**a**) Monophasic, (**b**) Biphasic, (**c**) Triphasic, and (**d**) Gradient matrix structures. Biphasic scaffolds have been widely investigated, however, recent studies have focused on developing gradient scaffolds with integrated bone and cartilage phases in a single structure. Some of the preliminary results regarding gradient OC scaffolds have been presented in this chapter

## 17.3.1 Monophasic

Scaffolds with no variation in composition or structure are referred to as monophasic scaffolds. This approach is centered on the application of one material, or multiple materials, so long as they are constructed in a homogenous arrangement throughout. Architecture and porosity are also spatially distributed uniformly throughout this category of construct. In vivo studies have shown evidence of cell recruitment by such scaffolds when implanted in an osteochondral defect model. However, since material composition and its arrangement are the same throughout the entirety of the scaffold structure, the scaffold is the least able to appropriately regenerate the respective tissue layers found in the native tissue. By not addressing the prevalent differences within the microenvironments of cartilage and bone, there are inherent limitations in promoting site-specific cellular differentiation and matrix deposition. It has been found that the regenerated tissue when used monophasic scaffolds is often incomplete and homogeneous. For instance, Jeong et al. employed polycaprolactone (PCL) scaffolds loaded with BMP-2, and found that these seeded biological constructs produced significant cartilaginous tissue. However, in vitro testing of the scaffolds showed higher expressions of hypertrophic chondrocyte markers, indicating a drift of the seeded primary chondrocytes towards endochondral ossification [11]. A relatively long term in vivo evaluation of monophasic PLA scaffolds in the rabbit osteochondral defect model conducted by Chu et al., showed that after a year, majority of the defect had been filled with regenerated cartilaginous growth lacking in glycosaminoglycans, while the bone regeneration, if any, was inconsistent [12]. These results establish that monophasic scaffolds do not have the appropriate architecture to support regeneration of a complex tissue such as the osteochondral tissue, often resulting in the regeneration of one type of tissue throughout its entirety.

## 17.3.2 Biphasic

A biphasic scaffold employs up to two different materials or two opposing architectural arrangements [5]. A scaffold can still be considered biphasic despite being composed of one material if there is a significant amount of structural disparity amongst its composition. These defining construct differences enable the scaffold to often obtain two opposing regions of differing physical or chemical makeup to more appropriately address the unique physiological characteristics found within the osteochondral region. Biphasic scaffolds have dedicated zones or phases for cartilage and bone layer growth. Since cartilage is a soft tissue, soft polymer matrices or hydrogels are commonly used to support cartilage layer regeneration. On the other hand, stiff matrices are proposed for bone layer regeneration. Together, biphasic matrices are designed to provide structural and mechanical flexibility with respect to cartilage and bone scaffold layer selection.

The use of biphasic scaffold design has been studied extensively and has shown promising results in a wide range of studies, with many designs making it to the clinical trial stage (Table 17.2). Previously, Jiang et al. constructed a biphasic composite scaffold with PLGA and partially seeded in the lower osseous phase with β-tricalcium phosphate, to promote osteogenesis [13]. The upper chondral phase of the graft was seeded with autologous porcine chondrocytes prior to implantation in a porcine model. Six months post implantation, histology data showed hyaline cartilage formation in the top and mineralization in the bottom layer of the scaffold. Moreover, there was integration of the bone phase with the surrounding host bone. In a separate study by Dresing et al., osteochondral scaffolds composed of poly(ester-urethane) and hydroxyapatite were press fit into defects within the femoral trochlear ridge of rabbits [14]. Following 12 weeks of implantation, histology revealed that subchondral bone generation rendered within the experimental groups was less than that seen with the control group, an empty defect without the scaffold. Frenkel et al. have also developed and tested a biphasic osteochondral design in a rabbit model, with focus on the regenerated cartilage layer [15]. The graft consisted of a PLA and hyaluronic acid osseous phase and fibrillar collagen I or hyaluronic acid-chitosan composite for its cartilage phase. While both graft types displayed hyaline-like regeneration of the articulating surface, with expression of collagen II and GAGs, an abnormal spatial dissociation of GAG expression and Col II was noticed, indicating regeneration of imperfect hyaline cartilage. These results demonstrated that one could not conclusively deduce if the biphasic unit had a profound effect on the regenerative process. One limitation with the biphasic design is that it is limited in its ability to produce conditions favorable for regeneration of the bone–cartilage interface that exists in native tissue.

### 17.3.3 Triphasic

Triphasic scaffolds consist of three different compositions or three opposing architectural arrangements. This configuration is proposed to have dedicated layers to support cartilage, mineralized cartilage, and subchondral bone regeneration, which are the three layers of an osteochondral tissue. Compared to biphasic structure, triphasic scaffold arrangements include a transitional zone between the cartilaginous and bone layers [16]. This would render an interface area for transfer of loading forces as well as selective nutrients and cells from compartment to compartment. Thus, the zones could essentially communicate with one another as it occurs in the native osteochondral tissue. Through seeding distinct cell types within their respective layers, one can more accurately mimic the microenvironments found within native osteochondral tissue, resulting in more successful cell-specific proliferation upon implantation.

A study by Marquass et al. investigated a triphasic scaffold, composed of an osseous phase constructed of β-tricalcium phosphate to support osteogenesis, a collagen I hydrogel phase for cartilaginous tissue formation, and an interfacial space filled by autologous plasma [17]. Autologous MSCs collected from the ovine (sheep) subjects and seeded into the osseous and cartilage phase of the grafts and predifferentiated separately into the respective cell types, then combined with the autologous plasma phase. Scaffolds were placed within 4 mm knee defects and analyzed after periods of 6 and 12 months. The 6-month group showed superior cartilage and bone layer formation when compared to the control group, an osteochondral autograft. However, contribution of the third intermediate phase was not fully analyzed by the group. On the other hand, Da et al. performed a similar experiment utilizing a bovine cartilage derived upper phase and 3D printed PLGA/TCP lower phase, combined with a compacted PLGA/TCP interfacial layer [18]. In vivo analysis in a rabbit model revealed superior regeneration of articular cartilage and trabecular bone in the triphasic designs. The regenerated tissue displayed columnar cartilage structure like that of native tissue, as well as an oriented trabecular structure for the subchondral bone after 6 months. In vitro analysis of the scaffolds showed that the compact intermediate layer presented very fine <10 μm pores that prevented cell migration between the two phases as well as limited fluid flow. Due to the restriction in cell and nutrient transport, it was speculated that the intermediate phase served to isolate each phase into its own compartment during early healing of the defect area, allowing the regenerating cells to mature in its own optimal microenvironment. However, as the intermediate layer degraded in vivo, the resulting cell and fluid access between the two phases would allow for integration of the two tissue types. These results indicate that the triphasic designs are advantageous over the monophasic and biphasic designs in terms of spatial arrangement, and the availability of dedicated layers to support bone and cartilage, while also promoting some integration between the two layers.

### 17.3.4 Gradient

To further replicate the native tissue characteristics, gradient designs have also garnered much attention in osteochondral regeneration. As mentioned previously, native osteochondral tissue presents a gradient structure of chondrocytes, hypertrophic chondrocytes, calcified cartilage, and subchondral bone, and several grafts were developed to fully recapitulate such an environment. A gradient scaffold aims to mimic the complex gradient structure of native osteochondral tissue, utilizing physical architecture of the scaffold, compositional variations, and growth factor concentrations. Physical gradient scaffolds aim to address the architectural and mechanical variations in the osteochondral tissue, these scaffolds may vary in mechanical properties along its axis to mimic the transition from soft cartilage tissue to the calcified cartilage, and ultimately subchondral bone. As the local environment stiffness can drastically influence cell behavior and differentiation, this variation in mechanical properties can help regenerate the osteochondral tissue, while also providing similar response to mechanical stress as the native tissue. Porosity of the graft can also be used in a graded manner to promote site specific differentiation of chondrocytes, hypertrophic chondrocytes, and osteoblasts. It was seen previously that larger pore sizes can significantly promote chondrocyte development [19, 20]. As the majority of the articular cartilage is relatively sparse and is mainly composed of extracellular matrix, the larger pore sizes allow the chondrocytes to reproduce matrix similar to that of native cartilage [3]. On the other hand, the subchondral phase requires a mechanically stable structure to support mechanical loading, which constrains the pore size of the lower region [21]. Compositional gradients are also used, where multiple osteoinductive and chondroinductive materials are used as a composite with opposing gradients to specifically drive the proliferation/differentiation of the relevant cell populations. The variety of known osteoinductive/chondroinductive materials allows for diverse compositions to be explored. Lastly, gradients of growth factors are used to directly induce desired cell behavior. Gradients of biochemical factors are prevalent in the body, with it being used for migration, proliferation, and differentiation in natural development and normal function of the body [22, 23]. Members of the transforming growth factor beta (TGF-$\beta$) superfamily have been used heavily in osteochondral tissue engineering. These factors, which include bone morphogenetic proteins (BMP), have been implicated in both osteogenesis and chondrogenesis, and have been used in gradient scaffold designs previously [24, 25]. The three types of gradient scaffolds are also often combined to develop more complex gradient scaffold systems. The main challenge associated with gradient scaffold arrangement is that it is an extremely difficult undertaking to construct a continuously and uniformly distributed structure. It requires highly specified and sophisticated fabrication techniques to achieve continuously gradient matrices. Some of the gradient scaffold fabrication methods are listed in Table 17.1.

Our group has recently developed a gradient matrix system for bone–cartilage interfacial tissue engineering [8, 26, 27]. The matrix was fabricated using an in house developed scaffold fabrication method, "particle sintering and porogen leaching." In this method, PLGA microparticles were mixed with a desired amount of porogen (NaCl particles) and sintered to form a three-dimensional and porous structure [10, 26]. Post-sintering, the porogen was leached to form a scaffold structure with interconnected pore structure. By varying the porogen content along the matrix length—increased porogen content from bottom to top—it was possible to develop a gradient porous matrix [8, 26, 28]. The physical gradient of the matrix was also combined with compositional gradient by infusing the upper cartilaginous portion with a hyaluronic hydrogel to form an inverse gradient matrix, with opposing gel polymer contents [8, 27]. The matrix's ability to support bone and cartilage layer regeneration was tested, in vitro, by seeding with human bone marrow derived and CD 271 positive mesenchymal stem cells [8]. Cell seeded constructs demonstrated chondrogenesis and osteogenesis when cultured separately in the respective media. Gradient grafts were cultured in a coculture media (1:1 ratio of osteogenic and chondrogenic media) to achieve simultaneous osteogenesis and chondrogenesis. In vitro culture of the gradient graft in the coculture media showed preferential chondrogenesis in the top of the matrix and osteogenesis in the bottom part of the matrix, as shown in Fig. 17.3. Through these studies, our group established the

**Table 17.1** Selected examples of published physical, compositional, and growth factor gradient grafts for osteochondral tissue regeneration

| Gradient type | | | Details | References |
|---|---|---|---|---|
| Physical | | Porosity | Nukavarapu et al. PLGA microsphere scaffold with gradient pore structure formed using porogen leaching method. Highly porous upper region for cartilage regeneration and dense lower region for bone regeneration.<br>Sherwood et al. 3D printed PLGA scaffold with highly porous cartilage region and denser bone region. Also modified using tricalcium phosphate for denser bone regeneration regions | [26, 28, 29] |
| | | Mechanical properties | Singh et al. Macroscopic gradient utilizing PLGA microsphere of different stiffness, controlled using incorporation of nanoscale $TiO_2$ or $CaCO_3$ | [30] |
| Compositional | | PLGA, Hyaluronic acid | Majumdar et al., Dorcemus et al. Opposing gradients for PLGA microspheres for bone tissue regeneration, and hyaluronic acid for cartilage tissue regeneration. Also modified with gradient growth factor and/or hydroxyapatite inclusion | [8, 27] |
| | | Silk, Silica nanoparticles | Guo et al. Silk fibroin matrix conjugated with varying concentrations of peptide to bind with silica nanoparticles. High conjugation with peptide and silica at subchondral regions, decreasing as it approaches the cartilage regions | [31] |
| | | Collagen, Hydroxyapatite | Liu et al. Collagen I matrix was created by chemically crosslinking bovine collagen. The matrix was then placed in solution of calcium and phosphate ion solution, then mineralized through diffusion. The resultant matrix showed graduated hydroxyapatite deposition | [32] |
| Growth factors | | TGF-β, BMP-2 | Dormer et al. Growth factor incorporated PLGA microsphere graft, with opposing gradients of TGF-β loaded microspheres (chondrogenesis) and BMP-2 loaded microspheres (osteogenesis)<br>Di Luca et al. 3D printed PCL scaffold with functionalized PEG brushes to covalently bind with TGF-β at upper cartilage regions and BMP-2 at lower bone regions, with opposing gradients | [33–35] |
| | | β-Glycerophosphate (β-GP), Insulin | Erisken et al. Insulin, which promotes chondrogenesis, and β-GP, which promotes matrix mineralization, were encapsulated in polycaprolactone fibers, and arranged in opposing gradients, with more Insulin at upper cartilage regions and more β-GP at lower subchondral regions | [36] |

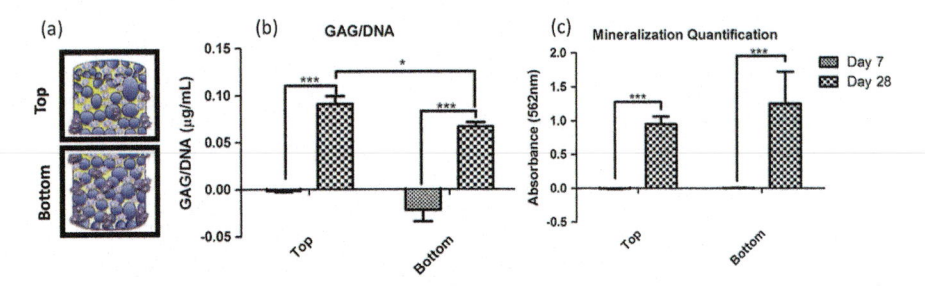

**Fig. 17.3** Gradient OC scaffold and stem cell differentiation in vitro. Gradient graft with cells in the gel phase were primed for chondrogenesis for 7 days in chondrogenic media followed by 21 days of culture in osteochondrogenic media with MSCs to attach to the bottom part of the matrix. (**a**) Pictorial representation of the top and bottom parts of the matrix, (**b**) GAG amount normalized to DNA, and (**c**) mineralization amount via alizarin red staining and quantification. GAG and mineralization quantifications were performed at days 7 and 28. Reprinted with permission from Tissue Engineering, Part A, Volume 23, Issue 15-16, by Dorcemus et al, published by Mary Ann Liebert, Inc., New Rochelle, NY [8]

potential of gradient matrix for osteochondral tissue engineering. Further studies may be needed to understand how the biomaterial physical cues such as matrix porosity and stiffness, in addition to gradient structure, play a role in achieving osteochondral tissue engineering. Such studies, in addition to in vivo analysis of the gradient scaffolds, may lead to better understanding of the scaffold architecture, composition, and local matrix stiffness/elasticity required for regeneration and the development of new regenerative strategies where the biomaterials alone guide tissue repair/regeneration.

## 17.4 Osteochondral Tissue Engineering: Bench to Bedside Translation

Based on preclinical success of multiphasic, many tissue engineering constructs have progressed to the clinical trial stage. Many clinical trials have been conducted using the biphasic design, composed of an upper cartilaginous phase and a lower osseous phase. Several biphasic scaffolds have been approved for use, mostly in the European Union, with trials still being conducted in the USA for many of the designs (Table 17.2).

One product currently undergoing clinical trials is Agili-C™ by CatriHeal. Agili-C™ is approved for use for osteochondral defects in the European Union, and is currently undergoing clinical trials in the USA. Agili-C™ aims to treat cartilage and osteochondral defects in both degenerative and nondegenerative lesions in the knee. The implant itself is biodegradable and biphasic, with coral aragonite, a crystalline calcium carbonate derived from coral matrix, serving as the osseous phase, and a hyaluronic acid impregnated aragonite cartilage phase [37, 38]. Coralline ara-

**Table 17.2** Selected examples of osteochondral grafts and treatment options currently in clinical trials or approved

| Product name | Company name | Material | Status |
|---|---|---|---|
| Agili-C Biphasic Implant | Cartiheal LTD | Biphasic construct: Aragonite and trace elements like strontium and magnesium (Bone); Modified aragonite and polymers (Cartilage) | Approved in EU; Clinical Trials in USA (NCT03299959) |
| BioMatrix Cartilage Repair Device | Kensey Nash Corporation | Bioresorbable scaffold with collagen fibril layer and calcium mineral layer held together with biodegradable polymer | Approved in EU |
| ChondroMimetic | TiGenix | Biphasic graft composed of unmineralized Type I collagen/chondritin sulfate cartilaginous phase and mineralized Type I collagen/GAG with brushite calcium phosphate | Approved in EU |
| MaioRegen | Fin-Ceramica Faenza Spa | Bioceramic, multilayered scaffold in a single gradient structure; deantigenated Type I equine collagen and magnesium enriched hydroxyapatite | Approved in EU |
| Chondrofix Osteochondral Allograft | Zimmer Biomet | Decellularized human hyaline cartilage and subchondral bone | Premarket approval not required in USA. |
| TruFit Plug | Smith & Nephew | Biphasic osteochondral graft with porous PLGA with addition of calcium sulfate for bone regeneration | Approval in EU; Limited approval in USA. |
| CartiFill | Sewon Cellontech | Modified Type I porcine collagen, stabilized with removal of telopeptides. For use in conjunction with microfracture technique to stabilize the mesenchymal stem cell population | Clinical Trials (NCT02685917) |

Taken and modified from Nukavarapu et al. 2012 [3]

gonite has been known to be osteogenic, being used in various clinical products, and owes its osteogenic abilities to the natural porous, microstructure that promotes osteoblast attachment and function [39–42]. The calcium carbonate that makes up aragonite is also readily resorbed by multinucleated giant cells (presumably osteoclasts), which then increases the concentrations of free calcium and carbonate ions that can promote new bone formation by osteoblasts [42]. Meanwhile, the cartilage regeneration of the graft is achieved with the use of hyaluronic acid (HA), which is extensively used in cartilage tissue engineering applications, and is shown to possess some degree of chondroinductive abilities [8, 43, 44]. Using these materials, in vivo data from preclinical trials was obtained in a goat model [37, 38]. The results showed that the aragonite-HA based graft was able to regenerate two distinct cartilaginous and osseous phases. Immunohistochemistry also showed spatially distinct expression of type I collagen (bone matrix protein) and type II collagen (hyaline cartilage protein), indicating proper regeneration of the articular cartilage without evidence of fibroblast or hypertrophic chondrocyte formation. A further 1 year follow-up of the study also showed similar results. Significant GAG deposition in the cartilage layer, along with type II collagen, and cell morphology suggests that the regenerated cartilage layer maintained its hyaline phenotype for a relatively extended period. However, the study does note that the 1 year follow-up does not necessarily confirm the maintenance of the hyaline cartilage, and further follow-ups would be needed to demonstrate a permanent hyaline cartilage regeneration. With these preclinical success of the graft, the Agili-C™ grafts are currently ongoing clinical trials, and are slated to be completed in 2019. While the complete clinical results are awaited, preliminary results published by CartiHeal show promising results for osteochondral repair in humans [45].

One product which is licensed and commercially available is the TruFit CB plug by Smith & Nephew. This plug is licensed to treat both chondral and osteochondral defects in Europe, but however is only approved for back filling of donor sites during autologous transplantation procedures in the USA [46, 47]. The TruFit CB plug is entirely synthetic, the plug is biphasic, composed of PLGA, with addition of calcium sulfate in the lower osseous phase, which allow for the plug to be resorbable once implanted [48]. The plug was designed to improve upon the microfracture technique, compensating for the loss in mechanical stability of the defect site during the procedure, using calcium sulfate to enhance bone growth, and allowing natural growth factors from bone marrow to infiltrate and initiate chondrogenesis [49]. Initial studies of the plug in a goat model showed favorable histological results, with evidence of hyaline-like cartilage regeneration, as well as integration with host tissue and between the regenerated tissues after 12 months [48]. Preliminary clinical results of the plug showed promising results, with all treated patients showing signs of improvement [50]. MRI analysis of the plugs also showed that the cartilage and bone phases of the plugs showed similar appearances to the native tissue [50].

However, questions of the plug's efficacy have been raised following long-term studies of the TruFit CB graft. Although initial results of the graft showed favorable regeneration of the target tissues, further studies and long-term follow-ups showed varied results [47, 51, 52]. In clinical study conducted by Joshi et al., 80% of the patients treated with TruFit plug reported improvements in conditions after 12 months, but patient satisfaction decreased to 10% after 24 months, and 70% of the patients required revision surgeries due to pain and loss of joint function [51]. Radiological evaluation of the defect site also showed conflicting and negative results for both cartilage and bone layers [49, 51, 53, 54]. The cartilage layer showed varied results, ranging from damaged surfaces featuring fissures, and fibrillation after 12 months [49, 51] to evidence of fibrocartilage rather than hyaline cartilage formation after 12 months [54]. The subchondral bone of defect site also showed overly negative results, displaying signs of granulation tissue, cyst formation, and sclerosis rather than bone regeneration [49, 51, 53, 54]. These results call to question the benefits of the TruFit plug in its use as an osteochondral graft, and raises further questions regarding use of acellular, biphasic designs for osteochondral defect repair.

MaioRegen is another multiphasic osteochondral scaffold that has obtained European Union market approval for osteochondral regeneration. In contrast to previously mentioned osteochondral grafts, MaioRegen is a triphasic graft, mainly composed of Type I equine collagen and hydroxyapatite. The lowest layer of the graft is composed of 70% Mg-hydroxyapatite—30% Collagen I, the middle layer is 40% Mg-hydroxyapatite—60% Collagen I, and the upper cartilaginous layer is made entirely with collagen I [55]. The incorporation of the middle layer was thought to enhance bonding between the regenerated cartilage and bone tissue, as well as recreate the graded structure of native tissue to an extent. In vitro and in vivo studies showed promising results, leading into extensive clinical study of the graft. Results from the clinical trials also showed marked improvements to the defect area [55–59]. MRI assessment of the surgical site showed integration with the host tissue by 6 months, and complete filling of the cartilage layer and integration by 2 years [58, 59].

With these promising results, MaioRegen was able to obtain approval to market in the European Union. However, issues regarding MaioRegen have also been brought up. In one case, 8 patients were implanted with MaioRegen graft in the femoral condyle, and observed via MRI after 18 months [60]. The MRI results indicated that defect filling in the subchondral and cartilaginous region did occur, but presence of newly formed bone was not detected. Also, a semiquantitative MRI method also showed differing zonal structure of the regenerated cartilage, which was confirmed through arthroscopy. The newly formed cartilage seemed to show more hypertrophic or fibroblastic characteristics in comparison to the native cartilage. Another set of clinical studies published by Christensen et al. also showed negative results of the MaioRegen graft [61]. The study was conducted on 10 patients, but two of the patients experienced increased pain and swelling soon after surgery, deemed to be due to the implant, and required removal of the implant. Furthermore, computed tomography (CT) scan images 3 years post-surgery showed complete absence of mineralized tissue within the subchondral phase of the graft, only displaying defect filling with soft tissue, evident through MRI analysis. These results also question the efficacy of the MaioRegen graft and its previous claims to regenerate both articular cartilage and the underlying bone. However, it is worth noting the small sample size of the two trials, in comparison to the previous trials done.

Nevertheless, regeneration of osteochondral defects and maintenance of phenotype with the use of synthetic grafts remain a significant challenge, but other methods of osteochondral repair also present shortcomings. For instance, Chondrofix Osteochondral Allograft from Zimmer is a human derived decellularized hyaline cartilage and subchondral bone graft. This graft is thoroughly processed to remove possible contaminants and disease vectors, while maintaining much of its native mechanical properties. Also, as an allograft, Chondrofix is not required to undergo the FDA premarket approval process, and claims to regenerate hyaline cartilage as well as the subchondral bone. However, after showing initial success, the Chondrofix Osteochondral Allograft has come under scrutiny following a 2016 study led by Farr et al. [62]. Chondrofix allografts were implanted on 32 patients and 23 implants were considered failures, with implant survivorship dipping to 19.6% after 2 years. Implant failure was defined as any need for reoperation of the implant or MRI/arthroscopic analysis showing subchondral collapse or cartilage delamination. Much of the implants resulted in increased pain and needed reoperation due to the delamination of the articular cartilage phase of the implant and the subsequent irregular cartilage development. Histological analysis of the failed implants showed that the cartilage phase of the graft was largely acellular, and the formation of vascularized tissue formation in the delaminated implants. Therefore, it is clear that there still remains a clinical need for a synthetic graft that can successfully regenerate hyaline cartilage as well as the subchondral bone, and maintain stability of the regenerated tissue.

## 17.5 Conclusions

Current treatment options for OC defect repair are primarily palliative as opposed to restorative, and often results in inadequate healing of the defect area. The defect repair is complicated by its complex tissue structure and heterogeneity. However, studies to mimic the native OC tissue structure with cartilage and bone layers with an integrating interface led to the development of various synthetic grafts. Some of the OC grafts developed in biphasic and triphasic configurations have advanced to clinical trials in the European Union. Although initial studies show promising results, long-term studies are needed to ensure (1) regeneration of cartilage/subchondral bone layers, (2) cartilage–bone interface development, and (3) no phenotypic drift of the regenerated hyaline cartilage into the fibroblastic or hypertrophic lineages. Current efforts are to develop novel OC grafts with local-microstructural control to regulate chondrogenesis/osteogenesis to form osteochondral tissue in structure and function similar to the native tissue and repair the defect fully.

**Acknowledgements** The authors acknowledge support from AO Foundation, Musculoskeletal Transplant Foundation, and NSF (EFRI and AIR). Dr. Nukavarapu acknowledges funding from Bioscience Connecticut through Technology Translation Pipe-line program and University of Connecticut through SPARK technology commercialization program. Dr. Nukavarapu also acknowledges funding support from the National Institute of Biomedical Imaging and Bioengineering of the National Institutes of Health (Award Number R01EB020640).

# References

1. Murphy L, Helmick CG (2012) The impact of osteoarthritis in the United States: a population-health perspective. Orthop Nurs 31(2):85–91
2. Pappas AM (1981) Osteochondrosis dissecans. Clin Orthop 158:59–69
3. Nukavarapu SP, Dorcemus DL (2013) Osteochondral tissue engineering: current strategies and challenges. Biotechnol Adv 31(5):706–721
4. Athanasiou KA, Zhu CF, Wang X, Agrawal CM (2000) Effects of aging and dietary restriction on the structural integrity of rat articular cartilage. Ann Biomed Eng 28(2):143–149
5. Martin I, Miot S, Barbero A, Jakob M, Wendt D (2007) Osteochondral tissue engineering. J Biomech 40(4):750–765
6. Fazzalari NL (2008) Bone remodeling: a review of the bone microenvironment perspective for fragility fracture (osteoporosis) of the hip. Semin Cell Dev Biol 19(5):467–472
7. Brandt KD, Dieppe P, Radin EL (2008) Etiopathogenesis of osteoarthritis. Rheum Dis Clin North Am 34(3):531–559
8. Dorcemus DL, George EO, Dealy CN, Nukavarapu SP (2017) Harnessing external cues: development and evaluation of an in vitro culture system for osteochondral tissue engineering. Tissue Eng Part A 23(15–16):719–737
9. Nukavarapu S, Freeman J, Laurencin C (2015) Regenerative engineering of musculoskeletal tissues and interfaces. Elsevier Science & Technology, Amsterdam
10. Amini AR, Laurencin CT, Nukavarapu SP (2012) Bone tissue engineering: recent advances and challenges. Crit Rev Biomed Eng 40(5):363–408
11. Jeong CG, Zhang H, Hollister SJ (2012) Three-dimensional polycaprolactone scaffold-conjugated bone morphogenetic protein-2 promotes cartilage regeneration from primary chondrocytes in vitro and in vivo without accelerated endochondral ossification. J Biomed Mater Res A 100A(8):2088–2096
12. Chu CR, Dounchis JS, Yoshioka M, Sah RL, Coutts RD, Amiel D (1997) Osteochondral repair using perichondrial cells. A 1-year study in rabbits. Clin Orthop 340:220–229
13. Jiang C-C et al (2007) Repair of porcine articular cartilage defect with a biphasic osteochondral composite. J Orthop Res 25(10):1277–1290
14. Dresing I, Zeiter S, Auer J, Alini M, Eglin D (2014) Evaluation of a press-fit osteochondral poly(ester-urethane) scaffold in a rabbit defect model. J Mater Sci Mater Med 25(7):1691–1700
15. Frenkel SR et al (2005) Regeneration of articular cartilage—evaluation of osteochondral defect repair in the rabbit using multiphasic implants. Osteoarthritis Cartilage 13(9):798–807
16. Jeon JE, Vaquette C, Klein TJ, Hutmacher DW (2014) Perspectives in multiphasic osteochondral tissue engineering. Anat Rec 297(1):26–35
17. Marquass B et al (2010) A novel MSC-seeded triphasic construct for the repair of osteochondral defects. J Orthop Res Off Publ Orthop Res Soc 28(12):1586–1599
18. Da H et al (2013) The impact of compact layer in biphasic scaffold on osteochondral tissue engineering. PLoS One 8(1):e54838
19. Woodfield T b f, Blitterswijk CAV, Wijn JD, Sims T j, Hollander A p, Riesle J (2005) Polymer scaffolds fabricated with pore-size gradients as a model for studying the zonal organization within tissue-engineered cartilage constructs. Tissue Eng 11(9–10):1297–1311
20. Oh SH, Kim TH, Im GI, Lee JH (2010) Investigation of pore size effect on chondrogenic differentiation of adipose stem cells using a pore size gradient scaffold. Biomacromolecules 11(8):1948–1955
21. Karageorgiou V, Kaplan D (2005) Porosity of 3D biomaterial scaffolds and osteogenesis. Biomaterials 26(27):5474–5491
22. Eichmann A, Le Noble F, Autiero M, Carmeliet P (2005) Guidance of vascular and neural network formation. Curr Opin Neurobiol 15(1):108–115
23. Parent CA, Devreotes PN (1999) A cell's sense of direction. Science 284(5415):765–770
24. Chen G, Deng C, Li Y-P (2012) TGF-β and BMP signaling in osteoblast differentiation and bone formation. Int J Biol Sci 8(2):272–288

25. Wozney JM (1992) The bone morphogenetic protein family and osteogenesis. Mol Reprod Dev 32(2):160–167
26. Nukavarapu SP, Laurencin CT, Amini AR, Dorcemus DL (2017) Gradient porous scaffolds. US 9,707,322 B2
27. Majumdar S, Pothirajan P, Dorcemus D, Nukavarapu S, Kotecha M (2016) High field sodium MRI assessment of stem cell chondrogenesis in a tissue-engineered matrix. Ann Biomed Eng 44(4):1120–1127
28. Dorcemus DL, Nukavarapu SP (2014) Novel and unique matrix design for osteochondral tissue engineering. MRS Online Proc Libr Arch 1621:17–23
29. Sherwood JK et al (2002) A three-dimensional osteochondral composite scaffold for articular cartilage repair. Biomaterials 23(24):4739–4751
30. Singh M et al (2010) Three-dimensional macroscopic scaffolds with a gradient in stiffness for functional regeneration of interfacial tissues. J Biomed Mater Res A 94(3):870–876
31. Guo J, Li C, Ling S, Huang W, Chen Y, Kaplan DL (2017) Multiscale design and synthesis of biomimetic gradient protein/biosilica composites for interfacial tissue engineering. Biomaterials 145(Supplement C):44–55
32. Liu C, Han Z, Czernuszka JT (2009) Gradient collagen/nanohydroxyapatite composite scaffold: development and characterization. Acta Biomater 5(2):661–669
33. Dormer NH, Singh M, Wang L, Berkland CJ, Detamore MS (2010) Osteochondral interface tissue engineering using macroscopic gradients of bioactive signals. Ann Biomed Eng 38(6):2167–2182
34. Dormer NH, Busaidy K, Berkland CJ, Detamore MS (2011) Osteochondral interface regeneration of rabbit mandibular condyle with bioactive signal gradients. J Oral Maxillofac Surg 69(6):e50–e57
35. Di Luca A, Klein-Gunnewiek M, Vancso JG, van Blitterswijk CA, Benetti EM, Moroni L (2017) Covalent binding of bone morphogenetic protein-2 and transforming growth factor-β3 to 3D plotted scaffolds for osteochondral tissue regeneration. Biotechnol J
36. Erisken C, Kalyon DM, Wang H, Örnek-Ballanco C, Xu J (2010) Osteochondral tissue formation through adipose-derived stromal cell differentiation on biomimetic polycaprolactone nanofibrous scaffolds with graded insulin and beta-glycerophosphate concentrations. Tissue Eng Part A 17(9–10):1239–1252
37. Kon E et al (2015) Osteochondral regeneration with a novel aragonite-hyaluronate biphasic scaffold: up to 12-month follow-up study in a goat model. J Orthop Surg 10
38. Kon E et al (2014) Osteochondral regeneration using a novel aragonite-hyaluronate bi-phasic scaffold in a goat model. Knee Surg Sports Traumatol Arthrosc 22(6):1452–1464
39. Demers C, Hamdy CR, Corsi K, Chellat F, Tabrizian M, Yahia L (2002) Natural coral exoskeleton as a bone graft substitute: a review. Biomed Mater Eng 12(1):15–35
40. Roudier M et al (1995) The resorption of bone-implanted corals varies with porosity but also with the host reaction. J Biomed Mater Res 29(8):909–915
41. Doherty MJ, Schlag G, Schwarz N, Mollan RA, Nolan PC, Wilson DJ (1994) Biocompatibility of xenogeneic bone, commercially available coral, a bioceramic and tissue sealant for human osteoblasts. Biomaterials 15(8):601–608
42. Petite H et al (2000) Tissue-engineered bone regeneration. Nat Biotechnol 18(9):959–963
43. Chung C, Burdick JA (2009) Influence of 3D hyaluronic acid microenvironments on mesenchymal stem cell chondrogenesis. Tissue Eng Part A 15(2):243–254
44. Amann E, Wolff P, Breel E, van Griensven M, Balmayor ER (2017) Hyaluronic acid facilitates chondrogenesis and matrix deposition of human adipose derived mesenchymal stem cells and human chondrocytes co-cultures. Acta Biomater 52(Supplement C):130–144
45. Kon E, Drobnic M, Davidson PA, Levy A, Zaslav K, Robinson D (2014) Chronic posttraumatic cartilage lesion of the knee treated with an acellular osteochondral-regenerating implant: case history with rehabilitation guidelines. J Sport Rehabil 23(3):270–275

46. Melton JTK, Wilson AJ, Chapman-Sheath P, Cossey AJ (2010) TruFit CB bone plug: chondral repair, scaffold design, surgical technique and early experiences. Expert Rev Med Devices 7(3):333–341
47. Verhaegen J, Clockaerts S, Van Osch GJVM, Somville J, Verdonk P, Mertens P (2015) TruFit plug for repair of osteochondral defects—where is the evidence? Systematic review of literature. Cartilage 6(1):12–19
48. Williams RJ, Gamradt SC (2008) Articular cartilage repair using a resorbable matrix scaffold. Instr Course Lect 57:563–571
49. Dhollander AAM et al (2012) A pilot study of the use of an osteochondral scaffold plug for cartilage repair in the knee and how to deal with early clinical failures. Arthrosc J Arthrosc Relat Surg 28(2):225–233
50. Saithna A, Dunne K, Kuchenbecker T, Thompson P, Dhillon M, Spalding T (2010) Qualitative MRI related to clinical results following cartilage repair using Trufit plugs: a two year follow up study. Orthop Proc 92-B(SUPP III):423
51. Joshi N, Reverte-Vinaixa M, Díaz-Ferreiro EW, Domínguez-Oronoz R (2012) Synthetic resorbable scaffolds for the treatment of isolated patellofemoral cartilage defects in young patients: magnetic resonance imaging and clinical evaluation. Am J Sports Med 40(6):1289–1295
52. Carmont MR, Carey-Smith R, Saithna A, Dhillon M, Thompson P, Spalding T (2009) Delayed incorporation of a TruFit plug: perseverance is recommended. Arthrosc J Arthrosc Relat Surg 25(7):810–814
53. Barber FA, Dockery WD (2011) A computed tomography scan assessment of synthetic multi-phase polymer scaffolds used for osteochondral defect repair. Arthrosc J Arthrosc Relat Surg 27(1):60–64
54. Pearce CJ, Gartner LE, Mitchell A, Calder JD (2012) Synthetic osteochondral grafting of ankle osteochondral lesions. Foot Ankle Surg 18(2):114–118
55. Kon E, Filardo G, Perdisa F, Venieri G, Marcacci M (2014) Clinical results of multilayered biomaterials for osteochondral regeneration. J Exp Orthop 1:10
56. Delcogliano M et al (2014) Use of innovative biomimetic scaffold in the treatment for large osteochondral lesions of the knee. Knee Surg Sports Traumatol Arthrosc Off J ESSKA 22(6):1260–1269
57. Winthrop Z, Pinkowsky G, Hennrikus W (2015) Surgical treatment for osteochondritis dessicans of the knee. Curr Rev Musculoskelet Med 8(4):467–475
58. Kon E et al (2014) Clinical results and MRI evolution of a nano-composite multilayered biomaterial for osteochondral regeneration at 5 years. Am J Sports Med 42(1):158–165
59. Kon E, Delcogliano M, Filardo G, Busacca M, Di Martino A, Marcacci M (2011) Novel nano-composite multilayered biomaterial for osteochondral regeneration: a pilot clinical trial. Am J Sports Med 39(6):1180–1190
60. Brix M et al (2016) Successful osteoconduction but limited cartilage tissue quality following osteochondral repair by a cell-free multilayered nano-composite scaffold at the knee. Int Orthop 40(3):625–632
61. Christensen BB, Foldager CB, Jensen J, Jensen NC, Lind M (2016) Poor osteochondral repair by a biomimetic collagen scaffold: 1- to 3-year clinical and radiological follow-up. Knee Surg Sports Traumatol Arthrosc Off J ESSKA 24(7):2380–2387
62. Farr J, Gracitelli GC, Shah N, Chang EY, Gomoll AH (2016) High failure rate of a decellularized osteochondral allograft for the treatment of cartilage lesions. Am J Sports Med 44(8):2015–2022

# Chapter 18
# Clinical Trials and Management of Osteochondral Lesions

Carlos A. Vilela, Alain da Silva Morais, Sandra Pina, J. Miguel Oliveira, Vitor M. Correlo, Rui L. Reis, and João Espregueira-Mendes

**Abstract** Osteochondral lesions are frequent and important causes of pain and disability. These lesions are induced by traumatic injuries or by diseases that affect both the cartilage surface and the subchondral bone. Due to the limited cartilage ability to regenerate and self-repair, these lesions tend to gradually worsen and progress towards osteoarthritis. The clinical, social, and economic impact of the osteochondral lesions is impressive and although therapeutic alternatives are under discussion, a consensus is not yet been achieved. Over the previous decade, new strategies based on innovative tissue engineering approaches have been developed with promising results. However, in order those products reach the market and help the actual patient in an effective manner, there is still a lot of work to be done. The current state of the implications, clinical aspects, and available treatments for this pathology, as well as the ongoing preclinical and clinical trials are presented in this chapter.

---

C. A. Vilela (✉)
Life and Health Sciences Research Institute (ICVS), School of Health Sciences, University of Minho, Braga, Portugal

Orthopedic Department, Centro Hospitalar do Alto Ave, Guimarães, Portugal

Dom Henrique Research Centre, Porto, Portugal

3B's Research Group, European Institute of Excellence on Tissue Engineering and Regenerative Medicine, University of Minho, Guimarães, Portugal

ICVS/3B's–PT Government Associate Laboratory, Braga/Guimarães, Portugal

A. da Silva Morais · S. Pina
3B's Research Group, European Institute of Excellence on Tissue Engineering and Regenerative Medicine, University of Minho, Guimarães, Portugal

ICVS/3B's–PT Government Associate Laboratory, Braga/Guimarães, Portugal

J. M. Oliveira · V. M. Correlo · R. L. Reis
3B's Research Group, European Institute of Excellence on Tissue Engineering and Regenerative Medicine, University of Minho, Guimarães, Portugal

ICVS/3B's–PT Government Associate Laboratory, Braga/Guimarães, Portugal

The Discoveries Centre for Regenerative and Precision Medicine, University of Minho, Barco/Guimarães, Portugal

© Springer International Publishing AG, part of Springer Nature 2018
J. M. Oliveira et al. (eds.), *Osteochondral Tissue Engineering*,
Advances in Experimental Medicine and Biology 1058,
https://doi.org/10.1007/978-3-319-76711-6_18

**Keywords** Musculoskeletal injuries · Bone repair · Cartilage repair · Tissue engineering · Osteochondral · Clinical studies

## 18.1 Introduction

Articular cartilage (AC) originated from the mesenchymal, more precisely from the skeletal blastemal, is a unique and highly specialized connective tissue, with 2–4 mm thickness, that covers the surface of diarthrodial joints [1]. Forming a very smooth, bright and sliding surface that facilitates pain-free movements during skeletal motion, this hyaline tissue has special mechanical properties. Composed primarily of water (65–80% of the wet weight), cells and macromolecules, AC possesses the unique ability to absorb shock impacts, support heavy and repetitive loads, and withstand wear and tear over the course of a lifetime. Water is mainly dispersed in the interfibrillar space of the matrix, thus assuring the diffusion of the nutrients and creating a moving load-dependent phase that provides flexibility, deformability and resilient strength [1]. Chondrocytes (1–5% of total tissue volume) are the sole cells found in the lacunae of the cartilage and are responsible for the synthesis of the extracellular matrix (ECM) environment [2, 3]. The ECM is primarily composed of collagens (10–20%), glycosaminoglycans (GAGs) and proteoglycans (PGs) [3]. There are others constituents that can be found in the matrix, namely, non-collagen proteins, glycoproteins, monosaccharides, and oligosaccharides [4].

Collagen type II represents the most abundant (90–95%) form of collagen in the AC. Collagen type X, XI and, although in minor quantity, collagen type V, VI, and IX are also present in the AC [4]. Collagen is a reinforcing structure of the water–proteoglycans gel phase of the matrix that increases the tensile strength, and facilitates the anchorage of others macromolecules and the mineralization process. Depending on the mechanical demands, the collagen orientation varies through the thickness of the AC. The specific GAGs of physiological significance in the AC are hyaluronic acid (also known as hyaluronan), chondroitin sulfate, keratan sulfate, and dermatan sulfate, all forming unbranched chains of repeating disaccharides.

---

J. Espregueira-Mendes
Clínica do Dragão—Espregueira-Mendes Sports Centre, FIFA Medical Centre of Excellence, Porto, Portugal

3B's Research Group, European Institute of Excellence on Tissue Engineering and Regenerative Medicine, University of Minho, Guimarães, Portugal

ICVS/3B's–PT Government Associate Laboratory, Braga/Guimarães, Portugal

The Discoveries Centre for Regenerative and Precision Medicine, University of Minho, Barco/Guimarães, Portugal

Orthopedic Department, Centro Hospitalar do Alto Ave, Guimarães, Portugal

The PGs, which represent the second-largest group of macromolecules in the ECM of the AC, are macromolecules produced by the chondrocytes that consist of a "core protein" with one or more covalently attached glycosaminoglycan (GAG) chain(s). The AC contains a diversity of PGs that are essential for normal function, namely, decorin, biglycan, fibromodulin, and aggrecan. The largest in size and most abundant by weight is aggrecan, a proteoglycan that possesses over 100 chondroitin sulfate and keratan sulfate chains. Aggrecan is able to interact with hyaluronan chains and fill the interfibrillar space of the AC matrix and constitutes approximately 90% of the total cartilage matrix [4]. Four different zones or layers could be easily identified in a histological examination of AC: (1) the superficial zone (10–20% of AC thickness), which contains collagen fibers packed tightly and aligned parallel to the articular surface, and a high water content and low PG concentration; (2) the middle or transitional zone (40–60% of AC thickness) with spherical and probably metabolic active chondrocytes showing collagen fibers in a less organized pattern; (3) the deep zone (30% of AC thickness), which contains ellipsoid cells that line up in a columnar fashion perpendicular to the joint surface and presents the highest PG concentration and the lowest water content; and (4) the calcified zone (5% of AC thickness), which plays an integral role in securing the cartilage to bone, by anchoring the collagen fibrils of the deep zone to the subchondral bone, and blocks the blood, neural or nutrients passage. Therefore, the nutrition to the AC is provided by the synovial and articular fluid in a very low oxygen tension environment [1, 3, 4]. Due to the absence of vasculature and nerve supply, cartilage has a low regenerative capacity. Thus, once injured, cartilage is much more difficult to repair.

Underneath the cartilage is the subchondral bone. Together, the AC and the subchondral bone form the osteochondral unit, which is a functional unit uniquely adapted to assure the transfer of loads across the diarthrodial joint. The subchondral bone plate constitutes the more superficial layer of compact cortical bone of the subchondral bone. Under the subchondral bone plate, a transitional subchondral spongiosa zone completes the osteochondral unit [5]. Supplementary to its role as a mechanical shock absorber, the subchondral bone possesses another important metabolic function. It is richly perforated by hollow spaces that allow the invasion of arterial and venous vessels, as well as of nerves, up to the calcified cartilage from the spongiosa. Therefore, the subchondral bone also nourishes the deeper cartilage layers, providing more than 50% of its glucose, oxygen and water requirements [6].

Chondral and osteochondral lesions are frequently observed by arthroscopy and can be diagnosed in 60–66% of all patients submitted to an arthroscopic procedure [7–11]. A retrospective analysis of 25,124 knee arthroscopies found chondral lesion in 60% of the patients [7]. Of these chondral lesions, 67% were classified as localized focal osteochondral or chondral lesions [7], 29% as osteoarthritis (OA), and, in 2% of the cases, an osteochondritis dissecans was diagnosed [7]. The medial condyle (34–58%) and the patella (11–36%) are the most frequent localizations of cartilage lesions in the knee [7, 8]. The most common cartilage lesion is a traumatic single lesion affecting a patient over 50 years old with a mean area of 2.1 cm$^2$ [8]. In the eighteenth century, Hunter observed, when talking about articular cartilage, that "once destroyed, is not repaired" and that the evolution to degenerative changes and OA is to be expected [12–15]. OA affects approximately 15% of the population and this number will double by the year 2020 due to population ageing and obesity [15]. The lower extremity is more commonly affected and researchers estimated the lifetime risk of developing symptomatic knee OA to be about 40% in men and 47% in women [15]. The direct and indirect costs of OA have been appraised at $65 billion per year [16]. Although a common location of osteochondral defects is the knee, any other joint could be affected by an osteochondral defect [17].

Several therapeutic alternatives are available for the treatment of chondral and osteochondral lesions, but a definitive and consensual treatment for the cartilage regeneration has not yet been found. In order to repair an osteochondral defect, the needs of the bone, cartilage, and the bone–cartilage interface must be taken into account [17]. A better understanding and knowledge of the cartilage and bone structure, of the biological and mechanical properties, and of the bone–cartilage interface will improve the therapeutic alternatives. Tissue engineering has been proposing single, biphasic and multiphasic scaffolds, cell therapies and bioactive molecules as advance therapies for cartilage regeneration and repair [17].

## 18.2 Clinical Management of Osteochondral Lesions

A traumatic chondral or osteochondral lesion usually involves the repeated application of a torsional and shear force or of a high energy force [18]. Advanced osteoarthritis and diseases involving necrosis of the subchondral bone due to different causes, including ischemia or repetitive trauma, can also be responsible for the development of osteochondral lesions [19]. Osteochondral lesions affect mostly the knee, particularly the medial femoral condyle (69%) (Fig. 18.1), the weight-bearing portion of the lateral femoral condyle (15%), the inferomedial pole of the patella (5%), and the trochlear fossa (1%) [19]. The other most commonly affected sites are the talar dome, the elbow, the shoulder, the wrist, and the hip [19]. The diagnosis is difficult. The symptoms are unspecific and the radiologic examination is negative in an early stage of the disease. Only a magnetic resonance imaging (MRI) is helpful as a diagnosis procedure for a cartilage lesion in an initial stage. The most important complaint is pain. It can be described in many ways and may have a mechanical rhythm that worsens gradually after the start of a new activity, especially when a previous trauma is involved [9]. Pain can be exacerbated for numerous reasons, from walking to more intense sport activities [9]. Associated symptoms, such as locking, pseudo-locking, and giving-way can be reported when loose bodies or associated lesions as meniscal tears or ligamentous injuries are present [20]. A methodic and exhaustive physical examination of the affected joint and of the contralateral joint is mandatory. Joint swelling or joint effusion, crepitation, lameness, and limitations of the range of motion of the affected joint are often present [21].

Although radiographic studies are not very helpful in diagnosing cartilage lesions in an early stage, they can be useful for detecting osteochondral lesions, osteoarthritis, osteochonditis dissecans, loose bodies, and limb malalignment [21]. In special

**Fig. 18.1** Chondral lesion

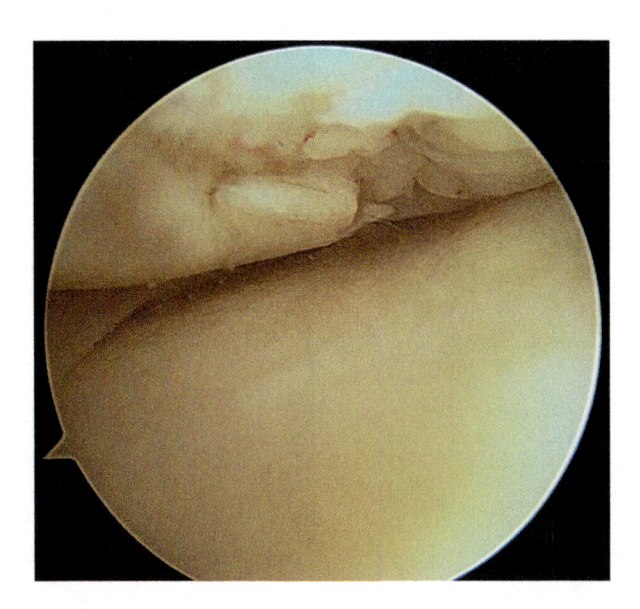

views, such as a 45° posteroanterior weight-bearing view of the knee (none as a "tunnel" view), a narrowing of the joint space of more than 2 mm when compared with the contralateral knee can diagnose a major cartilage lesion of the affected knee [22, 23].

MRI confirms the clinical diagnosis and is a valuable method not only for the characterization of the lesion, but also for the evaluation of the stability and viability of the osteochondral fragments. Moreover, it is useful for staging the disease, for the follow-up evaluation, for monitoring the effects of chondral pharmacologic and surgical therapies, and in cartilage scientific research, namely, for the quantitative or semiquantitative assessment of cartilage [19, 24, 25]. MRI is helpful in the differential diagnosis of osteochondral fractures and stress fractures, which must be differentiated from osteochondral lesions [19], and to understand all the associated lesions, such as subchondral cysts and meniscal or ligamentous injuries [21]. MRI is a noninvasive procedure able to identify a cartilage lesion in a very early stage due to its ability to detect metabolic water content and structural defects. Therefore, this technique can diagnose cartilage lesions before arthroscopic observation [24, 25]. Despite the relevance of the information provided by MRI, arthroscopy remains the chosen technique for diagnosis and validation procedure of new therapeutic approaches [24].

For the correct evaluation of the severity of the cartilage lesion and its clinical implications in the treatment and outcomes, several classification scores were developed [26]. The popular Outerbridge classification score intended to classify the cartilage lesions according to macroscopic aspects, depth and extension of the cartilage injury [27]. Grade 1 included cartilage lesions presenting softening and swelling. Lesions with fissuring or fragmentation of the cartilage surface that do not exceed 1.5 cm in diameter were classified as Outerbridge grade 2. In grade 3, fissures and cracks of the cartilage surface are present in an area of more than 1.5 cm. Grade 4

**Fig. 18.2** Grade IV chondral lesion

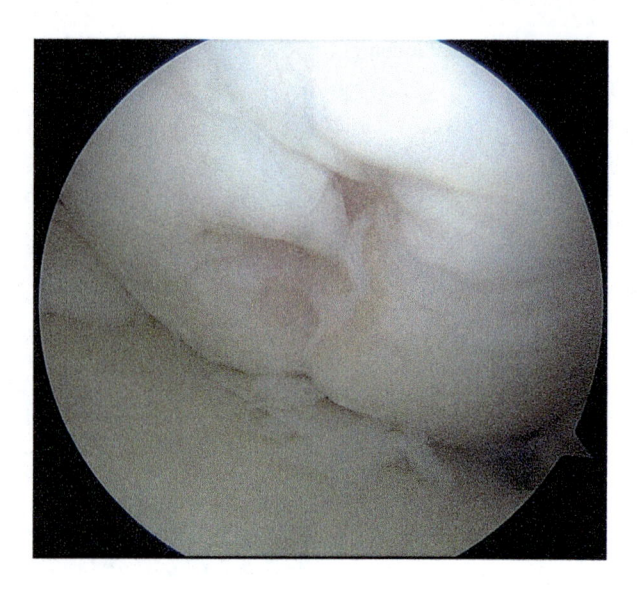

of the Outerbridge classification score included the lesions where the cartilage is eroded down to the bone [27]. Despite the simplicity and the dissemination of the Outerbridge classification score, some limitations are present [28]. Outerbridge classification score is mainly focused on the depth of the lesion with very little attention to its size. For example, a narrow deep cartilage lesion is classified as a grade 4, according to the Outerbridge classification score, but a lesion with a potentially worse prognosis, like an extensive partial thickness, is also classified as a grade 4 in this grading system (Fig. 18.2).

Due to the limitations of the Outerbridge classification score aforementioned, the International Cartilage Repair Society (ICRS) developed a five grade cartilage lesion classification system based on the macroscopic evaluation and the depth of the cartilage defect [29]. In grade 0, the cartilage is normal. Grade 1A includes superficial cartilage lesions with a cartilage softening and/or superficial fissures. Grade 1B includes superficial lesions where fissures and cracks are present. In grade 2, cartilage lesions present a defect extending down to less than 50% of the cartilage thickness and fraying is present. In grade 3, the cartilage defect has to extend down to more than 50% of the cartilage thickness as well as down to the calcified layer. When there is complete loss of cartilage thickness and the underlying bone is exposed, the cartilage lesion will be classified as a grade 4 (Fig. 18.3).

Osteochondritis Dissecans (OCD) is a disorder characterized by the degeneration or aseptic necrosis of the subchondral bone followed by fragmentation of the overlying cartilage and that can progress to osteoarthritis [19, 30]. As repetitive trauma is thought to be the most frequent cause, the term osteochondral lesion seemed to be more appropriate than the term "osteochondritis [19, 30]. Other possible causes for this disorder are epiphyseal ossification abnormalities, endocrine imbalances, ischemia or genetic predisposition [19, 30]. The ICRS developed an OCD evaluation score system: in grade 0, the lesion is stable and the overlying cartilage is normal and

**Fig. 18.3** Chondral lesion with the underlying bone exposed

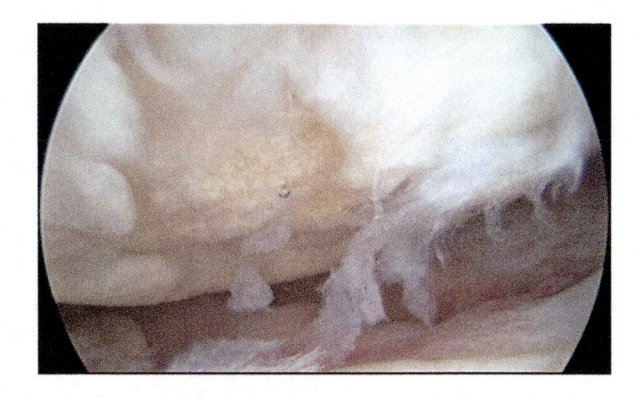

intact; grade I includes stable lesions with some softening of the cartilage surface; grade II refers to lesions with partial discontinuity of the cartilage surface; the lesion is classified as grade III when the defect is unstable due to a complete discontinuity of the osteochondral defect; a grade IV lesion is an empty defect or a defect with loose fragments [31]. According to this grading system, grade III and IV lesions are unstable and, therefore, may have indication for surgical orthopedic treatment [19, 30].

Cartilage treatment seeks to: (1) restore a smooth articular cartilage surface and match the biomechanical/biochemical properties of normal hyaline cartilage; (2) relieve patients' symptoms, namely, pain, swelling, and limping; (3) prevent or slow the progress to osteoarthritis; (4) lessen the morbidity of the disease and of the treatment techniques; (5) reduce the direct and indirect costs of the disease and expenses of the treatments [21].

Palliative treatments, such as articular lavage or articular debridement, are therapeutic approaches for the relief of the patient's symptoms. An example that could clinically justify the use of a palliative treatment is a case of a symptomatic patient with pain and blocking that needs the removal of a degenerated meniscus, a loose body (Fig. 18.4), a chondral fragment or a redundant synovia. Those procedures are in decline and their effectiveness has not been proven [32, 33].

Reparative procedures, such as arthroscopic abrasion arthroplasty, abrasion chondroplasty, Pridie drilling, microfracture (MF), or spongialization, are bone marrow stimulating techniques that seek a spontaneous and natural cartilage healing by perforating through the subchondral bone to promote bleeding and consequent recruitment of bone marrow cells [34, 35]. The MF technique was initially described by Stedman. It consists in the removal of all the instable cartilage using a small awl, after which holes of 3–4 mm depth separated by 3–4 mm are done by perforating the subchondral bone [36]. The fibrocartilaginous repair tissue formed showed weaker mechanical properties when compared to the natural hyaline cartilage [34, 35]. MF became a very popular cartilage restoration procedure, but although good short-term clinical outcomes were reported, in the long run, or when compared with other techniques, the impact on cartilage repair remained controversial [35, 37]. Future improvements in MF results could be achieved by administering growth factors (GF), platelet-rich plasma (PRP) or genetic engineering products [38, 39], and

**Fig. 18.4** Chondral lesion with a loose body

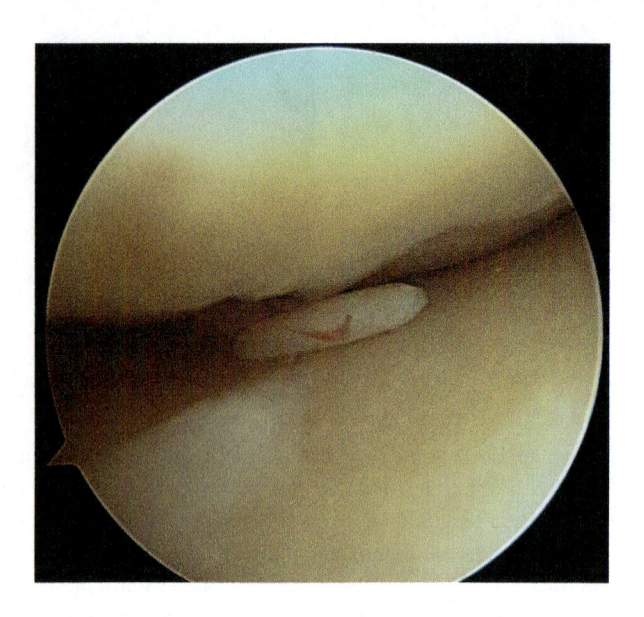

by using collagen membranes to cover the area treated with MF (AMIC- autologous matrix-induced chondrogenesis or BMAC—bone marrow aspirate concentrate) [37, 40, 41]. Another cartilage repair system that can be used with the MF procedure is the ChonDux™ (Cartilix, USA). This system consists of a biological adhesive and a photopolymerized hydrogel that are used combined with microfracture to enhance the stem cells migration from the bone marrow to the cartilage lesion [42].

The osteochondral autograft transplantation (OAT) is a demanding surgical procedure that consists in harvesting an osteochondral graft from the same joint or from another joint to repair the cartilage defect [34, 43, 44]. Regarding knee lesions, an osteochondral graft is harvested from a non-weight bearing area, the medial and lateral border of the condyles, the intercondylar notch, or the sulcus terminalis of the lateral femoral condyle being the preferred areas [34, 43]. Mosaicplasty (MP) is a similar repair cartilage procedure, but it uses more than one graft [35, 45]. The graft presents good bone-to-bone healing interface by contrast with cartilage-to-cartilage interface [35]. Others limitations regarding these procedures concern the morbidity of the donor site, the quantity of graft that can be harvested and the congruity of the repaired articular surface [35]. The OAT technique presents its best results in small lesions (between 2.5 and 4 $cm^2$) located on the condyle in the weight bearing area of a young patient [11, 43]. This technique could be an alternative to MF and other regenerative cartilage failed procedures [43]. The Cartilage Autograft Implantation System—CAIS® (DePuy/Mitek, USA) is a cartilage repair procedure that uses a minced autograft cartilage dispersed in a three-dimensional scaffold based on an absorbable copolymer foam (35% polycaprolactone and 65% polyglycolic acid reinforced with a polydioxanone mesh) [46, 47].

Osteochondral allograft (OAG) transplantation could be indicated (1) for the treatment of an active patient with a massive lesion (larger than 2.5 $cm^2$) [11, 48],

(2) for the treatment of an older patient when the alternative procedure is the arthroplasty; (3) in patients previously submitted to other cartilage repair techniques; (4) in osteonecrosis or osteochondritis dissecans cases, or (5) for the reconstructive repair of an extensive traumatic osteochondral lesion [48]. However, the reported outcomes are confusing [10, 49, 50]. This procedure presents serious limitations: (1) difficulty to obtain, preserve and manage the allografts, (2) risk of potential disease transmission, (3) immunogenicity of the allograft, and (4) surgical difficulties in fixating the allograft and achieving a good congruity with the joint cartilage surface. Chondrofix® allograft (Zimmer, USA) is considered the first off-the-shelf osteochondral allograft. It combines donated human cancellous bone and decellularized hyaline cartilage and is recommended to repair grade III and IV osteochondral lesions in a single, less-invasive procedure [51]. DeNovo NT (Zimmer, USA) is another allograft cartilage repair procedure that consists in securing allogeneic fetal chondrocytes into the cartilage defect with fibrin glue [46].

When the lesion has a viable part and the detached fragment presents no serious damage signs, the fixation in situ of the fragment could be the most appropriate therapeutic alternative, particularly in a young and/or active patient. After debridement of the lesion site and removal of blood clots and fibrous tissue, the fragment is then fixated in place. Several fixation devices are available: Kirshnerwires, compression screws, Herbert screws, osteochondral plugs, biodegradable pins and screws with acceptable clinical results [52, 53].

More than 20 years ago, Peterson and Brittberg, gave way to a new era in cartilage repair [54]. Their innovative two-step regenerative procedure consisted in collecting approximately 200–300 mg of a patient's articular cartilage in order to harvest chondrocytes that were then cultured and expanded. During the second stage of the process, the expanded chondrocytes were implanted into the defect and subsequently covered with a periosteal flap. The periosteal patch must be sutured water-tight to the surrounding cartilage to contain the injected suspension [11, 18, 43]. This technique is named as the autologous chondrocyte implantation (ACI) procedure. Since Brittberg et al. first described the ACI procedure, in 1994, as a treatment for chondral knee lesions [54], modifications have been introduced, thus, resulting in an evolution from a first-generation to a second-generation and third-generation ACI. This technique became one of the most important surgical alternatives in the treatment of chondral lesions of the knee and its use has extended to the treatment of chondral lesions in other joints, such as the ankle, shoulder, hip, and wrist [18, 21, 55]. ACI is recommended for younger patients who present symptoms of joint pain and swelling related to a chondral articular lesion. The first-generation ACI procedure was associated with a series of complications, such as periosteal graft hypertrophy, periosteal delamination, immune reaction, technical difficulties in fixating the periosteal flap, large surgical exposition, time consuming and joint stiffness [18, 55]. Carticel® (Genzyme Biosurgery, USA), an autologous cellular product commercialized since 1995 [56], is one of the techniques included in the first-generation ACI, defined as a two-stage procedure with implantation of cultured chondrocytes under a periosteal ACI (PACI).

In the second-generation ACI, a membrane of porcine type I/III collagen was used (ACI-C) to cover the treated lesion. This technique claimed a decrease operating time and reduced graft complications [18, 55]. One system used in both first- and second-generation ACI is ChondroCelect® (TiGenix, Belgium), a unique cell-based cartilage repair technique that involves the combination of cultured cells with a biodegradable type I-III collagen patch. In this technique, a gene expression score was used during the isolation and expansion of chondrocytes to identify and predict the cells ability to form hyaline cartilage. ChondroCelect® was the first cartilage repair system commercialized in Europe, but it was withdrawn from the market in 2016 [47]. Chondro-Gide® (Geistlich Pharma, Switzerland), also used in first- and second-generation ACI, is a porcine collagen type I/III bilayer matrix that promotes support and adhesion of autologous cultured chondrocytes in cartilage repair [41, 57].

In the third-generation ACI, a matrix seeded with autologous chondrocytes is used for cartilage repair [12]. This matrix, commonly referred to as matrix-assisted chondrocyte implantation or MACI (a trademark of Genzyme), has a specific and adapted mechanical profile with chondroinductive or chondroconductive properties that allows for a more homogenous distribution of the chondrocytes and that does not require a fixation procedure [18].

There are different types of scaffolds currently used in clinical settings [58].

1. BioSeed®-C (Biotissue Technologies, Germany) is a fibrin and polymer-based scaffold composed of polyglycolic/polylactic acid and polydioxanone [59];
2. Cartipatch® (Tissue Bank of France, France) is amonolayer agarose-alginate hydrogel scaffold [60, 61];
3. Hyalograft®-C (Fidia Advanced Biopolymers, Italy) is a hyaluronic acid-based scaffold [62];
4. MACI® (Genzyme, USA) uses a purified porcine collagen matrix to build the matrix scaffold [58];
5. Novocart®3D (TETEC Tissue Engineering Technologies AG, Germany) utilizes a collagen-chondroitin-sulfate based membrane [63];
6. NeoCart® (Histogenics, USA) implant is produced using a patient's own chondrocytes, which are then expanded and embedded in a type I collagen scaffold and incubated under low oxygen tension and variable mechanical pressure [64];
7. BioCart II™ (ProchonBioteK, Israel) is a matrix-assisted autologous chondrocytes implant made of human fibrin and hyaluronic acid that contains a patient's own cartilage cells after being expanded in a medium supplemented with a specific fibroblast growth factor [65];
8. CaRes®—Cartilage Regeneration System (Arthro-Kinetics, Germany) is a collagen type I matrix colonized with autologous cartilage cells for cartilage repair [66].

In the fourth-generation of ACI procedures, also known as MASI procedures (matrix-induced autologous stem-cell implantation), a matrix seeded with stem cells is used to treat the cartilage lesion. DeNovoET® (Zimmer, USA) and CARTYSTEM® (Medipost, Korea) are matrices seeded with allogeneic cells (juvenile allograft chondrocytes and allogeneic umbilical cord blood-derived mesenchymal stem cells, respectively) that are already available and with ongoing clinical trials [67, 68]. Although ACI procedures are linked to better clinical outcomes, especially when compared to other techniques, such as MF or MP [18, 43, 67, 69–71], these clinical improvements are not always statistically significant [11, 18, 55, 67, 69, 71–73].

Instead of cellular-based strategies, acellular scaffolds are becoming a feasible alternative for repair of cartilage defects: TruFit® (Smith & Nephew, USA), Maioregen® (FinCeramic, Italy), BST-CarGEL® (Smith & Nephew,England), CaRes-1S® (Arthro-Kinetics, Austria), CRD-Cartilage Repair Device (Kensey Nash Corp, USA), ChondroMimetic™ (TiGenix, Belgium), Vericart™ (Histogenics, USA), and Agili-C™(CartiHeal, Israel).

TruFit® is a bilayer porous poly(lactic-co-glycolic) acid–calcium sulfate biopolymer scaffold used in the treatment of OC defects. Its effectiveness, however, was not proven [74, 75] and the product is no longer available. Maioregen® is a scaffold with a porous three-dimensional tri-layer composite structure: the upper layer, consisting of Type I collagen, reproduces the cartilage surface; the intermediate layer is composed of magnesium-enriched hydroxyapatite and collagen and simulates the tide-mark; and the lower layer consists of magnesium-enriched hydroxyapatite, mimicking the subchondral bone [76]. BST-CarGEL®, composed of a mixture of chitosan and glycerol phosphate, was developed as a soluble polymer scaffold to stabilize the blood clot in the cartilage defect [77, 78]. CaRes-1S® is a collagen type I matrix cell-free/"cell-catcher" scaffold developed for extensive cartilage and osteochondral lesions [79, 80]. Cartilage Repair Device (Kensey Nash Corp.) is a biphasic scaffold intended to be implanted at the site of focal articular cartilage lesions or OC lesions. The chondral phase consists of a unique bovine collagen type I matrix. The subchondral phase consists of beta-tricalcium phosphate ($\beta$-TCP) mineral suspended within a porous bioresorbable synthetic polymer scaffold [51]. ChondroMimetic™ (TiGenix, Belgium) is a biocompatible, bilayered off-the-shelf scaffold composed of collagen, glycosaminoglycan and calcium phosphate that is used for the arthroscopic repair of small lesions [51]. Vericart™ is an acellular double-structured collagen scaffold that seeks to attract chondrocytes and stem cells to promote cartilage repair [42]. Agili-C™ is a biphasic osteochondral implant consisting of two layers: a bone phase made of calcium carbonate in the aragonite crystalline form, and a superficial cartilage phase composed of modified aragonite and hyaluronic acid [51]. The last solution when in the presence of an advanced osteochondral lesion that affects the patient in a painful and restrictive way is a partial or total prosthetic replacement [81]. The treatment for chondral and osteochondral lesions depends on (1) the type, grade, location, thickness and size of the lesion, (2) the age and activity of the patient, (3) previous treatment, (4) associated lesions, and (5) systemic diseases [82]. Some orientations and guidelines for the treatment of osteochondral lesions were presented by several authors. In Fig. 18.5, Vaquero et al. proposed an algorithm for the treatment of cartilage lesions. A consensus regarding the treatment of cartilage lesions has not yet been achieved [21], and the cost–benefit ratio of the current techniques is not entirely known [11]. The algorithm proposed by the ICRS committee for the treatment of osteochondral lesions is described in Fig. 18.6.

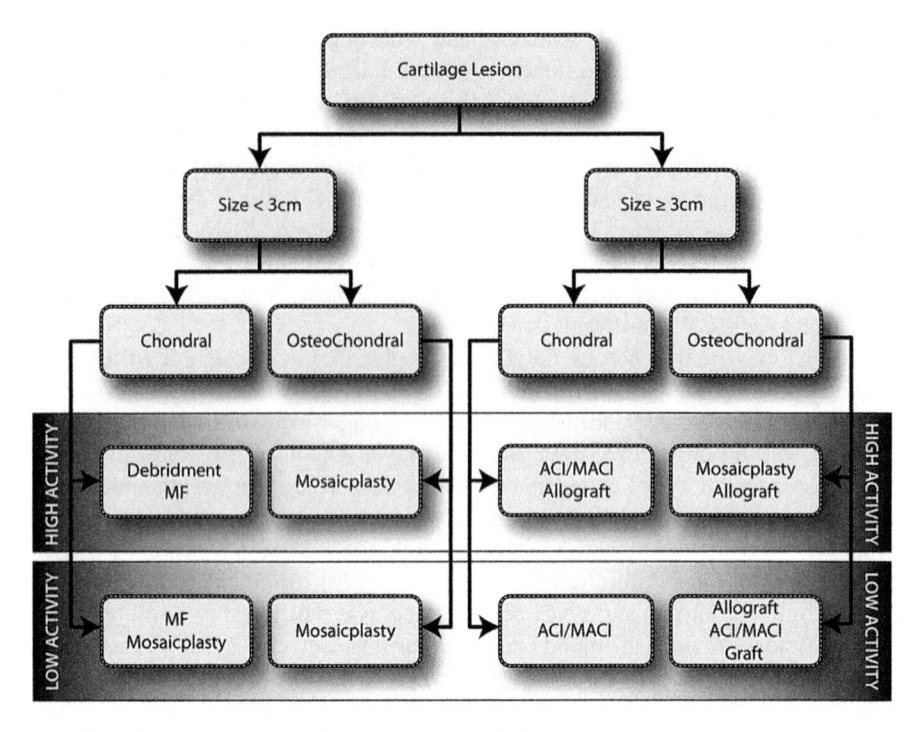

Fig. 18.5 Algorithm for chondral lesions treatment [21]

## 18.3 Preclinical and Clinical Trials

Preclinical and clinical trials are very complex and time-consuming steps prior to the introduction of a new therapeutic formulation in the market. During preclinical study, special attention must be given to the choice of the animal model. For cartilage repair, these tests should be conducted in a large animal model, such as goats, horses, or sheep, and should last enough time and have sufficient dimension to obtain the necessary evidences and allow for a robust analysis [83, 84]. Smaller animal models could be of interest in proof-of-concept studies, but its use presents important translational limitations: (1) the smaller volume of the cartilage defect, (2) the smaller thickness of the cartilage, and (3) the high degree of flexion in those small animals and consequent partial weight-bearing condition, are important drawbacks when compared to the human condition [26]. The biocompatibility and sterilization procedures of all the material during the process, including the manufacturing process, as well as the quality and the correct amount of the collected cells needed to be appropriate to the size of the lesion and to respect the limits of the population doubling/passage number. Tests regarding the biomechanical properties, tissue integrity and morphological characteristics of the repaired cartilage are also demanded. Biodistribution and toxicity studies could be necessary depending on the specific characteristics of the therapeutic product [83, 84].

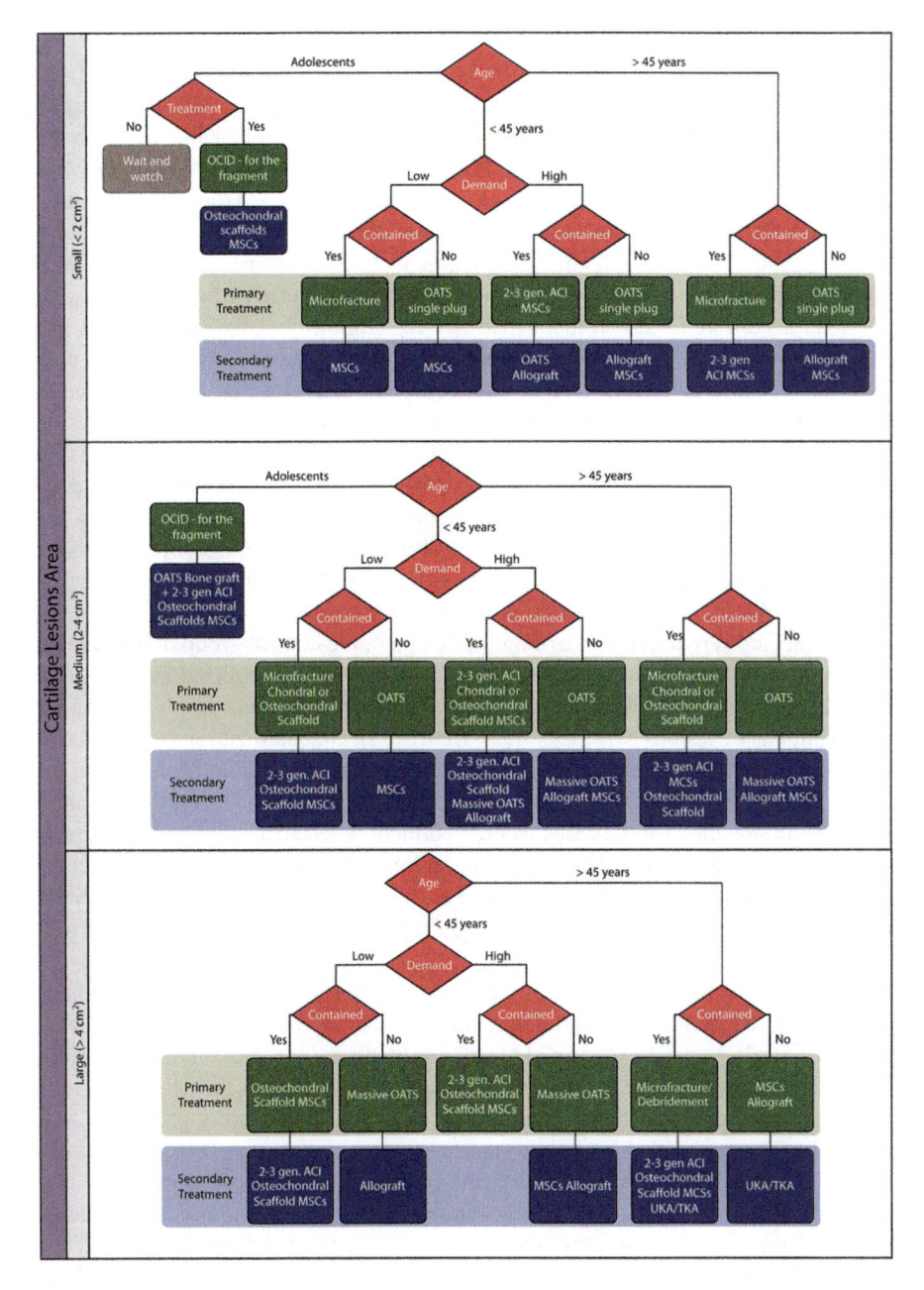

Fig. 18.6 Algorithm for chondral lesions treatment according ICRS

The clinical trial should be held in hospitals, universities, doctors' offices, or other locations under the direction of a principal investigator leading a multidisciplinary team and reviewed by an institutional review board [85, 86]. The research plan or protocol is designed to find answers to specific questions while assuring the health safety of the participants [85, 86]. The protocol should contain complete information about the study, namely, (1) data related to the reasons for promoting the study, (2) length and schedule of the clinical trial, (3) number of participants involved, and (4) drugs, techniques or treatments tested. Regarding the selection of the participants, the exclusion and inclusion criteria must be well-defined and recorded according to the protocol to assure the eligibility of the participants. The participants should be thoroughly informed and provided with answers to all possible questions in the informed consent document [85]. The trial should present a clear definition of the selected population; associated pathologies; cartilage lesion etiologies; treatments indications; type, size, localization, and grade of the lesion (using a score such as the ICRS score system); and the previous failed treatments [83]. Scoring scales, such as the Knee injury and Osteoarthritis outcome score (KOOS), the IKDC subjective scale, the Lysholm score, the ICRS objective scale, an MRI with or without histological evaluation, an X-ray, or an arthroscopic evaluation, are strongly recommended as confirmatory trials [83].

Special care is required concerning the design of the control group of the clinical trial. Standard therapy or the best standard of care with a centralized authorization should be used in this group. For example, microfracture remains the best option for the treatment of lesions with less than 4 $cm^2$ [83]. Standardization of associated therapies is strongly advised: surgical protocols, symptomatic treatments, perisurgical procedures, rehabilitation protocols and follow-up programs. A root-cause analysis is recommended for the cases where treatment failed so as to identify the cause of failure. Special attention has to be paid on long-term structural changes, such as local histological or MRI detectable changes, rates of treatment failures, as defined through relevant investigation techniques, including reoperation for revision purpose [83].

Clinical trials or interventional studies are often needed for the investigation of new approaches in the field of chondral and osteochondral repair studies. In clinical trials, the participants are submitted to specific treatments or interventions that have been previously established in the research plan. Different points have to be included in the plan, namely, changes in participants' behavior and recovery protocols [85]. Clinical trials are divided into different phases according to the stage of development of the drug or treatment. The five different phases are described in Table 18.1. In observational studies, researchers assess the result of interventions or procedures as part of the patient's standard medical routine treatment [85]. In Table 18.2, the ongoing clinical studies are presented in accordance with the Federal Drug Administration (FDA) data based on the use of the term osteochondral [86]. Open studies could be recruiting participants or not yet recruiting. Closed studies status could be active not recruiting, completed, terminated, or withdrawn [86].

**Table 18.1** Five phases of clinical trials description

| Phase | Description |
|---|---|
| Phase 0 | Exploratory study involving very limited human exposure to the drug, with no therapeutic or diagnostic goals (e.g., screening studies, microdose studies) |
| Phase 1 | Studies that are usually conducted with healthy volunteers and that emphasize safety. The goal is to find out what the drug's most frequent and serious adverse events are and, often, how the drug is metabolized and excreted. |
| Phase 2 | Studies that gather preliminary data on effectiveness (whether the drug works in people who have a certain disease or condition). For example, participants receiving the drug may be compared to similar participants receiving a different treatment, usually an inactive substance, called a placebo, or a different drug. Safety continues to be evaluated, and short-term adverse events are studied. |
| Phase 3 | Studies that gather more information about safety and effectiveness by studying different populations and different dosages and by using the drug in combination with other drugs. |
| Phase 4 | Studies occurring after FDA has approved a drug for marketing. These include postmarket requirement and commitment studies that are required of or agreed to by the study sponsor. These studies gather additional information about a drug's safety, efficacy, or optimal use. |

https://clinicaltrials.gov/ct2/help/glossary/phase

**Table 18.2** Ongoing clinical trials (Searching in https://clinicaltrials.gov using the term osteochondral)

| Identifier/status | Study name/Sponsor | Description study |
|---|---|---|
| NCT02736318<br>Not yet recruiting | OD-PHOENIX in Talus Osteochondral Lesion<br>SPONSOR: TBF Genie Tissulaire | Osteochondral allograft<br>Interventional<br>Phase 1/2 |
| NCT01209390<br>Terminated | A Prospective, Post-marketing Registry on the Use of ChondroMimetic for the Repair of Osteochondral Defects<br>SPONSOR: TiGenix n.v. | Device: Chondromimetic<br>Observational |
| NCT02345564<br>Active, not recruiting | Clinical and Radiological Results of Osteochondral Repair Using MaioRegen in Knee and Ankle Surgery<br>SPONSOR: Barmherzige Brüder Eisenstadt | MaioRegen<br>Interventional |
| NCT01282034<br>Completed | Study for the Treatment of Knee Chondral and Osteochondral Lesions<br>SPONSOR: Fin-Ceramica Faenza Spa | MaioRegen Surgery<br>Interventional<br>Phase 4 |
| NCT01409447<br>Unknown | Repair of Articular Osteochondral Defect<br>SPONSOR: National Taiwan University Hospital | Biphasic osteochondral Composite<br>Interventional |
| NCT02430558<br>Not Yet recruting | Second Line Treatment of Knee Osteochondral Lesion With Treated Osteochondral Graft<br>SPONSOR: TBF Genie Tissulaire | OD-PHOENIX<br>Interventional<br>Phase 1/2 |

(continued)

**Table 18.2** (continued)

| Identifier/status | Study name/Sponsor | Description study |
|---|---|---|
| NCT01410136 Terminated | Chondrofix Osteochondral Allograft Prospective Study SPONSOR: Zimmer Orthobiologics, Inc. | Chondrofix Osteochondral Allograft Interventional |
| NCT02308358 Withdrawn | Long Term Outcomes of Osteochondral Allografts for Osteochondral Defects of the Knee SPONSOR: University of Missouri-Columbia | Outcomes of Allograft Observational |
| NCT02011295 Recruting | Bone Marrow Aspirate Concentrate (BMAC) Supplementation for Osteochondral Lesions SPONSOR: Duke University | Microfracture + Bone Marrow Interventional Phase 4 |
| NCT02423629 Recruting | Agili-C™ Implant Performance Evaluation in the Repair of Cartilage and Osteochondral Defects SPONSOR: Cartiheal (2009) Ltd | AgiliC™ implantation Interventional Phase 4 |
| NCT02503228 Recruting | Clinical Assessment of the Missouri Osteochondral Allograft Preservation System—MOPS SPONSOR: James Cook, University of Missouri-Columbia | Receiving MOPS-Preserved Cartilage Observational |
| NCT01554878 Completed | Observational Study on the Treatment of Knee Osteochondral Lesions of Grade III–IV SPONSOR: Ettore Sansavini Health Science Foundation | Knee Surgery Observational |
| NCT03036878 Recruting | ReNu™ Marrow Stimulation Augmentation SPONSOR: NuTech Medical, Inc | ReNu Interventional |
| NCT01290991 Completed | A Study to Evaluate the Safety of Augment™ Bone Graft SPONSOR: William Stanish, Capital District Health Authority, Canada | Augment Bone Graft |
| NCT02005861 Recruting | "One-step" Bone Marrow Mononuclear Cell Transplantation in Talar Osteochondral Lesions SPONSOR: Istituto Ortopedico Rizzoli | Bone marrow cells + Collagen scaffold Interventional |
| NCT02338375 Unknown | Safety and Efficacy of Allogenic Umbilical Cord Blood-derived Mesenchymal Stem Cell Product SPONSOR: | Cartistem Interventional Phase 1 |
| NCT02309957 Recruting | EAGLE European Post Market Study SPONSOR: Kensey Nash Corporation | BioMatrix CD Interventional |
| NCT01477008 Recruting | BiPhasic Cartilage Repair Implant (BiCRI) IDE Clinical Trial - Taiwan SPONSOR: Exactech Taiwan, Ltd | BiPhasic Cartilage Repair Implant Interventional Phase 3 |

(continued)

**Table 18.2** (continued)

| Identifier/status | Study name/Sponsor | Description study |
|---|---|---|
| NCT00560664 Completed | Comparison of Autologous Chondrocyte Implantation vs. Mosaicoplasty: a Randomized Trial SPONSOR: University Hospital, Brest | Mosaicoplasty Interventional Phase 3 |
| NCT00984594 Terminated has results | Evaluation of a Composite Cancellous and Demineralized Bone Plug (CR-Plug) for Repair of Knee Osteochondral Defects SPONSOR: RTI Surgical | Composite of Cancellous and Demineralized Bone Plug (CR-Plug) Interventional Phase 3 |
| NCT00891501 Unknown | The Use of Autologous Bone Marrow Mesenchymal Stem Cells in the Treatment of Articular Cartilage Defects SPONSOR: Cairo University | Bone marrow mesenchymal stem cell implantation Interventional Phase 2/3 |
| NCT00821873 Completed has results | Evaluation of the CR Plug for Repair of Defects Created at the Harvest Site From an Autograft in the Knee. SPONSOR: RTI Surgical | CR Plug Interventional Phase 3 |
| NCT01183637 Terminated | Evaluation of an Acellular Osteochondral Graft for Cartilage LEsions Pilot Trial SPONSOR: Kensey Nash Corporation | Kensey Nash Corp. Cartilage Repair Device Interventional Phase 2 |
| NCT01159899 Unknown | Transplantation of Bone Marrow Stem Cells Stimulated by Proteins Scaffold to Heal Defects Articular Cartilage of the Knee SPONSOR: Michel Assor, MD, University of Marseille | Transplantation of Bone Marrow Stem Cells Activated in Knee Arthrosis Interventional Phase1 |
| NCT01471236 Active, not recruiting | Evaluation of the Agili-C Biphasic Implant in the Knee Joint SPONSOR: BioPoly LLC | BioPoly RS Partial Resurfacing Knee Implant Interventional Phase 4 |
| NCT00945399 Terminated | Comparison of Microfracture Treatment and CARTIPATCH® Chondrocyte Graft Treatment in Femoral Condyle Lesions SPONSOR: TBF Genie Tissulaire | CARTIPATCH® procedure Interventional Phase 3 |
| NCT01473199 Completed | BioPoly RS Knee Registry Study for Cartilage Defect Replacement SPONSOR: BioPoly LLC | BioPoly RS Partial Resurfacing Knee Implant Interventional |
| NCT01799876 Active, not recruiting | Use of Cell Therapy to Enhance Arthroscopic Knee Cartilage Surgery SPONSOR: Fondren Orthopedic Group L.L.P. | Autologous Cell/microfracture Interventional |
| NCT01747681 Completed | Results at 10 to 14 Years After Microfracture in the Knee SPONSOR: Bergen Orthopedic Study Group | Microfracture Observational |

(continued)

**Table 18.2** (continued)

| Identifier/status | Study name/Sponsor | Description study |
|---|---|---|
| NCT01347892 Active, not recruiting | DeNovo NT Ankle LDC Study SPONSOR: Zimmer Orthobiologics, Inc. | DeNovo NT Natural Tissue Graft Interventional |
| NCT01920373 Withdrawn | Platelet-Rich Plasma vs. Corticosteroid Injection as Treatment for Degenerative Pathology of the Temporomandibular Joint SPONSOR: Kaiser Permanente | Corticosteroid injection/ platelet-rich plasma injection Interventional Phase 1 |
| NCT02991300 Recruiting | BioPoly® RS Partial Resurfacing Patella Registry Study SPONSOR: BioPoly LLC | BioPoly RS Partial Resurfacing Patella Implant Interventional |
| NCT00793104 Terminated has results | Evaluation of the CR Plug (Allograft) for the Treatment of a Cartilage Injury in the Knee SPONSOR: RTI Surgical | CR Plug Interventional Phase 3 |

## 18.4 Conclusions

Despite all the therapeutic approaches for the treatment of chondral and osteochondral lesions, a consensus regarding a definitive treatment has not yet been achieved. Allografts, autografts, or substitution arthroplasties have already proven their value in cartilage repair. Emerging tissue engineering and regenerative approaches have shown promising results and could advance new solutions for osteochondral lesions treatment, but further studies and research need to be conducted. Although tissue engineering has already shown tremendous progress, a long and difficult road in the regulatory and legal path has to be travelled in order to transform new therapeutic approaches into a clinical reality.

**Acknowledgments** A. da Silva Morais acknowledges ERC-2012-ADG 20120216–321266 (ComplexiTE) for his Postdoc scholarship. Thanks to the project FROnTHERA (NORTE-01-0145-FEDER-000023), supported by Norte Portugal Regional Operational Programme (NORTE 2020), under the PORTUGAL 2020 Partnership Agreement, through the European Regional Development Fund (ERDF). The financial support from the Portuguese Foundation for Science and Technology for the M-ERA-NET/0001/2014 "HierarchiTech" project and for the funds provided under the program Investigador FCT 2012, 2014, and 2015 (IF/00423/2012, IF/01214/2014, and IF/01285/2015) is also greatly acknowledged.

# References

1. Ondrésik M, Oliveira JM, Reis RL (2017) Knee articular cartilage. In: Oliveira JM, Reis RL (eds) Regenerative strategies for the treatment of knee joint disabilities. Springer International, Cham, pp 3–20. https://doi.org/10.1007/978-3-319-44785-8_1
2. Pearle AD, Warren RF, Rodeo SA Basic science of articular cartilage and osteoarthritis. Clin Sports Med 24(1):1–12. https://doi.org/10.1016/j.csm.2004.08.007

3. Bolog NV, Andreisek G, Ulbrich EJ (2015) Articular cartilage and subchondral bone. In: MRI of the knee: a guide to evaluation and reporting. Springer International, Cham, pp 95–112. https://doi.org/10.1007/978-3-319-08165-6_6
4. Flik KR, Verma N, Cole BJ, Bach BR (2007) Articular cartilage. In: Williams RJ (ed) Cartilage repair strategies. Humana Press, Totowa, NJ, pp 1–12. https://doi.org/10.1007/978-1-59745-343-1_1
5. Goldring SR (2012) Alterations in periarticular bone and cross talk between subchondral bone and articular cartilage in osteoarthritis. Ther Adv Musculoskelet Dis 4(4):249–258. https://doi.org/10.1177/1759720X12437353
6. Imhof H, Sulzbacher I, Grampp S, Czerny C, Youssefzadeh S, Kainberger F (2000) Subchondral bone and cartilage disease: a rediscovered functional unit. Invest Radiol 35(10):581–588
7. Widuchowski W, Widuchowski J, Trzaska T (2007) Articular cartilage defects: study of 25,124 knee arthroscopies. Knee 14(3):177–182. https://doi.org/10.1016/j.knee.2007.02.001
8. Hjelle K, Solheim E, Strand T, Muri R, Brittberg M (2002) Articular cartilage defects in 1,000 knee arthroscopies. Arthroscopy 18(7):730–734
9. Aroen A, Loken S, Heir S, Alvik E, Ekeland A, Granlund OG, Engebretsen L (2004) Articular cartilage lesions in 993 consecutive knee arthroscopies. Am J Sports Med 32(1):211–215
10. Farr J, Cole B, Dhawan A, Kercher J, Sherman S (2011) Clinical cartilage restoration: evolution and overview. Clin Orthop Relat Res 469(10):2696–2705. https://doi.org/10.1007/s11999-010-1764-z
11. Vaquero J, Forriol F (2012) Knee chondral injuries: clinical treatment strategies and experimental models. Injury 43(6):694–705. https://doi.org/10.1016/j.injury.2011.06.033
12. Foldager CB (2013) Advances in autologous chondrocyte implantation and related techniques for cartilage repair. Dan Med J 60(4):B4600
13. Hunziker EB (1999) Articular cartilage repair: are the intrinsic biological constraints undermining this process insuperable? Osteoarthritis Cartilage 7(1):15–28. https://doi.org/10.1053/joca.1998.0159
14. Hunziker EB (2002) Articular cartilage repair: basic science and clinical progress. A review of the current status and prospects. Osteoarthritis Cartilage 10(6):432–463. https://doi.org/10.1053/joca.2002.0801
15. Johnson VL, Hunter DJ (2014) The epidemiology of osteoarthritis. Best Pract Res Clin Rheumatol 28(1):5–15. https://doi.org/10.1016/j.berh.2014.01.004
16. Jackson DW, Simon TM, Aberman HM (2001) Symptomatic articular cartilage degeneration: the impact in the new millennium. Clin Orthop Relat Res (391 Suppl):S14–25
17. Nukavarapu SP, Dorcemus DL (2013) Osteochondral tissue engineering: current strategies and challenges. Biotechnol Adv 31(5):706–721. https://doi.org/10.1016/j.biotechadv.2012.11.004
18. Batty L, Dance S, Bajaj S, Cole BJ (2011) Autologous chondrocyte implantation: an overview of technique and outcomes. ANZ J Surg 81(1–2):18–25. https://doi.org/10.1111/j.1445-2197.2010.05495.x
19. Durur-Subasi I, Durur-Karakaya A, Yildirim OS (2015) Osteochondral lesions of major joints. Eur J Med 47(2):138–144. https://doi.org/10.5152/eurasianjmed.2015.50
20. Lewandrowski KU, Muller J, Schollmeier G (1997) Concomitant meniscal and articular cartilage lesions in the femorotibial joint. Am J Sports Med 25(4):486–494
21. Vilela CA, Correia C, Oliveira JM, Sousa RA, Reis RL, Espregueira-Mendes J (2017) Clinical management of articular cartilage lesions. In: Oliveira JM, Reis RL (eds) Regenerative strategies for the treatment of knee joint disabilities. Springer International, Cham, pp 29–53. https://doi.org/10.1007/978-3-319-44785-8_3
22. Resnick D, Vint V (1980) The "Tunnel" view in assessment of cartilage loss in osteoarthritis of the knee. Radiology 137(2):547–548. https://doi.org/10.1148/radiology.137.2.7433690
23. Rosenberg TD, Paulos LE, Parker RD, Coward DB, Scott SM (1988) The forty-five-degree posteroanterior flexion weight-bearing radiograph of the knee. J Bone Joint Surg Am 70(10):1479–1483
24. Casula V, Hirvasniemi J, Lehenkari P, Ojala R, Haapea M, Saarakkala S, Lammentausta E, Nieminen MT (2016) Association between quantitative MRI and ICRS arthroscopic grading

of articular cartilage. Knee Surg Sports Traumatol Arthrosc 24(6):2046–2054. https://doi.org/10.1007/s00167-014-3286-9
25. Chan DD, Neu CP (2013) Probing articular cartilage damage and disease by quantitative magnetic resonance imaging. J R Soc Interface 10(78):20120608. https://doi.org/10.1098/rsif.2012.0608
26. Vilela CA, Correia C, Oliveira JM, Sousa RA, Espregueira-Mendes J, Reis RL (2015) Cartilage repair using hydrogels: a critical review of in vivo experimental designs. ACS Biomater Sci Eng 1(9):726–739. https://doi.org/10.1021/acsbiomaterials.5b00245
27. Outerbridge RE (1961) The etiology of chondromalacia patellae. J Bone Joint Surg Br 43-B:752–757
28. Casscells SW (1990) Outerbridge's ridges. Arthroscopy 6(4):253
29. Brittberg M, Winalski CS (2003) Evaluation of cartilage injuries and repair. J Bone Joint Surg Am 85-A(Suppl 2):58–69
30. Chen CH, Liu YS, Chou PH, Hsieh CC, Wang CK (2013) MR grading system of osteochondritis dissecans lesions: comparison with arthroscopy. Eur J Radiol 82(3):518–525. https://doi.org/10.1016/j.ejrad.2012.09.026
31. Ellermann JM, Donald B, Rohr S, Takahashi T, Tompkins M, Nelson B, Crawford A, Rud C, Macalena J (2016) Magnetic resonance imaging of osteochondritis dissecans: validation study for the ICRS Classification System. Acad Radiol 23(6):724–729. https://doi.org/10.1016/j.acra.2016.01.015
32. Lazic S, Boughton O, Hing C, Bernard J (2014) Arthroscopic washout of the knee: a procedure in decline. Knee 21(2):631–634. https://doi.org/10.1016/j.knee.2014.02.014
33. Katz JN, Brownlee SA, Jones MH (2014) The role of arthroscopy in the management of knee osteoarthritis. Best Pract Res Clin Rheumatol 28(1):143–156. https://doi.org/10.1016/j.berh.2014.01.008
34. Marcacci M, Filardo G, Kon E (2013) Treatment of cartilage lesions: what works and why? Injury 44(Suppl 1):S11–S15. https://doi.org/10.1016/S0020-1383(13)70004-4
35. Bentley G, Bhamra JS, Gikas PD, Skinner JA, Carrington R, Briggs TW (2013) Repair of osteochondral defects in joints—how to achieve success. Injury 44(Suppl 1):S3–10. https://doi.org/10.1016/S0020-1383(13)70003-2
36. Steadman JR, Briggs KK, Rodrigo JJ, Kocher MS, Gill TJ, Rodkey WG (2003) Outcomes of microfracture for traumatic chondral defects of the knee: average 11-year follow-up. Arthroscopy 19(5):477–484. https://doi.org/10.1053/jars.2003.50112
37. Versier G, Dubrana F, French Arthroscopy S (2011) Treatment of knee cartilage defect in 2010. Orthop Traumatol Surg Res 97(8 Suppl):S140–S153. https://doi.org/10.1016/j.otsr.2011.09.007
38. Lee GW, Son JH, Kim JD, Jung GH (2013) Is platelet-rich plasma able to enhance the results of arthroscopic microfracture in early osteoarthritis and cartilage lesion over 40 years of age? Eur J Orthopaed Surg Traumatol 23(5):581–587. https://doi.org/10.1007/s00590-012-1038-4
39. Tuan RS, Chen AF, Klatt BA (2013) Cartilage regeneration. J Am Acad Orthop Surg 21(5):303–311. https://doi.org/10.5435/JAAOS-21-05-303
40. Chung JY, Lee DH, Kim TH, Kwack KS, Yoon KH, Min BH (2014) Cartilage extra-cellular matrix biomembrane for the enhancement of microfractured defects. Knee Surg Sports Traumatol Arthrosc 22(6):1249–1259. https://doi.org/10.1007/s00167-013-2716-4
41. Usuelli FG, de Girolamo L, Grassi M, D'Ambrosi R, Montrasio UA, Boga M All-arthroscopic autologous matrix-induced chondrogenesis for the treatment of osteochondral lesions of the talus. Arthrosc Tech 4(3):e255–e259. https://doi.org/10.1016/j.eats.2015.02.010
42. Cascio BM, Sharma B The future of cartilage repair. Oper Tech Sports Med 16(4):221–224. https://doi.org/10.1053/j.otsm.2009.01.001
43. Gomoll AH, Filardo G, de Girolamo L, Espregueira-Mendes J, Marcacci M, Rodkey WG, Steadman JR, Zaffagnini S, Kon E (2012) Surgical treatment for early osteoarthritis. Part I: cartilage repair procedures. Knee Surg Sports Traumatol Arthrosc 20(3):450–466. https://doi.org/10.1007/s00167-011-1780-x
44. Espregueira-Mendes J, Pereira H, Sevivas N, Varanda P, da Silva MV, Monteiro A, Oliveira JM, Reis RL (2012) Osteochondral transplantation using autografts from the upper tibio-fibular

joint for the treatment of knee cartilage lesions. Knee Surg Sports Traumatol Arthrosc 20(6): 1136–1142. https://doi.org/10.1007/s00167-012-1910-0
45. Robert H (2011) Chondral repair of the knee joint using mosaicplasty. Orthop Traumatol Surg Res 97(4):418–429. https://doi.org/10.1016/j.otsr.2011.04.001
46. Farr J, Cole BJ, Sherman S, Karas V (2012) Particulated articular cartilage: CAIS and DeNovo NT. J Knee Surg 25(1):23–29
47. Stein S, Strauss E, Bosco J 3rd (2013) Advances in the surgical management of articular cartilage defects: autologous chondrocyte implantation techniques in the pipeline. Cartilage 4(1):12–19. https://doi.org/10.1177/1947603512463226
48. Gomoll AH, Filardo G, Almqvist FK, Bugbee WD, Jelic M, Monllau JC, Puddu G, Rodkey WG, Verdonk P, Verdonk R, Zaffagnini S, Marcacci M (2012) Surgical treatment for early osteoarthritis. Part II: allografts and concurrent procedures. Knee Surg Sports Traumatol Arthrosc 20(3):468–486. https://doi.org/10.1007/s00167-011-1714-7
49. Haene R, Qamirani E, Story RA, Pinsker E, Daniels TR (2012) Intermediate outcomes of fresh talar osteochondral allografts for treatment of large osteochondral lesions of the talus. J Bone Joint Surg Am 94(12):1105–1110. https://doi.org/10.2106/JBJS.J.02010
50. Giorgini A, Donati D, Cevolani L, Frisoni T, Zambianchi F, Catani F (2013) Fresh osteochondral allograft is a suitable alternative for wide cartilage defect in the knee. Injury 44(Suppl 1):S16–S20. https://doi.org/10.1016/S0020-1383(13)70005-6
51. Pina S, Ribeiro V, Oliveira JM, Reis RL (2017) Pre-clinical and clinical management of osteochondral lesions. In: Oliveira JM, Reis RL (eds) Regenerative strategies for the treatment of knee joint disabilities. Springer International, Cham, pp 147–161. https://doi.org/10.1007/978-3-319-44785-8_8
52. Barrett I, King AH, Riester S, van Wijnen A, Levy BA, Stuart MJ, Krych AJ (2016) Internal fixation of unstable osteochondritis dissecans in the skeletally mature knee with metal screws. Cartilage 7(2):157–162. https://doi.org/10.1177/1947603515622662
53. Grimm NL, Ewing CK, Ganley TJ (2014) The knee internal fixation techniques for osteochondritis dissecans. Clin Sports Med 33(2):313–319. https://doi.org/10.1016/j.csm.2013.12.001
54. Brittberg M, Lindahl A, Nilsson A, Ohlsson C, Isaksson O, Peterson L (1994) Treatment of deep cartilage defects in the knee with autologous chondrocyte transplantation. N Engl J Med 331(14):889–895. https://doi.org/10.1056/NEJM199410063311401
55. Dhollander AA, Verdonk PC, Lambrecht S, Verdonk R, Elewaut D, Verbruggen G, Almqvist KF (2012) Short-term outcome of the second generation characterized chondrocyte implantation for the treatment of cartilage lesions in the knee. Knee Surg Sports Traumatol Arthrosc 20(6):1118–1127. https://doi.org/10.1007/s00167-011-1759-7
56. Zaslav K, Cole B, Brewster R, DeBerardino T, Farr J, Fowler P, Nissen C, Investigators SSP (2009) A prospective study of autologous chondrocyte implantation in patients with failed prior treatment for articular cartilage defect of the knee: results of the Study of the Treatment of Articular Repair (STAR) clinical trial. Am J Sports Med 37(1):42–55. https://doi.org/10.1177/0363546508322897
57. D'Ambrosi R, Maccario C, Ursino C, Serra N, Usuelli FG (2017) Combining microfractures, autologous bone graft, and autologous matrix-induced chondrogenesis for the treatment of juvenile osteochondral talar lesions. Foot Ankle Int 38(5):485–495. https://doi.org/10.1177/1071100716687367
58. Jacobi M, Villa V, Magnussen RA, Neyret P (2011) MACI—a new era? Sports Med Arthrosc Rehabil Ther Technol 3(1):10. https://doi.org/10.1186/1758-2555-3-10
59. Kreuz PC, Müller S, Ossendorf C, Kaps C, Erggelet C (2009) Treatment of focal degenerative cartilage defects with polymer-based autologous chondrocyte grafts: four-year clinical results. Arthritis Res Ther 11(2):R33. https://doi.org/10.1186/ar2638
60. Selmi TA, Verdonk P, Chambat P, Dubrana F, Potel JF, Barnouin L, Neyret P (2008) Autologous chondrocyte implantation in a novel alginate-agarose hydrogel: outcome at two years. J Bone Joint Surg Br 90(5):597–604. https://doi.org/10.1302/0301-620X.90B5.20360

61. Clavé A, Potel J-F, Servien E, Neyret P, Dubrana F, Stindel E (2016) Third-generation autologous chondrocyte implantation versus mosaicplasty for knee cartilage injury: 2-year randomized trial. J Orthop Res 4(4):658–665. https://doi.org/10.1002/jor.23152
62. Brix MO, Stelzeneder D, Chiari C, Koller U, Nehrer S, Dorotka R, Windhager R, Domayer SE (2014) Treatment of full-thickness chondral defects with Hyalograft C in the knee. Am J Sports Med 42(6):1426–1432. https://doi.org/10.1177/0363546514526695
63. Zak L, Albrecht C, Wondrasch B, Widhalm H, Vekszler G, Trattnig S, Marlovits S, Aldrian S (2014) Results 2 years after matrix-associated autologous chondrocyte transplantation using the Novocart 3D scaffold. Am J Sports Med 42(7):1618–1627. https://doi.org/10.1177/0363546514532337
64. Crawford DC, DeBerardino TM, Williams RJ 3rd (2012) NeoCart, an autologous cartilage tissue implant, compared with microfracture for treatment of distal femoral cartilage lesions: an FDA phase-II prospective, randomized clinical trial after two years. J Bone Joint Surg Am 94(11):979–989. https://doi.org/10.2106/JBJS.K.00533
65. Yayon A, Neria E, Blumenstein S, Stern B, Barkai H, Zak R, Yaniv Y (2006) BIOCART™II a novel implant for 3D reconstruction of articular cartilage. J Bone Jt Surg Br 88-B(SUPP II):344
66. Zeifang F, Oberle D, Nierhoff C, Richter W, Moradi B, Schmitt H (2010) Autologous chondrocyte implantation using the original periosteum-cover technique versus matrix-associated autologous chondrocyte implantation: a randomized clinical trial. Am J Sports Med 38(5):924–933. https://doi.org/10.1177/0363546509351499
67. Dewan AK, Gibson MA, Elisseeff JH, Trice ME (2014) Evolution of autologous chondrocyte repair and comparison to other cartilage repair techniques. Biomed Res Int 2014:272481. https://doi.org/10.1155/2014/272481
68. McCormick F, Cole BJ, Nwachukwu B, Harris JD, Adkisson HDIV, Farr J Treatment of focal cartilage defects with a juvenile allogeneic 3-dimensional articular cartilage graft. Oper Tech Sports Med 21(2):95–99. https://doi.org/10.1053/j.otsm.2013.03.007
69. Rodriguez-Merchan EC (2013) Regeneration of articular cartilage of the knee. Rheumatol Int 33(4):837–845. https://doi.org/10.1007/s00296-012-2601-3
70. Schüttler KF, Schenker H, Theisen C, Schofer MD, Getgood A, Roessler PP, Struewer J, Rominger MB, Efe T (2014) Use of cell-free collagen type I matrix implants for the treatment of small cartilage defects in the knee: clinical and magnetic resonance imaging evaluation. Knee Surg Sports Traumatol Arthrosc 22(6):1270–1276. https://doi.org/10.1007/s00167-013-2747-x
71. Perera JR, Gikas PD, Bentley G (2012) The present state of treatments for articular cartilage defects in the knee. Ann R Coll Surg Engl 94(6):381–387. https://doi.org/10.1308/0035884 12X13171221592573
72. Negrin LL, Vecsei V (2013) Do meta-analyses reveal time-dependent differences between the clinical outcomes achieved by microfracture and autologous chondrocyte implantation in the treatment of cartilage defects of the knee? J Orthop Sci 18(6):940–948. https://doi.org/10.1007/s00776-013-0449-3
73. Petri M, Broese M, Simon A, Liodakis E, Ettinger M, Guenther D, Zeichen J, Krettek C, Jagodzinski M, Haasper C (2013) CaReS (MACT) versus microfracture in treating symptomatic patellofemoral cartilage defects: a retrospective matched-pair analysis. J Orthop Sci 18(1):38–44. https://doi.org/10.1007/s00776-012-0305-x
74. Gelber PE, Batista J, Millan-Billi A, Patthauer L, Vera S, Gomez-Masdeu M, Monllau JC (2014) Magnetic resonance evaluation of TruFit(R) plugs for the treatment of osteochondral lesions of the knee shows the poor characteristics of the repair tissue. Knee 21(4):827–832. https://doi.org/10.1016/j.knee.2014.04.013
75. Hindle P, Hendry JL, Keating JF, Biant LC (2014) Autologous osteochondral mosaicplasty or TruFit plugs for cartilage repair. Knee Surg Sports Traumatol Arthrosc 22(6):1235–1240. https://doi.org/10.1007/s00167-013-2493-0
76. Delcogliano M, de Caro F, Scaravella E, Ziveri G, De Biase CF, Marotta D, Marenghi P, Delcogliano A (2014) Use of innovative biomimetic scaffold in the treatment for large osteo-

chondral lesions of the knee. Knee Surg Sports Traumatol Arthrosc 22(6):1260–1269. https://doi.org/10.1007/s00167-013-2717-3
77. Stanish WD, McCormack R, Forriol F, Mohtadi N, Pelet S, Desnoyers J, Restrepo A, Shive MS (2013) Novel scaffold-based BST-CarGel treatment results in superior cartilage repair compared with microfracture in a randomized controlled trial. J Bone Joint Surg Am 95(18):1640–1650. https://doi.org/10.2106/JBJS.L.01345
78. Shive MS, Hoemann CD, Restrepo A, Hurtig MB, Duval N, Ranger P, Stanish W, Buschmann MD (2006) BST-CarGel: in situ chondroinduction for cartilage repair. Oper Tech Orthopaed 16(4):271–278. https://doi.org/10.1053/j.oto.2006.08.001
79. Schüettler KF, Struewer J, Rominger MB, Rexin P, Efe T (2013) Repair of a chondral defect using a cell free scaffold in a young patient—a case report of successful scaffold transformation and colonisation. BMC Surg 13(1):11. https://doi.org/10.1186/1471-2482-13-11
80. Roessler PP, Pfister B, Gesslein M, Figiel J, Heyse TJ, Colcuc C, Lorbach O, Efe T, Schüttler KF (2015) Short-term follow up after implantation of a cell-free collagen type I matrix for the treatment of large cartilage defects of the knee. Int Orthop 39(12):2473–2479. https://doi.org/10.1007/s00264-015-2695-9
81. Panseri S, Russo A, Cunha C, Bondi A, Di Martino A, Patella S, Kon E (2012) Osteochondral tissue engineering approaches for articular cartilage and subchondral bone regeneration. Knee Surg Sports Traumatol Arthrosc 20(6):1182–1191. https://doi.org/10.1007/s00167-011-1655-1
82. Mall NA, Harris JD, Cole BJ (2015) Clinical evaluation and preoperative planning of articular cartilage lesions of the knee. J Am Acad Orthop Surg 23(10):633–640. https://doi.org/10.5435/JAAOS-D-14-00241
83. Agency E-EM (2009) Reflection paper on in-vitro cultured chondrocyte containing products for cartilage repair of the knee—Doc. Ref. EMEA/CAT/CPWP/288934/2009. http://www.ema.europa.eu/docs/en_GB/document_library/Scientific_guideline/2009/10/WC500004223.pdf. Acessed Mar 2017
84. Administration F-FD (2011) Guidance for Industry—preparation of IDEs and INDs dor products intended to repair or replace knee cartilage. http://www.fda.gov/BiologicsBloodVaccines/GuidanceComplianceRegulatoryInformation/Guidances/default.htm. Accessed Mar 2017
85. Administration F-FD (2017) Learn about clinical studies. https://clinicaltrials.gov/ct2/about-studies/learn#WhatIs. Accessed Mar 2017
86. Administratioon F-FD (2017) Clinical trias. https://clinicaltrials.gov/ct2/home. Accessed Mar 2017

# Chapter 19
# Commercial Products for Osteochondral Tissue Repair and Regeneration

Diana Bicho, Sandra Pina, Rui L. Reis, and J. Miguel Oliveira

**Abstract** The osteochondral tissue represents a complex structure composed of four interconnected structures, namely hyaline cartilage, a thin layer of calcified cartilage, subchondral bone, and cancellous bone. Due to the several difficulties associated with its repair and regeneration, researchers have developed several studies aiming to restore the native tissue, some of which had led to tissue-engineered commercial products. In this sense, this chapter discusses the good manufacturing practices, regulatory medical conditions and challenges on clinical translations that should be fulfilled regarding the safety and efficacy of the new commercialized products. Furthermore, we review the current osteochondral products that are currently being marketed and applied in the clinical setting, emphasizing the advantages and difficulties of each one.

**Keywords** Commercial products · Bone, cartilage, and osteochondral regeneration

---

D. Bicho (✉) · S. Pina
3B's Research Group—Biomaterials, Biodegradables and Biomimetics, European Institute of Excellence on Tissue Engineering and Regenerative Medicine, University of Minho, Barco GMR, Portugal

ICVS/3B's—PT Government Associate Laboratory, Braga/Guimarães, Portugal
e-mail: dianabicho@dep.uminho.pt

R. L. Reis · J. M. Oliveira
3B's Research Group—Biomaterials, Biodegradables and Biomimetics, European Institute of Excellence on Tissue Engineering and Regenerative Medicine, University of Minho, Barco GMR, Portugal

ICVS/3B's—PT Government Associate Laboratory, Braga/Guimarães, Portugal

The Discoveries Centre for Regenerative and Precision Medicine, University of Minho, Barco, Guimarães, Portugal

© Springer International Publishing AG, part of Springer Nature 2018
J. M. Oliveira et al. (eds.), *Osteochondral Tissue Engineering*,
Advances in Experimental Medicine and Biology 1058,
https://doi.org/10.1007/978-3-319-76711-6_19

## 19.1 Introduction

Treatment of bone and cartilage defects represents a current problem that needs to be solved. Although the present therapies are well established and effective for reducing pain, thus improving the patients' quality of life, the hyaline or articular cartilage has a limited regeneration capacity, demanding new therapeutic options for complete healing of the osteochondral (OC) lesions. In this sense, tissue engineered biomaterials for OC application present some key challenges regarding their biocompatibility, bioactivity, biodegradation, and biomechanical properties. Additionally, the ion release of metallic materials and the reproducibility of the techniques are also fundamental aspects to address [1]. Furthermore, the strategies applied need to present biodegradable with nontoxic degradation products of easy metabolization and excretion, well-regulated degradation kinetics and similar rate to native tissue. Similarly, the expected local or systemic immune responses should be controlled since it will affect significantly the implant–host integration. Most importantly, OC biomaterials should be able to mimic the extracellular matrix (ECM) and the complex mechanisms involved in the surrounding cells where the biomaterial will be applied [2]. The biomaterial chosen should be able to aid the cells to grow and proliferate at a similar natural rate, with an efficient gas and nutrients exchange [3]. Thus, the choice of the biomaterial is extremely important and needs to consider not only its chemical composition, but also its physical properties [4]. Specifically, in OC tissue engineering the mechanical properties must be able to bear the daily stress to which this tissue is subjected, as well as to support integration of the cells involved. Equally, the microstructure of the scaffold is essential for chondrogenesis and osteogenesis. Normally, it is believed that a superior cell ingrowth, improved transport of nutrient and vascular formation is related with high porosity (> 300 μm) [5] and interconnectivity (>100 μm) of scaffolds to allow a proper cell colonization [6]. In regards to osteoinductive potential, there has been extensive research with some good synthetic materials having emerged, for example hydroxyapatite (HA), octacalcium phosphate (OCP), and β-tricalcium phosphate (β-TCP). However, regardless of their ability to be integrated into host tissues, they present poor mechanical properties, being therefore necessary to mix them with different materials that could overcome such limitations, and improve integration in OC lesions [7]. In addition, different natural and synthetic materials have been employed to engineer OC repair, presenting advantages and limitations. For instance, natural polymers are normally biocompatible and allow the interaction with cell receptors. However, safety concerns are usually an issue. In contrast, synthetic materials are more easily controllable and reproducible but lack the cell-recognition signals [8]. Researchers have suggested these materials to be especially used in the regeneration of large OC defects and sometimes combined with cells, growth factors, and tissue grafts [9].

Beside these scientific advances, when trying to launch new medical products and technologies to the market, several regulatory medical conditions should be fulfilled regarding their safety and efficacy. Moreover, regulatory hurdles associated with the commercialization of new products are critical. Therefore, laboratory facilities, manufacturing practices and documentation related to products development should follow strict requirements to ensure both the welfare of the individuals involved in the process and the reproducibility of the procedures. It is important to stress that all the directives involved in the manufacture and propagation of medical products and devices, can cause disagreements depending on the governmental administration involved. Among the multitude of options, Federal Food and Drug Administration (FDA) in the USA, and regional or centralized regulatory bodies like the European Medicines Agency (EMA) in the European Union (EU) are the most used. They are responsible for the development, assessment, and supervision of medicines [10]. Beyond this fact, it is also important to point out that the translations of the medical technologies into the clinic also rely on its nature, because cellular and acellular devices face different regulatory scrutiny in each country [11].

This chapter covers the general cares required for manufacturing tissue engineering and regenerative medicine (TERM) products. Particularly, it is focused the existing marketed products for OC tissue engineering and regeneration.

## 19.2 Good Manufacturing Practices and Regulatory Hurdles in Tissue-Engineered Products

Over the recent years, TERM has witnessed a rapid development that has motivated the marketing of novel products and with it some quality control procedures, and consistency and reproducibility guarantees. To assure that these conditions are being implemented, the FDA, EMA, and other world organizational committees are responsible to inspect the developers of commercial products. These organizations allow the implementation of standard procedures that extend away from individuals, research groups, and organizational procedures. It is their intention to have protocols in place that are independent of the operators and/or equipment guaranteeing uniformity of the data [12].

Good manufacturing practices (GMP) are a series of regulations that ensure that diagnostics, the production of pharmaceutical and medical devices, are controlled according to defined quality standards. GMP refers to all up-to-date aspects of the production from materials and equipment to the training staff and hygiene [13]. Moreover, the use of cells and tissues of human origin for TERM products also need to answer the good tissue practices (GTP). Particularly, GTP focuses its requirements on the prevention of the initiation, diffusion, and spread of contagious diseases besides ensuring uniformity, consistency, reliability, and reproducibility [14]. Following these lines, the description of a task or operation has to be performed in an identical manner and in compliance with appropriate regulation, as an approved standard operation procedure (SOP). Highly specific SOPs are usually required during all phases of a manufacturing process, and therefore are used to control the manufacturing of TERM products. Companies must agree to operate under harmonized guidelines across different geographic locations to assure that the best practices exist in every corner of the world. For instance, regarding TERM products using patient samples, companies need to pay attention to the appropriate ways to transport them once the classification of the shipment is the key in defining the level of containment required [13]. Fortunately, the TERM field is at the front line in terms of harmonization of international regulatory agencies, mostly because of the use of human cells in therapy, which has lead a worldwide joint effort [15]. Nevertheless, it should be kept in mind that the introduction of cells in tissue-engineered materials has associated hazards such as teratoma formation, contamination, immunogenicity, and insufficient cell adaptation. Thus, even though the materials used for tissue engineered strategies affect the regulatory process, cellular scaffolds pass through more regulatory hurdles to assure their safety.

## 19.3 Challenges in Clinical Translation and FDA Regulation

The programs for safety regulation vary widely by the type of product, its potential risks, and the regulatory powers granted to the agency. For example, the FDA regulates almost every facet of prescription drugs, including testing, manufacturing, labeling, advertising, marketing, efficacy, and safety, yet FDA regulation of cosmetics focuses primarily on labeling and safety. The FDA regulates most products with a set of published standards enforced by a modest number of facility inspections [16]. Each type of material is subjected to a different type of regulation based on its classification. Therefore, it is possible to characterize devices as "substantially equivalent" to currently accepted devices, allowing them to be more easily commercialized [17]. Contrariwise, when using novel bioactive scaffolds it is necessary to deeply describe their degradation and safety profile in preclinical and clinical studies, which normally results in a 30% of the costs increment [11]. However, if possible, new materials with more refined features should be created for TERM applications despite the innumerous hurdles that ought to be addressed for eventual clinical success. The first aspect to keep in mind is the scientific basis and the patentability of the technology. Furthermore, clinical studies should be carefully conducted. Then, the company where the product has been developed needs to assure not only enough financial support but also the regulatory requirements for GMP in order to have reproducible products [12]. Finally, the market potential of the therapeutic solution, possible competitors, and target audience may be considered. Nevertheless, even with these concerns, TERM is experiencing a boost in the development of new therapies for the treatment of chronic diseases and damaged tissues.

## 19.4 Commercially Available Products for Osteochondral Regeneration

The commercialization process of the scaffolds for implantation involves multiple stages of R&D replications before reaching the final approval from the government. R&D stages ensure safety and efficacy of the implants, which involve the production of medical grade scaffolds followed by animal testing, under regulatory approved conditions. Over the recent years, the concentrated research on TERM has resulted in few clinically approved therapeutics. Biomaterials applied in tissue regeneration are normally composed of a temporary three-dimensional (3D) support for the growth of cells that will regenerate a given injury, being then biodegraded and substituted by the new tissue. In OC regeneration, different materials have been employed as templates for cell interactions and formation of ECM to support the newly formed tissue. However, the most commonly used technique consists in designing bilayered scaffolds able to regenerate both cartilage and subchondral bone [18]. Normally, autologous chondrocytes are seeded at the top of the scaffold and allows the application of a cell–scaffold implantation [19]. An alternative approach to this procedure uses two scaffolds from cartilage and bone assembled either before or during implantation to assure OC regeneration [20].

At the present time, some commercial products have appeared. For example, Collagraft® (Nuecoll Inc.) consists of a mixture of collagen with HA and β-TCP in the form of granules. In previous studies, this product was used as subchondral support using chondrocytes harvested from rabbit articular cartilage. The animal model used survived through the regenerative process which occurred after 6 months presenting the adequate features for bone integration, but not for cartilage [21]. On the other hand, the treatment of small OC defects is also possible with the collagen-based implant ChondroMimetic™ (TiGenix NV). This product is an off-the-shelf bilayer implant launched in the European market. The chondral layer is made of collagen and glycosaminoglycan while the osseous layer is composed of calcium phosphates. It showed to support the simultaneous natural repair mechanism of both articular cartilage and bone, following by implantation in patients [22]. Another collagen-based 3D scaffold to treat knee chondral or OC defects is MaioRegen® (Med&Care). This matrix mimics the entire osteo-cartilaginous tissue and is composed of deantigenated type I equine collagen that resembles the cartilaginous tissue, and magnesium enriched-HA for the subchondral bone structure [23]. Preclinical studies using 12 sheep proved that this biomaterial is able to promote bone and hyaline-like cartilage tissue restoration. Quantitative macroscopic analysis showed absence of inflammation with some hyperemic synovium, but no synovial hypertrophy or fibrosis was noted. The histological score evaluations confirmed the presence of a newly formed tissue and a good integration of scaffolds [24]. Also, clinical evaluation of knee chondral and OC lesions in 27 patients, during a 5-year follow-up scored clinical improvements. Magnetic resonance imaging (MRI) results demonstrated a complete graft integration in 78.3% of patients offering a good clinical outcome for MaioRegen®. However, in another clinical investigation using knee and talar OC injuries, the biological response in vivo evaluated over 2.5 years showing no improvements after the implantation of MaioRegen®. Radiographic results of computed tomography and MRI presented a complete defect filling, integration, and an intact articular surface after 2.5 years of implantation [25].

One of the products used as an injectable material, in the treatment of OA of the knee, is the Gel-One®, which is composed of a cross-linked hyaluronic acid hydrogel through a photo-gelation process [26]. This product was applied in a clinical study using 379 patients with OA, and a single injection allowed both the relief of the pain associated to this condition for 13 weeks, as well as physical improvements [27].

BST-Cargel® (Piramal Life Sciences) has emerged as an advanced biodegradable and injectable chitosan hydrogel mixed with glycerophosphate and autologous blood to improve cartilage regeneration [28]. This off-the-shelf product can be used in conjunction with bone marrow cell stimulation by directly mixing blood from the patient with the biopolymer. This product has proved to be efficient in the initiation and amplification of the intrinsic wound healing processes of subchondral bone, as well as of the cartilage repair [28, 29]. An international randomized controlled trial with 80 patients was performed to compare the BST-Cargel® treatment with microfractured untreated patients. The results showed to be effective in the mid-term cartilage repair, and after 5-year follow up the treatment resulted in a sustained and significantly superior quantity and quality of repaired tissue against the microfracture alone [30]. Another commercial approach used in the treatment of cartilage defects, consists of a resorbable and textile polyglycolic acid–hyaluronan (PGA/hyaluronan) implant named Chondrotissue® (Biotissue). In preclinical studies with an ovine animal model it has shown improved tissue formation [31]. Clinical results with 5-year follow-up registered that this product had a good safety profile and provided a good filling of the chondral defects of the knee [32]. The effects of Chondrotissue® in patella defects of the cartilage were also evaluated by first debriding the damaged cartilage down to the subchondral bone. Then, the immersion of the implant in venous blood allows an enhanced MSC recruitment and integration leading to the improvement of the symptoms and no intraoperative or postoperative complications [33]. Chondrocushion (Advanced Bio Surfaces, Inc) made of biphasic polyurethane cylinders is a synthethic product evaluated for cartilage application. This biomaterial presents some potential disadvantages related with the lack of porosity which impedes tissue ingrowth and replacement [34]. Other drawbacks of this type of product include the displacement of the implant site and the release of potentially toxic by-products that can cause inflammation and cell death [22, 35]. Another product within this category is SaluCartilage™ (SaluMedica), a biocompatible and hydrophilic cylindrical device consisting of a polyvinyl alcohol hydrogel. This material mimics human cartilage in terms of water proportions and has been evaluated as a synthetic surface for the replacement of damaged cartilage. There is no evidence of inflammatory reaction or osteolysis associated with this implant [35]. Correspondingly, clinical results showed improvement of the chondral defects, but the hydrogel showed inadequate connection to the bone and risk of dislocation [36].

TruGraft™ (Osteobiologics) is a poly(lactic-co-glycolic) acid (PLGA) granulate used as a bone void filler, and has shown to support osteoblast proliferation and differentiation proved by high alkaline phosphatase activity and deposition of a mineralized matrix used in OC repair [37]. Bioseed®-C (Biotissue Technologies GmbH) is a bioresorbable scaffold for OC composed of fibrin, PLA and polyglycoic acid (PGA) copolymer, and polydioxanone embedding chondrocytes [38]. This product was preclinically evaluated on an equine animal model of full thickness cartilage defects and showed capacity to be integrated while promoting the formation of cartilaginous tissue [39]. These promising results made the testing of this product proceed into humans with posttraumatic and degenerative cartilage defects of the knee

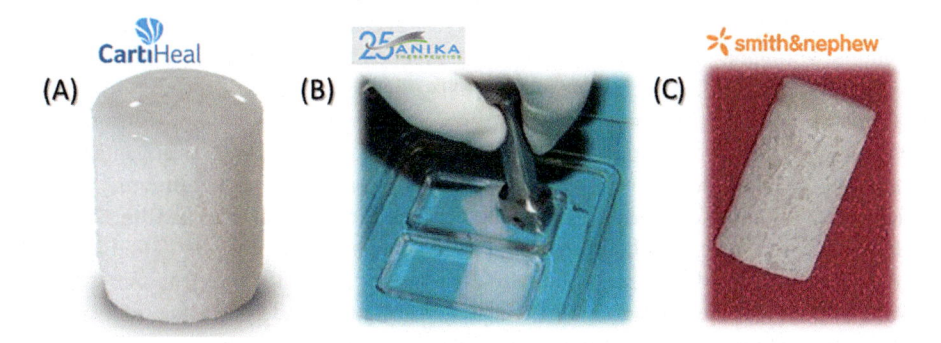

# 19 Commercial Products for Osteochondral Tissue Repair and Regeneration

**Fig. 19.1** Commercial biphasic scaffolds for osteochondral repair/regeneration: (**a**) Agili-C™ (reprinted with permission from [50]); (**b**) HYAFF™ (reprinted with permission from [51]. Copyright© 2015, Springer Nature) and (**c**) Trufit™

[40, 41]. It was reported that most of the grafts were able of completely fill the defects and formed tough hyaline-like cartilage, being well integrated into the tissue with good connection with the articular cartilage and the subchondral bone.

Feeding the importance of subchondral bone in the maintenance of articular cartilage, researchers have been developing some biphasic products in order to mimic both structures. These bilayered scaffolds have different characteristics to address different biological and functional requirements of both bone and cartilage, which are essential in the treatment of OC defects [23]. Accordingly, some commercial biphasic scaffolds are available in the market. For example, a product derived from natural sources for OC application, is Agili-C™ (CartiHeal) composed of calcium carbonate for the bone region, and aragonite and hyaluronic acid for the cartilage part (Fig. 19.1a). Results, after 12 months of implantation in a caprine animal model, showed an improvement in terms of cartilage repair and osteointegration with a reduction of the symptoms (limited motion, swelling of the joint, and pain). The bone phase of the implant shows a structure similar to natural bone presenting high pore interconnectivity essential for blood vessels [42, 43]. In contrast, the chondral phase rich in hyaluronic acid helps the ECM of the cartilage to be maintained with their proper characteristics. Kensey Nash Corporation started developing an acellular Cartilage Repair Device (CRD) to tackle primary defects of the articular cartilage in the joints, the OsseoFit® plug. This product was indicated to support the regeneration of hyaline cartilage and subchondral bone by promoting the correct cellular morphology, and structural organization during the healing process. The OsseoFit® plug is composed of a bioresorbable biphasic scaffold of collagen type-I fibrils, to simulate the cartilage, and 80% $\beta$-TCP + 20% polylactic acid (PLA) for the subchondral bone phase. A study using OsseoFit® applied in 10 plugs on the medial femoral condyle and on the lateral femoral condyle displayed a reduction in height of the material and the integration of the product on the surrounding native cartilage [19]. HYAFF® (Fidia Advanced Biopolymers) is a biodegradable scaffold used for repair of chondral and OC lesions. It is composed of purified hyaluronan esterified in its glucuronic acid group with distinct types of alcohols (Fig. 19.1b).

In vitro degradation profile of HYAFF® 11 suggested that the hyaluronan esters undergo spontaneous deesterification in an aqueous environment, meaning a good integration in the biological tissues [44]. Also, in vivo studies showed a minimal response after the first month of implantation, and no evidence of toxicity during the 1-year study following implantation. HYAFF® 11 scaffolds present the advantage of having good cell adhesiveness even without coating and surface conditioning [45]. TruFit® (Smith & Nephew) is another product used in OC applications (Fig. 19.1c) that consists of calcium phosphate and PLGA. This product presented good filling of OC defects, good integration in the native cartilage, and histological assays that showed a high percentage of hyaline cartilage formation, and good bone renovation after implantation in the femoral condyles and trochleae of goat defects [46]. Originally, this plug was designed as an alternative treatment for OC autologous transplantation, but in Europe has also been applied for acute focal articular cartilage and OC defects [47, 48]. Conversely, another work showed a comparative study between patients undergoing mosaicplasty (harvest and transplant of plugs of bone and cartilage from one place to another) and patients implanted with TruFit® showed no improvement in applying this product in comparison to the other procedure [49].

In order to establish a complete regenerative engineered strategy for TERM application, the use of cellular scaffolds has been proposed. The inclusion of living cells into the scaffolds enables real time growth factors, cytokines and matrix proteins which will accelerate the regenerative process. However, this technology also brings many complications besides the inevitable high costs, complexity of manufacture, sterility, preservation and regulatory issues. Currently, the application of cells without a scaffold has undergone several clinical studies showing that integration with the host tissue remains a problem [52]. This fact had led to the development of grafts combining living cells with biomaterials ex vivo allowing the construction of a 3D structure to be implanted in the living organism. Nevertheless, the type of cells and their quantity will help to set the mechanical and biochemical characteristics of the graft [53, 54]. The marketed product Osteocel® Plus (NuVasive®) is an example of a cell-based bone graft. It contains native MSCs and osteoprogenitor cells combined with an osteoconductive demineralized bone matrix (DBM), and cancellous bone, presenting osteogenic capacity. Promising experiences with this material have been reported for several applications, including lumbar [55], as well as periodontal [56], foot and ankle defects [57].

On the other side, demineralized matrices from human donors are also applied as optimal biomaterials. An example is Dynagraft®, a DBM combined with poloxamer, the nonionic triblock copolymer (GenSci Regeneration Sciences). This graft is moldable, packable and can be mixed with grafting materials being resistant to irrigation. Additionally, hyaluronate (i.e., DBX®, Synthes), glycerol (Grafton®, Osteotech, USA), and calcium sulfate (Allomatrix®, Wright Medical Technology) have also been applied with DBM and are also being commercialized [58, 59].

Commercially available cartilage graft BioCartilage® (Arthrex) provides a simple and inexpensive method to use extracellular cartilage matrix that has been dehydrated and micronized. It provides a proper scaffold to correct microfracture defects of articular cartilage, providing the appropriate biochemical signals, including collagen type II, cartilage matrix elements, and growth factors [60]. One of the few FDA approved ACT products is Carticel® (genzyme), which a cellular graft of autologous cultured cells derived from in vitro expansion of chondrocytes of femoral articular cartilage from the patient. The application of Carticel® in young patients has originated a cartilage tissue containing predominately collagen type II but lacking total host integration and alignment with the surrounding cartilage [61]. Hyalograft® C is another commercially available matrix to assist chondrocyte implantation. This product is composed of a hyaluronic acid-based cartilage graft in combination with autologous isolated and enriched chondrocytes. The clinical data of the application of this product showed improvements in 91.5% of the patients [62, 63]. The use of minced articular cartilage (autologous or allogeneic) defects is being explored for cartilage repair in OC defects. The off-the-shelf human tissue allograft DeNovo NT (Zimmer), consisting of juvenile hyaline cartilage pieces with viable chondrocytes, has emerged. This commercial product is intended to finds uses in lesions of articular cartilage. When implanting DeNovo NT, the debridement of the fibrous and calcified tissue of the defect should be performed without violating the subchondral bone layer. Then the implant is added to the lesion site and using fibrin glue will help the tissue to be maintained in place [64, 65]. Recently, this juvenile particulate cartilage has been employed in patellar lesions with success. It has been demonstrated that the repair hyaline cartilage is performed by integration with the surrounding tissue showing a good recovery and improvement of the movements and reducing the pain associated with OC lesions [66]. Another alternative is the OC allograft Chondrofix® (Zimmer®) composed of decellularized cadaveric human joints consisting of hyaline cartilage and cancellous bone [67]. A case report of Chondrofix® implanted in a large full-thickness OC defect demonstrated to be completely incorporated by the bone without articular cartilage margins and restoring the native femoral condylar radius of curvature [68]. Nevertheless, when working with these types of materials, the risk of disease transmission or immunogenicity still remains. Moreover, they can be quite brittle leading to the accumulation of microfractures during the remodeling phase [69].

The available commercial products aforementioned show that there is still much research that needs to be done to create new therapies to significantly increase the regenerative capacity of OC structure. A summary of these commercialized products for bone, cartilage, and OC tissue regeneration is presented in Table 19.1.

**Table 19.1** Commercial products for the repair and regeneration of bone, cartilage, and OC defects

| Product | Manufacturer | Composition | Bioresorbable | Applications |
|---|---|---|---|---|
| Chondromimetic™ | TiGenix NV | Collagen, GAG, and CaP | ✓ | OC |
| Trufit™ | Smith & Nephew | Calcium sulfate, PLGA/PGA | ✓ | OC |
| MaioRegen® | Med&Care | Collagen type I and magnesium enriched-HA | ✓ | OC |
| OsseoFit® plug |  | Type I collagen, and 80% β-TCP + 20% PLA | ✓ | OC |
| BST-Cargel® | Piramal Life Sciences | Chitosan gel, glycerophosphate, and autologous blood | n.d. | OC |
| Bioseed®-C | Biotissue Technologies GmbH | PLA/PGA | n.d. | OC |
| Agili-C™ | CartiHeal | Calcium carbonate and aragonite with hyaluronic acid. | ✓ | OC |
| Collagraft® | Nuecoll Inc | Collagen with granules of HA and β-TCP | ✓ | OC |
| Chondrotissue® | Biotissue | PGA/hyaluronan | ✓ | Cartilage |
| HYAFF® 11 | Anika Therapeutics | Hyaluronan | ✓ | Cartilage and OC |
| Chondrocushion | Advanced Bio Surfaces, Inc | Polyurethane | No | Cartilage |
| BST-Cargel® | Piramal Life Sciences | Chitosan with glycerol phosphate and autologous blood | ✓ | Cartilage |
| SaluCartilage™ | SaluMedica | Polyvinyl alcohol hydrogel | ✓ | Cartilage |
| Gel-One® | Zimmer | Hyaluronic acid | n.d. | Knee OA |

*n.d* not defined, *GAG* glycosaminoglycans, *CaP* calcium phosphates, *OC* osteochondral, *PLGA* poly(lactic-co-glycolic) acid, *HA* hydroxyapatite, *PLA* polylactic acid, *PGA* polyglycolic acid, *OA* osteoarthritis

## 19.5 Concluding Remarks and Future Trends

Several attempts are being made to mimic in vivo situations, and in fact enormous advances as regards not only to OC tissues but also to other tissues of the human body have been made. Herein, in this chapter we describe some commercial OC approaches, particularly based on 3D scaffolds envisioned to support newly formed tissues. The presented scaffolds are either biphasic, injectable hydrogels or decellularized matrices, and their outcomes reinforce the ideal basic requirements for the design of OC constructs aiming at tissue repair and regeneration. Such necessities include porosity, mechanical strength, biocompatibility, bioactivity, biodegradability, bio-integration, and proper cell proliferation and differentiation. Besides, due to its medical nature, new commercial products must undergo laborious testing, as demanded by regulatory approval bodies, before their use in humans. Therefore, some prospective improvements are under investigation to create better OC products. In this front, researchers are trying to enhance cell attachment to scaffolds using cell-adhesive ligands, and changing cell morphology, alignment, and phenotype, by varying the topographic surface of the scaffolds or even by mechanobiological stimulation of cells. Growth factors can also be incorporated directly into the scaffolds to help cells to differentiate, but immunomodulatory molecules are also an option to help to control inflammation towards the regenerative process. In the end, the final purpose is to create a scaffold that entirely mimics the ECM of OC tissue, having simultaneously proper mechanical properties, biochemical cues and the appropriate degradation profile, providing the ideal conditions for tissue growth.

**Acknowledgments** The authors acknowledge the project FROnTHERA (NORTE-01-0145-FEDER-000023), supported by Norte Portugal Regional Operational Programme (NORTE 2020), under the PORTUGAL 2020 Partnership Agreement, through the European Regional Development Fund (ERDF). The authors would also like to acknowledge H2020-MSCA-RISE program, as this work is part of developments carried out in BAMOS project, funded by the European Union's Horizon 2020 research and innovation program under grant agreement N° 734156. The financial support from the Portuguese Foundation for Science and Technology under the program Investigador FCT 2012 and 2015 (IF/00423/2012 and IF/01285/2015) is also greatly acknowledged.

# References

1. Bose S, Roy M, Bandyopadhyay A (2012) Recent advances in bone tissue engineering scaffolds. Trends Biotechnol 30:546–554. https://doi.org/10.1016/j.tibtech.2012.07.005
2. Benders KEM, van Weeren PR, Badylak SF et al (2013) Extracellular matrix scaffolds for cartilage and bone regeneration. Trends Biotechnol 31:169–176. https://doi.org/10.1016/j.tibtech.2012.12.004
3. Oliveira JT, Reis RL (2011) Polysaccharide-based materials for cartilage tissue engineering applications. J Tissue Eng Regen Med 5:421–436. https://doi.org/10.1002/term.335
4. Ge Z, Jin Z, Cao T (2008) Manufacture of degradable polymeric scaffolds for bone regeneration. Biomed Mater 3:22001. https://doi.org/10.1088/1748-6041/3/2/022001
5. Karageorgiou V, Kaplan D (2005) Porosity of 3D biomaterial scaffolds and osteogenesis. Biomaterials 26:5474–5491. https://doi.org/10.1016/j.biomaterials.2005.02.002
6. Habibovic P, Yuan H, van der Valk CM et al (2005) 3D microenvironment as essential element for osteoinduction by biomaterials. Biomaterials 26:3565–3575. https://doi.org/10.1016/j.biomaterials.2004.09.056
7. Pina S, Oliveira JM, Reis RL (2015) Natural-based nanocomposites for bone tissue engineering and regenerative medicine: a review. Adv Mater 27:1143–1169. https://doi.org/10.1002/adma.201403354
8. Canadas RF, Pina S, Marques AP et al (2016) Cartilage and bone regeneration—how close are we to bedside? In: Transl. Regen. Med. to Clin. Elsevier, Amsterdam, pp 89–106
9. Giannoudis PV, Dinopoulos H, Tsiridis E (2005) Bone substitutes: an update. Injury 36:S20–S27. https://doi.org/10.1016/j.injury.2005.07.029
10. Van Norman GA (2016) Drugs and devices: comparison of European and U.S. approval processes. JACC Basic Transl Sci 1:399–412. https://doi.org/10.1016/j.jacbts.2016.06.003

11. Webber MJ, Khan OF, Sydlik SA et al (2014) A perspective on the clinical translation of scaffolds for tissue engineering. Ann Biomed Eng 43:641–656. https://doi.org/10.1007/s10439-014-1104-7
12. Dodson BP, Levine AD (2015) Challenges in the translation and commercialization of cell therapies. BMC Biotechnol 15:70. https://doi.org/10.1186/s12896-015-0190-4
13. Basu J, Ludlow JW (2012) Regulatory and quality control. Dev Tissue Eng Regen Med Prod A Pract Approach 125–148. https://doi.org/10.1533/9781908818119.125
14. Idowu B, Di Silvio L (2013) Principles of good laboratory practice (GLP) for in vitro cell culture applications. Stand Cell Tissue Eng Methods Protoc 127–147. https://doi.org/10.1533/9780857098726.2.127
15. Gálvez P, Clares B, Hmadcha A et al (2013) Development of a cell-based medicinal product: regulatory structures in the European Union. Br Med Bull 105:85–105. https://doi.org/10.1093/bmb/lds036
16. Tyler RS (2013) The goals of FDA regulation and the challenges of meeting them. Health Matrix Clevel 22:423–431
17. Lewin A (2012) Medical device innovation in America: tensions between food and drug law and patent law. Harv J Law Technol 26
18. Dormer NH, Berkland CJ, Detamore MS (2010) Emerging techniques in stratified designs and continuous gradients for tissue engineering of interfaces. Ann Biomed Eng 38:2121–2141. https://doi.org/10.1007/s10439-010-0033-3
19. Elguizaoui S, Flanigan DC, Harris JD et al (2012) Proud osteochondral autograft versus synthetic plugs—contact pressures with cyclical loading in a bovine knee model. Knee 19:812–817. https://doi.org/10.1016/j.knee.2012.03.008
20. Swieszkowski W, Tuan BHS, Kurzydlowski KJ, Hutmacher DW (2007) Repair and regeneration of osteochondral defects in the articular joints. Biomol Eng 24:489–495. https://doi.org/10.1016/j.bioeng.2007.07.014
21. Schaefer D, Martin I, Jundt G et al (2002) Tissue-engineered composites for the repair of large osteochondral defects. Arthritis Rheum 46:2524–2534. https://doi.org/10.1002/art.10493
22. Getgood A, Henson F, Skelton C et al (2014) Osteochondral tissue engineering using a biphasic collagen/GAG scaffold containing rhFGF18 or BMP-7 in an ovine model. J Exp Orthop 1:13. https://doi.org/10.1186/s40634-014-0013-x
23. Kon E, Delcogliano M, Filardo G et al (2009) Novel nano-composite multi-layered biomaterial for the treatment of multifocal degenerative cartilage lesions. Knee Surg Sport Traumatol Arthrosc 17:1312–1315. https://doi.org/10.1007/s00167-009-0819-8
24. Kon E, Delcogliano M, Filardo G et al (2010) Orderly osteochondral regeneration in a sheep model using a novel nano-composite multilayered biomaterial. J Orthop Res 28:n/a. https://doi.org/10.1002/jor.20958
25. Christensen BB, Foldager CB, Jensen J et al (2016) Poor osteochondral repair by a biomimetic collagen scaffold: 1- to 3-year clinical and radiological follow-up. Knee Surg Sport Traumatol Arthrosc 24:2380–2387. https://doi.org/10.1007/s00167-015-3538-3
26. Ishikawa M, Yoshioka K, Urano K et al (2014) Biocompatibility of cross-linked hyaluronate (Gel-200) for the treatment of knee osteoarthritis. Osteoarthr Cartil 22:1902–1909. https://doi.org/10.1016/j.joca.2014.08.002
27. Strand V, Baraf HSB, Lavin PT et al (2012) A multicenter, randomized controlled trial comparing a single intra-articular injection of Gel-200,?a?new cross-linked formulation of hyaluronic acid, to phosphate buffered saline for treatment of osteoarthritis of the knee. Osteoarthr Cartil 20:350–356. https://doi.org/10.1016/j.joca.2012.01.013
28. Stanish WD, McCormack R, Forriol F et al (2013) Novel scaffold-based BST-CarGel treatment results in superior cartilage repair compared with microfracture in a randomized controlled trial. J Bone Jt Surg Am 95:1640–1650. https://doi.org/10.2106/JBJS.L.01345
29. Shive MS, Hoemann CD, Restrepo A et al (2006) BST-CarGel: in situ chondroinduction for cartilage repair. Oper Tech Orthop 16:271–278. https://doi.org/10.1053/j.oto.2006.08.001
30. Shive MS, Stanish WD, McCormack R et al (2015) BST-CarGel® treatment maintains cartilage repair superiority over microfracture at 5 years in a multicenter randomized controlled trial. Cartilage 6:62–72. https://doi.org/10.1177/1947603514562064

31. Erggelet C, Endres M, Neumann K et al (2009) Formation of cartilage repair tissue in articular cartilage defects pretreated with microfracture and covered with cell-free polymer-based implants. J Orthop Res 27:1353–1360. https://doi.org/10.1002/jor.20879
32. Siclari A, Mascaro G, Gentili C et al (2014) Cartilage repair in the knee with subchondral drilling augmented with a platelet-rich plasma-immersed polymer-based implant. Knee Surg Sport Traumatol Arthrosc 22:1225–1234. https://doi.org/10.1007/s00167-013-2484-1
33. Becher C, Ettinger M, Ezechieli M et al (2015) Repair of retropatellar cartilage defects in the knee with microfracture and a cell-free polymer-based implant. Arch Orthop Trauma Surg 135:1003–1010. https://doi.org/10.1007/s00402-015-2235-5
34. McNickle AG, Provencher MT, Cole BJ (2008) Overview of Existing Cartilage Repair Technology. Sports Med Arthrosc 16:196–201. https://doi.org/10.1097/JSA.0b013e31818cdb82
35. Falez F, Sciarretta FV (2015) Treatment of osteochondral symptomatic defects of the knee with salucartilage. Orthop Proc 87-B
36. Lange J, Follak N, Nowotny T, Merk H (2006) Ergebnisse der SaluCartilage-implantation bei viertgradigen Knorpelschäden im Bereich des Kniegelenks. Unfallchirurg 109:193–199. https://doi.org/10.1007/s00113-005-1025-x
37. Ishaug-Riley SL, Crane-Kruger GM, Yaszemski MJ, Mikos AG (1998) Three-dimensional culture of rat calvarial osteoblasts in porous biodegradable polymers. Biomaterials 19:1405–1412. https://doi.org/10.1016/S0142-9612(98)00021-0
38. Demoor M, Ollitrault D, Gomez-Leduc T et al (2014) Cartilage tissue engineering: Molecular control of chondrocyte differentiation for proper cartilage matrix reconstruction. Biochim Biophys Acta Gen Subj 1840:2414–2440. https://doi.org/10.1016/j.bbagen.2014.02.030
39. Barnewitz D, Endres M, Krüger I et al (2006) Treatment of articular cartilage defects in horses with polymer-based cartilage tissue engineering grafts. Biomaterials 27:2882–2889. https://doi.org/10.1016/j.biomaterials.2006.01.008
40. Kreuz PC, Müller S, Ossendorf C et al (2009) Treatment of focal degenerative cartilage defects with polymer-based autologous chondrocyte grafts: four-year clinical results. Arthritis Res Ther 11:R33. https://doi.org/10.1186/ar2638
41. Ossendorf C, Kaps C, Kreuz PC et al (2007) Treatment of posttraumatic and focal osteoarthritic cartilage defects of the knee with autologous polymer-based three-dimensional chondrocyte grafts: 2-year clinical results. Arthritis Res Ther 9:R41. https://doi.org/10.1186/ar2180
42. Guillemin G, Patat J-L, Fournie J, Chetail M (1987) The use of coral as a bone graft substitute. J Biomed Mater Res 21:557–567. https://doi.org/10.1002/jbm.820210503
43. Kon E, Filardo G, Shani J et al (2015) Osteochondral regeneration with a novel aragonite-hyaluronate biphasic scaffold: up to 12-month follow-up study in a goat model. J Orthop Surg Res 10:81. https://doi.org/10.1186/s13018-015-0211-y
44. Zhong SP, Campoccia D, Doherty PJ et al (1994) Biodegradation of hyaluronic acid derivatives by hyaluronidase. Biomaterials 15:359–365
45. Campoccia D, Doherty P, Radice M et al (1998) Semisynthetic resorbable materials from hyaluronan esterification. Biomaterials 19:2101–2127. https://doi.org/10.1016/S0142-9612(98)00042-8
46. Iii RJW, Gamradt SC, Williams RJ (2008) Articular cartilage repair using a resorbable matrix scaffold. Instr Course Lect 57:563–571
47. Melton JT, Wilson AJ, Chapman-Sheath P, Cossey AJ (2010) TruFit CB ® bone plug: chondral repair, scaffold design, surgical technique and early experiences. Expert Rev Med Devices 7:333–341. https://doi.org/10.1586/erd.10.15
48. Carmont MR, Carey-Smith R, Saithna A et al (2009) Delayed Incorporation of a TruFit Plug: perseverance is recommended. Arthrosc J Arthrosc Relat Surg 25:810–814. https://doi.org/10.1016/j.arthro.2009.01.023
49. Hindle P, Hendry JL, Keating JF, Biant LC (2014) Autologous osteochondral mosaicplasty or TruFit® plugs for cartilage repair. Knee Surg Sport Traumatol Arthrosc 22:1235–1240. https://doi.org/10.1007/s00167-013-2493-0
50. Kon E, Filardo G, Perdisa F et al (2014) Clinical results of multilayered biomaterials for osteochondral regeneration. J Exp Orthop 1:10. https://doi.org/10.1186/s40634-014-0010-0

51. Buda R, Vannini F, Castagnini F et al (2015) Regenerative treatment in osteochondral lesions of the talus: autologous chondrocyte implantation versus one-step bone marrow derived cells transplantation. Int Orthop 39:893–900. https://doi.org/10.1007/s00264-015-2685-y
52. Worthen J, Waterman BR, Davidson PA, Lubowitz JH (2012) Limitations and sources of bias in clinical knee cartilage research. Arthrosc J Arthrosc Relat Surg 28:1315–1325. https://doi.org/10.1016/j.arthro.2012.02.022
53. Bertolo A, Mehr M, Aebli N et al (2012) Influence of different commercial scaffolds on the in vitro differentiation of human mesenchymal stem cells to nucleus pulposus-like cells. Eur Spine J 21(Suppl 6):S826–S838. https://doi.org/10.1007/s00586-011-1975-3
54. Chen RR, Mooney DJ (2003) Polymeric growth factor delivery strategies for tissue engineering. Pharm Res 20:1103–1112. https://doi.org/10.1023/A:1025034925152
55. Ammerman JM, Libricz J, Ammerman MD (2013) The role of Osteocel Plus as a fusion substrate in minimally invasive instrumented transforaminal lumbar interbody fusion. Clin Neurol Neurosurg 115:991–994. https://doi.org/10.1016/j.clineuro.2012.10.013
56. McAllister BS (2011) Stem cell-containing allograft matrix enhances periodontal regeneration: case presentations. Int J Periodontics Restorative Dent 31:149–155
57. Scott RT, Hyer CF (2013) Role of cellular allograft containing mesenchymal stem cells in high-risk foot and ankle reconstructions. J Foot Ankle Surg 52:32–35. https://doi.org/10.1053/j.jfas.2012.09.004
58. Dhandayuthapani B, Yoshida Y, Maekawa T, Kumar DS (2011) Polymeric scaffolds in tissue engineering application: a review. Int J Polym Sci. https://doi.org/10.1155/2011/290602
59. Drosos GI (2015) Use of demineralized bone matrix in the extremities. World J Orthop 6:269. https://doi.org/10.5312/wjo.v6.i2.269
60. Abrams GD, Mall NA, Fortier LA et al (2013) BioCartilage: background and operative technique. Oper Tech Sports Med 21:116–124. https://doi.org/10.1053/j.otsm.2013.03.008
61. Kurkijärvi JE, Mattila L, Ojala RO et al (2007) Evaluation of cartilage repair in the distal femur after autologous chondrocyte transplantation using T2 relaxation time and dGEMRIC. Osteoarthr Cartil 15:372–378. https://doi.org/10.1016/j.joca.2006.10.001
62. Gobbi A, Kon E, Berruto M et al (2009) Patellofemoral full-thickness chondral defects treated with second-generation autologous chondrocyte implantation. Am J Sports Med 37:1083–1092. https://doi.org/10.1177/0363546509331419
63. Marcacci M, Berruto M, Brocchetta D, et al (2005) Articular cartilage engineering with hyalograft(R) C: 3-year clinical results. [Report]. Clin Orthop Relat Res 96–105
64. Tompkins M, Adkisson HD, Bonner KF (2013) DeNovo NT allograft. Oper Tech Sports Med 21:82–89. https://doi.org/10.1053/j.otsm.2013.03.005
65. Kruse DL, Ng A, Paden M, Stone PA (2012) Arthroscopic de novo NT? juvenile allograft cartilage implantation in the talus: a case presentation. J Foot Ankle Surg 51:218–221. https://doi.org/10.1053/j.jfas.2011.10.027
66. Buckwalter JA, Bowman GN, Albright JP et al (2014) Clinical outcomes of patellar chondral lesions treated with juvenile particulated cartilage allografts. Iowa Orthop J 34:44–49
67. Gomoll AH (2013) Osteochondral allograft transplantation using the chondrofix implant. Oper Tech Sports Med 21:90–94. https://doi.org/10.1053/j.otsm.2013.03.002
68. Reynolds KL, Bishai SK (2014) In situ evaluation of chondrofix(registered trademark) osteochondral allograft 25 months following implantation: a case report. Osteoarthr Cartil 22:S155–S156. https://doi.org/10.1016/j.joca.2014.02.288
69. Horton MT, Pulido PA, McCauley JC, Bugbee WD (2013) Revision osteochondral allograft transplantations. Am J Sports Med 41:2507–2511. https://doi.org/10.1177/0363546513500628

# Index

**A**

Absorbing film-assisted laser-induced forward transfer (AFA-LIFT), 227
Acellular bioprinting techniques
　advantages, 228
　cellular constructs, 228, 231, 232
　disadvantage, 231
　FDM, 228, 229
　hybrid chitosan–PCL scaffolds, 232
　hydrogel-based constructs, 231
　MEW, 230
　MHDS, 232
　sacrificial templates, 231
　SLS, 230, 231
　t-FDM, 232
　thermoplastic polymer-natural polymer, 232
Actifit®, 45
Additive manufacturing (AM), 130
Adipose-derived stem cells (ADSCs), 135, 148, 319, 362, 363
Agili-C™, 383, 401
Alginate, 14, 15, 338–339
Alginate–TGF-β–BMMSCs prints, 236
α,β-poly-(*N*-hydroxyethyl)-DL-aspartamide, 341
α-elastin-grafted HA-EDA (HA-EDA-g-α-elastin), 334
α-tricalcium phosphate (α-TCP), 296
Alumina-based bioceramics, 56
Anchorage dependent cells (ADCs), 289
Anti-inflammatory, 362
Aragonite, 19
Arthroscopy
　ACI, 360
　anteromedial and anterolateral accesses, 366
　BMS, 368
　focal osteochondral lesions, 363
　subchondral bone bed, 366
Articular cartilage tissue engineering, 32, 33
Autologous chondrocyte implantation (ACI), 36, 145, 196, 197, 208, 210, 376, 399
Autologous matrix-induced chondrogenesis (AMIC), 398

**B**

Basic fibroblast growth factor (bFGF), 256
Bench to bedside translation, 383–387
β-D-glucuronic acid (GlcA), 283
β-sheet crystalline content, 313
β-tricalcium phosphate (β-TCP), 199, 231, 379, 401
Bilayered scaffolds, 200–203, 298, 299, 316
Bioactive factor, 190
BioCart II™, 400
BioCartilage®, 294
Bioceramics
　CaPs, 19
　classification, 54
　clinical trials, 66–69
　fabrication method, 54
　hydroxyapatite, 19, 20
　ionic elements, 55
　limitations, 54
　natural (*see also* Natural bioceramics)
　in OTE
　　biphasic composite, 63
　　clinical use, 63
　　diverse materials, 65, 66

Bioceramics (cont.)
    growth factors, 63
    HAp porous scaffold, 64
    OC repair/regeneration, 63, 65
    silk and nano CaP layers, 65
    sodium alginate, 64
    synthetic (see also Synthetic bioceramics)
Biochemical gradients, 136
Biodegradable synthetic polymer, 180, 190
Bioglass nanofibers, 113
Bioglass®, 267, 272
Bioink
    cellular and acellular scaffold-based bioprinting, 224
    cross-linking mechanisms, 235
    3D bioprinting, 221, 235
    dispensation, 225
    GelMA and GelMA-HAMA, 236
    LBB, 226
    nanocellulose, 225
    rheological and biological properties, 223
    silk–gelatin, 234
BioINK™, 235
Biomaterials
    biodegradability, 85
    biphasic biomaterials, 34
    chemical, topological, mechanical and structural cues, 42, 43
    classes, 85
    composite, 86
    ECM functions, 32
    features, 34, 47
    hydrogels, 34
    minimally invasive surgery, 34
    in OTE
        ACI, 36
        challenge, 37
        defect management and repair, 36
        MACI, 36
        multi-layer acellular scaffolds, 37
        patients quality of life, 36
    PLGA and PVA, synthetic, 32–34
    scaffolds, 32
    self-reinforced and injectable, 34
Biomimetic OC TE
    aims, 144
    anatomical shape, 145
    bioreactors, 154, 155
    chondrocytes, 146–148
    components, 144, 145
    ECM, 144
    growth factors, 155, 157
    hESC and hiPSCs, 149, 150
    immunomodulation, 157, 158

    interface engineering, 160, 161
    low oxygen tension conditioning, 158, 159
    mesenchymal condensation, 150–151
    osteoblast-like cells, 148
    pluripotent stem cells, 148
    primary cells, 145
    purinergic signaling, 160
    scaffold-less techniques, 151–154
    stem/stromal cells, 145
    vascularization, 157
Biopolymers
    PHAs, 18, 19
    polysaccharides
        alginate, 14, 15
        chitin and chitosan, 13, 14
        gellan gum, 17, 18
        hyaluronic acid, 15–17
    proteins
        collagen, 7, 8, 10–19
        fibrin, 12
        gelatin, 8, 10
        keratin, 13
        silk proteins, 10, 11
Bioreactor
    chondrogenesis, 155
    interstitial flow, 155
    OC TE, 154
    predifferentiated osteoblasts, 155
    principles, 155
    undifferentiated hMSCs, 155
BioSeed®-C, 400
Biphasic scaffold, 378, 379
Bone–cartilage interface, 376, 379
Bone marrow
    aspiration, 364–365
    BMDCT, 360
    BMP-2, 293
    BMS, 368
    cartilage regeneration, 299
    concentration, 365
    fibroblasts, 361
    GG backbone, 284, 287
    mesenchymal stem cells, 361, 362
    regeneration, 295–298
    subchondral bone damage, 291
Bone marrow aspirate concentrate (BMAC), 398
Bone marrow-derived cll transplantation (BMDCT), 362, 364
Bone marrow-MSCs (BMSCs), 148, 150, 232, 315
Bone morphogenetic protein (BMPs), 63, 92, 183, 236, 237, 256, 293, 311, 362, 381

# Index

Bone tissue engineering
 collagen–CaP–DEX hybrid scaffolds, 185–186
 PLGA–collagen–BMP4 hybrid scaffolds, 183–185
BST-CarGel®, 333, 344, 345

## C

Calcified cartilage, 374, 375, 380
Calcium carbonate mineral ($CaCO_3$), 19
Calcium-deficient hydroxyapatite (CDHA), 296
Calcium glycerophosphate (CaGP) solution, 295
Calcium phosphates (CaPs), 19, 114, 131, 185–186, 267, 272, 274, 282
 α-TCP and β-TCP, 61
 CDHA, 62
 chemical compositions, 59, 60
 CPCs, 62, 63
 custom-designed forms, 61
 HAp, 61
 load-bearing applications, 60
 orthopedic and dental applications, 59
 types, 59
Calcium phosphates-based cements (CPCs), 62, 63
Carbon fibers, 117, 118
Carbon nanotubes (CNTs), 89
Carboxymethyl (CM) chitosan, 287
Cartilage
 articular TE, 32, 33, 83
 biomaterials, 32, 35
 bone–cartilage interface, 36
 chondrogenesis, 82
 collagen network, 81
 composites, 40
 CTE (see Cartilage tissue engineering (CTE))
 description, 80, 81
 diseases, 83
 GG, 284
 hydrogels, 34
 mechanical factors, 82
 meniscal fibrocartilage, 45
 OC, 292
 osteoarthritis, 83
 physical structure, 81
 PLGA, 34
 principal synthetic matrices, 33
 proteoglycan–water gel, 81
 regeneration, 292–295
 and subchondral bone damage, 291
 tissue engineering methods, 36
 treatments, osteochondral diseases, 83
 types, 80, 81
Cartilage Regeneration System, 400
Cartilage repair, 397–402, 408
 ICRS, 363
 in OLT, 369
Cartilage tissue engineering (CTE)
 free ice, 176–180
 funnel-like surface pore structures, 172–176
 hydrogels cells and growth factors, 38–40
 PLGA–collagen hybrid scaffolds, 180–183
Cartipatch®, 400
Cellular bioprinting
 components, 224
 DBB, 228
 EBB, 225, 226
 LBB, 226, 227
Cellulose fibers, 112
Ceramic fibers, 113
Chitin, 13, 14
Chitosan, 13, 14, 330–333
Chitosan–glycolic acid/tyrosine (Ch-GA/Tyr), 331
Chondral lesion, 393, 395, 397–399, 402, 403
ChondroCelect®, 271, 400
Chondrocytes, 271, 334, 336, 338, 375, 376, 378–381
Chondrofix Osteochondral Allograft, 387
Chondrogenesis, 82, 340
Chondro-Gide®, 400
Chondroitin sulfate amino ethyl methacrylate (CS-AEMA), 234
ChondroMimetic™, 401
ChonDux™, 398
Coaxial electrospinning, 255
Coil-helix transition temperature, 285
Collagen
 description, 7
 fibrils, 7
 fundamental building block, 7
 mammalian, 8
 marine-derived, 8
 type I and II, 7
Collagen binding domain (CBD), 184
Collagen–CaP–DEX hybrid scaffold, 186
Collagen/gelatin-based in situ hydrogels, 336–338
Collagen/PLGA–collagen biphasic sponge, 189
Collagen–PVA nanofibers, 116
Colony-forming capacity (CFU-F), 147
Commercial products, 417, 424

Composites
    fiber-reinforced material, 328
    injectable pullulan/chondroitin sulfate, 340
    nanocomposite hydrogel, 335
Computer-aided additive manufacturing (CAM), 153, 154, 161
Condensed mesenchymal bodies (CMBs), 150, 158
Continuous passive motion (CMP), 367

**D**
Decellularized extracellular matrix (dECM), 234
Denatured human fibrinogen (DHF), 343
DeNovoET®, 400
Dental pulp stem cells (DPSCs), 157
Desferoxamine (DFO), 134, 257
Diarthrodial joint, 393
Differential scanning calorimetry (DSC), 285, 286
3D printing
    acellular techniques (see Acellular bioprinting techniques)
    applications, 222
    bioactive compounds, 236–237
    biofabrication techniques, 223
    biological specifications, 223
    economic issues, 223
    heterogeneity and anisotropy, 237
    mechanical properties, 235–236
    mimetic scaffolds, 238
    optimal rheological and biological properties, 223
    osteochondral TE (see Osteochondral TE)
    physical, biological and economical specifications, 222
    preclinical studies, 238
    soft hydrogels, 223
4D printing, 44
Dulbecco's Modified Eagle Medium (DMEM), 234
Dynamical mechanical analysis (DMA), 273, 298

**E**
Elastin-based in situ hydrogels, 340
Elastin fibers, 117
Electrospinning (ES)
    benefit, 99
    bioactive glass, 113
    biomedical community, 247
    biphasic nanofiber scaffolds, 107
    calcium phosphate, 114
    cell proliferation and osteogenic differentiation, 99
    conventional arrangement, 249
    electrospun fibers, 248, 253–255
    electrostatic forces, 248
    environmental parameters, 253
    fibrous scaffolds, 107
    HA fibers, 114
    metallic collector base, 248
    microfiber- and nano-fibers production, 99
    nanofibers, 249
    nanofibrous meshes (see Nanofibrous meshes)
    natural fibrillar ECM, 258
    nonwoven mesh structure, 249
    parameters, 250–252
    PCL, 105
    principle, 248, 249
    pullulan and cellulose acetate, 112
    solution parameters, 252, 253
    Taylor cone, 248
    versatile and cost-effective polymer process, 248
Electrospraying, 248, 251
Electrospun nanofibrous structures, 254
Epithelial growth factor (EGF), 362
Ethylenediamine (EDA), 334
1-Ethyl-3-(3-dimethylaminopropyl) carbodiimide (EDAC), 289
European Medicines Agency (EMA), 417
Extracellular matrix (ECM), 126–128, 144, 146, 152, 160, 161, 208, 222, 232, 234, 282, 284, 292–294, 298, 299
Extrusion-based bioprinting (EBB), 225

**F**
Federal Food and Drug Administration (FDA), 417
Fibrin
    clot, 12
    description, 12
    fiber, 117
    and fibrinogen, 12
    hemostatic effects, 12
    hydrogels, 12
    in situ hydrogels, 339–340
Fibroblast growth factor 2 (FGF-2), 92
Fibrocartilaginous repair, 360, 397
Fibrous scaffolds
    application, 106
    biological assays, 108–111, 119
    drugs, 107

Index   433

electrospinning, 107
mechanical properties
    compressive tests, 100–102, 119
    tensile and flexural tests, 102–104, 119
PLGA, 106
polyamide fibers, 106
PVA-MA and chondroitin sulfate, 107
Fused deposition modeling (FDM), 131, 228–229
Fused filament fabrication (FFF), 228

## G

Gelatin methacrylamide (GelMA), 161, 225, 227, 230, 234–236, 293, 294
Gelatin–hydroxyphenylpropionic acid (Gtn-HPA), 337
Gel-Gro®, 284
Gellan gum (GG)-based hydrogels, 17, 18
    acetylation, 284
    in acyl groups' hydrolysis, 284
    advantages, biopolymer, 282
    bilayered scaffolds, 298, 299
    biocompatibility/non-cytotoxicity, 284
    biomedical field, 284
    bone, 295–298
    cartilage, 292–295
    chemical structure, deacylation, 283
    EPS, 282
    FDA and EU, 284
    functionalization strategies, 287–291
    gel-forming ability, 284
    mineral components, 282
    non-invasive approaches, 282
    OC tissue repair, 291, 292
    parameters, 283
    structure and properties, 285, 286
    substantial pre-clinical studies, 299
    TERM approaches, 282, 284
    versatile biomaterial, 300
GELRITE®, 284
Gelzan®, 284
Genipin
    crosslinking agent, 269
    morphological characterization, 270
    PL scaffolds, 267, 270, 275
Glass fibers, 113
Glycol chitosan, 331, 336
Glycoproteins, 128
Glycosaminoglycans (GAG), 132, 313, 318, 328, 334, 337, 338, 343, 344, 392
Gradient scaffolds, 206–207, 380–383
Gradient structures, 268

Growth factors
    BMP-2, 92
    IGF-1, 92
    TGF-β, 92

## H

HA–Ca–GG gelling system, 298
Hematoxylin, 175, 179
High acyl gellan gum (HAGG), 284, 285
Horseradish peroxidase (HRP), 152, 331, 337, 341
Human-adipose stem cells (hASCs), 297
Human articular chondrocytes (hACs), 315
Human dermal fibroblasts (HDFs), 290
Human epidermis fibroblasts (hEFBs), 287
Human foetal osteoblasts (hFOBs), 290
Human MSCs (hMSCs), 227
Human umbilical vein endothelial cells (HUVECs), 237
Hyalograft® C, 197, 400
Hyaluronic acid (HA), 15–17, 333–336
Hyaluronic acid methacrylate (HAMA) hydrogels, 225, 234, 236
Hybrid scaffold
    advantages, respective components, 190
    Collagen–CaP–DEX, 185–186
    PLGA, 180–183
    PLGA–Collagen–BMP4, 183–185
Hydrogels
    bilayered, 307
    HAMA, 225, 234, 236
    silk fibroin-based, 309–319
Hydrolyzed polyacrylamide (HPAM), 329, 341
Hydroxyapatite (HAp), 19, 20, 114, 115, 195, 199, 209, 210, 282
3-(4-hydroxyphenyl) propanamide (HPPA), 342
Hypoxia-inducible factors (HIF), 159

## I

Ice–collagen sponges, 176–179
Immunomodulatory capacity, 362
In situ hydrogels
    alginate, 338–339
    alginate-based, 339
    biocompatibility and biodegradability, 346
    chitosan, 330–333
    chondroitin sulfate, 340–341
    clinical applications, 342–345
    collagen/gelatin, 336–338
    elastin, 340

In situ hydrogels (*cont.*)
  fibrin, 339–340
  hyaluronic acid, 333–336
  mechanical properties and stability, 346
  physical/chemical cross-linking reactions, 346
  synthetic biomaterials, 341–342
Injectable hydrogels
  biomaterials and fabrication techniques, 329
  bone repair, 328
  cartilage and bone clinical trials, 328
  ECM, 328
  hydrophobic/ionic interactions, 329
  in situ hydrogels (*see* In situ hydrogels)
  natural and synthetic biopolymers, 329
  TE and regenerative medicine, 328
Inkjet bioprinting, 154
Insulin-like growth factors (IGFs), 134, 155, 236, 362
International Cartilage Repair Society classification (ICRS), 360
International Society for Cell Therapy (ISCT), 146, 363
Ionic- and photo-crosslinkable methacrylated GG (GG-MA), 288

## K
KELCOGEL®, 284
Keratin, 13
Knee injury and Osteoarthritis outcome score (KOOS), 404

## L
Laponite® XLG, 288
Laser-based bioprinting (LBB), 154, 226, 227
Laser-induced forward transfer (LIFT), 226, 227
Layer-by-layer deposition (LbL), 229
Layered scaffolds
  cartilage and subchondral bone, 210
  gradient scaffolds, 205–207
  graft–tissue interface, 211
  hierarchical structured scaffolds, 198–200
  OC tissue (*see* OC tissue)
  OCTE strategies, 194, 211
  trilayered scaffolds, 203–205
Low acyl gellan gum (LAGG), 284, 299

## M
MACI®, 400
MaioRegen®, 209, 210, 386
Matrix-assisted ACI (MACI), 36, 145, 197
Matrix-assisted autologous chondrocyte transplantation (MACT), 368
Matrix-assisted pulsed laser evaporation direct writing (MAPLE DW), 227
MC3T3-E1 osteoblast-like cells, 296
Medical grade PCL (mPCL) fibers, 230
Menaflex®, 45
Meniscal allografts, 47
Meniscal lesion, 45
Mesenchymal and Tissue Stem Cell Committee, 363
Mesenchymal stem cells (MSCs), 91, 147, 195, 196, 205, 209, 361–363
Mesenchymal stromal cells (MSCs), 146–150, 156, 160
Metalloproitenase-9 (MMP9), 157
Methacrylated glycol chitosan hydrogel, 334
Methacrylated hexanoyl glycol chitosan (M-HGC), 331
Methacrylated kappa carrageenan (MκCA), 161
1-(2-(2-Methoxyethoxy)ethoxy)-2-azidoethane, 342
Methoxy polyethylene glycol (MPEG), 342
Methoxy polyethylene glycol–poly(ε-caprolactone), 341
Micro-computed tomography (micro-CT), 270
Micro/nano scaffolds
  biological and chemical gradients, 126
  bone tissue, 129, 131
  bone–cartilage interface, 135
  cartilage tissue, 131, 132
  ECM, 126–128
  osteochondral regeneration, 132, 134, 135
  osteochondral TE, 126, 128, 129
  physiological properties, 135
  scientific literature, 136
Microextrusion bioprinting, 154
Microfibers
  advantages, 99
  carbon fiber, 117
  electrospinning, 99
  PLLA/PVA sheets, 100, 105
  submicron 45S5 bioglass fibers, 113
  TFG-β1 from PCL, 107
  titanium, 118
Mineral fibers, 113
Monophasic scaffolds, 378
Mosaicplasty (MP), 398

Index

Multi-head tissue/organ building system (MtoBS), 134
Multilayer/stratified scaffolds, 135, 211
Multiphasic calcium phosphate fibers, 114
Multiscale organization, 126–128

**N**
Nanobiomaterials, 89
Nanocomposite, 89–91, 93
Nanocrystalline hydroxyapatite (nHAp), 134, 341
Nanofibers
  advantages, 99
  bioglass, 113
  carboxymethylcellulose, 112
  collagen and electrospun PLLA, 105
  collagen–PVA, 116
  electrospinning, 99
  fibrous scaffolds, 107
  PDLLA, 105
  PLGA-based nanofibrous scaffolds, 106
Nanofibrous meshes
  bio-functionalization, 256
  chemical modification, 255, 256
  electrospun, 255
  osteochondral application, 257, 258
  surface modification methods, 255
Nanomaterials
  advantages, 91
  ceramic NPs, 89
  CNTs, 89
  fibrous scaffolds, 89
  multilayer scaffolds, 90
  PEG hydrogel, 90
  tissue engineering approach, 85
Nanostructured porous polycaprolactone (NSP-PCL) scaffold, 132
Natural biomaterials
  advantages, 86
  bioceramics
    calcium carbonate, 56
    CaPs, 19
    corals, 56
    *Coralline officinallis*, 56
    hydroxyapatite, 19, 20
    natural aragonite, 19
    scanning electron micrographs, 55
  biopolymers (*see* Biopolymers)
  characteristics and applications, 4–6
  chitosan, 86

chondrocytes and extracellular synthesis, 86
collagen and hyaluronic acid, 86
classes, 4
immunogenic incompatibility, 86
multicomponent and multilayered scaffolds, 4
properties, 4
TE scaffolds, 4
NeoCart®, 400
Nonsteroidal anti-inflammatory drug (NSAID), 288
Novocart®3D, 400
Nuclear magnetic resonance (NMR), 285, 288, 289

**O**
Osteochondral (OC) defect
  classification, 196
  clinical repair, 197
  magnesium-enriched HAp, 209
  repair, 44
  subchondral bone repair with TEC-HAp, 209
  tissue engineering strategies, 200
Osteochondral (OC) lesions
  articular cartilage, 392
  chondral lesions, 393, 394
  clinical management, 394–402
  collagen type II, 392
  ECM, 392
  histological examination, AC, 393
  hyaluronan, 392
  preclinical and clinical trials, 402–408
  tissue engineering, 408
  water, 392
Osteochondral (OC) tissue
  clinical applications, 209–210
  clinical repair of, 197
  complexity, 194–196
  development, 376
  injuries, 80
  integration, 207–209
Osteochondral allograft (OAG), 398
Osteochondral autograft transplantation (OAT), 398
Osteochondral lesions of the knee (OLKs), 368
Osteochondral lesions of the talus (OLTs), 368
Osteochondral regeneration, 66–69, 307, 313

Osteochondral tissue engineering (OCTE)
    autografts and allografts, 224
    biomimetic (see Biomimetic OC TE)
    categories, 224
    cellular bioprinting, 224–228
    challenges
        biocompatibility, 93
        ethical issues, 93
        immune response, 93
        scientific, 94
    collagen-based scaffolds, 10–9
    3D/4D printing, 44
    DBB, 228
    EBB, 225, 226
    hydrogels cells and growth
        factors, 38–40
    LBB, 226, 227
    natural origin materials (see Natural
        materials)
    PLGA–collagen hybrid mesh, 186–189
    PLGA–collagen/collagen biphasic
        scaffold, 189–190
    scaffolding design approaches, 308
    surgical technique, 47
    synthetic materials, 7
Osteochondral tissue regeneration
    Agili-C™ grafts, 383, 385
    aragonite-HA based graft, 385
    biphasic, 378–379
    calcium carbonate, 384
    chondrocytes, 374
    Chondrofix Osteochondral Allograft, 386
    clinical trials, 383
    GAG deposition, 385
    gradient, 380–383
    histological analysis, 387
    hyaline-like cartilage regeneration, 385
    ischemia and avascular necrosis, 374
    long-term studies, 387
    MaioRegen, 386
    monophasic, 378
    MRI method, 386
    non-calcified cartilage region, 374
    osteoarthritis, 374
    radiological evaluation, 385
    scaffold configurations, 377
    subchondral bone, 375
    tissue structure, 375
    treatment options, 376, 384, 387
    triphasic, 379, 380
    TruFit CB plug, 385
Osteochondritis dissecans (OCD), 374, 396
Outerbridge classification, 395, 396
Ovine MSCs (oMSCs), 228

**P**
Patello femoral joints, 360, 364
PCL-tricalcium phosphate (TCP), 131
PEG-poly(N-isopropyl acrylamide)
    (PNIPAAm), 341
Perfluoroalkoxy (PFA)-film, 172, 177
Phytagel®, 284, 290
Platelet lysates (PLs)
    BG-enriched region, 276
    chondral and osteochondral defects, 266
    development, 267, 276
    3D hydrogels, 275
    3D multifunctional architectures, 268–272
    immunogenic and disease transmission, 275
    natural and synthetic polymers, 266
    osteochondral interface, 266
    osteochondral TE, 266, 272–275
    supercritical fluid technology, 267–268
    tridimensional network, 267
Platelet–derived growth factor BB
    (PDGF-BB), 92, 156
Platelet-rich fibrin (PRF), 360, 364, 366, 367
PLGA–collagen hybrid mesh scaffold, 187
PLGA–collagen hybrid sponge layer, 190
PL$_{HXL}$ scaffolds, 271, 273, 274
Poly (ε-caprolactone) (PCL), 105, 106, 131,
    223, 229–233, 235, 236, 257, 378
Poly(ethylene glycol) (PEG), 152, 156, 329,
    334, 339, 341, 342
Poly(ethylene glycol)-terephthalate
    (PEGT), 229
Poly(hydroxybutyrate-co-hydroxyvalerate
    (PHBV), 231
Poly(hydroxymethylglycolide-co-ε-
    caprolactone)/PCL (pHMGCL/
    PCL), 232
Poly(lactic-co-glycolic acid) (PLGA), 34, 44,
    180, 229, 231, 233, 234, 236, 311
Poly(γ-propargyl-L-glutamate) (PPLG), 342
Polyamide (PA) fibers, 106
Polyethylene glycol diacrylate
    (PEG-DA), 343
Polyethylene glycol dimethacrylate
    (PEG-DMA), 288
Polyhydroxyalkanoates (PHAs), 18, 19
Polylactic acid (PLA), 105, 255
Polysaccharides
    alginate, 14, 15
    chitin and chitosan, 13, 14
    gellan gum, 17, 18
    hyaluronic acid, 15–17
Polyvinyl alcohol (PVA), 34, 36
Porcine mesenchymal stem cells
    (pMSCs), 255

Index 437

Porous scaffolds
  BMP4 and DEX, 190
  bone tissue engineering (see Bone tissue engineering)
  cartilage tissue engineering (see Cartilage tissue engineering)
  osteochondral tissue engineering (see Osteochondral tissue engineering)
Processing parameters, 250, 252
Proteins
  collagen, 7, 8, 10–19
  fibrin, 12
  gelatin, 8, 10
  keratin, 13
  silk proteins, 10, 11
PVA-DFO/shell PCL fibers, 135

**R**

Rabbit bone marrow mesenchymal stromal cells (RBMSCs), 316
Rat mesenchymal stem cells (rMSCs), 296
Recombinant human bone morphogenetic protein (rhBMP-2), 237
Regenerative medicine
  approaches, 84
  cellular and bioreactor components, 85
  degradation-regeneration approach, 85

**S**

Scaffolds
  artificial meniscus, substitution of, 47
  biphasic biomaterials, 34
  described, 32
  ECM, 32
  gradual development, 37
  multilayered osteochondral, 37
  PLGA, 34
  properties, 41
  single phase, 37
  Sr- and Mn-doped, 317
  stem cell- and scaffold-based OTE, 41, 42
  strategies, 36
  tissue-engineering strategy, 38
Scanning electron microscopy (SEM), 130, 173, 175, 177, 189, 270, 296, 316
Schiff-base reaction, 331
SEMI-type PLGA–collagen hybrid scaffolds, 181
Sericin, 11
Silica fibers, 115

Silk
  chondral and subchondral repair, 11
  described, 10
  fibers, 116, 117
  fibroin, 11
  sericin, 11, 307
  silkworm silk, 10
  spinning process, 11
  structure, 10
Silk fibroin (SF)
  bilayered scaffolds, 307
  characteristics, 308, 309
  ECM, 306, 320
  hydrogels formation, 309–313
  innovative strategies, 319
  large-scale processing, 306
  natural biopolymers, 320
  OC defects, 307
  osteochondral TE, 308
  scaffolds, 314–319
  self-healing and load-bearing, 320
  silk protein, 307
  spun fibers, 307
Silk-nanoCaP layer, 134
Simulated body fluid (SBF), 273–274
Solution parameters, 252, 253
Sonication-induced SF hydrogels, 312
Stem cells
  ACI, 360
  cartilage lesions, 360
  cBMA/BMS group, 368
  Hyaluronan-based scaffold, 368
  MSCs, 91, 360–363
  osteochondral lesions, 359
  postoperative treatment, 367, 368
  radiological analysis, 369
  surgical techniques
    approach, 366, 367
    bone marrow, 364–365
    indications, 363–364
    platelet-rich fibrin gel production, 364
  talar dome, 360
  talar osteochondral lesions, 368
Subchondral bone, 366, 374–376, 379, 380, 385–387
Supercritical fluid technology, 267–269, 275
Surface functionalization, 248
Synovial-derived stem cells (SDSC), 363
Synovium-derived mesenchymal stem cells (SMSC), 293
Synthetic bioceramics
  bioactive glasses and glass-ceramics, 58, 59
  bioinert materials, 56, 57
  CaPs, 59–62

Synthetic biomaterials, 34, 35, 40, 87, 88, 341–342
Synthetic polymeric fibers
  electrospinning, 99
  nanofibers, 99
  PA fibers, 106
  polycaprolactone, 105, 106
  polylactic acid, 105

**T**
Taylor cone, 249, 251
Thermal-induced phase separation (TIPS) technique, 89, 319
THIN-type PLGA–collagen hybrid scaffold, 181, 183
Thiol-ene PEG hydrogels, 342
Tissue-engineered products
  biomaterial, 416
  bone and cartilage defects, 416
  challenges, 416
  chondrogenesis and osteogenesis, 416
  clinical translation, 418
  3D scaffolds, 424
  FDA regulation, 418
  local/systemic immune responses, 416
  manufacturing practices, 417, 418
  medical technologies, 417
  OC lesions, 416
  osteochondral regeneration, 418–424
  regulatory medical conditions, 416
  researchers, 425
Tissue engineering (TE)
  aim of, 80
  approaches, 84
  articular cartilage repair treatment, 32, 33
  biomaterials
    natural, 86, 87
  scaffold designs, 84, 85
  synthetic, 87, 88
  medical implications, 92
  nanomaterials and nanocomposites, 85, 89–91
  natural biomaterials and applications, 87
  and regenerative approaches, 408
  scaffolds (*see also* Scaffolds)
  surgical treatments, 92
Tissue engineering and regenerative medicine (TERM), 282, 284, 291, 295, 299, 417
Tissue inhibitor of metalloprotease-3 (TIMP3), 157
Tissue integration, 194, 203, 209
Titanium fibers, 118, 119
Topographies, 253–255
Transcription activator-like effector nuclease (TALEN), 154
Transforming growth factor-β (TGF-β), 134, 150, 155, 157, 272, 293, 362, 381
Trilayered scaffolds, 203–205
Triphasic scaffolds, 379, 380
TruFit CB plug, 385
Trufit®, 209

**V**
Vascular endothelial growth factor (VEGF), 157, 159, 311
Vivostat® System, 364

**X**
XanoMatrix™, 112

**Y**
Young's modulus, 179, 180

Printed by Printforce, the Netherlands